ADAPTIVE APPROXIMATION BASED CONTROL

ADAPTIVE APPROXIMATION BASED CONTROL

Unifying Neural, Fuzzy and Traditional Adaptive Approximation Approaches

Jay A. Farrell
University of California Riverside

Marios M. Polycarpou
University of Cyprus and University of Cincinnati

WILEY-INTERSCIENCE

A JOHN WILEY & SONS, INC., PUBLICATION

Published by John Wiley & Sons, Inc., Hoboken, New Jersey.
Published simultaneously in Canada.

For general information on our other products and services or for technical support, please contact our Customer Care Department within the United States at (800) 762-2974, outside the United States at (317) 572-3993 or fax (317) 572-4002.

Wiley also publishes its books in a variety of electronic formats. Some content that appears in print may not be available in electronic formats. For more information about Wiley products, visit our web site at www.wiley.com.

Library of Congress Cataloging-in-Publication Data:

Farrell, Jay.
Adaptive approximation based control : unifying neural, fuzzy and traditional adaptive approximation approaches / Jay A. Farrell, Marios M. Polycarpou.
p. cm.
Includes bibliographical references and index.
ISBN-13 978-0-471-72788-0 (cloth)
ISBN-10 0-471-72788-1 (cloth)
1. Adaptive control systems. 2. Feedback control systems. I. Polycarpou, Marios. II. Title.
TJ217.F37 2006
629.8'36—dc22

2005021385

To our families and friends.

CONTENTS

Preface xiii

1 Introduction **1**

1.1 Systems and Control Terminology 1
1.2 Nonlinear Systems 3
1.3 Feedback Control Approaches 4
1.3.1 Linear Design 4
1.3.2 Adaptive Linear Design 6
1.3.3 Nonlinear Design 9
1.3.4 Adaptive Approximation Based Design 11
1.3.5 Example Summary 13
1.4 Components of Approximation Based Control 15
1.4.1 Control Architecture 15
1.4.2 Function Approximator 16
1.4.3 Stable Training Algorithm 17
1.5 Discussion and Philosophical Comments 18
1.6 Exercises and Design Problems 19

2 Approximation Theory **23**

2.1 Motivating Example 24
2.2 Interpolation 29

2.3 Function Approximation 30
- 2.3.1 Offline (Batch) Function Approximation 31
- 2.3.2 Adaptive Function Approximation 33

2.4 Approximator Properties 39
- 2.4.1 Parameter (Non)Linearity 39
- 2.4.2 Classical Approximation Results 43
- 2.4.3 Network Approximators 46
- 2.4.4 Nodal Processors 48
- 2.4.5 Universal Approximator 50
- 2.4.6 Best Approximator Property 52
- 2.4.7 Generalization 54
- 2.4.8 Extent of Influence Function Support 56
- 2.4.9 Approximator Transparency 65
- 2.4.10 Haar Conditions 66
- 2.4.11 Multivariable Approximation by Tensor Products 67

2.5 Summary 68

2.6 Exercises and Design Problems 69

3 Approximation Structures 71

3.1 Model Types 72
- 3.1.1 Physically Based Models 72
- 3.1.2 Structure (Model) Free Approximation 73
- 3.1.3 Function Approximation Structures 74

3.2 Polynomials 75
- 3.2.1 Description 75
- 3.2.2 Properties 77

3.3 Splines 78
- 3.3.1 Description 78
- 3.3.2 Properties 83

3.4 Radial Basis Functions 84
- 3.4.1 Description 84
- 3.4.2 Properties 86

3.5 Cerebellar Model Articulation Controller 87
- 3.5.1 Description 88
- 3.5.2 Properties 89

3.6 Multilayer Perceptron 93
- 3.6.1 Description 93
- 3.6.2 Properties 95

3.7 Fuzzy Approximation 96
- 3.7.1 Description 96
- 3.7.2 Takagi–Sugeno Fuzzy Systems 104
- 3.7.3 Properties 105

3.8 Wavelets 106
3.8.1 Multiresolution Analysis (MRA) 108
3.8.2 MRA Properties 110
3.9 Further Reading 112
3.10 Exercises and Design Problems 112

4 Parameter Estimation Methods 115

4.1 Formulation for Adaptive Approximation 116
4.1.1 Illustrative Example 116
4.1.2 Motivating Simulation Examples 118
4.1.3 Problem Statement 124
4.1.4 Discussion of Issues in Parametric Estimation 125
4.2 Derivation of Parametric Models 127
4.2.1 Problem Formulation for Full-State Measurement 128
4.2.2 Filtering Techniques 129
4.2.3 SPR Filtering 131
4.2.4 Linearly Parameterized Approximators 131
4.2.5 Parametric Models in State Space Form 133
4.2.6 Parametric Models of Discrete-Time Systems 134
4.2.7 Parametric Models of Input-Output Systems 136
4.3 Design of Online Learning Schemes 138
4.3.1 Error Filtering Online Learning (EFOL) Scheme 138
4.3.2 Regressor Filtering Online Learning (RFOL) Scheme 140
4.4 Continuous-Time Parameter Estimation 141
4.4.1 Lyapunov-Based Algorithms 143
4.4.2 Optimization Methods 148
4.4.3 Summary 154
4.5 Online Learning: Analysis 154
4.5.1 Analysis of LIP EFOL Scheme with Lyapunov Synthesis Method 155
4.5.2 Analysis of LIP RFOL Scheme with the Gradient Algorithm 158
4.5.3 Analysis of LIP RFOL Scheme with RLS Algorithm 160
4.5.4 Persistency of Excitation and Parameter Convergence 161
4.6 Robust Learning Algorithms 163
4.6.1 Projection Modification 165
4.6.2 σ-Modification 168
4.6.3 ϵ-Modification 169
4.6.4 Dead-Zone Modification 170
4.6.5 Discussion and Comparison 172
4.7 Concluding Summary 173
4.8 Exercises and Design Problems 173

5 Nonlinear Control Architectures 179

5.1 Small-Signal Linearization 180
5.1.1 Linearizing Around an Equilibrium Point 181
5.1.2 Linearizing Around a Trajectory 183
5.1.3 Gain Scheduling 186
5.2 Feedback Linearization 188
5.2.1 Scalar Input-State Linearization 188
5.2.2 Higher-Order Input-State Linearization 190
5.2.3 Coordinate Transformations and Diffeomorphisms 193
5.2.4 Input-Output Feedback Linearization 196
5.3 Backstepping 203
5.3.1 Second Order System 203
5.3.2 Higher Order Systems 205
5.3.3 Command Filtering Formulation 207
5.4 Robust Nonlinear Control Design Methods 211
5.4.1 Bounding Control 211
5.4.2 Sliding Mode Control 212
5.4.3 Lyapunov Redesign Method 215
5.4.4 Nonlinear Damping 219
5.4.5 Adaptive Bounding Control 220
5.5 Adaptive Nonlinear Control 222
5.6 Concluding Summary 225
5.7 Exercises and Design Problems 226

6 Adaptive Approximation: Motivation and Issues 231

6.1 Perspective for Adaptive Approximation Based Control 232
6.2 Stabilization of a Scalar System 236
6.2.1 Feedback Linearization 237
6.2.2 Small-Signal Linearization 238
6.2.3 Unknown Nonlinearity with Known Bounds 239
6.2.4 Adaptive Bounding Methods 241
6.2.5 Approximating the Unknown Nonlinearity 243
6.2.6 Combining Approximation with Bounding Methods 250
6.2.7 Combining Approximation with Adaptive Bounding Methods 252
6.2.8 Summary 252
6.3 Adaptive Approximation Based Tracking 253
6.3.1 Feedback Linearization 253
6.3.2 Tracking via Small-Signal Linearization 253
6.3.3 Unknown Nonlinearities with Known Bounds 256
6.3.4 Adaptive Bounding Design 258
6.3.5 Adaptive Approximation of the Unknown Nonlinearities 262

6.3.6 Robust Adaptive Approximation 264
6.3.7 Combining Adaptive Approximation with Adaptive Bounding 266
6.3.8 Advanced Adaptive Approximation Issues 271
6.4 Nonlinear Parameterized Adaptive Approximation 278
6.5 Concluding Summary 280
6.6 Exercises and Design Problems 281

7 Adaptive Approximation Based Control: General Theory 285

7.1 Problem Formulation 286
7.1.1 Trajectory Tracking 286
7.1.2 System 286
7.1.3 Approximator 287
7.1.4 Control Design 288
7.2 Approximation Based Feedback Linearization 288
7.2.1 Scalar System 289
7.2.2 Input-State 294
7.2.3 Input-Output 306
7.2.4 Control Design Outside the Approximation Region $\mathcal{D}$ 308
7.3 Approximation Based Backstepping 309
7.3.1 Second Order Systems 309
7.3.2 Higher Order Systems 316
7.3.3 Command Filtering Approach 323
7.3.4 Robustness Considerations 328
7.4 Concluding Summary 330
7.5 Exercises and Design Problems 331

8 Adaptive Approximation Based Control for Fixed-Wing Aircraft 333

8.1 Aircraft Model Introduction 334
8.1.1 Aircraft Dynamics 334
8.1.2 Nondimensional Coefficients 335
8.2 Angular Rate Control for Piloted Vehicles 336
8.2.1 Model Representation 337
8.2.2 Baseline Controller 337
8.2.3 Approximation Based Controller 338
8.2.4 Simulation Results 345
8.3 Full Control for Autonomous Aircraft 349
8.3.1 Airspeed and Flight Path Angle Control 350
8.3.2 Wind-Axes Angle Control 355
8.3.3 Body Axis Angular Rate Control 359
8.3.4 Control Law and Stability Properties 362
8.3.5 Approximator Definition 365

8.3.6 Simulation Analysis 367
8.3.7 Conclusions 371
8.4 Aircraft Notation 371

Appendix A: Systems and Stability Concepts 377

A.1 Systems Concepts 377
A.2 Stability Concepts 379
A.2.1 Stability Definitions 379
A.2.2 Stability Analysis Tools 381
A.2.3 Strictly Positive Real Transfer Functions 391
A.3 General Results 392
A.4 Trajectory Generation Filters 394
A.5 A Useful Inequality 397
A.6 Exercises and Design Problems 398

Appendix B: Recommended Implementation and Debugging Approach 399

References 401

Index 417

PREFACE

During the last few years there have been significant developments in the control of highly uncertain, nonlinear dynamical systems. For systems with parametric uncertainty, adaptive nonlinear control has evolved as a powerful methodology leading to global stability and tracking results for a class of nonlinear systems. Advances in geometric nonlinear control theory, in conjunction with the development and refinement of new techniques, such as the backstepping procedure and tuning functions, have brought about the design of control systems with proven stability properties. In addition, there has been a lot of research activity on robust nonlinear control design methods, such as sliding mode control, Lyapunov redesign method, nonlinear damping, and adaptive bounding control. These techniques are based on the assumption that the uncertainty in the nonlinear functions is within some known, or partially known, bounding functions.

In parallel with developments in adaptive nonlinear control, there has been a tremendous amount of activity in neural control and adaptive fuzzy approaches. In these studies, neural networks or fuzzy approximators are used to approximate unknown nonlinearities. The input/output response of the approximator is modified by adjusting the values of certain parameters, usually referred to as *weights*. From a mathematical control perspective, neural networks and fuzzy approximators represent just two classes of function approximators. Polynomials, splines, radial basis functions, and wavelets are examples of other function approximators that can be used—and have been used—in a similar setting. We refer to such approximation models with adaptivity features as *adaptive approximators*, and control methodologies that are based on them as *adaptive approximation based control*.

Adaptive approximation based control encompasses a variety of methods that appear in the literature: intelligent control, neural control, adaptive fuzzy control, memory-based control, knowledge-based control, adaptive nonlinear control, and adaptive linear control.

Researchers in these fields have diverse backgrounds: mathematicians, engineers, and computer scientists. Therefore, the perspective of the various papers in this area is also varied. However, the objective of the various practitioners is typically similar: to design a controller that can be guaranteed to be stable and achieve a high level of control performance for systems that contain poorly modeled nonlinear effects, or the dynamics of the system change during operation (for example, due to system faults). This objective is achieved by adaptively developing an approximating function to compensate the nonlinear effects during the operation of the system.

Many of the original papers on neural or adaptive fuzzy control were motivated by such concepts as ease of use, universal approximation, and fault tolerance. Often, ease of use meant that researchers without a control or systems background could experiment with and often succeed at controlling certain dynamics systems, at least in simulation. The rise of interest in the neural and adaptive fuzzy control approaches occurred at a time when desktop computers and dynamic simulation tools were becoming sufficiently cheap at reasonable levels of performance to support such research on a wide basis.

However, prior to application on systems of high economic value, the control system designer must carefully consider any new approach within a sound analytical framework that allows rigorous analysis of conditions for stability and robustness. This approach opens a variety of questions that have been of interest to various researchers: What properties should the function approximator have? Are certain families of approximators superior to others? How should the parameters of the approximator be estimated? What can be guaranteed about the properties of the signals within the control system? Can the stability of the approximator parameters be guaranteed? Can the convergence of the approximator parameters be guaranteed? Can such control systems be designed to be robust to noise, disturbances, and unmodeled effects. Can this approach handle significant changes in the dynamics due to, for example, a system failure. What types of nonlinear dynamic systems are amenable to the approach? What are the limitations? The objective of this textbook is to provide readers with a framework for rigorously considering such questions.

Adaptive approximation based control can be viewed as one of the available tools that a control designer should have in her/his control toolbox. Therefore, it is desirable for the reader not only to be able to apply, for example, neural network techniques to a certain class of systems, but more importantly to gain enough intuition and understanding about adaptive approximation so that she/he knows when it is a useful tool to be used and how to make necessary modifications or how to combine it with other control tools, so that it can be applied to a system that has not be encountered before.

The book has been written at the level of a first-year graduate student in any engineering field that includes an introduction to basic dynamic systems concepts such as state variables and Laplace transforms. We hope that this book has appeal to a wide audience. For use as a graduate text, we have included exercises, examples, and simulations. Sufficient detail is included in examples and exercises to allow students to replicate and extend results. Simulation implementation of the methods developed herein is a virtually necessary component of understanding implications of the approach. The book extensively uses ideas from stability theory. The advantage of this approach is that the adaptive law is derived based on the Lyapunov synthesis method and therefore the stability properties of the closed-loop system are more readily determined. Therefore, an appendix has been included as an aid to readers who are not familiar with the ideas of Lyapunov stability analysis. For theoretically oriented readers, the book includes complete stability analysis of the methods that are presented.

Organization. To understand and effectively implement adaptive approximation based control systems that have guaranteed stability properties, the designer must become familiar with concepts of dynamic systems, stability theory, function approximation, parameter estimation, nonlinear control methods, and the mechanisms to apply these various tools in a unified methodology.

Chapter 1 introduces the idea of adaptive approximation for addressing unknown nonlinear effects. This chapter includes a simple example comparing various control approaches and concludes with a discussion of components of an adaptive approximation based control system with pointers to the locations in the text where each topic is discussed.

Function approximation and data interpolation have long histories and are important fields in their own right. Many of the concepts and results from these fields are important relative to adaptive approximation based control. Chapter 2 discuss various properties of function approximators as they relate to adaptive function approximation for control purposes. Chapter 3 presents various function approximation structures that have been considered for implementation of adaptive approximation based controllers. All of the approximators of this chapter are presented using a single unifying notation. The presentation includes a comparative discussion of the approximators relative to the properties presented in Chapter 2.

Chapter 4 focuses on issues related to parameter estimation. First we study the formulation of parametric models for the approximation problem. Then we present the design of online learning schemes; and finally, we derive parameter estimation algorithms with certain stability and robustness properties. The parameter estimation problem is formulated in a continuous-time framework. The chapter includes a discussion of robust parameter estimation algorithms, which will prove to be critical to the design of stable adaptive approximation based control systems.

Chapter 5 reviews various nonlinear control system design methodologies. The objective of this chapter is to introduce the methods, analysis tools, and key issues of nonlinear control design. The chapter begins with a discussion of small-signal linearization and gain scheduling. Then we focus on feedback linearization and backstepping, which are two of the key design methods for nonlinear control design. The chapter presents a set of robust nonlinear control design techniques. These methods include bounding control, sliding mode control, Lyapunov redesign method, nonlinear damping, and adaptive bounding. Finally, we briefly study the adaptive nonlinear control methodology. For each approach we present the basic method, discuss necessary theoretical ideas related to each approach, and discuss the effect (and accommodation) of modeling error.

Chapters 6 and 7 bring together the ideas of Chapters 1–5 to design and analyze control systems using adaptive approximation to compensate for poorly modeled nonlinear effects. Chapter 6 considers scalar dynamic systems. The intent of this chapter is to allow a detailed discussion of important issues without the complications of working with higher numbers of state variables. The ideas, intuition, and methods developed in Chapter 6 are important to successful applications to higher order systems. Chapter 7 will augment feedback linearization and backstepping with adaptive approximation capabilities to achieve high-performance tracking for systems with significant unmodeled nonlinearities. The presentation of each approach includes a rigorous Lyapunov analysis.

Chapter 8 presents detailed design and analysis of adaptive approximation based controllers applied to fixed-wing aircraft. We study two control situations. First, an angular rate controller is designed and analyzed. This controller is applicable in piloted aircraft applications where the stick motion of the pilot is processed into body-frame angular rate commands. Then we develop a full vehicle controller suitable for uninhabited air vehicles

(UAVs). The control design is based on the approximation based backstepping methodology.

Acknowledgments. The authors would like to thank the various sponsors that have supported the research that has resulted in this book: the National Science Foundation (Paul Werbos), Air Force Wright-Patterson Laboratory (Mark Mears), Naval Air Development Center (Marc Steinberg), and the Research Promotion Foundation of Cyprus. We would like to thank our current and past employers who have directly and indirectly enabled this research: University of California, Riverside; University of Cyprus; University of Cincinnati; and Draper Laboratory. In addition, we wish to acknowledge the many colleagues, collaborators, and students who have contributed to the ideas presented herein, especially: P. Antsaklis, W. L. Baker, J.-Y. Choi, M. Demetriou, S. Ge, J. Harrison, P. A. Ioannou, H. K. Khalil, P. Kokotovic, F. L. Lewis, D. Liu, M. Mears, A. N. Michel, A. Minai, J. Nakanishi, K. Narendra, C. Panayiotou, T. Parisini, K. M. Passino, T. Samad, S. Schaal, M. Sharma, J.-J. Slotine, E. Sontag, G. Tao, A. Vemuri, H. Wang, S. Weaver, Y. Yang, X. Zhang, Y. Zhao, and P. Zufiria. Finally, we would like to thank our families for their constant support and encouragement throughout the long period that it took for this book to be completed.

Jay A. Farrell
Marios M. Polycarpou

Riverside, California and Nicosia, Cyprus
(10 hours time difference)
July 2005

CHAPTER 1

INTRODUCTION

This book presents adaptive function estimation and feedback control methodologies that develop and use approximations to portions of the nonlinear functions describing the system dynamics while the system is in online operation. Such methodologies have been proposed and analyzed under a variety of titles: neural control, adaptive fuzzy control, learning control, and approximation-based control. A primary objective of this text is to present the methods systematically in a unifying framework that will facilitate discussion of underlying properties and comparison of alternative techniques.

This introductory chapter discusses some fundamental issues such as: (i) motivations for using adaptive approximation-based control; (ii) when adaptive approximation-based control methods are appropriate; (iii) how the problem can be formulated; and (iv) what design decisions are required. These issues are illustrated through the use of a simple simulation example.

1.1 SYSTEMS AND CONTROL TERMINOLOGY

Researchers interested in this area come from a diverse set of backgrounds other than control; therefore, we start with a brief review of terminology standard to the field of control systems, as depicted in Figure 1.1. The *plant* is the system to be controlled. The plant will by modeled herein by a typically nonlinear set of ordinary differential equations. The plant model is assumed to include the actuator and sensor models. The *control system* is designed to achieve certain control objectives. As indicated in Figure 1.1, the inputs to the control system include the reference input $y_c(t)$ (which is possibly passed through

Adaptive Approximation Based Control: Unifying Neural, Fuzzy and Traditional Adaptive Approximation Approaches. By Jay A. Farrell and Marios M. Polycarpou

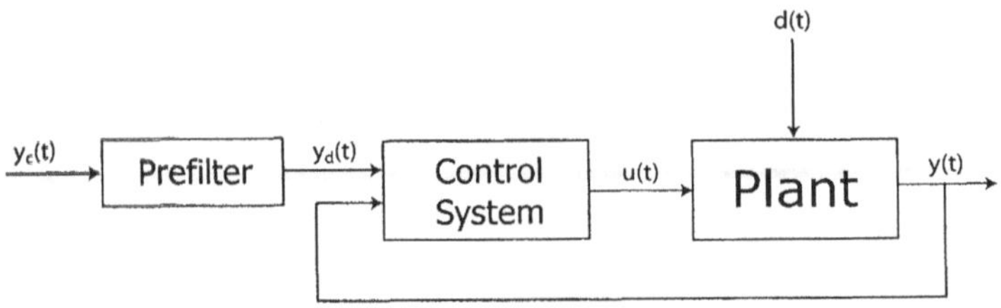

Figure 1.1: Standard control system block diagram.

a prefilter to yield a smoother function $y_d(t)$ and its first r time derivatives $y_d^{(i)}(t)$ for $i = 1, \ldots, r$) and a set of measurable plant outputs $y(t)$. The control system processes its inputs to produce the control system output $u(t)$ that is applied to the plant actuators to affect the desired change in the plant output. The control system output $u(t)$ is sometimes referred to as *control signal* or *plant input*. Figure 1.1 depicts as a block diagram a standard closed-loop control system configuration.

The control system determines the stability of the closed-loop system and the response to disturbances $d(t)$ and initial condition errors. A *disturbance* is any unmodeled physical effect on the plant state, usually caused by the environment. A disturbance is distinct from *measurement noise*. The former directly and physically affects the system to be controlled. The latter affects the measurement of the physical quantity without directly affecting the physical quantity. The physical quantity may be indirectly affected by the noise through the feedback control process.

Control design typically distinguishes *regulation* from *tracking* objectives. Regulation is concerned with designing a control system to achieve convergence of the system state, with a desirable transient response, from any initial condition within a desired domain of attraction, to a single operating point. In this case, the signal $y_c(t)$ is constant. Tracking is concerned with the design of a control system to cause the system output $y(t)$ to converge to and accurately follow the signal $y_d(t)$. Although the input signal $y_c(t)$ to a tracking controller could be a constant, it typically is time-varying in a manner that is not known at the time that the control system is designed. Therefore, the designer of a tracking controller must anticipate that the plant state may vary significantly on a persistent basis. It is reasonable to expect that the designer of the open-loop physical system and the designer of the feedback control system will agree on an allowable range of variation of the state of the system. Herein, we will denote this *operating envelope* by $\mathcal{D}$. The designer of the physical system ensures safe operation when the state of the system is in $\mathcal{D}$. The designer of the controller must ensure that the state the system remains in $\mathcal{D}$. Implicitly it is assumed that the state required to track y_d lies entirely in $\mathcal{D}$.

To illustrate the control terminology let us consider the example of a simple cruise control system for automobiles. In this case, the control objective is to make the vehicle follow a desired speed profile $y_c(t)$, which is set by the driver. The measured output $y(t)$ is the sensed vehicle speed and the control system output $u(t)$ is the throttle angle and/or fuel injection rate. The disturbance $d(t)$ may arise due to the wind or road incline. In addition to disturbances, which are external factors influencing the state, there may also be modeling errors. In the cruise control example, the plant model describes the effect of changing the throttle angle on the actual vehicle speed. Hence, modeling errors may arise from simplifications or inaccuracies in characterizing the effect of changing the throttle angle on the vehicle speed. Modeling errors (especially nonlinearities), whether they arise due to inaccuracies or intentional model simplifications, constitute one of the key motivations for employing adaptive approximation-based control, and thus are crucial to the techniques developed in this book.

In general, the objectives of a control system design are:

1. to stabilize the closed-loop system;
2. to achieve satisfactory reference input tracking in transient and at steady state;
3. to reduce the effect of disturbances;
4. to achieve the above in spite of modeling error;
5. to achieve the above in spite of noise introduced by sensors required to implement the feedback mechanism.

Introductory textbooks in control systems provide linear-based design and analysis techniques for achieving the above objectives and discuss some basic robustness and implementation issues [61, 66, 86, 140]. The theoretical foundations of linear systems analysis and design are presented in more advanced textbooks (see, for example, [10, 19, 39, 130]), where issues such as controllability, observability, and model reduction are examined.

1.2 NONLINEAR SYSTEMS

Most dynamic systems encountered in practice are inherently nonlinear. The control system design process builds on the concept of a model. Linear control design methods can sometimes be applied to nonlinear systems over limited operating regions (i.e., $\mathcal{D}$ is sufficiently small), through the process of small-signal linearization. However, the desired level of performance or tracking problems with a sufficiently large operating region $\mathcal{D}$ may require in which the nonlinearities be directly addressed in the control system design. Depending on the type of nonlinearity and the manner that the nonlinearity affects the system, various nonlinear control design methods are available [121, 134, 159, 234, 249, 279]. Some of these methods are reviewed in Chapter 5.

Nonlinearity and model accuracy directly affect the achievable control system performance. Nonlinearity can impose hard constraints on achievable performance. The challenge of addressing nonlinearities during the control design process is further complicated when the description of the nonlinearities involves significant uncertainty. When portions of the plant model are unknown or inaccurately defined, or they change during operation, the control performance may need to be severely limited to ensure safe operation. Therefore there is often an interest to improve the model accuracy. Especially in tracking applications this will typically necessitate the use of nonlinear models. The focus of this text is on adaptively improving models of nonlinear effects during online operation.

In such applications the level of achievable performance may be enhanced by using adaptive function approximation techniques to increase the accuracy of the model of the nonlinearities. Such adaptive approximation-based control methods include the popular areas of adaptive fuzzy and neural control. This chapter introduces various issues related to adaptive approximation-based control. This introductory discussion will direct the reader to the appropriate sections of the text where more detailed discussion of each issue can be found.

1.3 FEEDBACK CONTROL APPROACHES

To introduce the concept of adaptive approximation-based control, consider the following example, where the objective is to control the dynamic system

$$\dot{y}(t) = f(y(t)) + g(y(t))u(t) \tag{1.1}$$

in a manner such that $y(t)$ accurately tracks an externally generated reference input signal $y_d(t)$. Therefore, the control objective is achieved if the tracking error $\tilde{y}(t) = y(t) - y_d(t)$ is forced to zero. The performance specification is for the closed-loop system to have a rate of convergence corresponding to a linear system with a dominant time constant τ of about 5.0 s. With this time constant, tracking errors due to disturbances or initial conditions should decay to zero in approximately 15 s ($= 3\tau$). The system is expected to normally operate within $y \in [20, 60]$, but may safely operate on the region $\mathcal{D} = \{y \in [0, 100]\}$. Of course, all signals in the controller and plant must remain bounded during operation.

However, the plant model is not completely accurate. The best model available to the control system designer is given by

$$\dot{y} = f_o(y) + g_o(y)u, \tag{1.2}$$

where $f_o(y) = -y$ and $g_o(y) = 1.0 + 0.3y$. The actual system dynamics are not known or available to the designer. For implementation of the following simulation results, the actual dynamics will be

$$\begin{aligned} f(y) &= -1 - 0.01y^2 \\ g(y) &= \frac{40(y+1)}{40+y}. \end{aligned}$$

Therefore, there exists significant error between the design model and the actual dynamics over the desired domain of operation.

This section will consider four alternative control system design approaches. The example will allow a concrete, comparative discussion, but none of the designs have been optimized. The objective is to highlight the similarities, distinctions, complexity, and complicating factors of each approach. The details of each design have been removed from this discussion so as not to distract from the main focus. The details are included in the problem section of this chapter to allow further exploration. These methodologies and various others will be analyzed in substantially greater detail throughout the remainder of the book.

1.3.1 Linear Design

Given the design model and performance specification, the objective in this subsection is to design a linear controller for the system

$$\dot{y}(t) = h(y(t), u(t)) = -y(t) + (1.0 + 0.3y(t))u(t) \tag{1.3}$$

so that the linearized closed-loop system is stable (stability concepts are reviewed in Appendix A) and has the desired tracking error convergence rate. This controller is designed based on the idea of small-signal linearization and is approximate, even relative to the model. Section 1.3.3 will consider feedback linearization, which is a nonlinear design approach that exactly linearizes the model using the feedback control signal.

For the scalar system $\dot{y} = h(y, u)$, an *operating point* is a pair of real numbers (y^*, u^*) such that $h(y^*, u^*) = 0$. If $y = y^*$ and $u = u^*$, then $\dot{y} = 0$. In a more general setting, the designer may need to linearize around a time-varying nominal trajectory $(y^*(t), u^*(t))$. Note that operating points may be stable or unstable (see the discussion in Appendix A). An operating point analysis only indicates the values of y at which it is possible, by appropriate choice of u, for the system to be in steady state. For our example, the set of operating points is defined by (y^*, u^*) such that

$$u^* = \frac{y^*}{1 + 0.3y^*}.$$

Therefore, the design model indicates that the system can operate at any $y \in \mathcal{D}$.

The operating point analysis does not indicate how $u(t)$ should be selected to get convergence to any particular operating point. Convergence to a desired operating point is an objective for the control system design. In a linear control design, the best available model is linearized around an operating point and a linear controller is designed for that linearized model. If we choose the operating point $(y^*,\ u^*) = \left(40,\ \frac{40}{13}\right)$ as the design point, then the linearized dynamics are (see Exercise 1.1)

$$\delta\dot{y} = -\frac{1}{13}\delta y + 13\delta u,$$

where $\delta y = y - 40$ and $\delta u = u - \frac{40}{13}$. The linear controller

$$U(s) = \frac{40}{13} - \frac{0.2(s + \frac{1}{13})}{13s}\tilde{Y}(s) \tag{1.4}$$

used with the design model results in a stable system that achieves the specification at $y^* = 40$. In the above, s is the Laplace variable, $U(s)$ denotes the Laplace transform of $u(t)$, $\tilde{y}(t) = y(t) - y_d(t)$, and $y_d(t)$ is the reference input. Of course, $\mathcal{D}$ is large enough that a linear controller designed to achieve the specification at one operating point will probably not achieve the specification at all operating points in $\mathcal{D}$ or for $y_d(t)$ varying with time over the region $\mathcal{D}$.

Figure 1.2 shows the performance using the linear controller of eqn. (1.4) for a series of amplitude step inputs changing between $y_d = 20$ and $y_d = 60$. Note that the response exhibits two different convergence rates indicated by τ_1 and τ_2. One is significantly slower than the desired 5 s. Therefore, the linear controller does not operate as designed. There are two reasons for this. First, there is significant error between the design model and the actual dynamics of the system. Second, an inherent assumption of linear design is that the linear controller will only be used in a reasonably small neighborhood of the operating point for which the controller was designed. The degree of reasonableness depends on the nonlinear system of interest. For these two reasons, the *actual* linearized dynamics at the two points $y^* = 20$ and $y^* = 60$ are distinct from the linearized dynamics of the *design* model at the design point $y^* = 40$. The design methodology to determine eqn. (1.4) relied on cancelling the pole of the linearized dynamics. With modeling error, even for a linear system, the pole is not cancelled; instead, there are two poles. One near the desired pole and one near the origin. The second pole is dominant and yields the slowly converging error dynamics.

Improved performance using linear methods could be achieved by various methods. First, additional modeling efforts could decrease the error between the actual dynamics and the design model, but may be expensive and will not solve the problem of operating far from the linearization point. Second, high gain control will decrease the sensitivity to

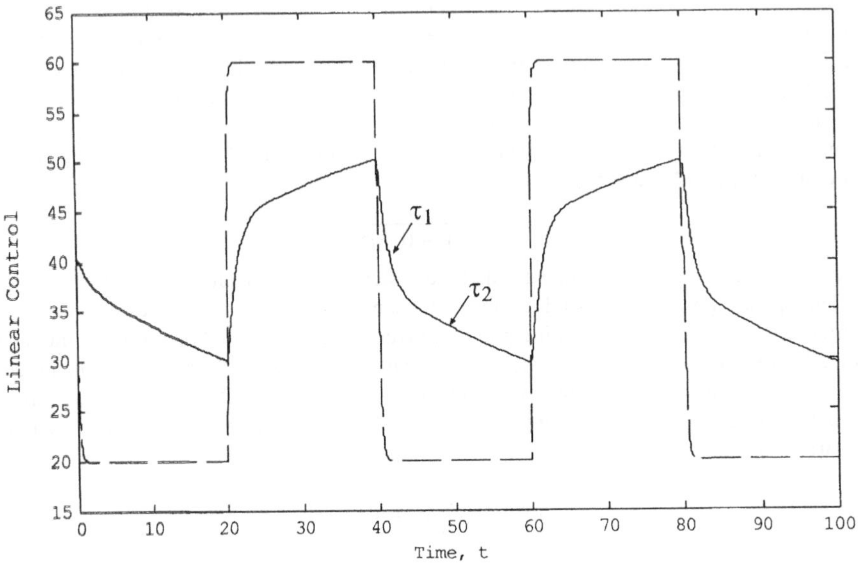

Figure 1.2: Performance of the linear control system of eqn. (1.4) with the dynamic system of eqn. (1.1). The solid curve is $y(t)$. The dashed curve is $y_d(t)$.

modeling error, but will result in a higher bandwidth closed-loop system as well as a large control effort. Third, *gain scheduling* methods (although not truly linear) address the issue of limiting the use of a linear controller to a region about its design point by switching between a set of linear controllers as a function of the system state. Each linear controller is designed to meet the performance specification (for the design model) on a small region of operation $\mathcal{D}_i$. The regions $\mathcal{D}_i$ are defined such that they cover the region of operation $\mathcal{D}$ (i.e., $\mathcal{D} \subset \cup_{i=1}^{N} \mathcal{D}_i$). Gain scheduling a set of linear controllers does not address the issue of error between the actual system and the design model.

1.3.2 Adaptive Linear Design

Through linearization, the dynamics near a fixed operating point (y^*, u^*) are approximated by

$$\dot{y}(t) = a^* + b^* y(t) + c^* u(t), \tag{1.5}$$

where a^*, b^*, and c^* are parameters that depend on (y^*, u^*). In one possible adaptive control approach, the control law is

$$u = \frac{1}{c}\left(-a - by + \dot{y}_d + 0.2(y_d - y)\right), \tag{1.6}$$

where $y_d \in \mathcal{C}^1(\mathcal{D})$ (i.e., the first derivative of y_d exists and is continuous within the region $\mathcal{D}$), and a, b, c are parameter estimates of a^*, b^*, and c^*, respectively. Note that if $(a, b, c) = (a^*, b^*, c^*)$, then exact cancellation occurs and the resulting error dynamics are

$$\dot{\tilde{y}} = -0.2\tilde{y},$$

where $\tilde{y} = y - y_d$. Therefore, the closed-loop error dynamics (with perfect modeling) achieve the performance specification. This closed-loop system has a time constant for rejecting disturbances and initial condition errors of 5.0 s, even though the feedforward term in eqn. (1.6) (i.e., $\frac{1}{c}\dot{y}_d$) will allow the system to track faster changes in the commanded input.

The differentiability constraint on $y_d(t)$ will be enforced by passing the reference input $y_c(t)$ through the first-order low pass prefilter

$$\frac{Y_d(s)}{Y_c(s)} = \frac{5}{s+5}, \tag{1.7}$$

where $Y_d(s)$, $Y_c(s)$ denote the Laplace transforms of the time signals $y_d(t)$ and $y_c(t)$ respectively. Therefore,

$$\dot{y}_d = -5(y_d - y_c),$$

which has the same bounded and continuous properties as y_c; whereas, the signal y_d will be bounded, continuous, and differentiable as long as y_c is bounded.

If (a^*, b^*, c^*) are assumed to be unknown constant parameters, then the corresponding parameter estimates (a, b, c) are derived from the following update laws

$$\begin{aligned} \dot{a} &= \gamma_1 \tilde{y}, & (1.8)\\ \dot{b} &= \gamma_2 \tilde{y} y, & (1.9)\\ \dot{c} &= \gamma_3 \tilde{y} u, & (1.10) \end{aligned}$$

where $\gamma_i > 0$ are design constants representing the adaptive gain of each parameter estimate. For the following simulation we select $\gamma_1 = \gamma_2 = \gamma_3 = 0.01$. In practice, the update law for $c(t)$ needs to be slightly modified in order to guarantee that $c(t)$ does not approach zero, which would cause $u(t)$ to become very large, or even infinite. The resulting error dynamic equations are

$$\begin{aligned} \dot{\tilde{y}} &= -0.2\tilde{y} + \tilde{a} + \tilde{b}y + \tilde{c}u, & (1.11)\\ \dot{\tilde{a}} &= -\gamma_1 \tilde{y}, & (1.12)\\ \dot{\tilde{b}} &= -\gamma_2 \tilde{y} y, & (1.13)\\ \dot{\tilde{c}} &= -\gamma_3 \tilde{y} u, & (1.14) \end{aligned}$$

where $\tilde{a} = a^* - a$, $\tilde{b} = b^* - b$, $\tilde{c} = c^* - c$. The adaptive control law is defined by eqns. (1.6) and (1.8 – 1.10). Note that this controller is not linear and that the controller implementation does not require knowledge of a^*, b^*, or c^* (other that the sign of c^*). If the above adaptive scheme is applied to the system model (1.5) (without noise, disturbances, and unmodeled states), it can be shown that the closed-loop system is stable, after some small modification to ensure that the parameter estimate c does not approach zero. It is noted that robustness issues are neglected at this point to simplify the presentation, but are addressed in Chapter 4.

Relative to (1.5), even if the tracking error $\tilde{y}(t)$ goes to zero, the adaptive parameters (a, b, c) may never converge to the "actual" parameters (a^*, b^*, c^*). Convergence (or not) of the parameter estimation error to zero depends on the nature of the signal $y_d(t)$. From eqn. (1.11), if $\tilde{a} + \tilde{b}y + \tilde{c}u = 0$, then $\tilde{y}$ will approach zero and parameter adaptation will stop. Since for any fixed values of y and u, the equation $\tilde{a} + \tilde{b}y + \tilde{c}u = 0$ defines a hyperplane of $(\tilde{a}, \tilde{b}, \tilde{c})$ values, there are many values of the parameter estimates that can result in $\tilde{y} = 0$. The hyperplane is distinct for different (y, u) and the only parameter estimates on all such hyperplanes satisfy $(\tilde{a}, \tilde{b}, \tilde{c}) = (0, 0, 0)$. Therefore, convergence of the parameter estimates would require that (y, u) change sufficiently in an appropriate sense, leading to the concept of *persistency of excitation* (see Chapter 4). An important fact to remember in the design of adaptive control systems is that convergence of the tracking error does not necessarily imply convergence (or even boundedness) of the parameter estimates.

Relative to (1.1), the parameters of (1.5) will be a function of the operating point (see Exercise 1.2). Each time that the operating point changes, the parameter estimates will adapt. If the operating point changed slowly, then a^*, b^*, and c^* could be considered as slowly time-varying. In such an approach, depending on the magnitude of the adaptive gains γ_i, the corresponding estimates may be able to change the adaptive parameters fast enough to maintain high performance. However, in this case the operating point would be restricted to vary slowly so that the control approach would behave properly. It is also important to note that increasing γ_i may create stability problems of the closed-loop system in the presence of measurement noise.

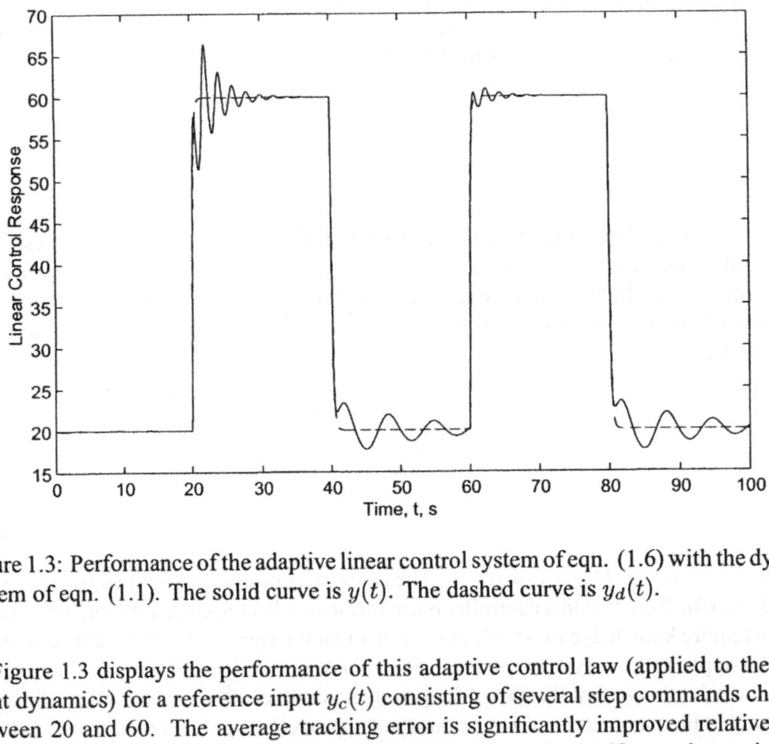

Figure 1.3: Performance of the adaptive linear control system of eqn. (1.6) with the dynamic system of eqn. (1.1). The solid curve is $y(t)$. The dashed curve is $y_d(t)$.

Figure 1.3 displays the performance of this adaptive control law (applied to the actual plant dynamics) for a reference input $y_c(t)$ consisting of several step commands changing between 20 and 60. The average tracking error is significantly improved relative to the linear control system. However, immediately following each significant change in $y_c(t)$, the tracking error is still large and oscillatory. Also, the estimated parameters that result in good performance at one operating point do not yield good performance at the other. Therefore, for this example, as the operating point is stepped back and forth, the estimated parameters step between the manifold of parameters (i.e., hyperplane) that yield good performance for $y \approx 20$ and the manifold of parameters that yield good performance for $y \approx 60$, see Figure 1.4. This is obviously inefficient. It would be convenient if the designer could devise a method to, in some sense, store the model (e.g., estimated parameters) as a function of the operating condition (e.g., y). Such ideas are the motivation for adaptive approximation-based control methods.

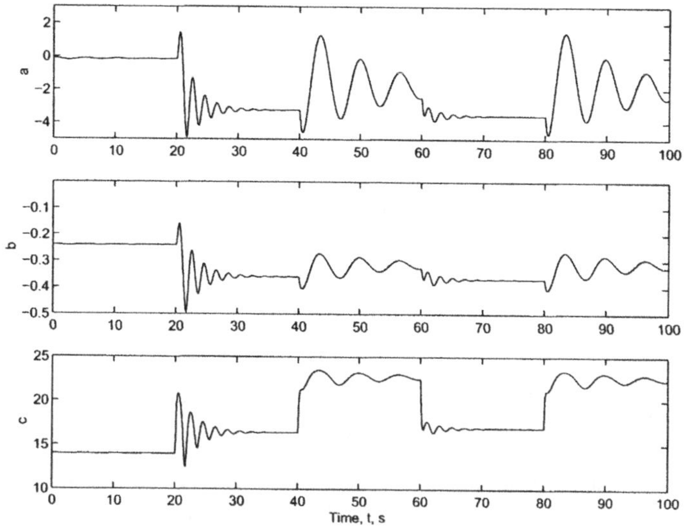

Figure 1.4: Time evolution of the estimated parameters $a(t)$, $b(t)$, $c(t)$ for the adaptive control system of eqn. (1.6) applied to the dynamic system of eqn. (1.1).

1.3.3 Nonlinear Design

Given the design model of eqn. (1.2), the feedback linearizing control law is

$$u(t) = \frac{1}{g_o(y(t))}\left(-f_o(y(t)) + \dot{y}_d(t) + K(y_d(t) - y(t))\right). \tag{1.15}$$

Combining the feedback linearizing control law with the design model and selecting $K = 0.2$, yields the following *nominal* closed-loop dynamics

$$\dot{\tilde{y}} = -0.2\tilde{y}, \tag{1.16}$$

where $\tilde{y} = y - y_d$. In contrast to the small signal linearization approach discussed in Section 1.3.1, the feedback linearizing controller is exact (for the design model). Therefore, the closed-loop tracking error dynamics based on the design model are asymptotically stable with the desired error convergence rate. Note also that (for the design model) the tracking is perfect in the sense that the initial condition $\tilde{y}(0)$ decays to zero with the linear dynamics of eqn. (1.16) and is completely unaffected by changes in $y_d(t)$.

However, since the design model is different from the actual plant dynamics, the performance of the actual closed-loop system will be affected by the modeling error. The dynamic model for the actual closed-loop system is

$$\dot{\tilde{y}} = -0.2\tilde{y} + (f(y) - f_o(y)) + (g(y) - g_o(y))\,u. \tag{1.17}$$

Accurate tracking will therefore depend on the accuracy of the design model.

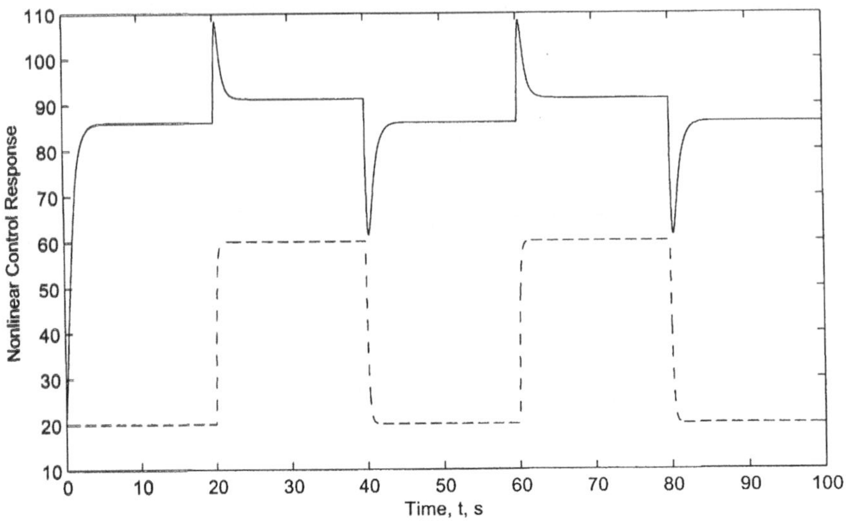

Figure 1.5: Performance of the nonlinear feedback linearizing control system of eqn. (1.15) with the dynamic system of eqn. (1.1). The dotted curve is the commanded response. The solid curve is the actual response.

Figure 1.5 displays the performance of the actual system compensated by the nonlinear feedback linearizing control law of eqn. (1.15) as a solid line. Again, the commanded state y_d (shown as a dashed line) and its derivative are generated by prefiltering y_c (a sequence of step changes) using the filter of eqn. (1.7). The actual response moves in the appropriate direction at the start of each step command, but the modeling error is significant enough that the steady state tracking error for each step is quite large. Since the feedback linearizing controller attempts to cancel the plant dynamics and insert the desired tracking error dynamics, the approach is very sensitive to model error. As shown in eqn. (1.17), the tracking error is directly affected by the error in the design model. An objective of adaptive approximation-based control methods is to adaptively decrease the amount of model error by using online data.

In addition to improving the model accuracy, either offline or online, the performance of the control law of eqn. (1.15) could be improved in a variety of other ways. The control gains could be increased, but this would change the rate of the error convergence relative to the specification, increase the magnitude of the control signal, and increase the effect of noise on the control signal. The linear portion of the controller, currently $K(y_d(t) - y(t))$ could be modified.[1] Also, additional robustifying terms could be added to the nonlinear control law to dominate the model error. These approaches will be described in Chapter 5.

[1] The difference in performance exhibited in Figs. 1.2 and 1.5 is worthy of comment, because the performance of the linear control is better even though both are based on the same design model. The major reason for the difference in performance is that the nonlinear controller is static whereas the linear controller is dynamic in the sense that it includes an integrator. The role of an integrator in a stable controller is to drive the steady state error to zero (see Exercise 1.3).

1.3.4 Adaptive Approximation Based Design

The performance of the feedback linearizing control law was significantly affected by the error between the design model and the actual dynamics. It is therefore of interest to consider whether the data accumulated online, in the process of controlling the system, can be used to decrease the modeling error and improve the control performance. This subsection discusses one such approach. The goal is to motivate various design issues relevant to generic adaptive approximation-based approaches. The remainder of this chapter will expand on these design issues and point the reader to the sections of the book that provide an in-depth discussion of both the issues and alternative design approaches.

In one method to implement such an approach, the designer assumes that the actual system dynamics can be represented as

$$\dot{y}(t) = f(y(t)) + g(y(t))u(t), \tag{1.18}$$

where $f(y) = (\theta_f^*)^\top \phi(y)$ and $g(v) = (\theta_g^*)^\top \phi(y)$ and $\phi(y)$ is a vector of basis functions selected by the designer during the offline design phase. Since f and g are unknown, the parameters θ_f^* and θ_g^* are also unknown and will be estimated online. Therefore, we define the approximated functions $\hat{f}(y) = \theta_f^\top \phi(y)$ and $\hat{g}(y) = \theta_g^\top \phi(y)$, where θ_f and θ_g are parameter vectors that will be estimated using the online data. One approach to using the design model (i.e., f_o and g_o of (1.2)) is to initialize the parameter vector estimates.

The adaptive feedback linearizing control law

$$\begin{aligned} u &= \frac{1}{\hat{g}(y)}\left(-\hat{f}(y) + \dot{y}_d + 0.2\,(y_d - y)\right) && (1.19)\\ \dot{\theta}_f &= \gamma_1 \tilde{y}\phi(y) && (1.20)\\ \dot{\theta}_g &= \gamma_1 u \tilde{y}\phi(y) && (1.21)\end{aligned}$$

results in the actual closed-loop system having error dynamics described by

$$\begin{aligned} \dot{\tilde{y}} &= -0.2\tilde{y} + \tilde{\theta}_f^\top \phi(y) + \tilde{\theta}_g^\top \phi(y)u + e_\phi(y,u) && (1.22)\\ \dot{\tilde{\theta}}_f &= -\gamma_1 \tilde{y}\phi(y) && (1.23)\\ \dot{\tilde{\theta}}_g &= -\gamma_1 u\tilde{y}\phi(y), && (1.24)\end{aligned}$$

where $\tilde{\theta}_f = \theta_f^* - \theta_f$, $\tilde{\theta}_g = \theta_g^* - \theta_g$, and $e_\phi(y,u)$ denotes the residual approximation error (i.e., the approximation error that may still exist even if the parameters of the adaptive approximators were set to their optimal values).[2] The $\tilde{y}$ error dynamics are very similar for the adaptive and nonadaptive feedback linearizing approaches. Relative to the nonadaptive feedback linearizing approach, the error dynamics are more complicated due to the presence of the dynamic equations for θ_f and θ_g. The expected payoff for this added complexity is higher performance (i.e., decreased tracking error). The designer must be careful to analyze the stability of the state of the adaptive feedback linearizing system (i.e., $\tilde{y}$, θ_f and θ_g) and to analyze the effect of $e_\phi(y,u)$. This term is rarely zero and the upper bound on its magnitude is a function of the designer's choice of approximation method (i.e., ϕ).

Figure 1.6 displays the performance of the approximation-based feedback linearizing control law using the basis functions defined by

$$\mu_i(v) = exp\left(\frac{(v-c_i)^2}{25}\right)$$

[2] Rigorous definitions of the optimal parameters and residual approximation error will be given in Section 1.4.2.

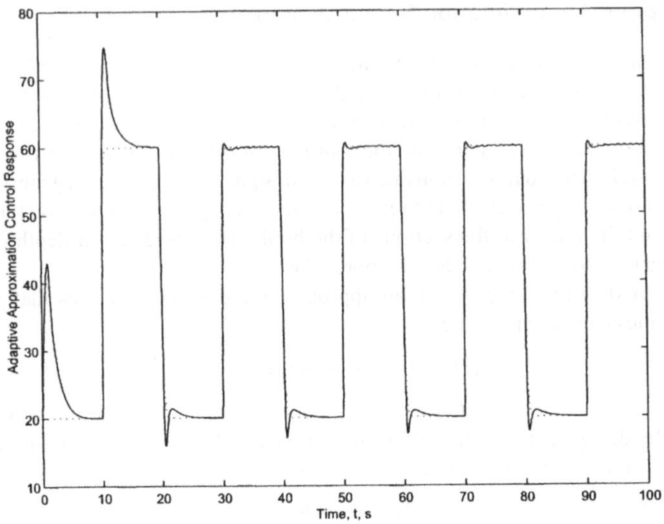

Figure 1.6: Performance of the approximation-based control system of eqn. (1.19)–(1.21) with the dynamic system of eqn. (1.1).

$$\phi_i(v) = \frac{\mu_i(v)}{\sum_{i=1}^{21} \mu_i(v)}$$
$$c_i = (i-1)5, \qquad \text{for } i = 1, \ldots, 21.$$

This simulation uses the actual plant dynamics. Initially, the tracking error is large, but as the online data is used to estimate the approximator parameters, the tracking performance improves significantly.

It is important that the designer understands the relationship between the tracking error and the function approximation error. It is possible for the tracking error to approach zero without the approximation error approaching zero. To see this, consider (1.22). If the last three terms sum to zero, then $\tilde{v}$ will converge to zero. The last three terms sum to zero across a manifold of parameter values, most of which do not necessarily represent accurate approximations over the region $\mathcal{D}$. If the designer is only interested in accurate tracking, then inaccurate function approximation over the entire region $\mathcal{D}$ may be unimportant. If the designer is interested in obtaining accurate function approximations, then conditions for function approximation error convergence must be considered.

Figure 1.7 displays the approximations at the initiation (dotted) and conclusion (solid) of the simulation evaluation, along with the actual functions (dashed). The simulation was concluded after 3000 s of simulated operation. The first 100 s of operation involved the filtered step commands displayed in Figure 1.6. The last 2900 s of operation involved filtered step commands, each with a 10-s duration, randomly distributed in a uniform manner with $y_c \in [20, 60]$. The initial conditions for the function approximation parameter vectors were defined to closely match the functions f_o and g_o of the design model. The bottom graph of Figure 1.8 displays the histogram of y_d at 0.1-s intervals. The top two graphes show the approximation error at the initial and final conditions. By 3000 s, both $\hat{f}$ and $\hat{g}$ have converged over the portion of $\mathcal{D}$ that contains a large amount of training data. Nothing can

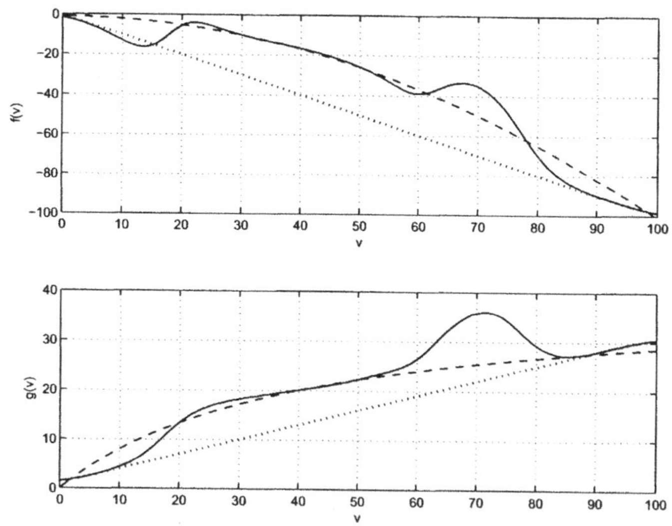

Figure 1.7: Approximations involved in the control system of eqn. (1.19)–(1.21) with the dynamic system of eqn. (1.1). Dotted lines represent initial conditions. Dashed lines represent the actual functions. Solid lines represent the approximation after 3000 s of operation.

be stated about convergence of the approximation outside this portion of $\mathcal{D}$. If the same plots are analyzed after the first 100 s of training, the approximation error is very small near $y = 20$ and $y = 60$, but not significantly improved elsewhere.

1.3.5 Example Summary

The four subsections 1.3.1 – 1.3.4 have each considered a different approach to feedback control design for a nonlinear system involving significant error between the design model (i.e., best available a priori model) and the actual dynamics. The four methods are closely related and all depend on cancelling the dynamics of the assumed model. The approximation-based method is closely related to the adaptive linear and feedback linearizing approaches discussed in the preceding sections. In fact, the approximation-based feedback linearizing approach can be conveniently considered as a combination of the preceding two methods. The differential equations for the parameter estimates of the approximation-based control approach have a structure identical to that for the adaptive linear approach while the control law is identical in structure to the feedback linearizing control approach.

Compared with the adaptive linear control approach, a more complex but more capable function approximation model is used. In the adaptive linear approach the parameter estimation routine attempted to track parameter changes as a function of the changing operation point. This is only feasible if the operating point changes slowly. Even then, tracking the changing model parameters is inefficient. If computer memory is not expensive, it would be more efficient to store the model information as a function of the operating point and recall the model information as needed when the operating point changes. This is a motivation for adaptive approximation-based methods.

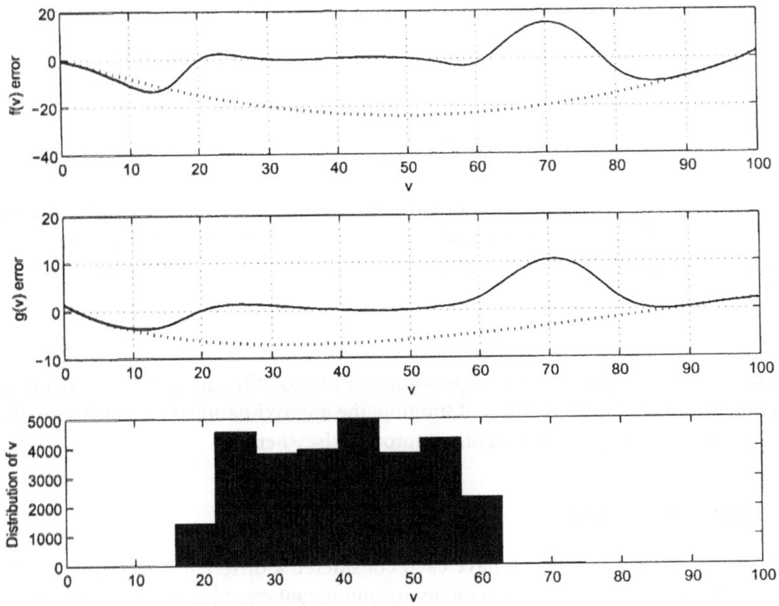

Figure 1.8: Approximation errors corresponding to Figure 1.7. Dotted lines represent initial approximation errors. Solid lines represent approximation errors after 3000 s of operation. The bottom figure shows a histogram of the values of v at 0.1-s increments.

Compared with the feedback linearizing approach, the approximation-based approach is more complex since the dimension of the parameter vectors may be quite large. The rapid increase in computational power and memory at reasonable cost over the last several decades has made the complexity feasible in an increasing array of applications. It is important to note that even though an adaptive approximator may have a very large number of adaptable parameters, with localized approximation models only a very small number of weights are adapted an any one time; therefore, while the memory requirements of adaptive approximation may be large, the computational requirements may be quite reasonable. Also, there is more risk in the approximation-based approach if the stability of the state and parameter estimates is not properly considered. On the positive side, the approximation-based approach has the potential for improved performance since the modeling or approximation error can be decreased online based on the measured control data. The extent to which performance improves will depend on several design choices: control design approach, approximator selection, parameter estimation algorithm, applications conditions, etc.

The following section discusses the major components of adaptive approximation-based control implementations. The discussion is broader than the example based discussion of this section and directs the reader to the appropriate sections of the book where each topic is discussed in depth.

1.4 COMPONENTS OF APPROXIMATION BASED CONTROL

Implementation or analysis of an adaptive approximation-based control system requires the designer to properly specify the problem and solution. This section discusses major aspects of the problem specification.

1.4.1 Control Architecture

Specification of the control architecture is one of the critical steps in the design process. Various nonlinear control methodologies and rigorous tools to analyze their performance have been developed in recent decades [121, 134, 139, 159, 234, 249, 279]. The choices made at this step will affect the complexity of the implementation, the type and level of performance that can be guaranteed, and the properties that the approximated function must satisfy. Major issues influencing the choice of control approach are the form of the system model and the manner in which the nonlinear model error appears in the dynamics. A few methods that are particularly appropriate for use with adaptive approximation are reviewed in Chapter 5.

Consider a dynamic system that can be described as

$$\begin{aligned}
\dot{x}_i &= x_{i+1}, \qquad \text{for } i = 1, \ldots, n-1 \\
\dot{x}_n &= (f_o(x) + f^*(x)) + (g_o(x) + g^*(x))\, u, \\
y &= x,
\end{aligned}$$

where $x(t)$ is the state of the system, $u(t)$ is the control input, f_o and $g_o > 0$ represent the known portions of the dynamics (i.e, the design model), and f^* and g^* are unknown nonlinear functions. Let $\hat{f}$ and $\hat{g}$ represent approximations to the unknown functions f^* and g^*. Then, a feedback linearizing control law can be defined as

$$u(t) = \frac{1}{g_o + \hat{g}} \left(\nu(t) - f_o - \hat{f} \right), \tag{1.25}$$

where $\hat{g}(x) > -g_o(x)$ and $\nu(t)$ can be specified as a function of the tracking error to meet the performance specification. If the approximations were exact (i.e., $f^* = \hat{f}$ and $g^* = \hat{g}$), then this control law would cancel the plant dynamics resulting in

$$\dot{x}_n(t) = \nu(t).$$

When the approximators are not exact, the tracking error dynamic equations are

$$\dot{x}_n(t) = \nu + (f^*(x) - \hat{f}(x)) + (g^*(x) - \hat{g}(x))\, u(t). \tag{1.26}$$

This simple example motivates a few issues that the designer should understand. First, if adaptive approximation is not used (i.e., $\hat{f}(x) = \hat{g}(x) = 0$), the tracking error will be determined by the n-th integral of the the interaction between the control law specified by ν and the model error, as expressed by eqn. (1.26). Second, adaptive approximation is not the only method capable of accomodating the unknown nonlinear effects. Alternative methods such as Lyapunov redesign, nonlinear damping, and sliding mode are reviewed in Section 5.4. These methods work by adding terms to the control law designed to dominate the worst case modeling error, therefore they may involve either large magnitude or high bandwidth control signals. Alternatively, adaptive approximation methods accumulate model information and attempt to remove the effects of a specific set of nonlinearities that fit the model information. These methods are compared, and in some cases combined, in Chapter 6. Third, it is not possible to approximate an arbitrary function over the entire $\Re^n$. Instead, we must restrict the class of functions, constrain the region over which the approximation is desired, or both. Since the operating envelope is already restricted for physical reasons, we will desire the ability to approximate the functions f^* and g^* only over the compact set denoted by $\mathcal{D}$. Note that $\mathcal{D}$ is a fixed compact set, but its size can be selected as large as need be at the design stage. Therefore, we are seeking to show that initial conditions outside $\mathcal{D}$ converge to $\mathcal{D}$ and that for trajectories in $\mathcal{D}$ the trajectory tracking error converges in a desired sense. Various techniques to achieve this are thoroughly discussed in Chapters 6, 7, and 8. The Lyapunov definitions of various forms of stability, and extensions to those definitions, are reviewed in Appendix A.

1.4.2 Function Approximator

Having analyzed the control problem and specified a control architecture capable of using an approximated function to improve the system control performance, the designer must specify the form of the approximating function. This specification includes the definition of the inputs and outputs of the function, the domain $\mathcal{D}$ over which the inputs can range, and the structure of the approximating function. This is a key performance limiting step. If the approximation capabilities are not sufficient over $\mathcal{D}$, then the approximator parameters will be adapted as the operating point changes with no long term retention of model accuracy.

For the discussion that follows, the approximating function will be denoted $\hat{f}(x;\theta,\sigma)$ where

$$\hat{f}(x;\theta,\sigma) = \theta^T \phi(x,\sigma). \tag{1.27}$$

In this notation x is a dummy variable representing the input vector to the approximation function. The actual function inputs may include elements of the plant state, control input, or outputs. The notation $\hat{f}(x;\theta,\sigma)$ implies that $\hat{f}$ is evaluated as a function of x when θ and σ are considered fixed for the purposes of function evaluation. In applications, the approximator parameters θ and σ will be adapted online to improve the accuracy of the approximating function — this is referred to as *training* in the neural network literature. The parameters θ are referred to in the (neural network) literature as the output layer parameters. The parameters σ are referred to as the input layer parameters. Note that the approximation of eqn. (1.27) is linear-in-the-parameters with respect to θ. The vector of basis functions ϕ will be referred to as the regressor vector. The regressor vector is typically a nonlinear function of x and the parameter vector σ. Specification of the structure of the approximating function includes selection of the basis elements of the regressor ϕ, the dimension of θ, and the dimension of σ. The values of θ and σ are determined through parameter estimation methods based on the online data.

Regardless of the choice of the function approximator and its structure, it will normally be the case that perfect approximation is not possible. The approximation error is denoted by $e(x;\theta,\sigma)$ where

$$e(x;\theta,\sigma) = f(x) - \hat{f}(x;\theta,\sigma). \tag{1.28}$$

If θ^* and σ^* denote parameters that minimize the ∞-norm of the approximating error over a compact region $\mathcal{D}$, then the Minimum Functional Approximation Error (MFAE) is defined as

$$e_\phi(x) = e(x;\theta^*,\sigma^*) = f(x) - \hat{f}(x;\theta^*,\sigma^*).$$

In practice, the quantities e_ϕ, θ^* and σ^* are not known, but are useful for the purposes of analysis. Note, as in eqn. (1.22), that $e_\phi(x)$ acts as a disturbance affecting the tracking error and therefore the parameter estimates. Therefore, the specification of the adaptive approximator $\hat{f}(x;\theta,\sigma)$ has a critical affect on the tracking performance that the approximation-based control system will be capable of achieving.

The approximator structure defined in eqn. (1.27) is sufficient to describe the various approximators used in the neural and fuzzy control literature, as well as many other approximators. Issues related to the adaptive approximation problem and approximator selection will be discussed in Chapter 2. Specific approximators will be discussed in Chapter 3.

1.4.3 Stable Training Algorithm

Given that the control architecture and approximator structure have been selected, the designer must specify the algorithm for adapting the adjustable parameters θ and σ of the approximating function based on the online data and control performance.

Parameter estimation can be designed for either a fixed batch of training data or for data that arrives incrementally at each control system sampling instant. The latter situation is typical for control applications; however, the batch situation is the focus for much of the traditional function approximation literature. In addition, much of the literature on function approximation is devoted to applications where the distribution of the training data in $\mathcal{D}$ can be specified by the designer. Since a control system is completing a task during the function approximation process, the distribution of training data usually cannot be specified by the control system designer. The portion of the function approximation literature concerned with batches of data where the data distribution is defined by the experiment and not the analyst is referred to as scattered data approximation methods [84]. Adaptive approximation-based control applications are distinct from traditional batch scattered data approximation problems in that:

- the data involved in the parameter estimation will become available incrementally (ad infinitum) while the approximated function is being used in the feedback loop;

- the training data might not be the direct output of the function to be approximated; and,
- the stability of the closed-loop system, which depends on the approximated function, must be ensured.

The main issue to be considered in the development of the parameter estimation algorithm is the overall stability of the closed-loop control system. The stability of the closed-loop system requires guarantees of the convergence of the system state and of (at least) the boundedness of the error in the approximator parameter vector. This analysis must be completed with caution, as it is possible to design a system for which the system state is asymptotically stable while

1. even when perfect approximation is possible (i.e., $e_\phi = 0$), the error in the estimated approximator parameters is bounded, but not convergent;
2. when perfect approximation is not possible, the error in the estimated approximator parameters may become unbounded.

In the first case, the lack of approximator convergence is due to lack of persistent excitation, which is further discussed in Chapter 4. This lack of approximator convergence may be acceptable, if the approximator is not needed for any other purpose, since the control performance is still achieved; however, control performance will improve as approximator accuracy increases. Also, the designer of a control system involving adaptive approximation sometimes has interest in the approximated function and is therefore interested in its accuracy. In such cases, the designer must ensure the convergence of the control state and approximator parameters. In the second case (the typical situation), the fact that e_ϕ cannot be forced to zero over $\mathcal{D}$ must be addressed in the design of the parameter estimation algorithm. Chapter 4 discusses the basic issues of adaptive (incremental) parameter estimation. Various methods including least squares and gradient descent (back-propagation) are derived and analyzed. Chapters 6 and 7 discuss the issues related to parameter estimation in the context of feedback control applications. Chapter 6 presents a detailed analysis of the issues related to stability of the state and parameter estimates. Robustness of parameter estimation algorithms to noise, disturbances, and $e_\phi(x)$ is discussed in Section 4.6 as well as in Chapter 7.

1.5 DISCUSSION AND PHILOSOPHICAL COMMENTS

The objective of adaptive approximation-based control methods is to achieve a higher level of control system performance than could be achieved based on the *a priori* model information. Such methods can be significantly more complicated (computationally and theoretically) than non-adaptive or even linear adaptive control methods. This extra complication can result in unexpected behavior (e.g., instability) if the design is not rigorously analyzed under realistic assumptions.

Adaptive function approximation has an important role to play in the development of advanced control systems. Adaptive approximation-based control, including neural and fuzzy approaches, have become feasible in recent decades due to the rapid advances that have occurred in computing technologies. Inexpensive desktop computing has inspired many ad hoc approximation-based control approaches. In addition, similar approaches in different communities (e.g., neural, fuzzy) have been derived and presented using different nomenclature yet nearly identical theoretical results. Our objective herein is to present such approaches rigorously within a unifying framework so that the resulting presentation encompasses both the adaptive fuzzy and neural control approaches, thereby allowing the discussion to focus on the underlying technical issues.

The three terms, adaptation, learning, and self-organization, are used with different meanings by different authors. In this text, we will use *adaptation* to refer to temporal changes. For example, adaptive control is applicable when the estimated parameters are slowly varying functions of time. We will use *learning* to refer to methods that retain information as a function of measured variables. Herein, learning is implemented via function approximation. Therefore, learning has a spatial connotation whereas adaptation refers to temporal effects. The process of learning requires adaptation, but the retention of information as a function of other variables in learning implies that learning is a higher level process than is adaptation.

Implementation of learning via function approximation requires specification of the function approximation structure. This specification is not straightforward, since the function to be approximated is assumed to be unknown and input-output samples of the function may not be available *a priori*. For the majority of this text, we assume that the designer is able to specify the approximation structure prior to online operation. However, an unsolved problem in the field is the online adaptation of the function approximation structure. We will refer to methods that adapt the function approximation structure during online operation as *self-organizing*.

Since most physical dynamic systems are described in continuous-time, while most advanced control systems are implemented via digital computer in discrete-time, the designer may consider at least two possible approaches. In one approach, the design and analysis would be performed in continuous-time with the resulting controller implemented in discrete-time by numeric integration. The alternative approach would be to transform the continuous-time ordinary differential equation to a discrete-time model that has equivalent state behavior at the sampling instants and then perform the control system design and analysis in discrete-time. Throughout this text, we will take the former approach. We do not pursue both approaches concurrently as the required significant increase in length and complexity would not provide a proportionate increase in understanding of the main design and analysis issues. Furthermore, the transformation of a continuous-time nonlinear system to a discrete-time equivalent model is not straightforward and often does not maintain certain useful properties of the continuous-time model (e.g., affine in the control).

1.6 EXERCISES AND DESIGN PROBLEMS

Exercise 1.1 This exercise steps through the design details for the linear controller of Section 1.3.1.

1. For the specified design model of eqn. (1.2), show that

$$\left.\frac{\partial h}{\partial y}\right|_{(y^*,u^*)} = -1.0 + 0.3u^* \qquad \left.\frac{\partial h}{\partial u}\right|_{(y^*,u^*)} = 1.0 + 0.3y^*,$$

and that the linearized system at $(y^*, u^*) = (40, \frac{40}{13})$ is

$$\delta\dot{y} = p\delta y + 13\delta u$$

with $p = \frac{-1}{13}$.

2. Analyze the linear control law of eqn. (1.4) and the linearized dynamics (above) to see that the nominal control design relies on cancelling the plant dynamics and replacing them with error dynamics of the desired bandwidth. Analyze the characteristic equation of the second-order, closed-loop linearized dynamics to see what happens to the closed-loop poles when p is near but not equal to $\frac{-1}{13}$.

3. Design a set of linear controllers and a switching mechanism (i.e., a gain scheduled controller) so that the closed-loop dynamics of the design model achieve the bandwidth specification over the region $v \in [20, 60]$. Test this in simulation. Analyze the performance of this gain scheduled controller using the actual dynamics.

Exercise 1.2 This exercise steps through the design details for the linear adaptive controller of Section 1.3.2.

1. Derive the error dynamics of eqns. (1.11)–(1.14) for the linear adaptive control law. (Hint: add $-\hat{c}u + \hat{c}u$ to eqn. (1.5) and substitute eqn. (1.6) for the latter term.)

2. Show that the correct values for the model of eqn. (1.5) to match eqn. (1.1) to first order are:

$$\begin{bmatrix} a \\ b \\ c \end{bmatrix} = \begin{bmatrix} f - \left(\frac{\partial f}{\partial y} + \frac{\partial g}{\partial y} u\right) y \\ \left(\frac{\partial f}{\partial y} + \frac{\partial g}{\partial y} u\right) \\ g \end{bmatrix}_{y=y^*, u=u^*}$$

3. Implement a simulation of the adaptive control system of Section 1.3.2. First, duplicate the results of the example. Do the estimated parameters converge to the same values each time v is commanded to the same operating point?

4. Using the Lyapunov function

$$V(\tilde{v}, \tilde{a}, \tilde{b}, \tilde{c}) = \frac{1}{2}\left(\tilde{v}^2 + \frac{1}{\gamma_1}\tilde{a}^2 + \frac{1}{\gamma_2}\tilde{b}^2 + \frac{1}{\gamma_3}\tilde{c}^2\right)$$

show that the time derivative of V evaluated along the error dynamics of the adaptive control system is negative semidefinite. Why can we only say that this derivative is semidefinite? What does this fact imply about each component of $(\tilde{v}, \tilde{a}, \tilde{b}, \tilde{c})$?

Exercise 1.3 This exercise steps through the design details of an extension to the feedback linearizing controller of Section 1.3.3.

Consider the dynamic feedback linearizing controller defined as

$$\begin{aligned} \dot{\tilde{x}} &= \tilde{y} \\ u &= \frac{1}{g_o(y)}\left(-f_o(y) + \dot{y}_d + K_1\tilde{x} + K_2\tilde{y}\right) \end{aligned}$$

where $\tilde{y} = (y - y_d)$. This controller includes an appended integrator with the goal of driving the tracking error to zero.

1. Show that the tracking error dynamics (relative to the design model) are

$$\begin{aligned} \dot{\tilde{x}} &= \tilde{y} \\ \dot{\tilde{y}} &= -K_1\tilde{x} - K_2\tilde{y}. \end{aligned}$$

2. For stability of the closed-loop system, relative to the design model, K_1 and K_2 must both be positive. If $K_1 = 0.04$ and $K_2 = 0.40$, then the linear tracking error dynamics have two poles at 0.2. If $K_1 = 1.00$ and $K_2 = 5.20$, then the poles are at 0.2 and 5.0. In each case, there is a dominant pole at 0.2. For each set of control gains:

   (a) Simulate the closed-loop system formed by this controller and the design model. Use this simulation to ensure that your controller is implemented correctly. The tracking should be perfect. That is, the tracking error states converge exponentially toward zero and are not affected by changes in y_d. If the tracking error states are initially zero, then they are permanently zero.

   (b) Simulate the closed-loop system formed by this controller and the actual dynamics.

   Discuss the effect of model error. Discuss the tradeoffs related to the choice of control gains.

Exercise 1.4 This exercise steps through the design details for the adaptive approximation-based feedback linearizing controller of Section 1.3.4.

1. Derive the error dynamics for the adaptive approximation-based control law.

2. Implement a simulation of the approximation-based control system of Section 1.3.4. First, duplicate the results of the example. Plot the approximation error versus v at $t = 100$. Discuss why it is small near $v = 20$ and $v = 100$, but not small elsewhere.

3. Using the Lyapunov function

$$V(\tilde{v}, \tilde{\theta}_f, \tilde{\theta}_g) = \frac{1}{2}\left(\tilde{v}^2 + \frac{1}{\gamma_1}\tilde{\theta}_f^\top \tilde{\theta}_f + \frac{1}{\gamma_2}\tilde{\theta}_g^\top \tilde{\theta}_g\right)$$

   show that the time derivative of V evaluated along the error dynamics of the approximation-based control system is negative semidefinite. Why can we only say that this derivative is semidefinite? What does this fact imply about each component of $(\tilde{v}, \tilde{\theta}_f, \tilde{\theta}_g)$?

CHAPTER 2

APPROXIMATION THEORY

This chapter formulates the numeric data processing issues of interpolation and function approximation, and then discusses function approximator properties that are relevant to the use of adaptive approximation for estimation and feedback control. Our interest in function approximation is derived from the hypothesis that online control performance could be improved if unknown nonlinear portions of the model are more accurately modeled. Although the data to improve the model may not be available *a priori*, additional data can be accumulated while the system is operating. Appropriate use of such data to guarantee performance improvement requires that the designer understand the areas of function approximation, control, stability, and parameter estimation. This chapter focuses on several aspects of approximation theory.

The discussion of function approximation is subdivided into offline and online approximation. Offline function approximation is concerned with the questions of selecting a family of approximators and parameters of a particular approximator to optimally fit a *given* set of data. The issue of the design of the set of data is also of interest when the acquisition of the data is under the control of the designer. An understanding of offline function approximation is necessary before delving into online approximation. The discussion of online approximation builds on the understanding of offline approximation, and also raises new issues motivated by the need to guarantee stability of the dynamic system and estimation process, the possible need to forget old stored information at a certain rate, and the inability to control the data distribution.

Section 2.1 presents an easy-to-understand (and replicate) example in order to motivate, in the context of online approximation based control, a few important issues that will

Adaptive Approximation Based Control: Unifying Neural, Fuzzy and Traditional Adaptive Approximation Approaches. By Jay A. Farrell and Marios M. Polycarpou

be discussed through the remainder of this chapter. Section 2.2 discusses the problem of function interpolation. Section 2.3 discusses the problem of function approximation. Section 2.4 discusses function approximator properties in the context of online function approximation.

2.1 MOTIVATING EXAMPLE

Consider the following simple example that illustrates some of the issues that arise in approximation based control applications.

■ EXAMPLE 2.1

Consider the control of the discrete-time system

$$\begin{aligned} x(k+1) &= f(x(k)) + u(k) \\ y(k) &= x(k), \end{aligned}$$

where $u(k)$ is the control variable at discrete-time k, $x(k)$ is the state, $y(k)$ is the measured output, the function $f(x)$ is not known to the designer, and the control law is given by

$$u(k) = y_d(k+1) - \beta\left[y_d(k) - y(k)\right] - \hat{f}(y(k)). \tag{2.1}$$

The above control law assumes that the reference trajectory y_d is known one step in advance. For the purposes of simulation in the example, we will use $f(x) = \sin(x)$.

If $\hat{f}(y) = \sin(y)$, then the closed-loop tracking error dynamics would be

$$e(k+1) = \beta e(k),$$

where $e(k) = y_d(k) - x(k)$, which is stable for $|\beta| < 1$ (in the following simulation example we use $\beta = 0.5$). If $\hat{f}(y) \neq f(x)$, then the closed-loop tracking error dynamics would be

$$e(k+1) = \beta e(k) - [f(x(k)) - \hat{f}(y(k))]. \tag{2.2}$$

Therefore, the tracking performance is directly affected by the accuracy of the design model $\hat{f}(x)$. The left hand column of Figure 2.1 shows the performance of this closed-loop system when $y_d(k) = \pi\sin(0.1k)$ and $\hat{f}(y) = 0$.

When $f(x)$ is not known *a priori*, the designer may attempt to improve the closed-loop performance by developing an online (i.e., adaptive) approximation to $f(x)$. In this section a straightforward database function approximation approach is used. At each time step k, the data

$$z(k) = [f(y(k-1)),\ y(k-1)]$$

will be stored. Note that the approach of this example requires that the function value $f(y(k-1))$ must be computable at each step from the measured variables. This assumed approach is referred to as *supervised learning*. This is a strict assumption that is not always applicable. Much more general control approaches that do not require this assumption are presented in Chapter 6. For this example, at time k, the information in $z(k)$ can be computed from available data according to

$$z(k) = [y(k) - u(k-1),\ y(k-1)].$$

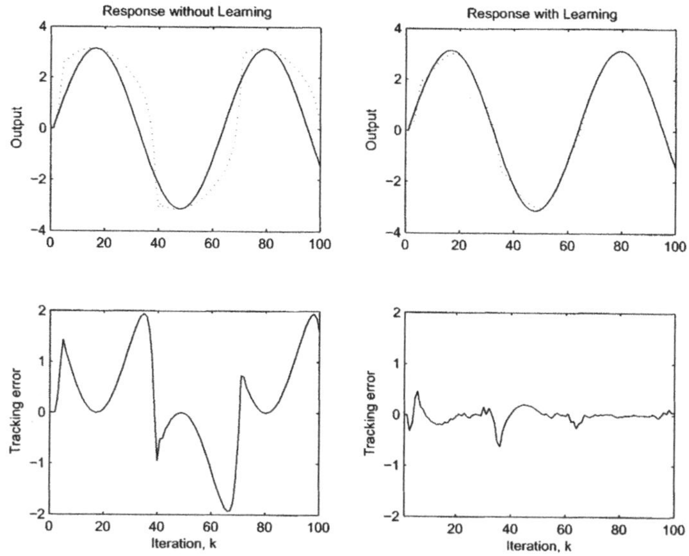

Figure 2.1: Closed-loop control performance for eqn. (2.1). Left column corresponds to $\hat{f} = 0$. Right column corresponds to $\hat{f}$ constructed via nearest neighbor matching. For the top row of graphs, the solid line is the reference trajectory. The dotted line is the system response. The tracking error is plotted in the bottom row of graphs.

At time step k with $y(k)$ available, $u(k)$ is calculated using eqn. (2.1) as follows: (1) search the second column of z for the row i that most closely matches $y(k)$ (i.e., $i = argmin_{0 \leq j \leq k} (\|z(j, 2) - y(k)\|)$, (2) use $\hat{f}(y(k)) = z(i, 1)$. The remaining terms in eqn. (2.1) can be directly calculated. The right-hand column of Figure 2.1 shows the performance of the closed-loop system using this adaptive approximation based method. Note that as the row dimension of z grows with k (i.e., more data values for $f(x)$ are stored) the tracking performance rapidly improves. However, both the memory required to store z and the computation required to search z increase at each iteration.[1]

The top graph of Figure 2.2 plots as discrete points the first column of z as a function of the second column of z. The approximate function used in the control law is piecewise constant with each piecewise section (of variable width) centered on one of the examples $y(i)$, as shown in the bottom graph of Figure 2.2. With noise-free data, the approximation becomes very good for large k. The approach defined above is referred to as nearest neighbor matching . Various other alternatives are possible such as k-nearest neighbor averaging, which perform better when noise is present in the measurement data. △

Note the following issues related to this example and the broader adaptive function approximation problem:

[1]Assuming a binary search of one-dimensional ordered data, the number of comparisons is on the order of $\log_2(k)$. In addition, as each new sample arrives, the stored data must be moved to maintain the ordering.

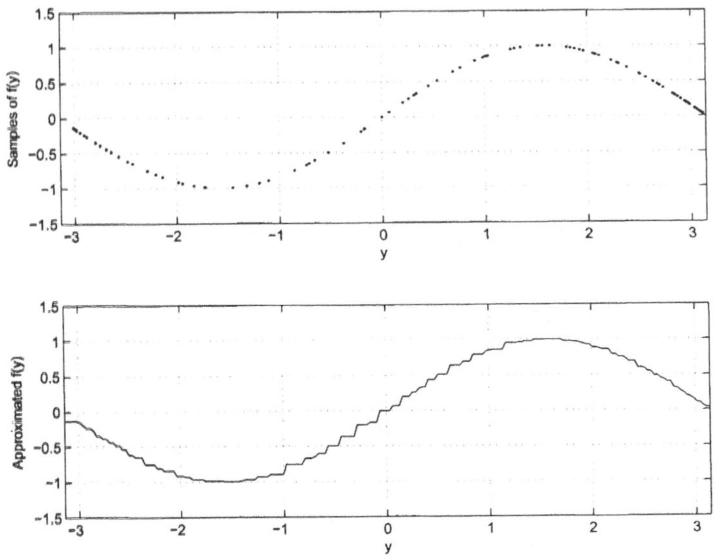

Figure 2.2: *Top* – Data for approximating $\hat{f}$ using nearest neighbor matching. *Bottom* – Approximated $\hat{f}$ resulting from nearest neighbor matching. Both graphs correspond to Example 2.1.

1. The input-output training data $(f(y(i)), y(i))$ cannot be expected to be distributed according to an analytic distribution. Instead, the training data will be defined by the control task that the system is performing. The distribution of training data over a fixed-duration window will typically be time varying. If control is operating well, then the training samples will cluster in the vicinity of a state trajectory (several may be possible) defined by the reference input. In particular, over short periods of time, the training data will not be uniformly distributed, but will cluster in some small subregion of the domain of approximation. For example, if the control objective is regulation to a certain fixed point (i.e., $y_d(k)$ = constant) then the training data may cluster around a single point.

2. When the raw training data are stored, as in this example, the approach will have growing memory and computational penalties. These can be overcome by the function approximation and recursive parameter estimation techniques to be described.

3. Consider the case of measurement data corrupted by noise. Direct storage of the data does not work as well as shown in Figures 2.1 and 2.2. Figure 2.3 shows performance[2] in the time domain when the measured $y(k)$ is corrupted with Gaussian random noise $n(k)$ with standard deviation $\sigma = 0.1$. In this case, $y(k) = x(k) + n(k)$ is stored in the database calculations and used in the control law. The actual tracking error $(x - y_d)$ is plotted. For $k > 100$, the tracking error has standard deviation of 0.16. So the approach has amplified the effects of noise. In this approach, noisy data are

[2]Note that the magnitude of the reference signal has also been decreased: $y_d = \frac{\pi}{2}\sin(0.1k)$. The reason for this will become clear in the subsequent item.

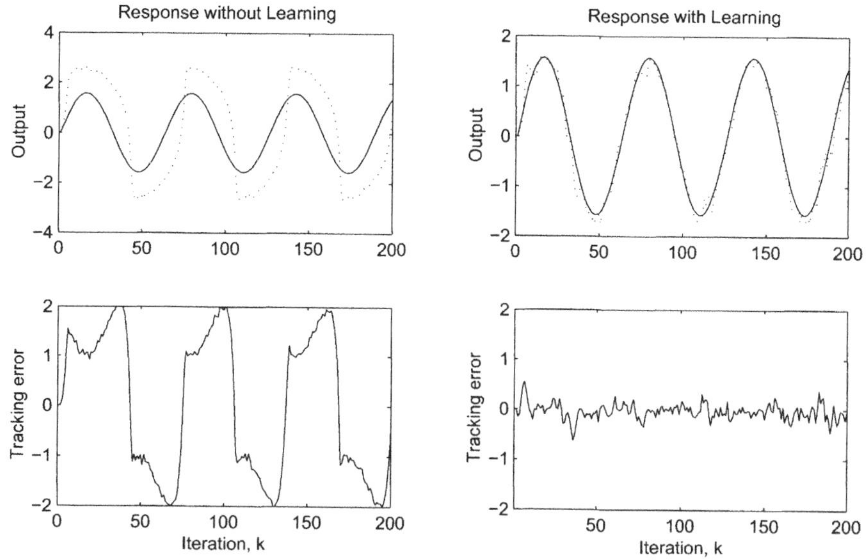

Figure 2.3: Closed-loop control performance for eqn. (2.1) with noisy measurement data. Left column corresponds to $\hat{f} = 0$. Right column corresponds to $\hat{f}$ constructed via nearest neighbor matching. In the top row of graphs, the solid line is the reference trajectory and the dotted line is the system response.

stored in the data vector without noise attenuation. It is important to note that, as we will see, methods to attenuate noise through averaging lead directly to function approximation methods.

4. Function approximation problems are not well defined. Consider Figure 2.4, which corresponds to the the data matrix z stored relative to Figure 2.3. If the domain of approximation that is of interest is $D = [-\pi, \pi]$, how should the approximation given the available data be extended to all of D (or should it?). A quick inspection of the data might lead to the conclusion that the function is linear. A more careful inspection, noting the apparent curvature near $\pm\frac{\pi}{2}$ might result in the use of a saturating function. From our knowledge of $f(x)$ neither of these is of course correct. Extreme care must be exercised in generalizing from available data in given regions to the form of the function in other regions. The manner in which data in one region affects the approximated function in another region is determined primarily by the specification of the function approximator structure. The assumed form of the approximation inserts the designer's bias into the approximation problem. The effect of this bias should be well understood.

5. From eqn. (2.2) the designer might expect that, as the database accumulates data, then the $(f - \hat{f})$ term and hence e should decrease; however, the control and function approximation approach of this example did not allow a rigorous stability analysis. The parametric function approximation methods that follow will enable a rigorous analysis of the stability properties of the closed-loop system.

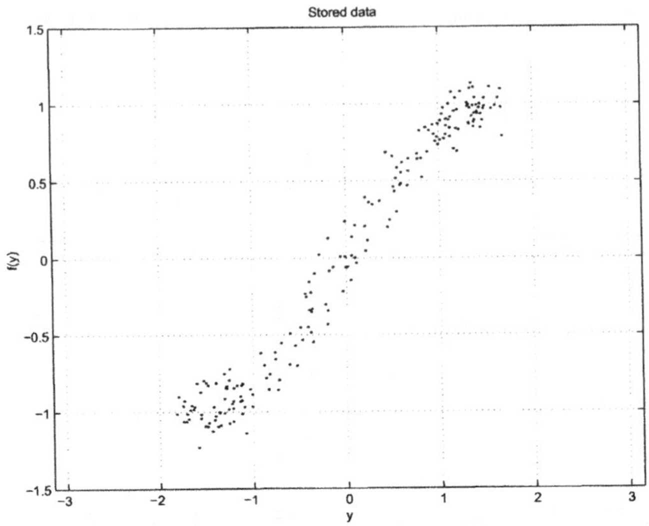

Figure 2.4: Data for approximating $\hat{f}$ corresponding to eqn. (2.1) with noisy measurement data.

Items 2 through 4 above naturally direct the attention of the designer to more general function interpolation and approximation issues. The above nearest neighbors approach can be represented as

$$\hat{f}(x : z(k)) = \sum_{i=1}^{k} z(i,1)\phi_i(x : z(k)) \tag{2.3}$$

where the notation $\hat{f}(x : z(k))$ means the value of $\hat{f}$ evaluated at x given the data in database matrix z at time k, and

$$\phi_i(x : z(k)) = \begin{cases} 1 & \text{if } |x - z(i,2)| = \min_{1 \le j \le k} |x - z(j,2)| \\ 0 & \text{otherwise,} \end{cases} \tag{2.4}$$

where we have assumed that no two entries (i.e., rows) have the same value for $z(j,2)$. Note that by its definition, this function passes exactly through each piece of measured data (i.e., $\hat{f}(z(i,2) : z(k)) = z(i,1)$). This is referred to as *interpolation.* Item 2 above points out the fact that this approximation structure has k basis elements that are redefined at each sampling instant. The computational complexity and memory requirements can be decreased and fixed by instead using a fixed number N of basis elements of the form

$$\hat{f}(x : \theta, \sigma, N) = \sum_{i=1}^{N} \theta_i \phi_i(x : \sigma_i) \tag{2.5}$$

where the data matrix z would be used to estimate $\theta = [\theta_1, \dots, \theta_N]$ and $\sigma = [\sigma_1, \dots, \sigma_N]$. With such a structure, it will eventually happen that there is more data than parameters, in which case interpolation may no longer be possible. After this instant in time, a well-designed parameter estimation algorithm will combine new and previous measurements to attenuate the affects of measurement noise on the approximated function. The choice of basis functions can affect the noise attenuation properties of the approximator. In addition, the choice of approximator will affect the accuracy of the approximation, the degree of approximator continuity and the extent of training *generalization*, as will be explained in Section 2.4.7.

2.2 INTERPOLATION

Given a set of input-output data $\{(x_j, y_j) \mid j = 1, \ldots, m;\ x_j \in \Re^n;\ y_j \in \Re^1\}$, function interpolation is the problem of defining a function $\hat{f}(x) : \Re^n \to \Re^1$ such that $\hat{f}(x_j) = y_j$ for all $j = 1, \ldots, m$. When $\hat{f}(x)$ is constrained to be an element of a finite dimensional linear space, this is called *Lagrange interpolation*. The interpolating function $\hat{f}(x)$ can then be used to estimate the value of $f(x)$ between the *known* values of $f(x_j)$.

In Lagrange interpolation with the basis functions $\{\phi_i(x)\}_{i=1}^N$,

$$\hat{f}(x) = \sum_{i=1}^{N} \theta_i \phi_i(x) = \theta^\top \phi(x) = \phi(x)^\top \theta, \tag{2.6}$$

where $\theta = [\theta_1, \ldots, \theta_N]^\top \in \Re^N$ and $\phi(x) = [\phi_1(x), \ldots, \phi_N(x)]^\top : \Re^n \to \Re^N$. The Lagrange interpolation condition can be expressed as the problem of finding θ such that

$$\begin{bmatrix} y_1 \\ \vdots \\ y_m \end{bmatrix} = \begin{bmatrix} \phi(x_1)^\top \\ \vdots \\ \phi(x_m)^\top \end{bmatrix} \begin{bmatrix} \theta_1 \\ \vdots \\ \theta_N \end{bmatrix} \tag{2.7}$$

$$Y = \Phi^\top \theta. \tag{2.8}$$

Note that $\Phi = [\phi(x_1), \ldots, \phi(x_m)] \in \Re^{N \times m}$. The matrix $\Phi^\top$ is referred to as the interpolation or collocation matrix. Much of the function approximation and interpolation literature focuses on the case where $n = 1$. When $n > 1$ and the data points are not defined on a grid, the problem is referred to as scattered data interpolation .

A necessary condition for interpolation to be possible is that $N \geq m$. In online applications, where m is unbounded (i.e., $x_k = x(kT)$), interpolation would eventually lead to both memory and computational problems.

If $N = m$ and Φ is nonsingular, the unique interpolated solution is

$$\theta = \left(\Phi^\top\right)^{-1} Y = \Phi^{-\top} Y. \tag{2.9}$$

Nonsingularity of Φ is equivalent to the column vectors $\phi(x_i), i = 1, \ldots, m$ being linearly independent. This requires (at least) that the x_i be distinct points. Once suitable N, $\phi(x)$, y_i, and x_i have been specified, the interpolation problem has a guaranteed unique solution.

When the basis set $\{\phi_j\}_{j=1}^N$ has the property that the matrix Φ is nonsingular for any distinct $\{x_i\}_{i=1}^N$, the linear space spanned by $\{\phi_j\}_{j=1}^N$ is referred to as a *Chebyshev space* or a *Haar space* [79, 155, 218]. The issue of how to select ϕ to form a Haar space has been widely studied. A brief discussion of related issues is presented in Section 2.4.10. Even if the theoretical conditions required for Φ to be invertible are satisfied, if x_i is near x_j for $i \neq j$, then Φ may be nearly singular. In this case, any measurement error in Y may be magnified in the determination of θ. In addition, the solution via eqn. (2.9) may be numerically unstable. Preferred methods of solution are by QR, UD, or singular value decompositions [99].

For a unique solution to exist, the number of free parameters (i.e., the dimension of θ) must be exactly equal to the number m of sample points x_i. Therefore, the dimension of the approximator parameter vector must increase linearly with the number of training points. Under these conditions, the number of computations involved in solving eqn. (2.9) is on the order of m^3 floating point operations (FLOPS) (see Section 5.5.9 in [99]). In addition to this large computational burden, the condition number of Φ often becomes small as m gets large.

As the number of data points m increases, there will eventually be more data (and for $m = N$ more degrees of freedom in the approximator) than degrees of freedom in the underlying function. In typical situations, the data y_i will not be measured perfectly, but will include errors from such effects as sensor measurement noise. The described interpolation solution attempts to fit this noisy data perfectly, which is not usually desirable. Approximators with $N < m$ parameters will be over-constrained (i.e., more constraints than degrees of freedom). In this case, the approximated function can be designed (in an appropriate sense) to attenuate the effects of noisy measurement data. An additional benefit of fixing N (independent of m) is that the computational complexity of the approximation and parameter estimation problems is fixed as a function of N and does not change as more data is accumulated.

■ **EXAMPLE 2.2**

Consider Figures 2.2 and 2.4. The former figure represents the underlying "true" function (i.e., noise-free data samples). The latter represents noisy samples of the underlying function. Interpolation of the data in Figure 2.4 would not generate a reliable representation of the desired function. In fact, depending on the choice of basis functions, interpolation of the noisy data may amplify the noise between the data points. △

2.3 FUNCTION APPROXIMATION

The linear in the parameters[3] (LIP) function approximation problem can be stated as: Given a basis set $\{\phi_i(x) : \Re^n \rightarrow \Re \text{ for } i = 1\dots, N\}$ and a function $f(x) : \Re^n \rightarrow \Re^1$ find a linear combination of the basis elements $\hat{f}(x) = \theta^\top \phi(x) : \Re^n \rightarrow \Re^1$ that is close to f. Key problems that arise are:

- How to select the basis set?
- How to measure closeness?
- How to determine the optimal parameter vector θ for the linear combination?

In the function approximation literature there are various broad classes of function approximation problems. The class of problems that will be of interest herein is the development of approximations to functions based on information related to input-output samples

[3] In general, the function approximation problem is not limited to LIP approaches; however, this introductory section will focus on LIP approaches to simplify the discussion.

of the function. The foundations of the results that follow are linear algebra and matrix theory [99].

2.3.1 Offline (Batch) Function Approximation

Given a set of input-output data $\{(x_i, y_i), i = 1, \ldots, m\}$ function approximation is the problem of defining a function $\hat{f}(x) : \Re^n \rightarrow \Re^1$ to minimize $\|\hat{Y} - Y\|$ where $Y = [y_1, \ldots, y_m]^\top$ and $\hat{Y} = [\hat{f}(x_1), \ldots, \hat{f}(x_m)]^\top$. The discussion of the following two sections will focus on the over and under constrained cases where $\|.\|$ denotes the $p = 2$ (Euclidean) norm. Solutions for other p norms are discussed, for example, in the references [54, 309].

2.3.1.1 Over-constrained Solution Consider the approximator structure of eqn. (2.6), which can be represented in matrix form as in eqn. (2.8). When $N < m$ the problem is over-specified (more constraints than degrees of freedom). In this case, the matrix Φ defined relative to eqn. (2.8) is not square and its inverse does not exist. In this case, there may be no solution to the corresponding interpolation problem. Since with the specified approximation structure the data cannot be fit perfectly, the designer may instead select the approximator parameters to minimize some measure of the function approximation error. If a weighted second-order cost function is specified then

$$J(\theta) = \frac{1}{2}(\hat{Y} - Y)^\top W(\hat{Y} - Y) \tag{2.10}$$

which corresponds to the norm $\|Y\|_W^2 = \frac{1}{2}Y^\top WY$ where W is symmetric and positive definite. In this case, the optimal vector θ^* can be found by differentiation:

$$\begin{aligned} J(\theta) &= \frac{1}{2}(\Phi^\top\theta - Y)^\top W(\Phi^\top\theta - Y) && (2.11)\\ \left.\frac{\partial J(\theta)}{\partial\theta}\right|_{\theta=\theta^*} &= (\Phi^\top\theta^* - Y)^\top W\Phi^\top = 0 && (2.12)\\ \theta^* &= \left(\Phi W\Phi^\top\right)^{-1}\Phi WY && (2.13) \end{aligned}$$

where it has been assumed that $rank(\Phi) = N$ (i.e., that Φ has N linearly independent rows and columns) so that $\Phi W\Phi^\top$ is nonsingular. When the rank of $\Phi < N$ (i.e., the N rows of Φ are not linearly independent), then either additional data are required or the under-constrained approach defined below must be used. Since the second derivative of $J(\theta)$ with respect to θ evaluated at θ^* (i.e., $\Phi W\Phi^\top$), is at least positive semidefinite, the solution of eqn. (2.13) is a minimum of the cost function.

Eqn. (2.13) is the *weighted least squares* solution. If W is a scalar multiple of the identity matrix, then the standard least squares solution results. Note from eqn. (2.12) that the weighted least squares approximation error $(\Phi^\top\theta^* - Y)$ has the property that it is orthogonal to all N columns of the weighted regressor $W\Phi^\top$.

Even when the $rank(\Phi)$ is N so that the inverse of $\left(\Phi W\Phi^\top\right)$ exists, the weighted least squares solution may still be poorly conditioned. In such a case, direct solution of eqn. (2.13) may not be the best numeric approach (see Ch. 5 in [99]). The condition number of the matrix A (i.e., $cond(A)$) provides an estimate of the sensitivity of the solution of the linear equation $Ax = b$ to errors in b. If $\bar{\sigma}(A)$ and $\underline{\sigma}(A)$ denote the maximum and minimum singular values of A, then $\log_{10}\left(\frac{\bar{\sigma}(A)}{\underline{\sigma}(A)}\right)$ provides an estimate in the number of decimal digits of accuracy that are lost in solving the linear equation. The function $C = \frac{\underline{\sigma}(A)}{\bar{\sigma}(A)}$ is an estimate of the distance between A and a singular matrix. Even if $(\Phi W \Phi^\top)$ has rank equal to N, if C is near zero, then the problem is not numerically well conditioned.

■ EXAMPLE 2.3

If a designer chooses to approximate a function $f(x)$ by an $(N-1)$-st order polynomial using the natural basis for polynomials $\{1, x, x^2, \ldots, x^{N-1}\}$ and the evaluation points are $\{\frac{1}{i}\}_{i=1:m}$ with $m \geq N$ then

$$\Phi = \begin{bmatrix} 1 & 1 & 1 & \cdots & 1 \\ 1 & \frac{1}{2} & \frac{1}{3} & \cdots & \frac{1}{m} \\ \vdots & \vdots & \vdots & \ddots & \vdots \\ 1 & \frac{1}{2^{N-1}} & \frac{1}{3^{N-1}} & \cdots & \frac{1}{m^{N-1}} \end{bmatrix}.$$

Although this Vandermonde matrix always has $rank(\Phi) = N$, it also has $cond(\Phi)$ increasing like 10^N. The condition of the matrix Φ will be affected by both the choice of basis functions and the distribution of evaluation points. △

2.3.1.2 ***Under-constrained Solution*** When $N > m$ the problem is under-specified (i.e., fewer constraints than degrees of freedom). This situation is typical at the initiation of an approximation based control implementation. In this case, the matrix Φ defined in eqn. (2.8) is not square and its inverse does not exist. Therefore, there will either be no solution (Y is not in the column space of $\Phi^\top$) or an infinite number of solutions. In the latter case, Y is in the column space of $\Phi^\top$; however, since the number of columns of $\Phi^\top$ is larger than the number of rows, the solution is not unique. The minimum norm solution can be found by application of Lagrange multipliers.

Define the cost function

$$J(\theta, \lambda) = \frac{1}{2}\theta^\top \theta + \lambda^\top (Y - \Phi^\top \theta) \tag{2.14}$$

which enforces the constraint of eqn. (2.8) and is minimized by the minimum norm solution. Taking derivatives with respect to θ and λ yields

$$\frac{\partial J}{\partial \theta} = \theta^\top - \lambda^\top \Phi^\top = 0 \quad \text{and} \quad \frac{\partial J}{\partial \lambda} = Y - \Phi^\top \theta = 0. \tag{2.15}$$

Combining these two equations and solving yields

$$\lambda = (\Phi^\top \Phi)^{-1} Y \qquad \theta = \Phi(\Phi^\top \Phi)^{-1} Y \tag{2.16}$$

where $(\Phi^\top \Phi)$ is an $m \times m$ matrix that is assumed to be nonsingular. The matrix $\Phi(\Phi^\top \Phi)^{-1}$ is the Moore-Penrose pseudo-inverse of $\Phi^\top$ [29, 99, 202].

Linear combinations of the rows of $\Phi^\top$, i.e. $\sum_{i=1}^m \beta_i \phi(x_i)$ for $\beta_i \in \Re^1$, form an linear space denoted L_Φ. L_Φ is a subspace of $\Re^N$. For simplicity, we will assume that the dimension of L_Φ is m. Let $L_\Phi^\perp$ denote the set of vectors perpendicular to L_Φ: $L_\Phi^\perp = \{v \in \Re^N | v^\top w = 0, \forall w \in L_\Phi\}$. The set $L_\Phi^\perp$ is also a linear subspace of $\Re^N$. Let $\{d_i\}$ for $i = 1, \ldots, N-m$ denote a basis for $L_\Phi^\perp$. The vector λ defines the unique linear combination of the $\phi(x_i)$ (i.e., $\theta = \Phi\lambda$) such that $\Phi^\top \theta = Y$. Every other solution v to $\Phi^\top v = Y$ can be expressed as

$$v = \theta + \sum_{i=1}^{N-m} \alpha_i d_i \quad \text{for some } \alpha_i \in \Re^1.$$

Since θ is orthogonal to $\sum_{i=1}^{N-m} \alpha_i d_i$ by construction, $\|v\| = \|\theta\| + \| \sum_{i=1}^{N-m} \alpha_i d_i\|$ which is always greater than $\|\theta\|$. For additional discussion, see Section 6.7 of [29].

2.3.1.3 Summary This section has discussed the offline problem of fitting a function to a fixed batch of data. In the process, we have introduced the topic of weighted least squares parameter estimation which is applicable when the number of data points exceeds the number of free parameters defined for the approximator. We have also discussed the under-constrained case when there is not sufficient data available to completely specify the parameters of the approximator. Normally in online control applications, the number of data samples m will eventually be much larger that the number of parameters N. This is true since additional training examples are accumulated at each sampling instant. The results for the under-constrained case are therefore mainly applicable during start-up conditions.

2.3.2 Adaptive Function Approximation

Section 2.3.1.1 derived a formula for the weighted least squares (WLS) parameter estimate. Given the first k samples, with $k \geq N$, the WLS estimate can be expressed as

$$\theta_k = \left(\Phi_k W_k \Phi_k^\top\right)^{-1} \Phi_k W_k Y_k$$

where $\Phi_k = [\phi(x_1), \ldots, \phi(x_k)] \in \Re^{N \times k}$, $Y_k = [y_1, \ldots, y_k]^\top$, and W_k is an appropriately dimensioned positive definite matrix. Solution of this equation requires inversion of an $N \times N$ matrix. When the $(k+1)$st sample becomes available, this expression requires the availability of all previous training samples and again requires inversion of a new $N \times N$ matrix. For a diagonal weighting matrix W_k, direct implementation of the WLS algorithm has storage and computational requirements that increase with k. This is not satisfactory, since k is increasing without bound.

A main goal of subsection 2.3.2.1 is to derive a recursive implementation of that algorithm. That subsection is technical and may be skipped by readers who are not interested in the algorithm derivation. Properties of the recursive weighted least squares (RWLS) algorithm will be discussed in subsection 2.3.2.2. Two properties that are critically important are that (given proper initialization) the WLS and RWLS provide identical parameter estimates and that the computational requirements of the RWLS solution method are determined by N instead of k.

2.3.2.1 Recursive WLS: Derivation The WLS parameter estimate can be expressed as

$$\theta_k = P_k^{-1} R_k, \text{ where } P_k = \frac{1}{k}\Phi_k W_k \Phi_k^\top \text{ and } R_k = \frac{1}{k}\Phi_k W_k Y_k. \tag{2.17}$$

In the case where $W_k = I$, P_k is the sample regressor autocorrelation matrix and R_k is the sample cross-correlation matrix between the regressor and the function output. For interpretations of these algorithms in a statistical setting, the interested reader should see, for example, [133, 164].

From the definitions of Φ, Y, and W, assuming that W is a diagonal matrix, we have that

$$Y_{k+1} = \begin{bmatrix} Y_k \\ y_{k+1} \end{bmatrix}, \quad \Phi_{k+1} = \begin{bmatrix} \Phi_k & \phi_{k+1} \end{bmatrix}, \text{ and } W_{k+1} = \begin{bmatrix} W_k & 0 \\ 0 & w_{k+1} \end{bmatrix}. \tag{2.18}$$

Therefore,

$$\Phi_{k+1} W_{k+1} \Phi_{k+1}^\top = \Phi_k W_k \Phi_k^\top + \phi_{k+1} w_{k+1} \phi_{k+1}^\top \tag{2.19}$$

$$\Phi_{k+1} W_{k+1} Y_{k+1} = \Phi_k W_k Y_k + \phi_{k+1} w_{k+1} y_{k+1}. \tag{2.20}$$

Calculation of the WLS parameter estimate after the $(k+1)$st sample is available will require inversion of the $\Phi_{k+1}W_{k+1}\Phi_{k+1}^{\mathsf{T}}$. The *Matrix Inversion Lemma* [99] will enable derivation of the desired recursive algorithm based on eqn. (2.19).

The Matrix Inversion Lemma states that if matrices A, C, and $(A+BCD)$ are invertible (and of appropriate dimension), then

$$(A+BCD)^{-1} = A^{-1} - A^{-1}B\left(DA^{-1}B + C^{-1}\right)^{-1} DA^{-1}.$$

The validity of this expression is demonstrated by multiplying $(A+BCD)$ by the right-hand side expression and showing that the result is the identity matrix.

Applying the Matrix Inversion Lemma to the task of inverting $\Phi_{k+1}W_{k+1}\Phi_{k+1}^{\mathsf{T}}$, with $A_k = \Phi_k W_k \Phi_k^{\mathsf{T}}$, $B = \phi_{k+1}$, $C = w_{k+1}$, and $D = \phi_{k+1}^{\mathsf{T}}$, yields

$$\begin{aligned} A_{k+1}^{-1} &= \left(\Phi_k W_k \Phi_k^{\mathsf{T}} + \phi_{k+1} w_{k+1} \phi_{k+1}^{\mathsf{T}}\right)^{-1} \\ A_{k+1}^{-1} &= A_k^{-1} - A_k^{-1}\phi_{k+1}\left(\phi_{k+1}^{\mathsf{T}} A_k^{-1}\phi_{k+1} + w_{k+1}^{-1}\right)^{-1} \phi_{k+1}^{\mathsf{T}} A_k^{-1}. \end{aligned} \tag{2.21}$$

Note that the WLS estimate after samples k and $(k+1)$ can respectively be espressed as

$$\theta_k = A_k^{-1}\Phi_k W_k Y_k \text{ and } \theta_{k+1} = A_{k+1}^{-1}\Phi_{k+1} W_{k+1} Y_{k+1}.$$

The recursive WLS update is derived, using eqns. (2.20) and (2.21), as follows:

$$\begin{aligned} \theta_{k+1} &= \left[A_k^{-1} - A_k^{-1}\phi_{k+1}\left(\phi_{k+1}^{\mathsf{T}} A_k^{-1}\phi_{k+1} + w_{k+1}^{-1}\right)^{-1} \phi_{k+1}^{\mathsf{T}} A_k^{-1}\right] \\ &\quad \left[\Phi_k W_k Y_k + \phi_{k+1} w_{k+1} y_{k+1}\right] \\ &= \theta_k - A_k^{-1}\phi_{k+1}\left(\phi_{k+1}^{\mathsf{T}} A_k^{-1}\phi_{k+1} + w_{k+1}^{-1}\right)^{-1} \phi_{k+1}^{\mathsf{T}}\theta_k \\ &\quad + A_k^{-1}\phi_{k+1} w_{k+1} y_{k+1} \\ &\quad - A_k^{-1}\phi_{k+1}\left(\phi_{k+1}^{\mathsf{T}} A_k^{-1}\phi_{k+1} + w_{k+1}^{-1}\right)^{-1} \phi_{k+1}^{\mathsf{T}} A_k^{-1}\phi_{k+1} w_{k+1} y_{k+1} \\ &= \theta_k - A_k^{-1}\phi_{k+1}\left(\phi_{k+1}^{\mathsf{T}} A_k^{-1}\phi_{k+1} + w_{k+1}^{-1}\right)^{-1} \phi_{k+1}^{\mathsf{T}}\theta_k \\ &\quad + A_k^{-1}\phi_{k+1}\left[I - \left(\phi_{k+1}^{\mathsf{T}} A_k^{-1}\phi_{k+1} + w_{k+1}^{-1}\right)^{-1} \phi_{k+1}^{\mathsf{T}} A_k^{-1}\phi_{k+1}\right] w_{k+1} y_{k+1} \\ &= \theta_k - A_k^{-1}\phi_{k+1}\left(\phi_{k+1}^{\mathsf{T}} A_k^{-1}\phi_{k+1} + w_{k+1}^{-1}\right)^{-1} \phi_{k+1}^{\mathsf{T}}\theta_k \\ &\quad + A_k^{-1}\phi_{k+1}\left(\phi_{k+1}^{\mathsf{T}} A_k^{-1}\phi_{k+1} + w_{k+1}^{-1}\right)^{-1} \\ &\quad \left[\phi_{k+1}^{\mathsf{T}} A_k^{-1}\phi_{k+1} + w_{k+1}^{-1} - \phi_{k+1}^{\mathsf{T}} A_k^{-1}\phi_{k+1}\right] w_{k+1} y_{k+1} \\ \theta_{k+1} &= \theta_k + A_k^{-1}\phi_{k+1}\left(\phi_{k+1}^{\mathsf{T}} A_k^{-1}\phi_{k+1} + w_{k+1}^{-1}\right)^{-1}\left(y_{k+1} - \phi_{k+1}^{\mathsf{T}}\theta_k\right), \\ \theta_{k+1} &= \theta_k + A_k^{-1}\left(\phi_{k+1}^{\mathsf{T}} A_k^{-1}\phi_{k+1} + w_{k+1}^{-1}\right)^{-1} \phi_{k+1}\left(y_{k+1} - \phi_{k+1}^{\mathsf{T}}\theta_k\right), \end{aligned} \tag{2.22}$$

where we have used the fact that $\left(\phi_{k+1}^{\mathsf{T}} A_k^{-1}\phi_{k+1} + w_{k+1}^{-1}\right)$ is a scalar. Shifting indices in eqn. (2.21) yields the recursive equation for A_k^{-1}:

$$A_k^{-1} = A_{k-1}^{-1} - A_{k-1}^{-1}\phi_k\left(\phi_k^{\mathsf{T}} A_{k-1}^{-1}\phi_k + w_k^{-1}\right)^{-1} \phi_k^{\mathsf{T}} A_{k-1}^{-1}. \tag{2.23}$$

2.3.2.2 *Recursive WLS: Properties* The RWLS algorithm is defined by eqns. (2.22) and (2.23). This algorithm has several features worth noting.

1. Eqn. (2.22) has a standard *predictor-corrector* format

$$\theta_{k+1} = \theta_k + \Omega_k \phi_{k+1}(y_{k+1} - \hat{y}_{k+1:k}) \tag{2.24}$$

where $\Omega_k = A_k^{-1}\left(\phi_{k+1}^\top A_k^{-1}\phi_{k+1} + w_{k+1}^{-1}\right)^{-1}$ and $\hat{y}_{k+1:k} = \phi_{k+1}^\top \theta_k$ is the estimate of y_{k+1} based on θ_k. The majority of computations for the RWLS algorithm are involved in the propagation of A_k^{-1} by eqn. (2.23).

2. The RWLS calculation only uses information from the last iteration (i.e., A_k^{-1} and θ_k) and the current sample (i.e., y_{k+1} and ϕ_{k+1}). The memory requirements of the RWLS algorithm are proportional to N, not k. Therefore, the memory requirements are fixed at the design stage.

3. The WLS calculation of eqn. (2.13) requires inversion of an $N \times N$ matrix. The RWLS algorithm only requires inversion of an $n \times n$ matrix where N is the number of basis functions and n is the output dimension of f, which we have assumed to be one. Therefore, the matrix inversion simplifies to a scalar division. Note that A_k is never required. Therefore, A_k^{-1} is propagated, but never inverted.

4. All vectors and matrices in eqns. (2.22) and (2.23) have dimensions related to N, not k. Therefore, the computational requirements of the RWLS algorithm are fixed at the design stage.

5. Since no approximations have been made, the recursive WLS parameter estimate is the same as the solution of eqn. (2.13), if the matrix A_k^{-1} is properly initialized. One approach is to accumulate enough samples that A_k is nonsingular before initializing the RWLS algorithm. An alternative common approach is to initialize A_0^{-1} as a large positive definite matrix. This approximate initialization introduces an error in θ_1 that is proportional to $\|A_0\|$. This error is small and decreases as k increases. For additional details see Section 2.2 in [154].

6. Due to the equivalence of the WLS and RWLS solutions, the RWLS estimate will not be the unique solution to the WLS cost function until the matrix $\Phi_k W_k \Phi_k^\top$ is not singular. This condition is referred to as Φ_k being *sufficiently exciting*.

Various alternative parameter estimation algorithms can be derived (see Chapter 4). These algorithms require substantially less memory and fewer computations since they do not propagate A_k^{-1}, the tradeoff is that the alternative algorithms converge asymptotically instead of yielding the optimal parameter estimate as soon as ϕ_k achieves sufficient excitation. In fact, if convergence of the parameter vector is desired for non-WLS algorithms, then the more stringent condition of *persistence of excitation* will be required.

■ EXAMPLE 2.4

Example 2.1 presented a control approach requiring the storage of all past data $z(k)$. That approach had the drawback of requiring memory and computational resources that increased with k. The present section has shown that use of a function approximation structure of the form

$$\hat{f}(x) = \phi(x)^\top \theta$$

and a parameter update law of the form eqn. (2.24) (e.g., the RWLS algorithm) results in an adaptive function approximation approach with fixed memory and computational

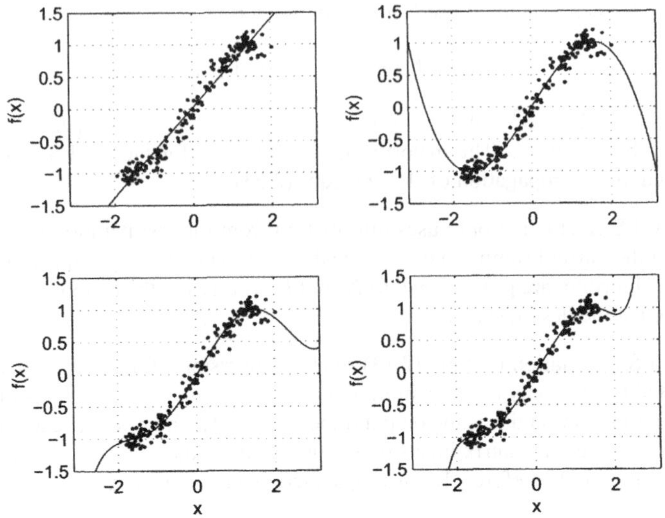

Figure 2.5: Least squares polynomial approximations to experimental data. The polynomial orders are 1 (*top left*), 3 (*top right*), 5 (*bottom left*), and 7 (*bottom right*).

requirements. This example further considers Example 2.1 to motivate additional issues related to the adaptive function approximation problem.

Let $\hat{f}$ be a polynomial of order m. Then, one possible choice of a basis for this approximator is (see Section 3.2) $\phi(x) = [1, x, \ldots, x^m]^\top$. Figure 2.5 displays the function approximation results for one set of experimental data (600 samples) and four different order polynomials. The x-axis of this figure corresponds to $\mathcal{D} = [-\pi, \pi]$ as specified in Example 2.1. Each of the polynomial approximations fits the data in the weighted least squares sense over the range of the data, which is approximately $\mathcal{B} = (-2, 2)$. Outside of the region $\mathcal{B}$, the behavior of each approximation is distinct.

The disparity of the behavior of the approximators on $\mathcal{D} - \mathcal{B}$ should motivate questions related to the idea of *generalization* relative to the training data. First, we dichotomize the problem into local and nonlocal generalization. Local generalization refers to the ability of the approximator to accurately compute $\hat{f}(x) = \hat{f}(x_i + dx)$ where x_i is the nearest training point and dx is small. Local generalization is a necessary and desirable characteristic of parametric approximators. Local generalization allows accurate function approximation with finite memory approximators and finite amounts of training data. The approximation and local generalization characteristics of an approximator will depend on the type and magnitude of the measurement noise and disturbances, the continuity characteristics of f and $\hat{f}$, and the type and number of elements in the regressor vector ϕ. Nonlocal generalization refers to the ability of an approximator to accurately compute $\hat{f}(x)$ for $x \in \mathcal{D} - \mathcal{B}$. Nonlocal generalization is always a somewhat risky proposition.

Although the designer would like to minimize the norm of the function approximation error $\int_{\mathcal{D}} \|f(x) - \hat{f}(x)\| dx$, this quantity is not able to be evaluated online, since $f(x)$ is not known. The norm of the sample data fit error $\frac{1}{k} \sum_{i=1}^{k} \|y_i - \hat{f}(x_i)\|$ can be evaluated and minimized. Figure 2.6 compares the minimum of these two quantities

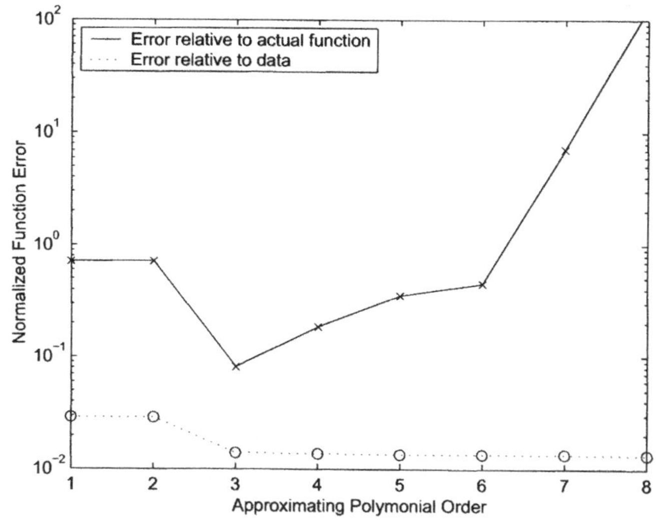

Figure 2.6: Data fit (dotted with circles) and function approximation (solid with x's) error versus polynomial order.

for the data of Figure 2.5 as the order m of the polynomial is increased. Both graphs decrease for small values of m until some critical regressor dimension m^* is attained. For $m > m^*$, the data fit error continues to decrease while the function approximation error actually increases. The data fit error decreases with m, since increasing the number of degrees of freedom of the approximator allows the measured data to be fit more accurately. The function approximation error increases with m for $m > m^*$, since the ability of the approximator to fit the measurement noise actually increases the error of the approximator relative to the true function. The value m^* is problem, data, and approximator dependent. In adaptive approximation problems where the data distribution and f are unknown, estimation of m^* prior to online operation is a difficult problem.

Since this example has used the RWLS method which propagates A^{-1} without data forgetting, the parameter estimate is independent of the order in which the data is presented. Generally, parameter estimation algorithms of the form eqn. (2.24) (e.g., gradient descent) are also trajectory (i.e., order of data presentation) dependent. △

Starting in Chapter 4, all derivations will be performed in continuous-time. In continuous-time, the analog of recursive parameter updates will be written as

$$\dot{\theta}(t) = \Gamma(t)\phi(t)(y(t) - \hat{y}(t)), \tag{2.25}$$

where $\Gamma(t)$ is the adaptive gain or learning rate. In discrete-time the corresponding adaptive gain $\Omega(t)$ (sometimes referred to as step size) needs to be sufficiently small in order to guarantee convergence; however, in continuous-time $\Gamma(t)$ simply needs to be positive definite (due to the infinitesimal change of the derivative $\dot{\theta}(t)$).

■ **EXAMPLE 2.5**

The continuous-time least squares problem estimates the vector θ such that $\hat{y}(t) = \phi(t)^\top\theta$ minimizes

$$J(\theta) = \int_0^t (y(\tau) - \hat{y}(\tau))^2\, d\tau = \int_0^t \left(y(\tau) - \phi(\tau)^\top\theta\right)^2 d\tau \tag{2.26}$$

where $y : \Re^+ \mapsto \Re^1$, $\theta \in \Re^N$, and $\phi : \Re^+ \mapsto \Re^N$. Setting the gradient of $J(\theta)$ with respect to θ to zero yields the following

$$\begin{aligned}
\int_0^t \phi(\tau)\left(y(\tau) - \phi(\tau)^\top\theta\right) d\tau &= 0\\
\int_0^t \phi(\tau)y(\tau)d\tau &= \int_0^t \phi(\tau)\phi(\tau)^\top d\tau\, \theta\\
R(t) &= P^{-1}(t)\,\theta\\
\theta(t) &= P(t)R(t)
\end{aligned} \tag{2.27}$$

where $R(t) = \int_0^t \phi(\tau)y(\tau)d\tau$ and $P^{-1}(t) = \int_0^t \phi(\tau)\phi(\tau)^\top d\tau$. Note by the definitions of P and R, that P^{-1} is symmetric and that

$$\begin{aligned}
\frac{d}{dt}[R(t)] &= \phi(t)y(t)\\
\frac{d}{dt}\left[P^{-1}(t)\right] &= \phi(t)\phi(t)^\top.
\end{aligned}$$

Since $P(t)P^{-1}(t) = I$, differentiation and rearrangement shows that in general the time derivative of a matrix and its inverse must satisfy $\dot{P} = -P\frac{d}{dt}\left[P^{-1}(t)\right]P$; therefore, in least squares estimation

$$\dot{P} = -P(t)\phi(t)\phi(t)^\top P(t). \tag{2.28}$$

Finally, to show that the continuous-time least squares estimate of θ satisfies eqn. (2.25) we differentiate both sides of eqn. (2.27):

$$\begin{aligned}
\dot{\theta}(t) &= \dot{P}(t)R(t) + P(t)\dot{R}(t)\\
&= -P(t)\phi(t)\phi(t)^\top P(t)R(t) + P(t)\phi(t)y(t)\\
&= P(t)\phi(t)\left(-\phi(t)^\top\theta(t) + y(t)\right)\\
\dot{\theta}(t) &= P(t)\phi(t)\left(y(t) - \hat{y}(t)\right).
\end{aligned} \tag{2.29}$$

Implementation of the continuous-time least squares estimation algorithm uses equations (2.28)–(2.29). Typically, the initial value of the matrix P is selected to be large. The initial matrix must be nonsingular. Often, it is initialized as $P(0) = \gamma I$ where γ is a large positive number. The implementation does not invert any matrix. △

Before concluding this section, we consider the problem of approximating a function over a compact region $\mathcal{D}$. The cost function of interest is

$$J(\theta) = \int_{\mathcal{D}} \left(f(x) - \phi(x)^\top\theta\right)^\top \left(f(x) - \phi(x)^\top\theta\right) dx.$$

Again, we find the gradient of J with respect to θ, set it to zero, and find the resulting parameter estimate. The final result is that θ must satisfy (see Exercise 2.9)

$$\left(\int_{\mathcal{D}} \phi(x)\phi(x)^\top dx\right)\theta = \int_{\mathcal{D}} \phi(x)f(x)^\top dx. \tag{2.30}$$

Computation of θ by eqn. (2.30) requires knowledge of the function f. For the applications of interest herein, we do not have this luxury. Instead, we will have measurements that are indirectly related to the unknown function. Nonetheless, eqn. (2.30) shows that the condition of the matrix $\int_{\mathcal{D}} \phi(x)\phi(x)^\top dx$ is important. When the elements of the ϕ are mutually orthonormal over $\mathcal{D}$, then $\int_{\mathcal{D}} \phi(x)\phi(x)^\top dx$ is an identity matrix. This is the optimal situation for solution of eqn. (2.30), but is often not practical in applications.

2.4 APPROXIMATOR PROPERTIES

This section discusses properties that families of function approximators may have. In each subsection, the technical meaning of each property is presented and the relevance and tradeoffs of the property in the applications of interest are discussed. Due to the technical nature of and the broad background that would be required for the proofs, in most cases the proofs are not presented. Literature sources for the proofs are cited.

2.4.1 Parameter (Non)Linearity

An initial decision that the designer must make is the form of the function approximator. A large class of function approximators (several are presented in Chapter 3) can be represented as

$$\hat{f}(x:\theta,\sigma) = \theta^\top \phi(x,\sigma) \tag{2.31}$$

where $x \in \Re^n$, $\theta \in \Re^N$, and the dimension of σ depends on the approximator of interest. The approximator has a linear dependence on θ, but a nonlinear dependence on σ.

■ EXAMPLE 2.6

The (N-1)-th order polynomial approximation $\hat{f}(x:\theta,N) = \sum_{i=0}^{N-1} \theta_i x^i$ for $x \in \Re^1$ has the form of eqn. (2.31) where $\phi(x,N) = [1, x, \ldots, x^{N-1}]^\top$. If N is fixed, then the polynomial approximation is linear in its adjustable parameter vector $\theta = [\theta_0, \ldots, \theta_{N-1}]$. See Section 3.2 for a more detailed discussion of polynomial approximators. △

■ EXAMPLE 2.7

The radial basis function approximator with Gaussian nodes:

$$\hat{f}(x:\theta,\sigma) = \sum_{i=1}^{N} \theta_i exp\left(-\frac{\|x-c_i\|^2}{\gamma_i^2}\right),$$

with x, $c_i \in \Re^n$ and θ_i, $\gamma_i \in \Re^1$, has the form of eqn. (2.31) where

$$\phi(x,\sigma) = \left[exp\left(-\frac{\|x-c_1\|^2}{\gamma_1^2}\right), \ldots, exp\left(-\frac{\|x-c_N\|^2}{\gamma_N^2}\right)\right]^\top,$$

$$\sigma = [c_1, \ldots, c_N, \gamma_1, \ldots, \gamma_N],$$

and

$$\theta = [\theta_1, \ldots, \theta_N].$$

This radial basis function approximator is only linear in its parameters when all elements of σ are fixed. See Section 3.4 for a more detailed discussion of radial basis function approximators. △

■ **EXAMPLE 2.8**

The sigmoidal neural network approximator

$$\hat{f}(x : \theta, \sigma)) = \sum_{i=1}^{N} \theta_i g(\lambda_i^\top x + b_i)$$

with *nodal processing function* g defined by the squashing function $g(u) = \frac{e^u - 1}{e^u + 1}$ has the form of eqn. (2.31) where

$$\phi(x, \sigma) = \left[g(\lambda_1^\top x + b_1), \ldots, g(\lambda_N^\top x + b_N)\right]$$

and

$$\sigma = [\lambda_1, \ldots, \lambda_N, b_1, \ldots, b_N].$$

The sigmoidal neural network approximator is again linear in its parameters if all elements of the vector σ are fixed *a priori*. Sigmoidal neural networks are discussed in detail in Section 3.6. △

In most articles and applications, the parameter N which is the dimension of ϕ is fixed prior to online usage of the approximator. When N is fixed prior to online operation, selection of its value should be carefully considered as N is one of the key parameters that determines the minimum approximation accuracy that can be achieved. All the uniform approximation results of Section 2.4.5 will contain a phrase to the effect "for N sufficiently large." Self-organizing approximators that adjust N online while ensuring stability of the closed-loop control system constitute an area of continuing research.

A second key design decision is whether σ will be fixed *a priori* (i.e., $\sigma(t) = \sigma(0)$ and $\dot{\sigma} = 0$) or adapted online (i.e., $\sigma(t)$ is a function of the online data and control performance). If σ is fixed during online operation, then the function approximator is linear in the remaining adjustable parameters θ so that the designer has a linear-in-the-parameter (LIP) adaptive function approximation problem. Proving theoretical issues, such as closed-loop system stability, is easier in the LIP case. In the case where the approximating parameters σ are fixed, these parameters will be dropped from the approximation notation, yielding

$$\hat{f}(x) = \theta^\top \phi(x). \tag{2.32}$$

Fixing σ is beneficial in terms of simplifying the analysis and online computations, but may limit the functions that can be accurately approximated and may require that $N = dim(\phi)$ be larger than would be required if σ were estimated online.

Example 2.8 has introduced the term *nodal processing function*. This terminology is used when each node in a network approximator uses the same function, but different nonlinear parameters. In Example 2.8, the i-th component of the ϕ can be written as $\phi_i(x) = g(x : \lambda_i, b_i)$. Using the idea of a nodal processor, the i-th element of the regressor vector in Example 2.7 can be written as $\phi_i(x) = g(x : c_i, \gamma_i)$ where for that example the nodal processor is $g(u) = exp\left(-u^2\right)$. Many of the other approximators defined in Chapter 3 can be written using the nodal processor notation.

To obtain a linear in the parameters function approximation problem, the designer must specify *a priori* values for (n, N, g, σ). If these parameters are not specified judiciously, then an approximator achieving a desired ϵ-accuracy may not be achievable for any value of θ. After (n, N, g, σ) are fixed, a family of linear in the parameter approximators results.

Definition 2.4.1 (Linear-in-Parameter Approximators) *The family of n input, N node, LIP approximators associated with nodal processor $g(\cdot)$ is defined by*

$$S_{n,N,g,\sigma} = \left\{ f : \Re^n \rightarrow \Re^1 \,\middle|\, f(x) = \sum_{i=1}^{N} \theta_i \phi_i(x) = \theta^\top \phi(x) \right\} \tag{2.33}$$

with $x \in \Re^n$, and $\theta \in \Re^N$ where $\phi_i(x) = g(x : \sigma_i)$ and σ_i is fixed at the design stage.

This family of LIP approximators defines a linear subspace of functions from $\Re^n$ to $\Re^1$. A basis for this linear subspace is $\{\phi_i(x)\}_{i=1}^{N}$.

The relative drawbacks of approximators that are linear in the adjustable parameters are discussed, for example, by Barron in [17]. Barron shows that under certain technical assumptions, approximators that are nonlinear in their parameters have squared approximation errors of order $O\left(\frac{1}{N}\right)$ while approximators that are linear in their parameters cannot have squared approximation errors smaller than order $O\left(\frac{1}{N^{2/n}}\right)$ (N is the number of nodal functions, n is the dimension of domain $\mathcal{D}$). Therefore, for $n > 2$ the approximation error for nonlinear in the parameter families of approximators can be significantly less than that for LIP approximators. This order of approximation advantage for nonlinear in the parameter approximators requires significant tradeoffs, that will be summarized at the end of this subsection. Note that this order of approximation advantage is a theoretical result, it does not provide a means of determining approximator parameters or an approximator structure that achieves the bound.

A cost function $J_e(e)$ is strictly convex in e if for $0 \leq \alpha \leq 1$ and for all $e_1, e_2 \in \Re^1$, the function J_e satisfies

$$J_e\left(\alpha e_1 + (1-\alpha) e_2\right) < \alpha J_e(e_1) + (1-\alpha) J_e(e_2).$$

If a continuous strictly convex function has a minimum e^*, then that minimum is a unique global minimum.

If J_e is strictly convex in e and $e(x) = \theta^\top \phi(x)$, then for an fixed value x_i the cost function $J(\theta) = J_e(\theta^\top \phi(x_i))$ is only convex in θ. This is important since some of the parameter estimation algorithms to be presented in Chapter 4 will be extensions of gradient following methods. For discussion, let $e = \phi(x_i)^\top \theta$ where $\phi(x_i)$ is a constant vector. Then, there is a linear space of parameter vectors Θ_i such that $e^* = \phi(x_i)^\top \theta$, $\forall \theta \in \Theta_i$. The fact that J is strictly convex in e and convex in θ for LIP approximators ensures that for any initial value of θ, gradient based parameter estimation will cause the parameter estimate to converge toward the space Θ_i (i.e., for LIP approximators, although there is a linear space of minima, there is a single basin of attraction). Alternatively when $J_e(e)$ is convex but the approximator is not linear in its parameters (i.e., $\hat{f}(x, \theta, \sigma) = \theta^\top \phi(x, \sigma)$), then the cost function $J(\theta, \sigma) = J_e(\theta^\top \phi(x, \sigma))$ may not be convex in θ and σ. If the cost

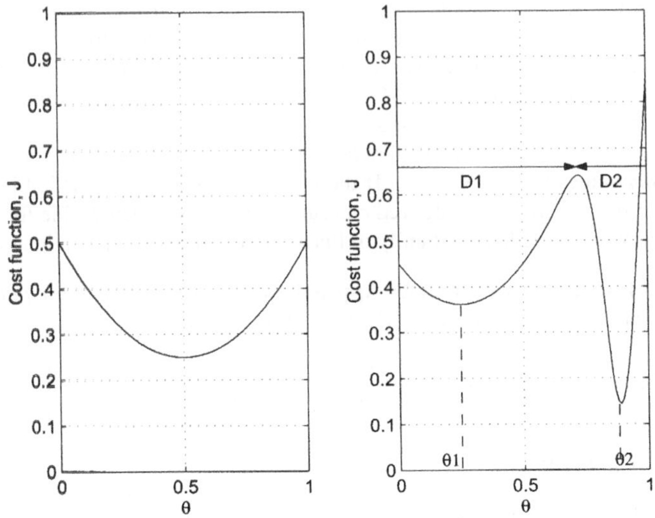

Figure 2.7: Convex (*left*) and nonconvex (*right*) cost functions.

function is not convex, then multiple local minima may exist. Each local minima could have its own basin of attraction.

Convex and nonconvex one dimensional cost functions are depicted in Figure 2.7. When the cost function is convex in the parameter error (as in the left graph of Figure 2.7), regardless of the initial parameter values, the gradient will point towards the θ^*. Therefore, LIP approximators allow *global* convergence results. For approximators that are not linear in their parameters, even if the cost function is convex in the approximation error, the cost function may not be convex in the parameter error. When the cost function is not convex in the parameter error, there may be saddle points or several values of θ that locally minimize the cost function. Each local minimum of the cost function will have associated with it a local domain of attraction (indicated by D1 and D2 in the figure). Therefore, when an approximator that is not linear in its parameters is used, only *local* convergence results may be possible. In this case it would be immaterial that the global minimizing parameter vector achieves a desired ϵ approximation accuracy if the parameter vector at the local minimum does not.

When multiple evaluation points $\{x_i\}_{i=1:m}$ are available, the cost function can be selected as

$$J(\theta) = \sum_{i=1}^{m} J_e(\theta^\top \phi(x_i)).$$

This cost function is minimized for $\theta \in \bigcap_{i=1}^m \Theta_i$. If $\phi(x_i)$ varies sufficiently, then $\bigcap_{i=1}^m \Theta_i$ will shrink to a single point θ^*. This condition is referred to as sufficiency of excitation.

In summary, the main advantage of nonlinear in their parameter approximators is that for a given accuracy of approximation, the minimum number of nodes or basis elements N will typically be less than for a LIP approximator. However, for the same value of N, the nonlinear in the parameter approximator will require much more computation due to the estimation of σ. When the LIP approximator is also a *lattice* approximator, the computation of the approximator is also significantly reduced, see Section 2.4.8.4. Additional advantages of LIP approximators are simplification of theoretical analysis, the existence of a global minimizing parameter value, the ability to prove global (in the parameter estimate) convergence results, and the ability (if desired) to initialize the approximation parameters based on prior data or model information using the methods of Section 2.3. An additional motivation for the use of LIP approximators is discussed in Section 2.4.6.

In the batch training of LIP approximators by the least squares methods of Section 2.3.1, unique determination of θ is possible once Φ is nonsingular. In the literature, this is sometimes referred to as a *guaranteed learning algorithm* [31]. This is in contrast to gradient descent learning algorithms (especially in the case of non-LIP approximators) that (possibly) converge asymptotically to the optimal parameter estimate.

2.4.2 Classical Approximation Results

This section reviews results from the classic theory of function approximation [155, 218] that will recur in later sections and that have direct relevance to the stated motivations for using certain classes of function approximators. The notation and technical concepts required to discuss these results will be introduced in this section and used throughout the reminder of the text.

2.4.2.1 Background and Notation The set of functions $\mathcal{F}(\mathcal{D})$ defined on a compact set[4] $\mathcal{D}$ is a linear space (i.e., if $f, g \in \mathcal{F}$, then $\alpha f + \beta g \in \mathcal{F}$.). The ∞-norm of $\mathcal{F}(\mathcal{D})$ is defined as

$$\|f\|_\infty = \sup_{x \in \mathcal{D}} |f(x)|.$$

The set $\mathcal{C}(\mathcal{D})$ of continuous functions defined on $\mathcal{D}$ is also a linear space of functions. Since $\mathcal{D}$ is compact,[5] for $f \in \mathcal{C}(\mathcal{D})$,

$$\|f\|_\infty = \max_{x \in \mathcal{D}} |f(x)|.$$

Since $\sup_{x \in \mathcal{D}} |f(x)|$ satisfies the properties of a norm, both $\mathcal{F}(\mathcal{D})$ and $\mathcal{C}(\mathcal{D})$ are normed linear spaces. Given a norm on $\mathcal{F}(\mathcal{D})$, the distance between $f, g \in \mathcal{F}(\mathcal{D})$ can be defined as $d(f,g) = \|f - g\|$. When f, g are elements of a space $\mathcal{S}$, $d(f,g)$ is a metric for $\mathcal{S}$ and the pair $\{\mathcal{S}, d\}$ is referred to as a *metric space*. When $\mathcal{S}(\mathcal{D})$ is a subset of $\mathcal{F}(\mathcal{D})$, the distance from $f \in \mathcal{F}(\mathcal{D})$ to $\mathcal{S}(\mathcal{D})$ is defined to be $d(f, \mathcal{S}) = \inf_{a \in \mathcal{S}} d(f, a)$.

A sequence $\{f_i\} \in \mathcal{X}$ is a *Cauchy* sequence if $\|f_i - f_j\| \to 0$ as $i, j \to \infty$. A space $\mathcal{X}$ is complete if every Cauchy sequence in $\mathcal{X}$ converges to an element of $\mathcal{X}$ (i.e., $\|f_i - f\| \to 0$ as $i \to \infty$ for some $f \in \mathcal{X}$). A *Banach space* is the name given to a complete normed linear space. Examples of Banach spaces include the $\mathcal{L}_p$ spaces for $p \geq 1$ where

$$\mathcal{L}_p = \left\{ f : \Re^1 \mapsto \Re^1 \,\middle|\, \|f\|_p = \left[\int |f(x)|^p \, dx \right]^{1/p} < \infty \right\}$$

or the set $\mathcal{C}(\mathcal{D})$ with norm $\|f\|_\infty$.

[4]The following properties are equivalent for a finite dimensional compact set $\mathcal{D} \subset \mathcal{X}$: (1) $\mathcal{D}$ is closed and bounded, (2) every infinite cover of $\mathcal{D}$ has a finite subcover (i.e., Given any $\{d_i\}_{i=1}^{\infty} \subset \mathcal{X}$ such that $\mathcal{D} \subset \cup_{i=1}^{\infty} d_i$, then there exist N such that $\mathcal{D} \subset \cup_{i=1}^{N} d_i$), (3) every infinite sequence in $\mathcal{D}$ has a convergent subsequence.

[5]If f is a continuous real function defined over a compact region $\mathcal{D}$, then f achieves both a maximum and a minimum value on $\mathcal{D}$.

EXAMPLE 2.9

Let $\mathcal{D} = [0, 1]$. Is $\mathcal{C}(\mathcal{D})$ with the $\mathcal{L}_2$ norm complete?

Consider the sequence of functions $\{x^n\}_{n=1}^{\infty}$ each of which is in $\mathcal{C}(\mathcal{D})$. Basic calculus and algebra (assuming without loss of generality that $m > n$) leads to the bound

$$\|x^n - x^m\|_2 = \left(\int_0^1 |x^n - x^m|^2 \, dx\right)^{1/2} < \left(\int_0^1 |x^n|^2 \, dx\right)^{1/2} = \left(\frac{1}{2n+1}\right)^{1/2}.$$

Since the right-hand side can be made arbitrarily small by choice on n, $\{x^n\}_{n=1}^{\infty}$ is a Cauchy sequence of function in $\mathcal{C}(\mathcal{D})$ with norm $\|\cdot\|_2$. The limit of this sequence is the function

$$g(x) = \begin{cases} 0, & x \in [0, 1) \\ 1, & x = 1. \end{cases}$$

which is not in $\mathcal{C}(\mathcal{D})$. Therefore, by counterexample, $\mathcal{C}(\mathcal{D})$ with the $\mathcal{L}_2$ norm is not complete.

Note that this sequence is not Cauchy with the ∞-norm. △

2.4.2.2 Weierstrass Results Given a Banach space $\mathcal{X}$ with elements f, norm $\|f\|$, and a sequence $\Phi_N = \{\phi_i\}_{i=1}^N \subset \mathcal{X}$ of basis elements, f is said to be *approximable* by linear combinations of Φ_N with respect to the norm $\|\cdot\|$ if for each $\epsilon > 0$ there exists N such that $\|f - P_N\| < \epsilon$ where

$$P_N(x) = \sum_{i=1}^{N} \theta_i \phi_i(x), \text{ for some } \theta_i \in \mathcal{R}. \tag{2.34}$$

The *N-th degree of approximation* of f by Φ_N is

$$E_N^{\Phi}(f) = d(f, P_N) = \inf_{\theta} \|f - P_N\|.$$

When the infimum is attained for some $P \in \mathcal{X}$, this P is referred to as the linear combination of *best approximation.*

Consider the theoretical problem of approximating a given function $f \in \mathcal{C}(\mathcal{D})$ relative to the two norm using P_N. The solution to eqn. (2.30) is

$$\theta = \left(\int_{\mathcal{D}} \Phi_N(x)\Phi_N^{\top}(x)dx\right)^{-1} \int_{\mathcal{D}} \Phi_N(x)f(x)dx \tag{2.35}$$

where the basis elements $\phi_i(x)$ are assumed to be linearly independent so that

$$\int_{\mathcal{D}} \Phi_N(x)\Phi_N^{\top}(x)dx$$

is not singular.[6] This solution shows that there is a unique set of coefficients for each N such that the two-norm of the approximation error is minimized by a linear combination of the basis vectors. This solution does not show that $f \in \mathcal{C}(\mathcal{D})$ is "approximable by linear combinations of Φ_N," since eqn. (2.35) does not show whether $E_N^{\Phi}(f)$ approaches zero as N increases.

[6]Note the similarity between eqns. (2.13) and (2.35). The properties of the matrix to be inverted in the latter equation are determined by $\mathcal{D}$ and the definition of the basis elements. The properties of the matrix to be inverted in the former equation depend on these same factors as will as the distribution of the samples used to define the matrix.

■ **EXAMPLE 2.10**

Let $\chi(x : a, b)$ be the characteristic function on the interval $[a, b]$:

$$\chi(x, a, b) = \begin{cases} 1 & \text{for } x \in [a, b], \\ 0 & \text{otherwise.} \end{cases}$$

If the designer selects the approximator basis elements to be $\phi_i(x) = \chi(x : 0, \frac{1}{i})$ where $\mathcal{D} = [0, 1]$, then

$$\int_{\mathcal{D}} \Phi_N(x)\Phi_N^\top(x)dx = \begin{bmatrix} 1 & \frac{1}{2} & \frac{1}{3} & \cdots & \frac{1}{N} \\ \frac{1}{2} & \frac{1}{2} & \frac{1}{3} & \cdots & \frac{1}{N} \\ \frac{1}{3} & \frac{1}{3} & \frac{1}{3} & \cdots & \frac{1}{N} \\ \vdots & \vdots & \vdots & & \vdots \\ \frac{1}{N} & \frac{1}{N} & \frac{1}{N} & \cdots & \frac{1}{N} \end{bmatrix}.$$

This matrix is nonsingular for all N. Therefore, for any continuous function f on $\mathcal{D}$, there exists a optimal set of parameters given by eqn. (2.35) such that $\sum_{i=1}^{N} \theta_i\phi_i(x)$ achieves $E_N^{\Phi}(f)$. However, this choice of basis function, even as N increases to infinity, cannot accurately approximate continuous functions that are nonconstant for $x \in [0.5, 1]$ (e.g., $f(x) = x$). △

The previous example shows that for a set of basis elements to be capable of uniform approximation of continuous functions over a compact region $\mathcal{D}$, conditions in addition to linear independence of the basis elements over $\mathcal{D}$ must be satisfied. The uniform approximation property of univariate polynomials

$$P_N(x) = \left\{ \sum_{k=0}^{N} a_k x^k, x, a_k \in \Re^1 \right\}$$

is addressed by the Weierstrass theorem.

Theorem 2.4.1 *Each real function f that is continuous on $D = [a, b]$ is approximable by algebraic polynomials with respect to the ∞-norm: $\forall \epsilon > 0$, $\exists M$ such that if $N > M$ there exists a polynomial $p \in P_N$ with $\|f(x) - p(x)\|_\infty < \epsilon$ for all $x \in D$.*

A set $\mathcal{S}$ being dense on a set $\mathcal{T}$ means that for any $\epsilon > 0$ and $T \in \mathcal{T}$, there exists $S \in \mathcal{S}$ such that $\|S - T\| < \epsilon$. A simple example is the set of rational numbers being dense on the set of real numbers. The Weierstrass theorem can be summarized by the statement that the linear space of polynomials is *dense* on the set of functions continuous on compact D. It is important to note that the Weierstass theorem shows existence, but is not constructive in the sense that it does not specify M or the parameters $[a_0, \ldots, a_M]$.

■ **EXAMPLE 2.11**

The Weierstrass theorem requires that the domain D be compact. This example motivates the necessity of this condition.

Let $D = (0, 1]$, which is not compact. Let $f = \frac{1}{x}$, which is continuous on D. Therefore, all preconditions of the Weierstrass theorem are met except for D being compact. Due to the lack of boundedness of f on D and the fact that the value of any element of P_N as $x \to 0$ is $a_0 < \infty$, ϵ accuracy over D cannot be achieved no matter how large N is selected. △

The remainder of this chapter will introduce the concept of network approximators and discuss the extension of the above approximation concepts to network approximators.

2.4.3 Network Approximators

Network approximators included some traditional (e.g., spline) and many recently introduced (e.g., wavelets, radial basis functions, sigmoidal neural networks) function approximation methods. The basic idea of a network approximator is to use a possibly large number of simple, identical, interconnected nodal processors. Because of the structure that results, matrix analysis methods are natural and parallel computation is possible.

Consider the family of affine functions.

Definition 2.4.2 (Affine Functions) *For any* $r \in \{1, 2, 3 \ldots\}$, $A^r : \Re^r \to \Re^1$ *denotes the set of* affine functions *of the form*

$$A(x) = w^\top x + b$$

where $w, x \in \Re^r$ *and* $b \in \Re^1$.

The affine function $A(x)$ defines a hyperplane that divides $\Re^r$ into two sets $\{x \in \Re^r | A(x) \geq 0\}$ and $\{x \in \Re^r | A(x) < 0\}$. In pattern recognition and classification applications, such hyperplane divisions can be used to subdivide an input space into classes of inputs [152].

Network approximators are defined by constructing linear combinations of processed affine functions [259].

Definition 2.4.3 (Single Hidden Layer (Σ) Networks) *The family of* r *input,* N *node, single hidden layer (Σ) network approximators associated with nodal processor* $g(\cdot)$ *is defined by*

$$S_{r,N} = \left\{ h : \Re^r \to \Re^1 \,\middle|\, h(x) = \sum_{i=1}^{N} \theta_i g\left(A_i(x)\right),\, x \in \Re^r, \theta \in \Re^N, \text{ and } A_i \in A^r \right\}$$

where $\theta = [\theta_1, \ldots, \theta_N]$.

The Σ designation in the title of this definition indicates that each nodal processor sums its scalar input variables. The type of nodal processor selected determines the form of the nodal processor $g(\cdot)$. In network function approximation structures x is the network input, w are the hidden layer weights, b is a bias, and θ are the output layer weights.

■ **EXAMPLE 2.12**

The well-known single-layer perceptron (which will be defined later in Section 3.6) is a Σ network there A_i would denote the input layer parameters of the i-th neuron.
△

Extending the Σ-network definition to allow nodal processors with outputs that are the product of Σ-network hidden layer outputs produces a wider class of network approximators.

Definition 2.4.4 (Single Hidden Layer ($\Sigma\Pi$) Networks) *The family of r input, N node, single hidden layer ($\Sigma\Pi$) network approximators associated with nodal processor $g(\cdot)$ is defined by*

$$S_{r,N} = \left\{ h : \Re^r \to \Re^1 \,\middle|\, h(x) = \sum_{i=1}^{N} \theta_i \Pi_{j=1}^{q} g\left(A_{ij}(x)\right),\, x \in \Re^r, \theta \in \Re^N, \textit{ and } A_{ij} \in A^r \right\}.$$

■ EXAMPLE 2.13

Radial basis functions (see Section 3.4) are defined by

$$y = \sum_{i=1}^{N} \theta_i \exp\left(-\left(x - c_i\right)^\top P_i \left(x - c_i\right)\right)$$

where $P_i \in \Re^{n \times n}$ is symmetric and positive semidefinite and $c_i \in \Re^n$. The matrix P_i can always be expressed as $P_i = V_i V_i^\top$ where $V_i \in \Re^{n \times q}$ and $q = rank(P_i)$.

In the special case that $q = 1$, where V_i is an n-dimensional vector, define

$$\begin{aligned} u_i &= V_i^\top \left(x - c_i\right) \\ &= V_i^\top x + b_i \end{aligned}$$

where $b_i = -V_i^\top \mu_i$. As a result,

$$y = \sum_{i=1}^{N} \theta_i \exp\left(-u_i^\top u_i\right).$$

Therefore, this special case of the radial basis function fits Definition 2.4.3 with the nodal processor defined as $g(u) = \exp(-u^\top u)$.

In the case that $q > 1$ (q is normally equal to n), V_i is a matrix so that u_i is a vector with components denoted by u_{ij}. Therefore,

$$\begin{aligned} y &= \sum_{i=1}^{N} \theta_i \exp(-u_i^\top u_i) \\ &= \sum_{i=1}^{N} \theta_i \exp\left(-\sum_{j=1}^{q} u_{ij}^2\right) \\ &= \sum_{i=1}^{N} \theta_i \Pi_{j=1}^{q} \exp\left(-u_{ij}^2\right) \\ &= \sum_{i=1}^{N} \theta_i \Pi_{j=1}^{q} g(u_{ij}) \end{aligned}$$

where u_{ij} is an affine function. Therefore, radial basis functions are $\Sigma\Pi$-networks.
$\triangle$

Any Σ-network can be written in the form of eqn. (2.31) by defining $\phi_i(x, \sigma)$ to be $g\left(A_i(x)\right)$, where σ is a vector composed of the elements of w and b. Similarly, any $\Sigma\Pi$-network can be written in the form of eqn. (2.31) by defining $\phi_i(x, \sigma)$ to be $\Pi_{j=1}^{q} g\left(A_{ij}(x)\right)$.

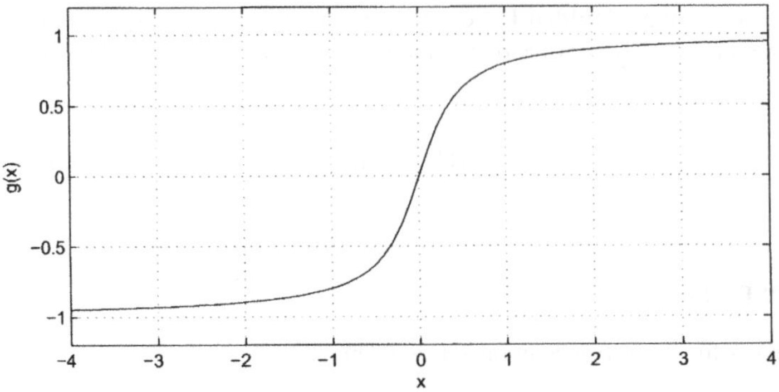

Figure 2.8: Example of a squashing function.

Specification of a unique single hidden layer network approximator requires definition of the following 5-tuple $\mathcal{F} = (r, N, g, \theta, \sigma)$. If all parameters except for θ are specified, then we have a linear-in-the-parameters approximator.

Definitions 2.4.3 and 2.4.4 explicitly define single output network functions. The definition of vector output network approximators is a direct extension of the definition, where each vector component is defined as in the definitions and θ is a matrix. With the definition of vector output single hidden layer networks, multi-hidden layer networks can be defined by using the vector output from one network as the vector input to another network.

The universal approximation results that follow utilize the concept of an algebra.

Definition 2.4.5 *A family of real functions S defined on $\mathcal{D}$ is an* algebra *if S is closed under the operations of addition, multiplication, and scalar multiplication.*

The set of functions in $\mathcal{L}_p(\mathcal{D})$ is an algebra. The set of functions in $C(\mathcal{D})$ is an algebra. The set of polynomial functions $\mathcal{P}$ is an algebra. The set $\mathcal{P}_m$ of polynomials of order m is not an algebra. The set of single hidden layer Σ-networks is not an algebra. Of particular note for the results to follow, the set of $\Sigma\Pi$-networks is an algebra as long as q and N are not fixed.

2.4.4 Nodal Processors

The universal approximation theorems of Section 2.4.5 will build on the Σ and $\Sigma\Pi$-networks of the previous section and the squashing and local functions defined below [110].

Definition 2.4.6 (Squashing functions) *The nodal processor $g(\cdot)$ is a* squashing function *if $g : \Re^1 \mapsto \Re^1$ is a non-constant, continuous, bounded, and monotone increasing function of its scalar argument.*

Definition 2.4.7 (Local functions) *The nodal processor $g(\cdot)$ is a local function if $g : \Re^1 \mapsto \Re^1$ is continuous, $g \in \mathcal{L}_1 \bigcap \mathcal{L}_p, 1 \leq p < \infty$ and $\int_{-\infty}^{\infty} g(x)dx \neq 0$.*

Figure 2.8 shows an example of a nodal function that satisfies Definition 2.4.6. Figure 2.9 shows three functions. The function g_1 is not a local function because $\int_{-\infty}^{\infty} g_1(x)dx = 0$. The functions g_2 and g_3 are local functions according to Definition 2.4.7.

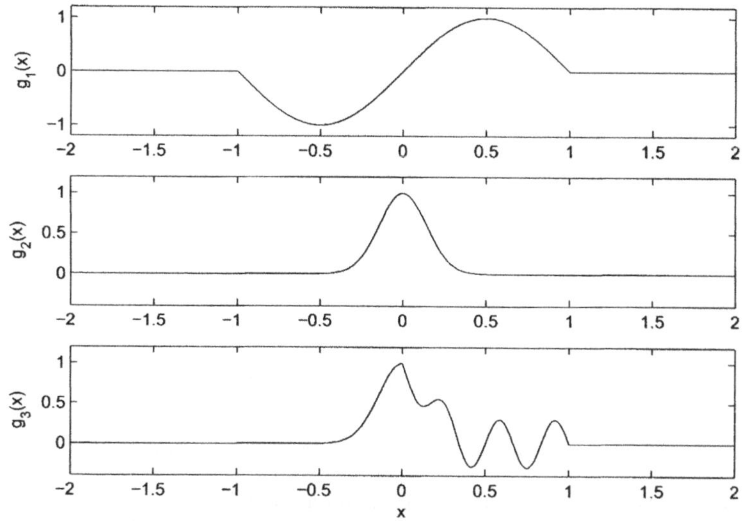

Figure 2.9: The function g_1 is not a local function because $\int g_1(x)dx = 0$. The functions g_2 and g_3 are local functions according to Definition 2.4.7.

To avoid difficulties such as those that occurred due to the choice of approximators in Example 2.10, we must introduce the following definition.

Definition 2.4.8 *A family of real functions $\mathcal{S}$ defined on $\mathcal{D}$ separates points on $\mathcal{D}$ if for any $x, y \in \mathcal{D}$ there exists $f \in \mathcal{S}$ such that $f(x) \neq f(y)$.*

If $\mathcal{S}$ did not separate points, then there would exist $x, y \in \mathcal{D}$ such that $f(x) = f(y)$ for all $f \in \mathcal{S}$. In this case, $\mathcal{S}$ could not approximate to arbitrary ϵ-accuracy any function for which $g(x) \neq g(y)$.

■ **EXAMPLE 2.14**

Consider a $\Sigma\Pi$-network with g satisfying either Definition 2.4.6 or 2.4.7. Pick $a, b \in \Re^1$ such that $g(a) \neq g(b)$. This is always possible since in both definitions g is nonconstant. For any $x, y \in \mathcal{D}$ such that $x \neq y$ it is possible to find $A \in A^r$ so that $A(x) = a$ and $A(y) = b$, which shows that $\Sigma\Pi$-networks with nodal processors satisfying either of these definitions separate points of $\mathcal{D}$. △

Definition 2.4.9 *A family of real functions $\mathcal{S}$ defined on $\mathcal{D}$ vanishes at no point of $\mathcal{D}$ if for any $x \in \mathcal{D}$ there exists $f \in \mathcal{S}$ such that $f(x) \neq 0$.*

If $\mathcal{S}$ did vanish at some point $x \in \mathcal{D}$, then $\mathcal{S}$ could not approximate functions with a nonzero value at x to arbitrary ϵ-accuracy.

■ **EXAMPLE 2.15**

Consider a $\Sigma\Pi$-network with g satisfying either Definition 2.4.6 or 2.4.7. By the definitions, there exist some b such that $g(b) \neq 0$. Choose $A(x) = 0x + b$. Then,

$g(A(x)) \neq 0$. Therefore, $\Sigma\Pi$-networks with nodal processors satisfying either of these definitions satisfy Definition 2.4.9. △

2.4.5 Universal Approximator

Consider the following theorem.

Theorem 2.4.2 *Given $f \in \mathcal{L}_2(\mathcal{D})$ and an approximator of the form eqn. (2.32), for any N if $\left(\int_{\mathcal{D}} \Phi_N(x)\Phi_N^\top(x)dx\right)$ is nonsingular, then there exists a unique $\theta^* \in \Re^N$ such that $f(x) = (\theta^*)^\top \phi(x) + e_f^*(x)$ where*

$$\theta^* = argmin_\theta \int_D \|f(x) - \hat{f}(x:\theta)\|_2^2 dx. \tag{2.36}$$

In addition, there are no local minima of the cost function (other than θ^*).

This theorem states the condition necessary that for a given N, there exists a unique parameter vector θ^* that minimizes the $\mathcal{L}_2$ error over $\mathcal{D}$. In spite of this, for $f \in \mathcal{L}_2(\mathcal{D})$, the error $e_f^*(x)$ may be unbounded pointwise (see Exercise 2.7). Since $\mathcal{D}$ is compact, if f and $\Phi_N \in C(\mathcal{D})$, then $e_f^*(x)$ is uniformily bounded on $\mathcal{D}$, but the theorem does not indicate how $e_f^*(x)$ changes as N increases. This is in contrast to results like the Weierstass theorem which showed polynomials could achieve arbitrary ϵ-accuracy approximation to continuous functions uniformly over a compact region, if the order of the polynomial was large enough. Development of results analogous to the Weierstrass theorem for more general classes of functions is the goal of this section.

For approximation based control applications, a fundamental question is whether a particular family of approximators is capable of providing a close approximation to the function $f(\mathbf{x})$. There are at least three interesting aspects of this question:

1. Is there some subset of a family of approximators that is capable of providing an ϵ-accurate approximation to $f(\mathbf{x})$ uniformly over $\mathcal{D}$.

2. If there exists some subset of the family of approximators that is capable of providing an ϵ-accurate approximation, can the designer specify an approximation structure in this subset *a priori*?

3. Given that an approximation structure can be specified, can appropriate parameter vectors θ and σ be estimated using data obtained during online system operation, while ensuring stable operation?

The first item is addressed by the universal approximation results of this subsection. The second item is largely unanswered, but easier for some approximation structures. Item 2 is discussed in Chapter 3. Item 3 which is a main focus of this text is discussed in Chapters 4-7. The discussion of this section focuses on single hidden layer networks. Similar results apply to multi-hidden layer networks [88, 259].

The N-th degree of approximation of f by $S_{r,N}$ is

$$E_N^\Phi(f) = \inf_{\theta_i, A_i, i=1\ldots,N} \left\| f - \left(\sum_{i=1}^{N} \theta_i g\left(A_i(\mathbf{x})\right) \right) \right\|.$$

Uniform Approximation is concerned with the question of whether for a particular family of approximators and f having certain properties (e.g., continuity), is it guaranteed to be true that for any $\epsilon > 0$, $E_N^{\Phi}(f) < \epsilon$ if N is large enough? Many such universal approximation results have been published (e.g.,[58, 88, 110, 146, 193, 259]). This section will present and prove one very general result for $\Sigma\Pi$-networks [110], and discuss interpretations and implications of this (and similar) results. Theorem 2.4.5 summarizes related results for Σ-networks.

The proof for $\Sigma\Pi$-networks uses the Stone-Weierstrass Theorem [48] which is stated below.

Theorem 2.4.3 (Stone-Weierstrass Theorem) *Let S be any algebra of real continuous functions on a compact set $\mathcal{D}$. If S separates points on $\mathcal{D}$ and vanishes at no point of $\mathcal{D}$, then for any $f \in C(\mathcal{D})$ and $\epsilon > 0$ there exists $\hat{f} \in S$ such that* $\sup_{\mathcal{D}} |f(x) - \hat{f}(x)| < \epsilon$.

Theorem 2.4.4 ([110]) *Let $\mathcal{D}$ be a compact subset of $\Re^r$ and $g : \Re^1 \mapsto \Re^1$ be any continuous, nonconstant function. The set S of $\Sigma\Pi$-networks with nodal processors specified by g has the property that for any $f \in C(\mathcal{D})$ and $\epsilon > 0$ there exists $\hat{f} \in S$ such that* $\sup_{\mathcal{D}} |f(x) - \hat{f}(x)| < \epsilon$.

Extensions of the examples of Section 2.4.4 show that for any continuous nonconstant g, $\Sigma\Pi$-networks satisfy the conditions of the Stone-Weierstrass Theorem. Therefore, the proof of Theorem 2.4.4 follows directly from the Stone-Weierstrass Theorem. An interesting and powerful feature of his theorem is that g is arbitrary in the set of continuous, nonconstant functions. The following theorem shows that Σ-networks with appropriate nodal functions also have the universal approximation property. The proof is not included due to the scope of the results that would be required to support it.

Theorem 2.4.5 *If g is either a squashing function or a local function (according to Definitions 2.4.6 or 2.4.7 respectively), f is continuous on the compact set $D \in \Re^r$, and S is the family of approximators defined as Σ network (according to Definition 2.4.3), then for a given ϵ there exist $\bar{N}(\epsilon)$ such that for $N > \bar{N}(\epsilon)$ there exist $\hat{f} \in S_{r,N}$ such that*

$$\rho(f, \hat{f}) < \epsilon$$

for an appropriately defined metric ρ for functions on D.

Approximators that satisfy theorems such as 2.4.4 and 2.4.5 are referred to as *universal approximators*. Universal Approximation Theorems such as this state that under reasonable assumptions on the nodal processor and the function to be approximated, if the (single hidden layer) network approximator has enough nodes, then an accurate network approximation can be constructed by selection of θ and σ. Such theorems do not provide constructive methods for determining appropriate values of $\bar{N}, \theta$, or σ.

Universal approximation results are one of the most typically cited reasons for applying neural or fuzzy techniques in control applications involving significant unmodeled nonlinear effects. The reasoning is along the following lines. *The dynamics involve a function $f(\mathbf{x}) = f_0(\mathbf{x}) + f^*(\mathbf{x})$ where $f^*(\mathbf{x})$ has a significant effect on the system performance and is known to have properties satisfying a universal approximation theorem, but $f^*(\mathbf{x})$ cannot be accurately modeled* a priori. *Based on universal approximation results, the designer knows that there exists some subset of S that approximates $f^*(\mathbf{x})$ to an accuracy ϵ for which the control specification can be achieved. Therefore, the approximation based* *control problem reduces to finding $\hat{f} \in S$ that satisfies the ϵ accuracy specification.* Most articles in the literature address the third question stated at the beginning of this section: selection of θ or (θ, σ) given that the remaining parameters of S have been specified. However, selection of N for a given choice of g and σ (or (N, σ) for a specified g) is the step in the design process that limits the approximation accuracy that can ultimately be achieved. To cite universal approximation results as a motivation and then select N as some arbitrary, small number are essentially contradictory.

Starting with the motivation stated in the previous paragraph, it is reasonable to derive stable algorithms for adaptive estimation of θ (or (θ, σ)) if N is specified large enough that it can be assumed larger than the unknown $\bar{N}$. Specification of too small of a value for N defeats the purpose of using a universal approximation based technique. When N is selected too small but a provably stable parameter estimation algorithm is used, stable (even satisfactory) control performance is still achievable; however, accurate approximation will not be achievable. Unfortunately, the parameter $\bar{N}$ is typically unknown, since $f^*(\mathbf{x})$ is not known. Therefore, the selection of N must be made overly large to ensure accurate approximation. The tradeoff for over estimating the value of N is the larger memory and computation time requirements of the implementation. In addition, if N is selected too large, then the approximator will be capable of fitting the measurement noise as well as the function. Fourier analysis based methods for selecting N are discussed in [232]. Online adjustment of N is an interesting area of research which tries to minimize the computational requirements while minimizing ϵ and ensuring stability [13, 37, 49, 72, 89, 178].

Results such as Theorems 2.4.4 and 2.4.5 provide sufficient conditions for the approximation of continuous functions over compact domains. Other approximation schemes exist that do not satisfy the conditions of these particular theorems but are capable of achieving ϵ approximation accuracy. For example, the Stone-Weierstrass Theorem shows this property for polynomial series. In addition, some classical approximation methods can be coerced into the form necessary to apply the universal approximation results. Therefore, there exist numerous approximators capable of achieving ϵ approximation accuracy when a sufficiently large number of basis elements is used. The decision among them should be made by considering other approximator properties and carefully weighing their relative advantages and disadvantages.

2.4.6 Best Approximator Property

Universal approximation theorems of the type discussed in Section 2.4.5 analyze the problem of whether for a family of function approximators $S_{r,N}$, there exists $a \in S_{r,N}$ that approximates a given function with at most ϵ error over a region D. Universal approximation results guarantee the existence of a sequence of approximators that achieve ϵ_i-accuracy, where $\{\epsilon_i\}$ is a sequence that converges to zero. Depending on the properties of the set $S_{r,N}$, the limit point of such a sequence may or may not exist in $S_{r,N}$.

This section considers an interesting related question: Given a convergent sequence of approximators $\{a_i\}$, $a_i \in S_{r,N}$, is the limit point of the sequence in the set $S_{r,N}$? If the limit point is guaranteed to be in $S_{r,N}$, then the family of approximators is said to have the *best approximator property*. Therefore, where universal approximation results seek approximators that satisfy a given accuracy requirement, best approximation results seek optimal approximation accuracy.

The best approximation problem [97, 155] can be stated as "Given $f \in C(D)$ and $S_{r,N} \subset C(D)$, find $a^* \in S_{r,N}$ such that $d(f, a^*) = d(f, S_{r,N})$." A set $S_{r,N}$ is called an *existence set* if for any $f \in C(D)$ there is at least one best approximation to f in $S_{r,N}$. A set

$S_{r,N}$ is called a *uniqueness set* if for any $f \in C(D)$ there is at most one best approximation to f in $S_{r,N}$. A set $S_{r,N}$ is called a *Tchebycheff set* if it is both a uniqueness set and an existence set. The results and discussion to follow are based on [48, 97].

Theorem 2.4.6 *Every existence set is closed.*

Proof. Assume that existence set $\mathcal{S} \subset C(\mathcal{D})$ is not closed. Then there exists a convergent sequence $\{s_i\} \subset \mathcal{S}$ such that the limit $f \notin \mathcal{S}$. Since f is a limit of $\{s_i\}$, $d(f, \mathcal{S}) = 0$. Since $\mathcal{S}$ is an existence set, there exists $g \in \mathcal{S}$ such that $d(f, g) = 0$. This implies that $f = g$ which is a contradiction. Therefore, $\mathcal{S}$ must be closed. ■

Theorem 2.4.7 *If $\mathcal{A}$ is a compact set in metric space $(\mathcal{S}, \|\cdot\|)$, then $\mathcal{A}$ is an existence set.*

Proof. Let $\mu = d(f, \mathcal{A})$ for $f \in \mathcal{S}$. By the definition of d as an infimum there exists a sequence $\{a_i\} \subset \mathcal{A}$ such that $d(f, a_i)$ converges to μ as $i \to \infty$. By the compactness of $\mathcal{A}$, the sequence $\{a_i\}$ has a limit $a^* \in \mathcal{A}$. By the triangle inequality, $d(f, a^*) \leq d(f, a_k) + d(a_k, a^*)$. Since the left side is independent of k and the right side converges to μ, $d(f, a^*) \leq \mu$. By the definition of μ as the infimum over all elements of $\mathcal{A}$, it is necessary that $d(f, a^*) \geq \mu$. Combining inequalities gives $d(f, a^*) = \mu$, which shows that the best approximation is achieved by an element of $\mathcal{A}$. ■

The above two theorems show that a set being closed is a necessary, but not sufficient condition for a set to be an existence set. Compactness is a sufficient condition.

Theorem 2.4.8 *For g continuous and nonconstant, let $S_{n,N,g,\sigma} \subset C(\mathcal{D})$ be defined as in Definition 2.4.1, then $S_{n,N,g,\sigma}$ is an existence set.*

Proof. Let f be an arbitrary fixed element of $C(\mathcal{D})$. Choose an arbitrary $h \in S_{n,N,g,\sigma}$. The set

$$\mathcal{H}_N = \{g \in S_{n,N,g,\sigma} \,|\, \|g - f\| \leq \|h - f\|\}$$

is closed and bounded. Therefore, the finite dimensional set $\mathcal{H}_N$ is compact. Theorem 2.4.7 implies that $\mathcal{H}$ (and therefore $S_{n,N,g,\sigma}$) is an existence set. ■

The set $\mathcal{H}_N$ being closed and bounded relies on the assumption that $S_{n,N,g,\sigma} \subset C(\mathcal{D})$ is defined by a finite dimensional LIP approximator. When the approximator $\hat{f}$ in not LIP, the proof will not typically go through, since the set $\mathcal{H}$ defined relative to

$$S_{r,N,g} = \left\{ f : \Re^r \to \Re^1 \,\middle|\, f(\mathbf{x}) = \sum_{i=1}^{N} \theta_i g_i(\mathbf{x}, \sigma) = \theta^\top \phi(\mathbf{x}, \sigma) \right\}$$

for $\mathbf{x} \in \Re^n$, $\sigma \in \Re^m$ and $\theta \in \Re^N$ is not usually closed. In particular, [97] shows that radial basis functions with adaptive centers and sigmoidal neural networks with an adaptive input layer (or multiple adaptive layers) do not have the best approximator property.

Although the best approximation property is a motivation to using LIP approximators, the motivation is not strong. If ϵ-accuracy approximation is required for satisfactory control performance and an approximator structure $S_{r,N,g}$ can be defined which is capable of achieving ϵ'-accuracy for some $\epsilon' < \epsilon$, then there exist a subset $\mathcal{A}$ of $S_{r,N,g}$ that achieves the desired ϵ-accuracy approximation. However, it may be quite difficult to specify the required approximation structure and find an element of the subset $\mathcal{A}$.

2.4.7 Generalization

Function approximation is the process of selecting a family of approximators, and the structure and parameters for a specific approximator in that family, to optimally fit a given set of training data. The subsequent process of generating reasonable outputs for inputs not in the training set is referred to as *generalization* [128, 226, 246, 300, 301]. Generalization is also closely related to *statistical learning theory* , which is a well-established field in machine learning [8, 239, 274].

The term *generalization* is often used to motivate the use of neural network/fuzzy methods. The motivational phrase is typically of the form "... neural networks have the ability to generalize from the training data." Analysis of such statements requires understanding of the term generalization.

Generalization refers to the ability of a function $\hat{f}(x;\theta)$ designed to approximate a given set of data $\{(x_i, y_i)\}_{i=1}^m$ also to provide accurate estimates of $y = f(x)$ for $x \notin \{x_i\}_{i=1}^m$. Generalization can be analyzed by considering whether the approximator that minimizes the sample cost function

$$J_m(\theta) = \frac{1}{m}\sum_{i=1}^{m} \|y_i - \hat{f}(x_i;\theta)\| \tag{2.37}$$

also minimizes the analytic cost function

$$J(\theta) = \int_{\mathcal{D}} \|f(x) - \hat{f}(x;\theta)\|dx. \tag{2.38}$$

Unfortunately, the cost function of eqn. (2.38) can only be evaluated if $f(x)$ is known. Therefore, implementations focus on the minimization of a sample cost function such as eqn. (2.37). This is a *scattered data approximation* problem. As $m \to \infty$, when $J_m(\theta)$ converges, its limit is

$$J_E(\theta) = \int_{\mathcal{D}} \|f(x) - \hat{f}(x;\theta)\|p(x)dx \tag{2.39}$$

where $p(x)$ is the distribution of training samples. If $p(x)$ is uniform, then the minima of the two cost functions will be the same; however, in general, the approximations that result from the two cost functions will be distinct.

If $x \in \mathcal{D}$ with $x \notin \{x_i\}_{i=1}^m$ and $\|x - x_i\| < \delta$ for some i. Then

$$\|f(x) - \hat{f}(x)\| \leq \|f(x) - f(x_i)\| + \|f(x_i) - \hat{f}(x_i)\| + \|\hat{f}(x_i) - \hat{f}(x)\|.$$

If the approximation is accurate at the points in the training set, then the middle right hand side term is small. If f and $\hat{f}$ are both continuous, then the outside terms on the right hand side are also small when δ is suitably small. Therefore, this expression yields two conclusions: (1) accurate approximation over the training set is a precondition to discussing generalization; and, (2) continuity of the function and approximator automatically give local generalization in the vicinity of the training points.

In offline training, the above analysis motivates the accumulation of a batch of data, with m large, that is uniformly distributed over $\mathcal{D}$. In adaptive approximation, the number of samples does eventually become large, but the distribution of samples is rarely uniform, is not known *a priori*, is time varying, and is usually not selectable by the designer. However, when the state is a continuous function of time, which is usually the case because the sample frequency is high relative to the system bandwidth and the state is the solution to a set of differential equations describing the evolution of a physical system, x_{i+1} is near x_i. If the approximator has been trained at x_i and is being evaluated at x_{i+1}, then

$$\|f(x_{i+1}) - \hat{f}(x_{i+1})\| \leq \|f(x_{i+1}) - f(x_i)\| + \|f(x_i) - \hat{f}(x_i)\| + \|\hat{f}(x_i) - \hat{f}(x_{i+1})\|.$$

The outside terms on the right-hand side are again small if f and $\hat{f}$ are continuous and $\|x_i - x_{i+1}\|$ is small. The middle right-hand side is small if the adaptive approximation algorithm has converged near x.

The ability of an approximator to "generalize from the training data" depends on (1) the properties of the function to be approximated, (2) the properties of the approximating function, (3) the amount and distribution of the training data, and (4) the method of evaluation of the generalization results. In particular, related to item (4), is localized generalization all that is expected or is the approximator expected to extrapolate from the training data to regions of $\mathcal{D}$ that are not represented by the training data?

Local generalization is the process of providing an estimate of $f(x)$ at a point x, where $x - x_i$ is small for some $i \leq 1 \leq m$. Conceptually, local generalization combines appropriately weighted training points in the vicinity of the evaluation point. Therefore, local generalization is desirable both for noise filtering and data reduction. The capability of the function approximator to generalize locally between training samples is necessary if the approximator is to make efficient use of memory and the training data. Based on the previous analysis, it is reasonable to expect local generalization when f and $\hat{f}$ are continuous in x.

Extrapolation is the process of providing an estimate of $f(x)$ at a point x, where $x - x_i$ is large for all $1 \leq i \leq m$. Therefore, extrapolation attempts to predict the value of the function in a region far from the available training data. In offline (batch) training scenarios, the set of training samples can be designed to be representative of the region D, so that extrapolation does not occur. In online control applications, operating conditions may force the designer to use whatever data the system generates even if the training data does not representatively cover all of D. Since the class of functions to be approximated is large (i.e., all continuous functions on D) and the training data will include measurement noise, accurate extrapolation should not be expected. In fact, the control methodology should include methods to accommodate regions of the state space for which adequate training has not occurred. Alternatively, the system should slowly move from regions for which accurate approximation has been achieved into regions still requiring exploration. Often, this is a natural result of the system dynamics, as discussed above.

■ EXAMPLE 2.16

Consider Figure 2.5 in the context of the discussion of this section. The figure shows polynomial approximations of various orders to a set of experimental data. The figure also shows the extrapolation of the function approximation to the portions of $\mathcal{D}$ that were not represented by the training data in that example. The extrapolation accuracy is dependent on both the approximator order and on the training data. Even the order of the polynomial that provides the "best" extrapolation relative to the true function is highly dependent on the elements of the training set. △

Since the control system performance is usually directly related to the approximation error, it is usually better for the approximator to be zero than possibly of the wrong sign (i.e., amplifying the approximation error) in a region not adequately represented by the training data. This constraint motivates the use of approximators with locally supported basis elements.

2.4.8 Extent of Influence Function Support

In the specification of the approximators of eqns. (2.31) or (2.32), a major factor in determining the ultimate performance that can be achieved is the selection of the functions $\phi(x)$. An important characteristic in the selection of ϕ is the extent of the support of the elements of ϕ, which is defined to be $S_i = Supp_{\phi_i} = \{x \in \mathcal{D} | \phi_i(x) \neq 0\}$. Let $\mu(A)$ be a function that measures the area of the set $A \subset \mathcal{D}$. Then, the functions ϕ_i will be referred to as *globally supported* functions if $\mu(Supp_{\phi_i}) = \mu(\mathcal{D})$. The functions ϕ_i will be referred to as *locally supported* functions if S_i is connected and $\mu(S_i) \ll \mu(\mathcal{D})$.

The solution of the theoretical least squares problem where f is a known function is given in eqn. 2.30. The accuracy of the solution depends on the condition of the matrix $\int_\mathcal{D} \phi(x)\phi(x)^\top dx$. The elements of this matrix are

$$\left[\int_\mathcal{D} \phi_i(x)\phi_j(x)dx\right] = \left[\int_{S_i \bigcap S_j} \phi_i(x)\phi_j(x)dx\right].$$

When the basis elements have local support, this matrix will be sparse and have a banded-diagonal structure. With careful design of the regressor vector the elements of each diagonal will each be of about the same size and the matrix $\int_\mathcal{D} \phi(x)\phi(x)^\top dx$ will be well conditioned.

The following subsections introduce a general representation for approximators with locally supported basis elements, contrast the advantages of locally and globally supported basis elements, and introduce the concept of a lattice network.

2.4.8.1 *Approximators with Local Influence Functions*

Several approximators with local influence functions have been proposed in the literature. This section analyzes such approximators in a general framework [73, 83, 85, 173, 175]. Specific approximators are discussed in Chapter 3.

Local and global approximation structures can be distinguished as follows.

Definition 2.4.10 (Local Approximation Structure) – *a function $\hat{f}(x, \hat{\theta})$ is a local approximation to $f(x)$ at x_0 if for any ε there exist $\hat{\theta}$ and δ such that $\| f(x) - \hat{f}(x, \hat{\theta}) \| \leq \varepsilon$ for all $x \in B(x_0, \delta) = \{x | \ \|x - x_0\| < \delta\}$.*

Two common examples of local approximation structures are constant and linear functions. It is well known that constant, linear, or higher order polynomial functions can be used to accurately approximate an arbitrary continuous function if the region of validity of the approximation is small enough.

Definition 2.4.11 (Global Approximation Structure) – *a parametric model $\hat{f}(x, \hat{\theta})$ is an ε-accurate global approximation to $f(x)$ over domain $\mathcal{D}$ if for the given ε there exists $\hat{\theta}$ such that $\| f(x) - \hat{f}(x, \hat{\theta}) \| \leq \varepsilon$ for all $x \in D$.*

Note the following issues related to the above definitions.

- Models derived from first principals are usually (expected to be) global approximation structures.
- Whether a given approximation structure is local or global is dependent on the system that is being modeled. For example, a linear approximating structure is global for linear plants, but only local for nonlinear plants.

- The set of global models is a strict subset of the set of local models. This is obvious, since if there exists a set of parameters $\hat{\theta}$ satisfying Definition 2.4.11 for a particular ε, then this $\hat{\theta}$ also satisfies Definition 2.4.10 for the same ε at each $x_0 \in \mathcal{D}$.

To maintain accuracy over domain $\mathcal{D}$, a local approximation structure can either adjust its parameter vector, through time, as the operating point x_0 changes; or store its parameter vector as a function of the operating point. The former approach is typical of adaptive control methodologies while the latter approach is being motivated herein as learning control. The latter approach can effectively construct a global approximation structure by connecting several local approximating structures.

A main objective of this subsection is to appropriately piece together a (large) set of local approximation structures to achieve a global approximation structure. The following definition of the class of *Basis-Influence Functions* [16, 76, 85, 122, 173] presents one means of achieving this objective.

Definition 2.4.12 (Basis-Influence (BI) Functions) – *A function approximator is of the BI Class if and only if it can be written as*

$$\hat{f}(x, \hat{\theta}) = \sum_i f_i(x, \hat{\theta})\Gamma_i(x) \tag{2.40}$$

where each $f_i(x, \hat{\theta})$ is a local approximation to $f(x)$ for all $x \in B(x_i, \delta)$, and $\Gamma_i(x)$ has local support S_i which is a subset of $B(x_i, \delta)$ such that $D \subseteq \bigcup_i S_i$.

Examples of Basis-Influence approximators include: Boxes [231], CMAC [2], Radial Basis Functions [205], splines, and several versions of fuzzy systems [198, 283]. In the traditional implementation of each of these approximators, the basis functions are constant on the support of the influence function. If more capable basis functions (e.g., linear functions) were implemented, then the designer should expect there to be a decrease in the number of required local approximation structures. An alternative definition of local influence, which also provides a measure of the degree of localization based on the learning algorithm, is given in [288].

The partition of unity is defined as follows [253, 293].

Definition 2.4.13 (Partition of Unity) – *The set of positive semidefinite influence functions $\{\Gamma_i\}$ form a Partition of Unity on $\mathcal{D}$ if for any $x \in \mathcal{D}$, $\sum_{i=1}^N \Gamma_i(x) = 1$.*

Influence functions that form a partition of unity have a variety of benefits. First, if $\{\Gamma_i\}$ form a Partition of Unity on $\mathcal{D}$, then there cannot be any $x \in \mathcal{D}$ such that $\sum_{i=1}^N \Gamma_i(x) = 0$. Also, when the approximator is defined by eqn. (2.40) with $\{\Gamma_i\}$ forming a Partition of Unity, then at any $x \in \mathcal{D}$, $\hat{f}(x, \hat{\theta})$ is a convex combination of $f_i(x, \hat{\theta})$.

If a set of positive semidefinite influence functions $\{\bar{\Gamma}_i\}$ do not form a partition of unity, but have the coverage property (i.e., for any $x \in \mathcal{D}$ there exists at least one i such that $\bar{\Gamma}_i(x) \neq 0$), then a partition of unity can be formed from $\{\bar{\Gamma}_i\}$ as

$$\Gamma_i(x) = \frac{\bar{\Gamma}_i(x)}{\sum_{i=1}^N \bar{\Gamma}_i(x)}. \tag{2.41}$$

This normalization operation should however be used cautiously [221]. Such normalization can yield $\Gamma_i(x)$ that have large flat areas. In addition, even when $\bar{\Gamma}_i(x)$ is unimodal, $\Gamma_i(x)$ may be multimodal. See Exercise 2.10. When the functions $\bar{\Gamma}_i(x)$ are fixed after the design stage, the designer can ensure that the $\Gamma_i(x)$ have desirable properties; however, when the centers and radii of the $\bar{\Gamma}_i(x)$ are adapted online (i.e., nonlinear in the parameter adaptive approximation), then such anomalous behaviors may occur.

Given Definition 2.4.12 it is possible to constructively prove a sufficient condition for Basis-Influence functions to be global approximators.

Theorem 2.4.9 *If $\hat{f}(x, \hat{\theta})$ is of Class BI with each $f_i(x, \theta)$ satisfying Definition 2.4.10 for a fixed $\epsilon > 0$, then*

$$\Gamma_i(x) \geq 0, \textit{and} \sum_i \Gamma_i(x) = 1, \textit{for any } x \in \mathcal{D}$$

are sufficient conditions for $\hat{f}(x, \hat{\theta})$ to be an ϵ accurate global approximation to $f \in C(\mathcal{D})$ for compact $\mathcal{D}$.

Proof. Fix $x \in \mathcal{D}$. Let $N_x = \{i \in I \,|\Gamma_i(x) \neq 0\}$. Then by Definition 2.4.12, $\sum_{i \in N_x} \Gamma_i(x) = 1$. For each $i \in N_x$, for $x \in S_i$, by Definitions 2.4.12 and 2.4.10, there exists $|\varepsilon_i(x)| \leq \varepsilon$ such that

$$\hat{f}_i(x, \hat{\theta}) = f(x) + \varepsilon_i(x). \tag{2.42}$$

Therefore,

$$\hat{f}(x, \hat{\theta}) = \sum_i \hat{f}_i(x, \hat{\theta}) \Gamma_i(x) = \sum_i (f(x) + \varepsilon_i(x)) \, \Gamma_i(x) = f(x) + \sum_i \Gamma_i(x) \varepsilon_i(x),$$

and

$$\begin{aligned} \left\| \hat{f}(x, \hat{\theta}) - f(x) \right\| &= \left\| \sum_i \varepsilon_i(x) \Gamma_i(x) \right\| \\ &\leq \left\| \sum_i |\varepsilon_i(x)| |\Gamma_i(x)| \right\| \\ &\leq \left\| \epsilon \sum_i \Gamma_i(x) \right\| \\ &= \epsilon. \end{aligned}$$

Since x is an arbitrary point in *D*, this completes the proof. ■

When a multivariable Basis-Influence approximator can be represented by taking the product of the influence functions for each single variable:

$$\hat{f}(x, y, \hat{\theta}) = \sum_i \sum_j f_{ij}(x, y, \hat{\theta}) \Gamma_{x_i}(x) \Gamma_{y_j}(y;) \tag{2.43}$$

the basis-influence approximator fits the definition of a $\Sigma\Pi$-network to which Theorem 2.4.4 applies.

■ **EXAMPLE 2.17**

A one input approximator that meets all the conditions of Theorem 2.4.9 is

$$\hat{f}(x,\hat{\theta}) = \sum_i f_i(x,\hat{\theta})\Gamma_i(x) \tag{2.44}$$

where $x_i = i\frac{\pi}{2\lambda}$

$$\Gamma_i(x) = \begin{cases} cos^2\left(\lambda\left(x - x_i\right)\right), & \text{for } -\frac{\pi}{2\lambda} \le (x - x_i) \le \frac{\pi}{2\lambda} \\ 0 & \text{otherwise} \end{cases} \tag{2.45}$$

and $f_i(x,\hat{\theta})$ can be any function capable of providing a local approximation to $f(x)$ at x_i. △

■ **EXAMPLE 2.18**

Figure 2.10 illustrates basis-influence function approximation. The routine for constructing this plot used Γ as defined in eqn. (2.45) with $\lambda = 0.785$. In the notation of Definition 2.4.12, for $i = 1, \ldots, 6$:

$$f_i(x,\theta) \quad = \quad a_i(x - c_i) + b_i = [a_i, b_i]\begin{bmatrix} (x - c_i) \\ 1 \end{bmatrix}$$

where $c_i = 0.2(i - 1)$ and $\mathcal{D} = [0, 1]$. For clarity, the influence functions are plotted at a 10% scale and only a portion of each linear approximation is plotted.

Note that the parameters of the approximator have been jointly optimized such that eqn. (2.44) has minimum least squares approximation error over $\mathcal{D}$. This does not imply that each f_i is least squares optimal over S_i. This is clearly evident from the figure. For example, f_5 is not least squares optimal over $S_5 = [0.6, 1.0]$. The least squared error of f_5 over S_5 would be decreased by shifting f_5 down. It is possible to improve the local accuracy of each f_i over S_i, but this will increase the approximation error of eqn. (2.44) over $\mathcal{D}$. Often, this increase is small and such receptive field weighted regression methods have other advantages in terms of computation and approximator structure adaptation (i.e., approximator self-organization) [13, 236, 237]. △

2.4.8.2 Retention of Training Experience Based on the discussion of Subsection 2.4.7, the designer should not expect $\hat{f}$ to accurately extrapolate training data from regions of $\mathcal{D}$ containing significant training data into other (unexplored) regions. In addition, it is desirable for training data in new regions to not affect the previously achieved approximation accuracy in distant regions. These two issues are tightly interrelated. The issues of localization and interference in learning algorithms were rigorously examined in [288, 289].

The online parameter estimation algorithms of Chapters 4, 6, and 7 will adapt the parameter vector estimate $\hat{\theta}(t)$ based on the current (possibly filtered) tracking error $e(t)$. The algorithms will have the generic forms of eqns. (2.24) and (2.25). If the regressor (i.e., $\phi(x)$) has global support, then changing the estimated parameter $\hat{\theta}_i$ affects the approximation accuracy throughout D. Alternatively, if ϕ_i has local support, then changing the estimated parameter $\hat{\theta}_i$ affects the approximation accuracy only on $Supp_{\phi_i}$ which by assumption is a small region of D containing the training point.

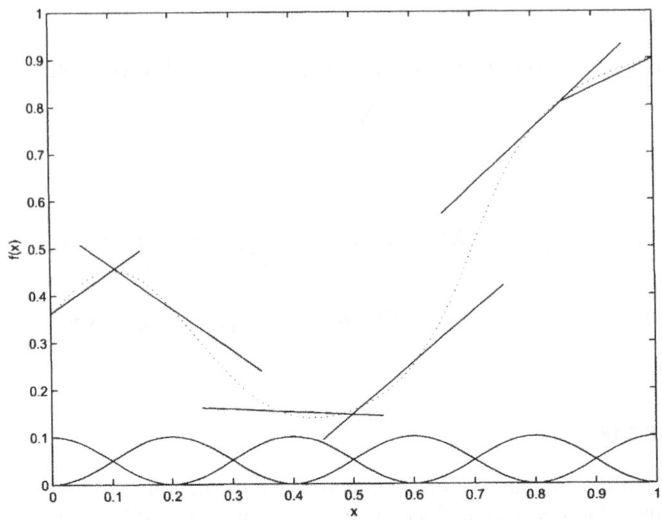

Figure 2.10: Basis-Influence Function Approximation of Example 2.18. The original function is shown as a dashed line. The local approximations (basis functions) are shown as solid lines. The influence functions (drawn at 10% scale) are shown as solid lines at the bottom of the figure.

■ **EXAMPLE 2.19**

Consider the task of estimating a function $f(x)$ by an approximator $\hat{f}(x) = \phi(x)^\top \theta$. As in a control application, assume that samples are obtained incrementally and the x_{k+1} is near x_k. This example considers how the support characteristics of the basis elements $\{\phi_i\}_{i=1}^N$ affects the convergence of the function approximation.

For computational purposes, assume that $f(x) = \sin(x)$ and the domain of approximation $\mathcal{D} = [-\pi, \pi]$. Also, let $x_k = -3.6 + 0.1k$ for $k = 0, \ldots, 72$. Consider two possible basis sets. The first set of basis elements is the first eight Legendre polynomials (see Section 3.2) with the input to the polynomial scaled so that $\mathcal{D} \mapsto [-1, 1]$. This basis sets has global support over $\mathcal{D}$. The approximator with the first eight Legendre polynomials as basis elements is capable of approximating the sin function with a maximum error over $\mathcal{D}$ of approximately 1.0×10^{-3}. The second set of basis elements is a set of Gaussian radial basis elements (see Section 3.4) with centers at $c_i = -4 + 0.5i$ for $i = 0, \ldots, 16$ and spread $\sigma = 0.5$. Although each Gaussian basis element is nonzero over all of $\mathcal{D}$, each basis element is effectively locally supported. This 17-element RBF approximator is capable of approximating the sin function with maximum error over $\mathcal{D}$ of approximately 0.5×10^{-3}. For both approximators, initially the parameter estimate is the zero vector.

Figure 2.11 shows the results of gradient descent based (normalized least mean squares) estimation of the sin function with each of the two approximators. The Legendre polynomial approximation process is illustrated in the top graph. The RBF approximation process is illustrated in the bottom graph. Each of the graphs contains three curves. The solid line indicates the function $f(x)$ that is to be approximated. The

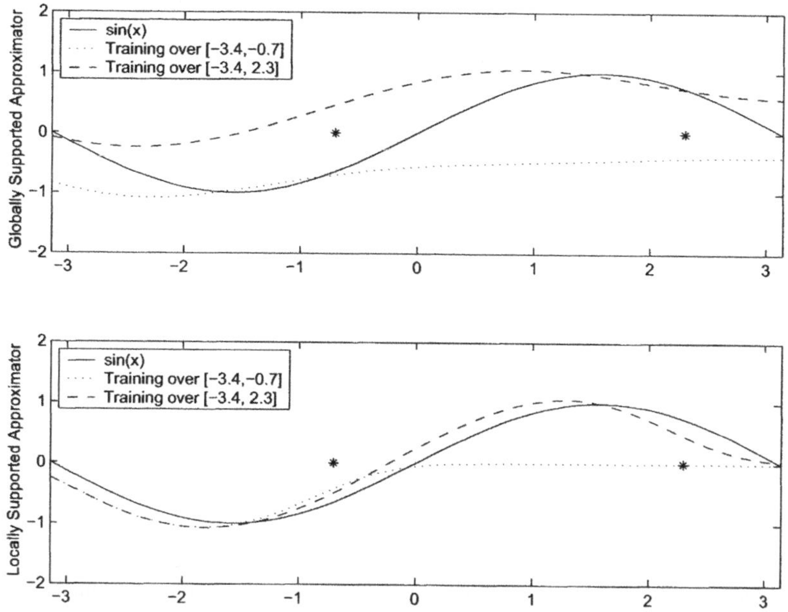

Figure 2.11: Incremental Approximations to a sin function. *Top* – Approximation by 8-th order Legendre Polynomials. *Bottom* – Approximation by normalized Radial Basis Functions. The asterisks indicate the rightmost training point for the two training periods discussed in the text.

dotted line is the approximation at $k = 29$. At this time, the approximation process has only incorporated training examples over the region $\mathcal{D}_{29} = [-3.6, -0.7]$. The left asterisk on the x-axis indicates the largest value of x in $\mathcal{D}_{29}$. Note that both approximators have partially converged over $\mathcal{D}_{29}$. The RBF approximation is more accurate over $\mathcal{D}_{29}$. The polynomial approximation has changed on $\mathcal{D} - \mathcal{D}_{29}$. The RBF approximation is largely unchanged on $\mathcal{D} - \mathcal{D}_{29}$. The dashed line is the approximation at $k = 59$. At this time, the approximation process has incorporated training examples over the region $\mathcal{D}_{29} = [-3.6, 2.3]$. The right asterisk on the x-axis indicates the largest value of x in $\mathcal{D}_{59}$. Note that while the polynomial approximation is now accurate near the current training point ($x = 2.3$), the approximation error has increased, relative to the dotted curve, on $\mathcal{D}_{29}$. Alternatively, the RBF approximator is not only accurate in the vicinity of the current training point, but is still accurate on $\mathcal{D} - \mathcal{D}_{29}$, even though no recent training data has been in that set. For both approximators, the norm of the parameter error is decreasing throughout the training.

This example has used polynomials and Gaussian RBFs for computational purposes, but the main idea can be more broadly stated. When the approximator uses locally supported basis elements, there is a close correspondence between parameters of the approximation and regions of $\mathcal{D}$. Therefore, the function can be adapted locally to learn new information, without affecting the function approximation in other regions of the domain of approximation. This fact facilitates the retention of past training data. When the basis elements have global support, retention of past training

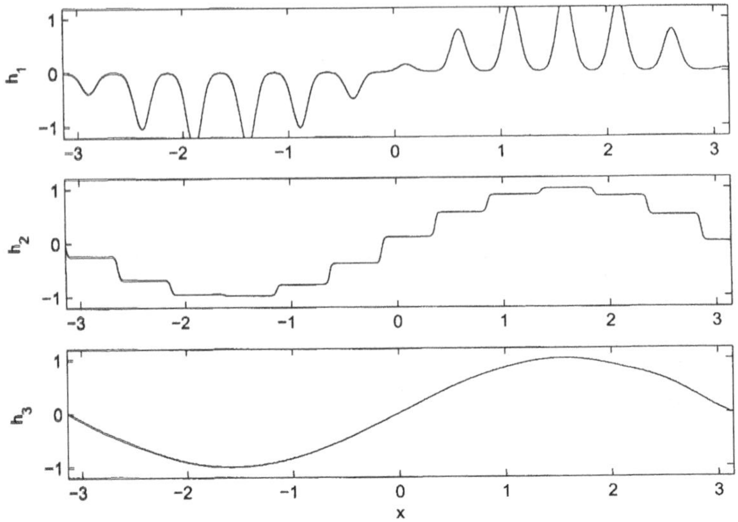

Figure 2.12: Three RBF approximations to a sine function using different values of σ. The basis elements of the middle and bottom approximations form partitions of unity.

data is much more complicated. It can be accomplished, for example, using recursive least squares, but only at significant computational expense. △

When an approximator uses influence functions that do not form a partition of unity and the influence functions are too narrow relative to their separation, the resulting approximation may be "spiky." Alternatively, when the influence functions do form a partition of unity and the influence functions are too narrow relative to their separation, the approximation may have flat spots.

EXAMPLE 2.20

Figure 2.12 shows three radial basis function approximations to a sine function. The top plot uses an approximation h_1 with unnormalized RBF functions every 0.5 units and $\sigma = 0.1$. Since σ is much less than the separation between the basis elements, the approximation is spiky. The middle approximation h_2 uses normalized RBF functions every 0.5 units with $\sigma = 0.1$. Since σ is much less than the separation between the basis elements, the normalization of the regressor vector results in an approximation that has flat regions. The bottom approximation h_3 uses normalized RBF functions every 0.5 units with $\sigma = 0.5$. Since σ is similar to the separation between the basis elements, the support of adjacent basis elements overlap. In this case, the approximation has neither spikes nor flat regions. △

The choice of the functions $f_i(x, \hat{\theta}; x_i)$ are important to application success and computational feasibility. Consider the case where $f_i(x, \hat{\theta}; x_i)$ are either zero or first order local Taylor series approximations:

$$f_i(x, \hat{\theta}) \quad = \quad A \tag{2.46}$$

$$\text{or } f_i(x, \hat{\theta}) = A + B(x - x_i). \tag{2.47}$$

In the first case, the basis functions are constants, as in the case of normalized radial basis functions. For a given desired approximation accuracy ε, many more basis-influence pairs may be required if constant basis functions are used instead of linear basis functions. Estimates of the magnitude of higher order derivatives can be used to estimate the number of Basis Influence (BI) function pairs required in a given application.

The linear basis functions hold two advantages in control applications.

1. Linear approximations are often known *a priori* (e.g., from previous gain scheduled designs or operating point experiments). It is straightforward to use this prior information to initialize the BI function parameters.

2. Linear approximations are often desired *a posteriori* either for analysis or design purposes. These linear approximations are easily derived from the BI model parameters.

See the related discussion in Section 2.4.9 of network transparency.

2.4.8.3 Curse of Dimensionality A well-known drawback [20] of function approximators with locally supported regressor elements is the "curse of dimensionality," which refers to the fact that the number of parameters required for localized approximators grows exponentially with the number of dimensions $\mathcal{D}$.

■ **EXAMPLE 2.21**

Let $d = dim(\mathcal{D})$. If $\mathcal{D}$ is partitioned into $\bar{N}$ divisions per dimension, then there will be $N = \bar{N}^d$ total partitions. △

This exponential increase in N with d is a problem if either the computation time or memory requirements of the approximator become too large. The embedding approach discussed in Section 3.5 is a method of allowing the number of partitions of $\mathcal{D}$ to increase exponentially without a corresponding increase in the number of approximator parameters. The lattice networks discussed in Section 2.4.8.4 illustrate a method by which the computational requirements grow much slower than the exponential growth in the number of parameters.

2.4.8.4 Lattice-Based Approximators Specification of locally supported basis functions requires specification of the type and support of each basis element. Typically, the support of a basis element is parameterized by the center and width parameters of each ϕ_i. This specification includes the choice as to whether the center and width parameters are fixed *a priori* or estimated based on the acquired data.

Adaptive estimation of the center and width parameters is a nonlinear estimation problem. Therefore, the resulting approximator would not have the best approximator property, but would have the beneficial "order of approximation" behavior as discussed in Section 2.4.1.

Prior specification of the centers on a grid of points results in a *lattice-based approximator* [32]. Lattice-based approximators result in significant computational simplification over adaptive center-based approximators for two reasons. First, the center adaptation calculations are not required. Second, the non-zero elements of the vector ϕ can be determined without direct calculation of ϕ (see below). If the width parameters are also fixed *a priori*, then a linear parameter estimation problem results with the corresponding benefits.

EXAMPLE 2.22

The purpose of this example [75] is to clarify how lattice-based approximators can reduce the amount of computation required per iteration. For clarity, the example discusses a two-dimensional region of approximation, as shown in Figure 2.13, but the discussion directly extends to $d > 2$ dimensions.

A function f is to be approximated over the region $\mathcal{D} = \{(x, y) \in [0,1]\times[0,1]\}$. If the approximator takes the form $\hat{f}(x) = \theta^\top \phi(x)$, where $\theta \in \Re^N$ and $\phi : \mathcal{R}^2 \rightarrow \Re^N$, then evaluation of $\hat{f}$ for a general approximator requires calculation of the N elements of $\phi(x)$ followed by an N-vector multiply (with the associated memory accesses). Assuming that $\phi(x)$ is maintained in memory between the approximator computation and parameter adaptation, then adaptation of θ requires (at the minimum) a scalar by N-vector multiply.

Alternatively, let the elements of $\phi(x)$ be locally supported with fixed centers defined on a lattice by

$$c_m = c_{i,j} = ((i-1)\cdot dx, (j-1)\cdot dy)$$

for $i = 1, \ldots, n_x$ and $j = 1, \ldots, n_y$, where $N = n_x n_y$, $m = i + n_x * (j-1)$, $dx = \frac{1}{n_x - 1}$, and $dy = \frac{1}{n_y - 1}$. Also, let $\phi_{i,j}(x) = g((x,y) - c_{i,j})$ be locally supported such $g((x,y) - c_{i,j}) = 0$ if $\|(x,y) - c_{i,j}\|_\infty > \lambda$. The parameter λ is referred to as the *generalization parameter*. To allow explicit discussion in the following, assume that $\lambda = 1.5dx$. Also, as depicted in Figure 2.13, assume that $n_x = n_y = 5$, so that $dx = dy = 0.25$. The figure indicates the nodal centers with x's and indicates the values of m on the lattice diagram. In general, these assumptions imply that although N may be quite large, at most 9 elements of the vector ϕ will

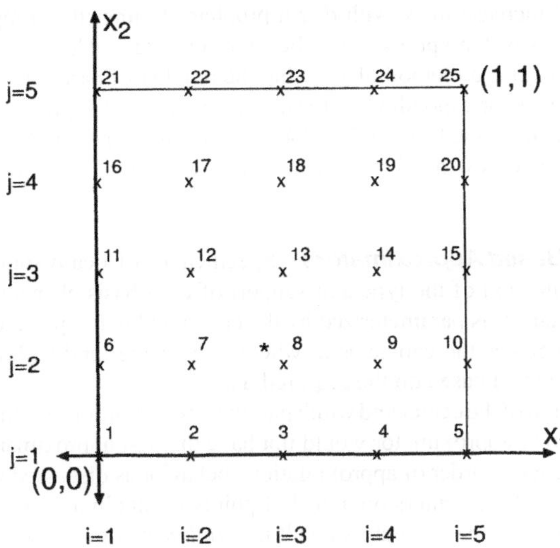

Figure 2.13: Lattice structure diagram for Example 2.22. The x's indicate locations of nodal centers. The integers near the x's indicate the nodal addresses m. The $*$ indicates an evaluation point.

be non-zero at a given value of x; therefore, calculation of $\hat{f}$ only requires a 9-element vector multiply (with the associated memory accesses). This computational simplification assumes that there is a simple method for determining the appropriate elements of ϕ and θ without search and without directly calculating all of $\phi(x)$.

The indices for the nonzero elements of ϕ and corresponding elements of θ (sometimes called *nodal addresses*) can be found by an algorithm such as

$$\begin{aligned} i_c(x) &= 1 + round\left(\frac{x}{dx}\right) \\ j_c(y) &= 1 + round\left(\frac{y}{dy}\right) \end{aligned}$$

where $round(z)$ is the function that returns the nearest integer to z. The set of indices corresponding to nonzero basis elements (neglecting evaluation points within $\frac{dx}{2}$ and $\frac{dy}{2}$ of the edges of $\mathcal{D}$) is then

$$\begin{array}{ccc} (i_c-1, j_c+1) & (i_c, j_c+1) & (i_c+1, j_c+1) \\ (i_c-1, j_c) & (i_c, j_c) & (i_c+1, j_c) \\ (i_c-1, j_c-1) & (i_c, j_c-1) & (i_c+1, j_c-1). \end{array}$$

At the evaluation point indicated by the $*$, $(i_c, j_c) = (3, 2)$, $m = 8$, and the nodal addresses of the nonzero basis elements are $\{2, 3, 4, 7, 8, 9, 12, 13, 14\}$. △

To summarize, if an approximator has locally supported basis elements defined on a lattice, then both the approximation at a point and the parameter estimation update can be performed (due to the sparseness of ϕ and the regularity of the centers) without calculating all of ϕ and without direct sorting of ϕ to find its non-zero elements.

Even if each element of ϕ is locally supported, if the centers are not defined on a lattice, then in general there is no method to find the nonzero elements of ϕ without direct calculation of and search over the vector ϕ.

A common argument against lattice networks is that fewer basis functions may be required if the centers are allowed to adapt their locations to optimize their distribution relative to the function being approximated. There is a tradeoff involved between the decrease in memory required (due to the potential decreased number of basis functions) and the increased per iteration computation (due to all of ϕ being calculated). In addition, online adaptation of the center locations optimizes the estimated center locations relative to the training data, which at any given time may not represent optimization relative to the actual function.

2.4.9 Approximator Transparency

Approximator transparency refers to the ability to preload *a priori* information into the function approximator and the ability to interpret the approximated function as it evolves in applications. Applications using fuzzy systems typically cite approximator transparency as a motivation. The fuzzy system can be interpreted as a rule base stating either the control value or control law applicable at a given system state [198, 283].

In any application, *a priori* information can be preloaded by at least two approaches. First, the function to be approximated can always be decomposed as

$$f(\mathbf{x}) = f_o(\mathbf{x}) + f^*(\mathbf{x}) \tag{2.48}$$

where $f_o(\mathbf{x})$ represents the known portion of the function and $f^*(\mathbf{x})$ represents the unknown portion for which an approximation will be developed online. In this case, the function approximator would approximate only $f^*(\mathbf{x})$. Second, if for some reason, the approach described in eqn. (2.48) is not satisfactory, then $\hat{f}(\mathbf{x})$ could be initialized by offline methods to accurately approximate the known portion of the function (i.e., $f_o(\mathbf{x})$). During online operation, the parameters of the approximator would be tuned to account also for the unknown portion of the function so that ultimately $\hat{f}(\mathbf{x}) = f_o(\mathbf{x}) + f^*(\mathbf{x})$.

Any approximator of the *basis-influence* class allows the user to interpret the approximated function. The influence functions dictate which of the basis functions are applicable (and the amount of applicability) at any given point.

The fuzzy logic (see Section 3.7) interpretation of approximator transparency is slightly more that the interpretation of the previous paragraph. In fuzzy logic approaches the influence variables are often associated with linguistic variables: "small," "medium," or "large." So that the ideas of the previous paragraph together with the linguistic variable can result in statements like: "If the ... is small, then use the control law ...." Similar ideas could be extended to any lattice based approximator, but when the number of influence functions per input dimension becomes large, the linguistic variables become awkward.

2.4.10 Haar Conditions

Section 2.2 introduced the idea of a Haar space: for unique function interpolation to be possible by a LIP approximator with N basis elements using training data from an arbitrary set of distinct locations $\{x_i\}_{i=1}^N$, the matrix $\Phi = [\phi_j(x_i)]$ must be nonsingular.

An example of a Haar subspace of $C[a,b]$ is the set of N-dimensional polynomials $P_N(x)$ defined on $[a,b]$. With the natural basis for polynomials, it can be shown that

$$det(\Phi) = \begin{vmatrix} 1 & 1 & \dots & 1 \\ x_1 & x_2 & \dots & x_{N+1} \\ \vdots & & & \vdots \\ x_1^N & x_2^N & \dots & x_{N+1}^N \end{vmatrix} = \Pi_{1 \leq i < j \leq N+1}(x_j - x_i)$$

which is positive if it is assumed that the x_i are sorted such that $x_1 < x_2 < \ldots < x_{N+1}$.

An N-dimensional Haar space (see Appendix A in [218]) can be considered as a generalized polynomial in the sense that the Haar space is a linear space of functions that retains the ability to interpolate a set of data defined at N arbitrary locations. For a Haar space $\mathcal{A} \subset C[a,b]$, the following conditions are equivalent:

1. If $\hat{f} \in \mathcal{A}$ and $\hat{f}$ is not identically zero, then the number of roots of the equation $\hat{f}(x) = 0$ in $[a,b]$ is less than N.

2. If $\hat{f} \in \mathcal{A}$ and $\hat{f}$ is not identically zero, if the number of roots of the equation $\hat{f}(x) = 0$ in [a,b] is j, and if k of these roots are interior points of $[a,b]$ at which $\hat{f}$ does not change sign, then $(j+k) < N$.

3. If $\{\phi_j, j = 1, \ldots, N\}$ is any basis for $\mathcal{A}$, and if $\{x_i, i = 1, \ldots, N\}$ is a set of any N distinct points in $[a,b]$ then the $N \times N$ matrix $[\phi_j(x_i)]$ is nonsingular.

The space $\mathcal{P}_N$ of N-th order polynomials is an example of a Haar space. It is straightforward to show that spline functions (see Section 3.3) with fixed knots that are not dependent on the data (or any approximator such that $Supp(\phi_j)$ is finite) is not a Haar space. This is

shown using item 1 or item 3 of the Haar conditions as follows.
Item 1: Fix j as an integer in $[1, N]$. Assume that $Supp(\phi_j) \subset \mathcal{D}$, $Supp(\phi_j) \neq \mathcal{D}$, and $\mathcal{D} \subset \cup_{i=1}^{N} Supp(\phi_i)$. Let $\hat{f}(x) = \theta^\top \phi(x)$ with $\theta_k = \begin{cases} 0 & k \neq j \\ 1 & k = j. \end{cases}$ This $\hat{f}$ is not identically zero, but has an infinite number of zeros since it is zero for all $x \in \mathcal{D} - Supp(\phi_j)$.
Item 3: If $\{x_i\}_{i=0}^{N}$ is selected such that $x_i \notin Supp(\phi_j)$ for any $i = 0, \ldots, N$, then the matrix $[\phi_j(x_i)]$ will have all zero elements in its j-th column. Therefore, this matrix is singular.

The fact that approximators using basis elements with finite support do not generate Haar spaces does not imply that such approximators are unsuitable for interpolation or adaptive approximation problems. Instead, it implies that the choice of the points $\{x_i\}$ affects the existence and uniqueness of a solution to the problem of interest. In offline data interpolation problems, the points $\{x_i\}$ are used to define the center or knot locations of the ϕ_j, in such a way that the matrix $[\phi_j(x_i)]$ is nonsingular.

In adaptive function approximation problems, defining the center or knot locations to match the first N data locations is typically not suitable, since these data locations will rarely be representative of all of $\mathcal{D}$. At least three alternative approaches to the definition of the center (or knot) locations are possible:

1. A set of experimental data representative of all expected system operating conditions could be accumulated and analyzed offline to determine appropriate center locations.

2. The center (or knot) locations could be altered during online operation as new data is received.

3. The center (or knot) locations could be defined, possibly on a lattice, such that the union of the support of the basis elements covers $\mathcal{D}$.

None of these three approaches will ensure that the interpolation problem is solvable after N samples, but that is not the objective. Instead, if appropriately implemented, these approaches will ensure that accurate approximation is possible over $\mathcal{D}$. Parameter estimation by the methods of Chapter 4 will result in convergence of the approximator locally in the neighborhood of each sample point. Because the sample points cover all of $\mathcal{D}$, global convergence can be achieved.

Note that the Haar condition ensures that the matrix $[\phi_j(x_i)]$ is nonsingular for solution of the interpolation problem. The Haar condition does not ensure that this matrix is well-conditioned.

2.4.11 Multivariable Approximation by Tensor Products

For dimensions greater than one, one means for constructing basis functions is as the product of basis functions defined separately for each dimension. This can be represented as a tensor product.

Let $G = span\{g_1, \ldots, g_p\}$ (i.e., $G = \{g | g(x) = \sum_{i=1}^{p} a_i g_i(x), a_i \in \Re^1, g_i : [a,b] \mapsto \Re^1\}$). Let $H = span\{h_1, \ldots, h_q\}$ where $h_i : [c,d] \mapsto \Re^1$. Then the *tensor product* of the spaces G and H is

$$f(x,y) = G \otimes H \ (x,y) = \sum_{i=1}^{p} \sum_{j=1}^{q} a_{ij} g_i(x) h_j(y) = \phi_g^\top(x) A \phi_h(y) \tag{2.49}$$

where $\phi_g^\top = [g_1, \ldots, g_p]$, $\phi_h^\top = [h_1, \ldots, h_p]$, and $A = [a_{ij}]$. The function f can be written in standard LIP form with

$$\theta^\top = [a_{11}, \ldots, a_{1q}, \ldots, a_{p1}, \ldots, a_{pq}]$$

and

$$\phi^\top = [g_1(x)\phi_h(y)^\top, \ldots, g_p(x)\phi_h(y)^\top].$$

If ϕ_g and ϕ_h are partitions of unity, then the ϕ corresponding to their tensor product is also also a partition of unity since

$$\sum_{k=1}^{pq} \phi_k(x,y) = \sum_{i=1}^{p}\sum_{j=1}^{q} g_i(x)h_j(y) \tag{2.50}$$

$$= \sum_{i=1}^{p} g_i(x) \sum_{j=1}^{q} h_j(y) = 1. \tag{2.51}$$

Assume that G and H vanish nowhere on their respective domains. If G separates points in $[a,b]$ and H separates points in $[c,d]$, then it is straightforward to show that $\phi(x,y)$ separates points in $[a,b] \times [c,d]$. Therefore, it is also straightforward to show that if G and H each satisfy the preconditions of the Stone-Weierstrass theorem, then the tensor product of G and H also satisfies the Stone-Weierstrass theorem. Therefore, $G_p \bigotimes H_q(x,y) = span\{g_i(x)h_j(y), i = 1, \ldots, p, j = 1, \ldots, q\}$ is a family of uniform approximators in $\mathcal{C}([a,b] \times [c,d])$.

This product of basis function approach can be directly extended to higher dimensions, but results in an exponential growth in the number of basis functions with the dimension of the domain of approximation. This approach is not restricted to locally supported basis elements. It can for example be applied to polynomial basis elements to produce multivariate polynomials.

2.5 SUMMARY

This chapter has introduced various function approximation issues that are important for adaptive approximation applications. In particular, this chapter has motivated why various issues should (or should not) be taken into account when selecting an appropriate approximator for a particular application.

Since the number of training samples will eventually become large, approximation by recursive parameter update eventually becomes important. All the data cannot be stored and a basis function cannot be associated with each training point. Due to noise on the measurements and the ever increasing number of samples, interpolation is neither desired nor practical.

Several factors influence the specification of the function approximator. Since the criteria for a family of approximators to be capable of uniform ϵ-accuracy approximation are actually quite loose, the existence of uniform approximation theorems for a particular family of approximators is not a key factor in the selection process. Important issues include the memory requirements, the computation required per function evaluation, the computation required for parameter update, and the numeric properties of the approximation problem. These issues are affected by whether or not the approximator is LIP, has locally supported basis elements, and is defined on a lattice. Various tradeoffs are possible.

The concept of a partition of unity has also been introduced. Advantages of approximators having the partition of unity property are (1) such approximators vanish nowhere and (2) such approximators are capable of exactly representing constant functions. The basis-influence function idea has been introduced to group together a set of approaches involving locally accurate approximations (i.e., basis functions) that are smoothly interpolated by the influence functions to generate an approximator capable of accurate approximation over the larger set $\mathcal{D}$. When the influence functions form a partition of unity, then the basis-influence approximator is formed as the convex combination of the local approximations.

Once a family of approximators has been selected, the designer must still specify the structure of the approximator, the parameter estimation algorithm, and the control architecture. Optimal selection of the structure of the approximator is currently an unanswered research question. The designer must be careful to ensure that the approximation structure that is specified is not too small or it will overly restrict the class of functions that can ultimately be represented. The parameter N should also not be too large or the approximated function may fit the noise on the measured data. Parameter estimation algorithms are discussed in Chapter 4. Control architectures and stability analysis are discussed in Chapters 5 – 7.

2.6 EXERCISES AND DESIGN PROBLEMS

Exercise 2.1 Implement a simulation to duplicate the results of Section 2.1.

Exercise 2.2 Show that the parameter vector that jointly minimizes the norm of the parameter vector and the approximation error is $\theta = \left(\lambda I + \Phi\Phi^\top\right)^{-1}\Phi Y$. Note that the cost function for this optimization problem is

$$J(\theta) = \frac{1}{2}\left(\lambda\theta^\top\theta + \left(Y - \Phi^\top\theta\right)^\top\left(Y - \Phi^\top\theta\right)\right).$$

Exercise 2.3 Perform the matrix algebraic manipulations to validate the Matrix Inversion Lemma.

Exercise 2.4 Show that if $J(e)$ is strictly convex in e and e is a linear function of θ, then J is convex in θ.

Exercise 2.5 Derive eqn. (2.35).

Exercise 2.6 Following Definition 2.4.5 a series of statements is made about whether or not given sets of functions are algebras. Prove each of these statements.

Exercise 2.7 Let $f(x) = x^{-1/3}$ and $\mathcal{D} = [0, 1]$.

1. Show that $f \in \mathcal{L}_2(\mathcal{D})$.
2. Is $f \in \mathcal{L}_\infty(\mathcal{D})$?
3. Use eqn. (2.35) to find the $\mathcal{L}_2$ optimal constant approximation (i.e., let $\phi(x) = [1]$) to f over $\mathcal{D}$.
4. Use eqn. (2.35) to find the $\mathcal{L}_2$ optimal linear approximation (i.e., let $\phi(x) = [1, x]^\top$) to f over $\mathcal{D}$.

For each of the constant and linear approximations, is the approximation error in $\mathcal{L}_2(\mathcal{D})$? $\mathcal{L}_\infty(\mathcal{D})$?

Exercise 2.8 Repeat Example 2.1 using recursive weighted least squares to estimate the parameters of the approximator $\hat{f} = \theta^\top \phi(x)$ where $\phi(x)$ is the Gaussian radial basis function described in Example 2.19.

Exercise 2.9 Show that eqn. (2.30) is true.

Exercise 2.10 The text following Definition 2.4.13 discussed normalization of the influence functions $\bar{\Gamma}_i$ to produce influence functions Γ_i forming a partition of unity. This problem further considers the cautions expressed in that text. Let $\mathcal{D} = [0, 1]$.

1. Let $\bar{\Gamma}_1(x) = \exp\left(-\left(\frac{x}{\sigma}\right)^2\right)$ and $\bar{\Gamma}_2(x) = \exp\left(-\left(\frac{x-1}{\sigma}\right)^2\right)$. Numerically computed and plot $\{\bar{\Gamma}_i(x)\}_{i=1:2}$ and $\{\Gamma_i(x)\}_{i=1:2}$ over $\mathcal{D}$ with $\sigma = 1$. Repeat for $\sigma = 0.1$ and $\sigma = 0.5$. Discuss the tradeoffs involved with choosing σ.

2. Let $\bar{\Gamma}_1(x) = \exp\left(-(x)^2\right)$ and $\bar{\Gamma}_2(x) = \exp\left(-\left(\frac{x-0.5}{0.1}\right)^2\right)$. Plot and discuss $\Gamma_1(x)$ and $\Gamma_2(x)$.

CHAPTER 3

APPROXIMATION STRUCTURES

The objective of this chapter is to present and discuss several neural, fuzzy, and traditional approximation structures in a unifying framework. The presentation will make direct references to the approximator properties presented in Chapter 2. In addition to introducing the reader to these various approximation structures, this chapter will be referenced throughout the remainder of the text.

Each section of this chapter discusses one type of function approximator, presents the motivation for the development of the approximator, and shows how the approximator can be represented in one of the standard nonlinearly and linearly parameterized forms:

$$NLIP: \quad \hat{f}(x:\theta,\sigma) = \theta^T \phi(x,\sigma) \tag{3.1}$$

$$LIP: \quad \hat{f}(x:\theta) = \theta^T \phi(x) \tag{3.2}$$

where $x \in D \subset \Re^n$, $\theta \in \Re^N$, $\sigma \in \Re^p$, $\hat{f} : D \mapsto \Re^1$, and D is assumed to be compact. Note that $\hat{f}$ is assumed to map a subset of $\Re^n$ onto $\Re^1$. This assumption that we are only concerned with scalar functions (i.e., single output) is made only for simplicity of notation. All the results extend to vector functions. Furthermore, vector functions will be used in several examples to motivate and exemplify this extension.

The ultimate objective is to adjust the approximator parameters θ and σ to encode information that will enable better control performance. Proper design requires selection of a family of function approximators, specification of the structure of the approximator, and estimation of appropriate approximator parameters. The latter process is referred to as parameter estimation, adaptation, or learning. Such processes are discussed in Chapter 4.

Adaptive Approximation Based Control: Unifying Neural, Fuzzy and Traditional Adaptive Approximation Approaches. By Jay A. Farrell and Marios M. Polycarpou

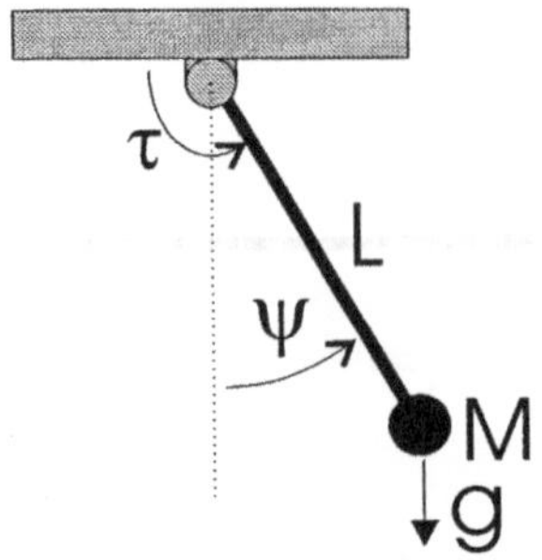

Figure 3.1: Simple pendulum.

3.1 MODEL TYPES

This section discusses three approaches to adaptive approximation. The first subsection discusses the use of a model structure derived from physical principles. The second subsection discusses the storage and use of the raw data without the intermediate step of function approximation. The third section discusses the use of generic function approximators. It is this third approach that will be the main focus of the majority of this text.

3.1.1 Physically Based Models

In some applications, the physics of the problem will provide a well-defined model structure where only parameters with well-defined physical interpretations are unknown. In such cases, the physically defined model may provide a structure appropriate for adaptive parameter identification.

EXAMPLE 3.1

The dynamics of the simple pendulum of Figure 3.1 are

$$\ddot{\psi}(t) = -\frac{g}{L}\sin(\psi(t)) + \frac{1}{M\,L^2}\tau(t)$$

where τ is the applied control torque. If the parameters M and L were unknown, they could be estimated based on the model structure

$$\ddot{\psi} = -\theta_1 \sin(\psi) + \theta_2 \tau = \theta^T \phi(x)$$

where $x = [\psi,\ \tau]^\top$ and $\phi(x) = [\sin(\psi),\ \tau]^\top$, while the parameters θ_1 and θ_2 are defined as $\theta_1 = \frac{g}{L}$ and $\theta_2 = \frac{1}{mL^2}$. △

When the physics of the problem provides a well-defined model structure, parameter estimation based on that model is often the most appropriate approach to pursue. However, even these applications must be designed with care to ensure stable operation and meaningful parameter estimates.

Alternatively, the physics of an application will often provide a model structure, but leave certain functions within the structure ill-defined. In these applications, adaptive approximation based approaches may be of interest.

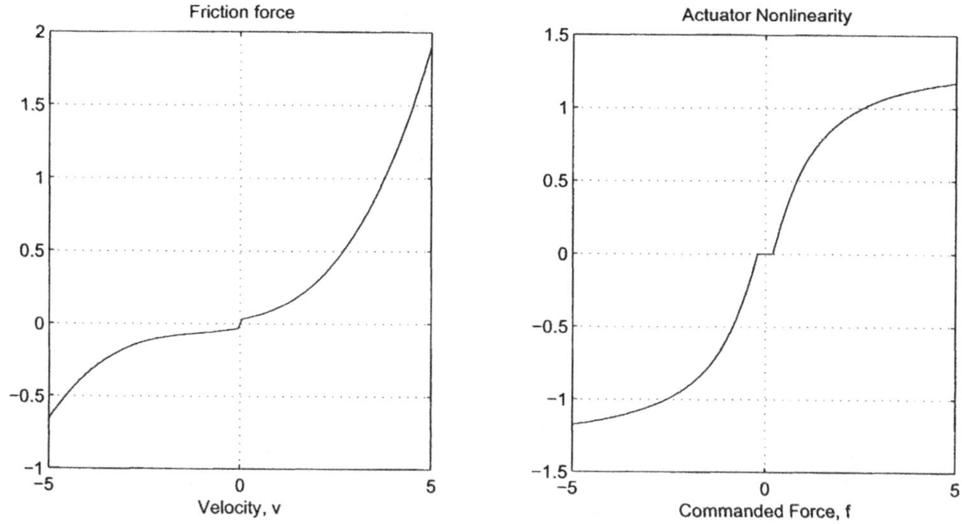

Figure 3.2: Friction and actuator nonlinearities.

EXAMPLE 3.2

The dynamics of a mass-spring-damper system are

$$\ddot{x}(t) = \frac{1}{m}\left[-h\left(\dot{x}(t)\right) - k(x(t)) + g\left(F(t)\right)\right]$$

where $x(t)$ is the distance from a reference point, $F(t)$ is the applied force (control input), $h(\cdot)$ represents friction, $k(\cdot)$ represents the nonlinear spring restoring force, and $g(\cdot)$ represents the actuator nonlinearity. Example friction and actuator nonlinearities are depicted in Figure 3.2. △

3.1.2 Structure (Model) Free Approximation

In applications where adaptive function approximation is of interest, the data necessary to perform the functions approximation will be supplied by the application itself and could arrive in a variety of formats. The easiest form of data to work with is samples of the input and output of the function to be approximated. Although this is often an unrealistic assumption, for this section we will assume availability of a set of data $\{z_i\}_{i=1}^{m}$, where each vector z_i can be decomposed as $z_i = [x_i, f(x_i)]$ with x_i being the function inputs and $f(x_i)$ being the function outputs. This set of data can be directly stored without further processing, as in Section 2.1. This is essentially a database approach. If the function value is required at x_j for some $1 \leq j \leq m$, then its value can be retrieved from the database. Note that there is no noise attenuation. However, in control applications, the chance of getting exactly the same evaluation point in the future as one of the sample points from the past is very small. Therefore, the exact input matching requirement would render the database useless.

Many extensions of the database type of approach are available to generate estimates of the functions values at evaluation points $x \notin \{x_i\}_{i=1}^m$, see for example Section 2.1 or [12, 222, 252]. In such approaches, the sample points $\{x_i\}_{i=1}^m$ affect the estimate of $f(x)$ at points $x \notin \{x_i\}_{i=1}^m$. Therefore, all such approaches cause generalization (appropriately or not) from the training data. If the function samples at several of the x_i are combined to produce the estimate of $f(x)$, then noise on individual samples might be attenuated.

When the designer does not have prior knowledge of a parametric description of function, then the basic function approximation problem is nonparametric. A complete description of an arbitrary function could require an infinite number of parameters, which is clearly not physically possible.

In the database approach of this section, the designer specifies a method to estimate $f(x)$ for $x \notin \{x_i\}_{i=1}^m$, but since all data is stored the approach is still infinite dimensional since $m \rightarrow \infty$. The label *structure free approximation* can be used to define the class of nonparametric approximation approaches that store all data as it becomes available and generate function estimates by combining the stored data. Since in such approaches all data is stored, the memory and computational requirements increase with time. Since online control applications theoretically run for infinite periods of time on computers with finite memory and computational resources, data reduction eventually becomes a requirement. Data reduction can be effectively implemented by specifying an approximation structure with unknown parameters and using the available data to estimate the parameter values. When the designer chooses such an approach, the problem is converted to one of parameter estimation for a finite dimensional parameter vector; however, the designer must expect that the approximated function will not perfectly match the actual function even for some optimal set of parameters. Therefore, the effect of *residual approximation error* must be considered.

Once the designer of a structure free approximator specifies a method to estimate $f(x)$ for $x \notin \{x_i\}_{i=1}^m$ the designer has specified a function approximation method. Therefore, the specified function approximator should be evaluated relative to existing approximation methods. Several traditional and recently developed parametric approximators are discussed in the subsequent sections of this chapter.

3.1.3 Function Approximation Structures

The design philosophy should be to use as much known information as is possible when constructing the dynamic model; however, when portions of a physically based model are either not accurately known or deemed inappropriate for online applications, then it is reasonable to use function approximation structures capable of approximating wide classes of functions. To make this point explicit, we will use the notation $f(z) = f_o(z) + f^*(x)$ to describe a partially known function f. In this notation, f_o is the known information about f and f^* represents the unknown portion of f. When there is no prior known information, the function f_o is set to zero.

Basic descriptions and properties of specific function approximation structures are discussed in the remaining sections of this chapter. Note that the choice of a family of approximators and the structure of a particular approximator is based on the implicit assumption by the designer that the selected approximator structure is sufficient for the application. Subsequent adaptive function approximation is constrained to the functions that can be implemented only by adjusting the parameters of the (now) fixed approximation structure.

Once the approximation structure and the compact region of approximation $\mathcal{D}$ are fixed, we can define an optimal parameter vector, a parameter error vector, and the *residual approximation error.* Given $f \in C(\mathcal{D})$, then by the properties of continuous functions on compact sets, we know that $f \in \mathcal{L}_\infty(\mathcal{D})$. For an approximator given by eqn. (3.2), we define

$$\theta^* = \arg\min_{\theta} \left(\sup_{x \in \mathcal{D}} \left| f^*(x) - \hat{f}(x:\theta) \right| \right). \tag{3.3}$$

For an approximator given by eqn. (3.1), we define

$$(\theta^*, \sigma^*) = \arg\min_{\theta,\sigma} \left(\sup_{x \in \mathcal{D}} \left| f^*(x) - \hat{f}(x:\theta,\sigma) \right| \right). \tag{3.4}$$

Given these definitions of the optimal parameters, the parameter error vector for LIP approximators is defined by

$$\tilde{\theta} = \theta - \theta^*. \tag{3.5}$$

For NLIP approximators, in addition of $\tilde{\theta}$, we also define

$$\tilde{\sigma} = \sigma - \sigma^*. \tag{3.6}$$

The residual or inherent approximation error (for the specified approximation structure) is defined as

$$e^*(x) = \hat{f}(x:\theta^*) - f^*(x) \tag{3.7}$$

for LIP approximators and as

$$e^*(x) = \hat{f}(x:\theta^*,\sigma^*) - f^*(x) \tag{3.8}$$

for NLIP approximators. This error will also sometimes be referred to as the Minimum Functional Approximation Error (MFAE).

Note that none of θ^*, σ^*, $\tilde{\theta}$, $\tilde{\sigma}$ or $e^*(x)$ are known. These are theoretical quantities that are necessary for analysis, but they cannot be used in implementation equations. When $f \in C(\mathcal{D})$ with $\mathcal{D}$ compact, then the quantities θ^* and $\sup_{x\in\mathcal{D}} |e^*(x)|$ are easily shown to be bounded.

3.2 POLYNOMIALS

Due to their long history in the field of approximation, polynomials are a natural starting point for a discussion of approximators. Examples of the use of polynomial approximators in control related applications can be found in [18, 200].

3.2.1 Description

The space $\mathcal{P}_N$ of polynomials of order N is

$$\mathcal{P}_N = \left\{ p(x) \middle| p(x) = \sum_{i=0}^{N} \theta_i x^i \text{ for } \theta_i, x \in \Re^1 \right\}. \tag{3.9}$$

The natural basis for this set of functions is $\{1, x, \ldots, x^N\}$. If, for example, the value of the function and its first N derivatives are known at a specific point x_o, then the well known Taylor series approximation is constructed as

$$p_T(x) = \sum_{i=0}^{N} \frac{f^{(i)}(x_o)}{i!} (x - x_o)^i$$

which is accurate for x near x_0. However, for interpolation or approximation problems, this basis set is not convenient. The basis elements are not orthogonal. In fact, their shapes are very similar over the standard interval $x \in [-1, 1]$. This choice of basis for $\mathcal{P}_N$ is well known to yield matrices with poor numeric properties.

An alternative choice of basis functions for $\mathcal{P}_N$ is the set of Legendre polynomials of degree N. The Legendre polynomials are generated by

$$\phi_j(x) = \frac{1}{2^j j!} \frac{d^j}{dx^j} \left[(x^2 - 1)^j \right]. \tag{3.10}$$

The first six Legendre polynomials are

$$\phi_0(x) = 1 \qquad \phi_1(x) = x$$
$$\phi_2(x) = \frac{1}{2}(3x^2 - 1) \qquad \phi_3(x) = \frac{1}{2}(5x^3 - 3x)$$
$$\phi_4(x) = \frac{1}{8}(35x^4 - 30x^2 + 3) \qquad \phi_5(x) = \frac{1}{8}(63x^5 - 70x^3 + 15x)$$

after scaling such that $\phi_i(1) = 1$. For $j > 1$, the Legendre polynomials can be generated using the recurrence relation

$$\phi_{j+1}(x) = \frac{2j+1}{j+1} x\phi_j(x) - \frac{j}{j+1}\phi_{j-1}(x).$$

This relation can also be used to compute recursively the values of the Legendre polynomials at a specific value of x.

Over the region $x \in [-1, 1]$, the Legendre polynomials are orthogonal, satisfying

$$\int_{-1}^{1} \phi_i(x)\phi_j(x)dx = \begin{cases} 0 & i \neq j \\ \dfrac{2}{2j+1} & i = j. \end{cases}$$

The fact that the Legendre polynomials are orthogonal over $[-1, 1]$ is the reason that they are a preferred basis set for performing function approximation over this interval.

■ EXAMPLE 3.3

If it is desired to find an N-th order polynomial approximation to the known function $f : \Re^1 \rightarrow \Re^1$ over the region $\mathcal{D} = [0, 1]$, we can select $g(x) = \sum_{i=0}^{N} \frac{2i+1}{2}\theta_i\phi_i(x)$ where $\theta_i = < \phi_i, f > = \int_{-1}^{1} \phi_i(x)f(x)dx$ and $< \phi_i, f >$ denotes the inner product between ϕ_i and f.

Let the error in this polynomial approximation be $h(x) = f(x) - g(x)$. For each $i \in [0, \ldots, N]$, $< h, \phi_i > = < f, \phi_i > - < g, \phi_i > = \theta_i - \theta_i = 0$. Therefore, the approximation error h is orthogonal to the space $\mathcal{P}_N$. This shows that g is in fact the optimal N-th order polynomial approximation to f.

It is due to the orthogonality of the ϕ_i that the coefficient θ_i can be computed independently of θ_j for $i \neq j$. Once the θ_i are available, if desired, they could be used to generate the coefficients for a polynomial as represented in the natural basis.
△

If an approximation is needed over $z \in [a, b]$ with $b > a$, then define $x = \frac{2z-a-b}{b-a}$ which maps $[a, b]$ to the interval $[-1, 1]$ where the standard Legendre polynomials can be used.

3.2.2 Properties

The space of polynomial approximators have several useful properties [240]:

1. $\mathcal{P}_N$ is a finite dimensional (i.e., $d = N + 1$) linear space with several convenient basis sets.

2. Polynomials are smooth (i.e., infinitely differentiable) functions.

3. Polynomials are easy to store, implement, and evaluate on a computer.

4. The derivative and integral of a polynomial are polynomials whose coefficients can be found algebraically.

5. Certain matrices involved in solving the interpolation or approximation problems can be guaranteed to be nonsingular (i.e., $\mathcal{P}_N$ on $\mathcal{D}$ is a Haar space).

6. Given any continuous function on an interval $[a, b]$, there exists a polynomial (for N sufficiently large) that is uniformly close to it (by the Weierstrass Theorem).

In contrast to these strong positive features, polynomial approximators have a few practical disadvantages.

Any basis $\{p_j(x)\}_{j=0}^N$ for $\mathcal{P}_N$ satisfies a Haar condition on $[a, b]$. This implies that if $\{x_i\}, i = 1, \ldots, N+1$ is a set of $N + 1$ distinct points on $[a, b]$, then the $(N+1) \times (N+1)$ collocation matrix with elements $\phi_{i,j} = p_j(x_i)$ is nonsingular. This fact is also true on any arbitrarily small subinterval of $[a, b]$. Therefore, the values of the polynomial at $N + 1$ distinct points on an arbitrarily small subinterval completely determine the polynomial coefficients; however, the condition number of this matrix can be arbitrarily bad. The fact that the matrix $[\phi_{i,j}]$ is nonsingular for $N + 1$ distinct evaluation points, is beneficial in the sense that the interpolation problem is guaranteed solvable and that the approximation problem has a solution once $m \geq N + 1$ distinct evaluation points (x_i, y_i) are available. The fact that the condition number of this matrix can be arbitrarily bad means that even small errors in the measurement of y_i or numeric errors in the algorithm implementation can greatly affect the estimated coefficients of the polynomial.

Any approximating polynomial can be manipulated into the form

$$p(x) = \sum_{i=0}^{N} \theta_i x^i.$$

The derivative of the approximation is

$$\frac{dp(x)}{dx} = \sum_{i=1}^{N} i\theta_i x^{i-1}.$$

Since i is greater than 1, the coefficients of the derivative are larger than the coefficients of the original polynomial. This fact becomes increasingly important as N increases. Therefore, higher order polynomials are likely to have steep derivatives. These steep derivatives may cause the approximating polynomial to exhibit large oscillations between interpolation points. Since the approximation accuracy of a polynomial is also directly related to N, polynomials are somewhat inflexible. To approximate the measured data more accurately, N must be increased; however, this may result in excessive oscillations between the points involved in the approximation (see Exercises 3.2, 3.4, and 3.6). Unfortunately, there are no parameters of the approximating structure other than N that can be manipulated to affect the approximating accuracy.

Finally, the polynomial basis elements are each globally supported over the interval of approximation I. Therefore, incremental training on a subinterval I_j will affect the approximation accuracy on all of I. This issue is further explored in Exercises 3.1.

These drawbacks have motivated researchers to develop alternative function approximators. The above text has discussed univariate polynomial approximation. Similar comments apply to multivariable polynomial approximation. In addition, the number of basis elements required to represent multivariable polynomials increases dramatically with the dimension of the input space n.

3.3 SPLINES

The previous section discussed the benefits and drawbacks of using polynomials as approximators. Although higher order polynomials had difficulties, low order polynomials are a reasonable approximator choice when the region of approximation is sufficiently small relative to the rate of change of the function f. This motivates the idea of subdividing a large region and using a low order approximating polynomial on each of the resulting subregions.

Numeric splines implement this idea by connecting in a continuous fashion a set of local, low order, piecewise polynomial functions to fit a function over a region $\mathcal{D}$. For example, given a set of data $\{(x_i, y_i)\}_{i=1}^{K+1}$ with $x_i < x_{i+1}$, if the data are drawn on a standard $x-y$ graph and connected with straight lines, this would be a spline of order two interpolation of the data set. If the data were connected using 2rd order polynomials between the data points in such a way that the graph had a continuous first derivative at these interconnection points, this would be a spline of order three. The name "spline" comes from drafting where flexible strips were used to aid the drafter to interpolate smoothly between points on paper.

Examples of the use of splines in control related applications can be found in [27, 38, 132, 142, 143, 174, 175, 294, 305].

3.3.1 Description

Various types of splines now exist in the literature. The types of splines differ in the properties that they are designed to optimize and in their implementation methods. In the following, natural splines will be discussed to allow a more complete discussion of the examples from the introduction to this section and to motivate B-splines. Then B-splines will be discussed in greater depth.

Natural Splines. In one dimension, a spline is constructed by subdividing the interval of approximation $I = (\underline{x}, \overline{x}]$ into K subintervals $I_j = (x_j, x_{j+1}]$ where the x_j, referred to as knots or break points, are assumed[1] to be ordered such that $\underline{x} = x_0 < x_1 < \ldots < x_K = \overline{x}$. For a spline of order k, a $(k-1)st$ order polynomial is defined on each subinterval I_j. Without additional constraints, each $(k-1)st$ order polynomial has k free parameters for a total of Kk free spline parameters. The spline functions are, however, usually defined so that the approximation is in $\mathcal{C}^{(k-2)}$ over the interior of I. For example, a $2nd$ order spline is composed of first order polynomials (i.e., lines) defined so that the approximation

[1] More generally, strict inequalities are not required. The entire spline theory goes through for $\underline{x} = x_0 \leq x_1 \leq \ldots \leq x_K = \overline{x}$. We use strict inequalities in our presentation as it simplifies the discussion.

is continuous over I including at the knots. With such continuity constraints, the spline has $Kk - (k-1)(K-1) = K + k - 1$ free parameters. With the constraint that the spline be continuous in $(k-2)$ derivatives, splines have the property of being continuous in as many derivatives as is possible without the spline degenerating into a single polynomial. In contrast to polynomial series approximation, the accuracy of a spline approximation can be improved by increasing either k or K. Therefore, splines approximations are more flexible than polynomials series approximators.

■ EXAMPLE 3.4

Consider the approximation of a function f by a spline of second order with continuity constraints using $K = 4$ subintervals defined over $[-1, 1]$. First, we define $\{x_j\}_{j=0}^{4}$ such that $x_0 = -1$, $x_4 = 1$, and $x_j < x_{j+1}$ for $j = 0, \ldots, 3$. The approximator can be expressed as

$$g(x) = \sum_{j=0}^{3} \left[(a_j + b_j(x - x_j))\, I_j(x) \right]$$

where I_j is an indicator function defined as

$$I_j(x) = \begin{cases} 1 & x_j < x \leq x_{j+1} \\ 0 & \text{otherwise.} \end{cases}$$

The eight unknown parameters can be arranged in a vector as $\theta = [a_0, b_0, \ldots, a_3, b_3]^\top$ with the basis vector for the approximator defined as

$$\phi(x) = [I_0(x), (x - x_0)I_0(x), \ldots, I_3(x), (x - x_3)I_3(x)]^\top,$$

so that $g(x) = \theta^\top \phi(x)$. Note that for arbitrary parameters, this approximation does not enforce the continuity constraint. To satisfy the continuity constraint, we must have

$$\begin{aligned} a_0 + b_0(x_1 - x_0) &= a_1 \\ a_1 + b_1(x_2 - x_1) &= a_2 \\ a_2 + b_2(x_3 - x_2) &= a_3 \end{aligned}$$

which can be written in matrix form as $G\theta = 0$ where

$$G = \begin{bmatrix} 1 & (x_1 - x_0) & -1 & 0 & 0 & 0 & 0 & 0 \\ 0 & 0 & 1 & (x_2 - x_1) & -1 & 0 & 0 & 0 \\ 0 & 0 & 0 & 0 & 1 & (x_3 - x_2) & -1 & 0 \end{bmatrix}.$$

If, given a dataset $\{(z_i, f(z_i))\}_{i=1}^{N}$, the objective is to find parameters θ to approximate f using the continuous spline of 2nd order denoted by g, then we have to solve a constrained optimization problem. If the optimization is being performed online (i.e., N is increasing), then the constraint must be accounted for each time that the parameters are adjusted. Due to the constraint, a change in the parameters for one interval can result in changes to the parameters in the other intervals. Constrained least squares parameter estimation is considered in Exercise 3.8. △

Note that in the previous example, the elements of the basis vector $\phi(x)$ are not themselves continuous. Therefore, the continuity of the approximator is enforced by additional constraints, resulting in the constrained optimization problem. An alternative approach is to generate a different set of basis elements for the set of splines of order k such that the new basis elements are themselves in $\mathcal{C}^{(k-2)}$. In this case, the adjustment of the coefficients of the approximator can be performed without "continuity constraint" equations. This approach results in the definition of B-splines, which are computationally efficient with good numeric properties [238].

Cardinal B-splines. When the B-splines are defined with the knots at

$$\{\ldots, -2, -1, 0, 1, 2, \ldots\},$$

they are called Cardinal B-splines. One of the common forms in which B-splines are used in adaptive approximation applications is by translation and dilation of the Cardinal B-splines.

Definition 3.3.1 *[59] The functions $g_k : \Re^1 \to \Re^1$ defined recursively, for $k > 1$, by*

$$g_k(x) = \int_0^1 g_{k-1}(x-\lambda)d\lambda = \int_{-\infty}^{\infty} g_{k-1}(x-\lambda)g_1(\lambda)d\lambda, \tag{3.11}$$

is the Cardinal B-spline of order k (for the knot at 0) where

$$g_1(x) = \begin{cases} 1, & \text{if } 0 \le x < 1 \\ 0, & \text{otherwise.} \end{cases}$$

The Cardinal B-splines of orders 2 and 3 are, respectively, for the knot at 0 given by

$$g_2(x) = \begin{cases} x & \text{for } 0 \le x < 1 \\ 2-x & \text{for } 1 \le x < 2 \\ 0 & \text{otherwise,} \end{cases} \tag{3.12}$$

$$g_3(x) = \begin{cases} \frac{x^2}{2} & \text{for } 0 \le x < 1 \\ -x^2 + 3x - \frac{3}{2} & \text{for } 1 \le x < 2 \\ \frac{(3-x)^2}{2} & \text{for } 2 \le x < 3 \\ 0 & \text{otherwise.} \end{cases} \tag{3.13}$$

Note that the Cardinal B-spline of order k is a piecewise polynomial of degree $k-1$. The piecewise polynomial is in $\mathcal{C}^{(k-2)}$ with points of discontinuity in the $(k-1)$ derivative at $x = 0, 1, 2, ..., k$.

The B-spline basis element of order k for the knot at $x = j$ is $g_{kj}(x) = g_k(x-j)$ and has support for $x \in (j, k+j)$. Conversely, for $x \in [0,1]$, the functions $g_k(x-j)$ are nonzero for $j \in [1-k, 0]$. The B-splines basis elements of order $k = 1, 2, 3$, and 4 are shown in Figure 3.3. This figure shows all the B-splines g_{kj} for $j = 1-k, \ldots, 0$ that would be necessary to form a partition of unity for $x \in (0,1)$.

The function $s_k(x) = \sum_{j=1-k}^{N-1} \theta_j g_k(x-j)$ is a spline of order k with $(N+k-1)$ knots at $x = 1-k, 1, 2, \ldots, N-1$. It is also a piecewise polynomial of degree $k-1$ with the same continuity properties as g_k. The function $s_k(x)$ is nonzero on $[1-k, k+N-1]$. For $N > k$, the set of basis elements $\{g_k(x-j)\}_{j=1-k}^{N-1}$ form a partition of unity on $[0, N]$.

If instead, the basis elements are selected as

$$\phi_j(x) = g_k\left(N\frac{x-a}{b-a} - j\right) \tag{3.14}$$

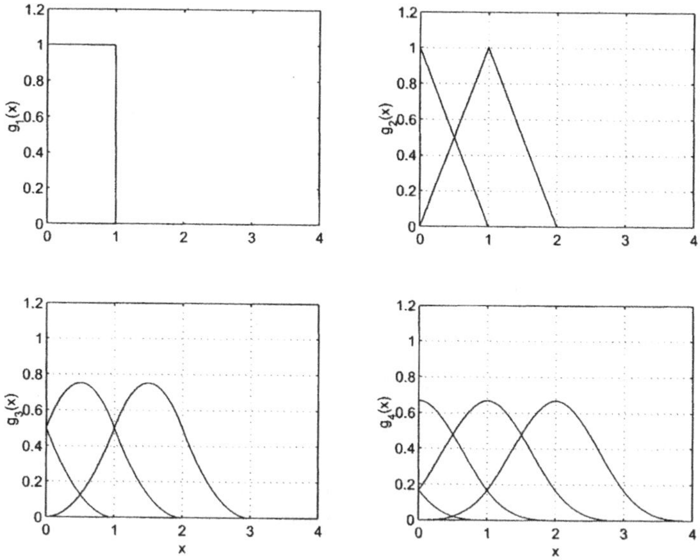

Figure 3.3: B-splines of order 1 thru 4 that are non-zero on $(0, 1)$.

for $j = 1 - k, \ldots, N - 1$, then this basis set $\{\phi_j\}_{j=1-k}^{N-1}$, formed by translating and dilating the k-th Cardinal B-spline, is a partition of unity on $[a, b]$. The span of this set of basis functions is a piecewise polynomial of degree $k - 1$ that is in $\mathcal{C}^{(k-2)}$. By using an approximator defined as

$$\hat{f}(x, \theta) = \theta^\top \phi(x) = \sum_{j=1-k}^{N-1} \theta_j \phi_j(x),$$

with $\phi_j(x)$ as defined in eqn. (3.14), we are able to adjust the parameters of the approximator without the explicit inclusion of continuity constraints in the parameter adjustment process, such as those that were required for the natural splines. We attain a piecewise polynomial of degree $k - 1$ in $\mathcal{C}^{(k-2)}$ because the basis elements have been selected to have these properties.

Nonuniformly spaced knots. Splines with uniformly spaced knots, as in the previous subsection, form lattice networks and are often used in online applications; however, B-splines are readily defined and implemented for nonuniformly spaced knots as well. In fact, the majority of the spline literature does not discriminate against nonuniform knot spacing or repeated knots.

Let there be $M + k + 1$ knots where $M > 0$. If the interval of approximation is (a, b), then the knots should be defined to satisfy the following conditions:

1. $x_j < x_{j+1}$ for $j \in [1 - k, M]$
2. $x_0 \leq a < x_1$
3. $x_M < b \leq x_{M+1}$.

When the knots satisfy these conditions, they are ordered as

$$x_{1-k} < x_{2-k} < \ldots < x_0 \leq a < x_1 < \ldots < x_M < b \leq x_{M+1}.$$

With these conditions, for $x \in (a, b)$, the B-splines of order k will provide an $M + k$ element basis for the set of splines of order k with knots at $\{x_j\}_{j=1-k}^{M+1}$. This basis will be a partition of unity on (a, b). Denote the B-spline basis functions as $\{B_{k,j}(x)\}_{j=1}^{M+k}$.

Define the interval index function

$$J(x) = i \quad \text{if } x \in (x_{i-1}, x_i]. \tag{3.15}$$

Note that $J(x) : (a, b) \mapsto [1, M + 1]$. This function simply provides the integer index for the interval containing the evaluation point x. For uniformly spaced knots (i.e., lattice approximation), $J(x)$ can be computed very efficiently. For nonuniformly spaced knots a search requiring on the order of $\log_2(K)$ comparisons will be required.

Given $J(x)$, the vector of first order splines is calculated as

$$B_{1,j}(x) = \begin{cases} 1, & \text{if } j = J(x), \\ 0, & \text{otherwise.} \end{cases} \tag{3.16}$$

For higher order splines (i.e., $k > 1$) it is computationally efficient to calculate the non-zero basis functions using the recursion relation

$$B_{k,j}(x) = \left(\frac{x - x_{j-k}}{x_{j-1} - x_{j-k}}\right) B_{k-1,j-1}(x) + \left(\frac{x_j - x}{x_j - x_{j-k+1}}\right) B_{k-1,j}(x) \tag{3.17}$$

for $j \in [J(x), J(x) + k - 1]$. For j outside this range, $B_{k,j}(x) = 0$. This requires about $\frac{3}{2}k^2$ multiplications. Derivatives and integrals of the spline can be calculated by related recursions [55, 142].

EXAMPLE 3.5

To clarify the above recursion, consider the following example. Let $I = (0, 2)$ and $M = 4$ with knots at

j =	-2	-1	0	1	2	3	4	5
x_j =	-0.75	-0.1	0.00	0.50	0.75	1.00	1.50	2.00.

For $x \in I$ and $k = 3$, $B_{3,j}$ can be nonzero for $j \in [1, 7]$. Consider the calculation of the third order spline basis at $x = 0.45$ and at $x = 1.95$.

Since $0.45 \in (0.00, 0.50]$, we have that $J(0.45) = 1$ and

$$B_{1,j}(0.45) = \begin{cases} 1, & \text{if } j = 1, \\ 0, & \text{for } j \neq 1. \end{cases} \tag{3.18}$$

The recursion of eqn. (3.17) defines (row-wise) the following array of values

$$\begin{array}{lll} B_{1,1} = 1.0000 & B_{1,2} = 0.0000 & B_{1,3} = 0.0000 \\ B_{2,1} = 0.1000 & B_{2,2} = 0.9000 & B_{2,3} = 0.0000 \\ B_{3,1} = 0.0083 & B_{3,2} = 0.4517 & B_{3,3} = 0.5400 \end{array}$$

and $B_{1,j} = B_{2,j} = B_{3,j} = 0$ for $j \geq 4$.

Since $1.95 \in (1.50, 2.00]$, we have that $J(1.95) = 5$ and

$$B_{1,j}(1.95) = \begin{cases} 1, & \text{if } j = 5, \\ 0, & \text{for } j \neq 5. \end{cases} \tag{3.19}$$

The recursion of eqn. (3.17) defines (row-wise) the following array of values

$$\begin{array}{lll} B_{1,5} = 1.000 & B_{1,6} = 0.000 & B_{1,7} = 0.000 \\ B_{2,5} = 0.100 & B_{2,6} = 0.900 & B_{2,7} = 0.000 \\ B_{3,5} = 0.005 & B_{3,6} = 0.320 & B_{3,7} = 0.675 \end{array}$$

and $B_{1,j} = B_{2,j} = B_{3,j} = 0$ for $j \leq 4$.

Note that each row sums to one. △

The order k B-spline approximator is

$$\hat{f}(x) = \sum_{j=1}^{M+k} \theta_j \phi_j(x)$$

where $\phi_j(x) = B_{k,j}(x)$. This approximator is a partition of unity for $x \in (a, b)$. Also, univariate B-splines of order k have support over k knot intervals. Each input x maps to k non-zero basis elements.

3.3.2 Properties

Splines can be defined as follows.

Definition 3.3.2 *The linear space of univariate spline functions of order k with knot sequence $X = \{x_j\}$ is*

$$S_{k,X} = \left\{ s(x) = \sum_j \theta_j g_{k,j}(x), \text{ for } \theta_j \in \Re^1 \right\} \tag{3.20}$$

where $g_{k,j}$ are the B-splines of order k corresponding to X.

Similarly, the space of approximations spanned by dilations and translations of the Cardinal B-splines has the following definition.

Definition 3.3.3 *The linear space of univariate spline functions with equally space knots is*

$$S_{k,\nu} = \left\{ s(x) = \sum_{j \in \mathcal{Z}} \theta_j g_k(\nu x - j - \mu), \text{ for } \theta_j, \mu, \nu \in \Re^1 \right\} \tag{3.21}$$

where ν is the dilation parameter, j counts over translations of the Cardinal B-spline, and μ is a phase shift constant.

Note that Definition 3.3.3 matches eqn. (3.14) if $\mu = \frac{Na}{b-a}$ and $\nu = \frac{N}{b-a}$.

When the region of approximation D is compact and the knots are defined by translation and dilation of the Cardinal B-splines, the summation will include a finite subset of $\mathcal{Z}$. The sets $S_{k,\nu}$ and $S_{k,X}$ are subsets of the Σ-networks and are linear in the parameter vector θ.

$S_{k,\nu}$ is also a lattice network. Splines have the uniform approximation property in the sense that any continuous function on a compact set can be approximated with arbitrary accuracy by decreasing the spacing between knots, which increases the number of basis elements. For nonuniformly spaced knots, if it is desired to add additional knots, there are available methods that can be found by searching for "knot insertion."

B-splines are locally supported, positive, normalized (i.e., $\int_0^k g_k(x)dx = 1$ where $g_k(\cdot)$ is the basis spline of order k), and form a partition of unity [59]. Each basis element is nonzero over the k intervals defined by the knots. Therefore, a change in the parameter θ_i only affects the approximation over the k intervals of its support. In addition, at any evaluation point, at most k of the basis elements are nonzero.

3.4 RADIAL BASIS FUNCTIONS

Radial basis functions (RBFs) were originally introduced as a solution method for *batch* multivariable scattered data *interpolation* problems [31, 83, 84, 104, 105, 204, 219]. Scattered data interpolation problems are the subset of interpolation problems, where the data samples are dictated not by some optimal criteria, but by the application or experimental conditions. Online control applications involve (non-batch) scattered data function *approximation*.

The main references for this section are [31, 79, 83, 84]. Examples of the use of RBFs in various control applications are presented in [43, 44, 46, 47, 74, 136, 156, 232, 272].

3.4.1 Description

A radial basis function approximator is defined as

$$\hat{f}(x) = \sum_{i=1}^{N} \theta_i g(\|x - c_i\|) + \sum_{i=1}^{\bar{k}} b_i p_i(x) \tag{3.22}$$

where $x \in \Re^n$, $\{c_i\}_{i=1}^N$ are a set of center locations, $\|x - c_i\|$ is the distance from the evaluation point to the i-th center, $g(\cdot) : \Re^+ \mapsto \Re^1$ is a radial function, and $\{p_i(x)\}_{i=1:\bar{k}}$ is a basis for the $\bar{k}$ dimensional linear space of polynomials of degree k in n variables

$$P_k = span\{x_1^{\alpha_1}, \cdots, x_n^{\alpha_n} | 0 \leq \alpha_1 + \ldots + \alpha_n \leq k\}.$$

The polynomial term in eqn. (3.22) is included so that the RBF approximator will have polynomial precision[2] k. Often in RBF applications, k is specified by the designer to be -1. In that case, the polynomial term does not appear in the approximator structure and the RBF does not have a guaranteed polynomial precision.

Some forms of the radial function that appear in the literature are

$$\text{Gaussian:} \quad g_1(\rho) = exp\left(-\frac{1}{2}\frac{\rho^2}{\gamma^2}\right) \tag{3.23}$$

$$\text{Multi-quadratic:} \quad g_2(\rho) = \left(\rho^2 + \gamma^2\right)^{\beta}, \ \beta \in (0,1) \tag{3.24}$$

$$\text{Inverse Mulit-quadratic:} \quad g_3(\rho) = \left(\rho^2 + \gamma^2\right)^{-\alpha}, \ \alpha > 0 \tag{3.25}$$

[2] An approximator having polynomial precision k means that the approximator is capable of exactly representing a polynomial of that order.

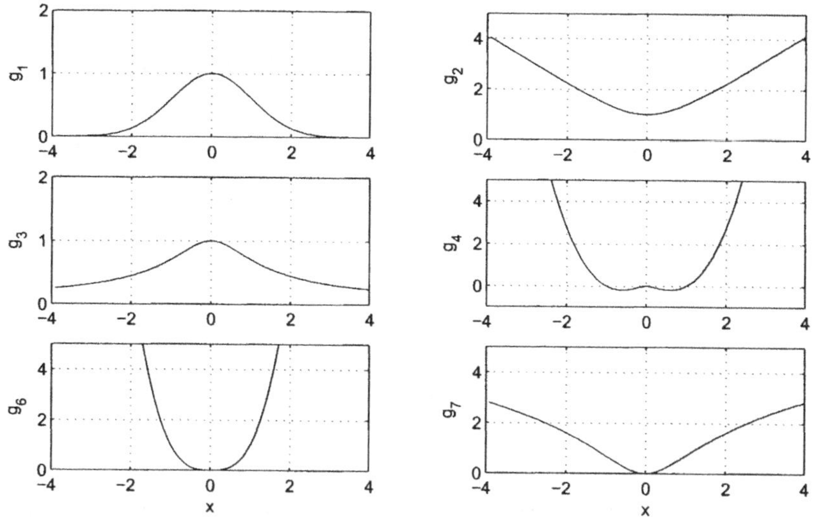

Figure 3.4: Radial basis nodal functions ($c = 0$). *Top left* – Gaussian g_1. *Top right* – Multi-quadratic g_2. *Middle left* – Inverse Multi-quadratic g_3. *Middle right* – Thin plate spline g_4. *Bottom left* – Cubic g_6. *Bottom right* – Shifted Logarithm g_7.

$$\text{Thin Plate Spline:} \quad g_4(\rho) = \rho^2 \log(\rho + \gamma) \tag{3.26}$$

$$\text{Linear:} \quad g_5(\rho) = \rho \tag{3.27}$$

$$\text{Cubic:} \quad g_6(\rho) = \rho^3 \tag{3.28}$$

$$\text{Shifted Logarithm:} \quad g_7(\rho) = \log\left(\rho^2 + \gamma^2\right) \tag{3.29}$$

where $\rho \in [0, \infty)$ and γ is a constant either defined by the designer prior to online application or a parameter to be estimated online. The multi-quadratic and inverse multi-quadratic are stated for specific ranges of β and α, but the names of the nodal functions relate explicitly to the case where $\alpha = \beta = 0.5$. Multi-quadratics were introduced by Hardy in 1971 [104]. Figure 3.4 displays plots of six radial functions with $\alpha = \beta = 0.5$ and $\gamma = 1$. Constraints on g for guaranteed solution of the interpolation problem will be discussed later.

Radial basis functions (with $\bar{k} = 0$) can be represented in the standard form

$$\hat{f}(x) = \theta^T \phi(x, c, \gamma) = \sum_{i=1}^{N} \theta_i \phi_i(x, c, \gamma)$$

where the i-th basis element is defined by

$$\phi_i(x, c, \gamma) = g\left(\|x - c_i\|, \gamma\right) \text{ for } x \in \Re^n \text{ and } i = 1, \ldots, m. \tag{3.30}$$

In the standard RBF, all the elements of ϕ are based on the same radial function $g(\cdot)$. The first argument of g is the radial distance from the input x to the i-th center c_i, $\rho_i(x) = \|c_i - x\|$.

When g is selected to be either the Gaussian function or the Inverse multi-quadratic, then the resulting basis function approximator will have localization properties determined by the parameter γ. In a more general approach, different values of γ can be used in different basis elements of the RBF approach.

3.4.2 Properties

Given a constant value for γ, three procedures are typical for the specification of the centers c_i.

1. For a fixed batch of sample data $\{(x_j, y_j)\}_{j=1}^{N}$, when the objective is interpolation of the data set, the centers are equated to the locations of the sample data: $c_i = x_i$ for $i = 1, \ldots, N$. Data interpolation for $j = 1, \ldots, N$ provides a set of N constraints

$$y_j = \sum_{i=1}^{N} \theta_i g(\|x_j - c_i\|) + \sum_{i=1}^{\bar{k}} b_i p_i(x_j) \tag{3.31}$$

leaving $\bar{k}$ degrees of freedom. These interpolation constraints can be written as

$$Y = [\Phi^\top, P^\top] \begin{bmatrix} \theta \\ b \end{bmatrix}$$

where Φ and Y are defined in eqn. (2.8), $P = [p(x_1), \ldots, p(x_N)]$, $p(x_j) = [p_1(x_j), \ldots, p_{\bar{k}}(x_j)]^\top$, and $b = [b_1, \ldots, b_{\bar{k}}]^\top$. Since g is a radial function, $\phi_i(x_j) = g(\|x_j - x_i\|) = \phi_j(x_i)$; therefore, the matrix Φ is symmetric. The RBF approximator of eqn. (3.22) still allows an additional $\bar{k}$ degrees of freedom. The additional constraint that $\sum_{i=1}^{N} \theta_i p_j(x_i) = 0$ for $j = 1, \ldots, \bar{k}$ is typically imposed. The resulting linear set of equations that must be solved for θ and b is

$$\begin{bmatrix} Y \\ 0 \end{bmatrix} = \begin{bmatrix} \Phi^\top & P^\top \\ P & 0 \end{bmatrix} \begin{bmatrix} \theta \\ b \end{bmatrix}.$$

This is a fully determined set of $N + \bar{k}$ equations with the same number of unknowns. It can be shown that when g is appropriately selected, this set of equations is well-posed [79]. The choice of g is further discussed below.

2. The c_i are specified on a lattice covering $\mathcal{D}$. Such specification results in a LIP approximation problem with memory requirements that grow exponentially with the dimension of x, but very efficient computation (see Section 2.4.8.4). Theorem 2.4.5 shows that this type of RBF is a universal approximator.

3. The c_i are estimated online as training data are accumulated. This results in a NLIP approximation problem. Theorem 2.4.4 shows that this type of RBF is a universal approximator. The resulting approximator may have fewer parameters than case 2, but the approach must address the difficulties inherent in nonlinear parameter estimation. In addition, the computation saving methods of Section 2.4.8.4 will not be applicable.

Although our main interest will be in approximation problems, the use of RBFs for data interpolation has an interesting history. The analysis of the interpolation has implications for the choice of g in approximation problems.

As described in Section 2.2, the LIP RBF interpolation problem with $c_i = x_i$ is solvable if the regressor matrix Φ with elements $\phi_{ij} = g(\|x_i - x_j\|)$ is nonsingular (assuming that $\bar{k} = 0$). Therefore, conditions on g such that Φ is nonsingular are of interest. An obvious necessary condition is that the points $\{x_i, i = 1 \ldots, N\}$ be distinct. This will be assumed throughout the following.

EXAMPLE 3.6

Let $g(r) = r^2$ with r defined as the Euclidean norm, then in two dimensions ($n = 2$), $\phi_i(r) = (x-x_i)^2+(y-y_i)^2$. For this nodal function, the approximator $\sum_{i=1}^N \theta_i\phi_i(r)$ does not define an N dimensional linear space of functions. Instead, it defines a subset of the linear space of functions spanned by $(1, x, y, x^2 + y^2)$. To see this, consider that

$$\sum_{i=1}^{N} \theta_i\phi_i(r) = \sum_{i=1}^{N} \theta_i \left((x-x_i)^2 + (y-y_i)^2\right) \tag{3.32}$$

$$= \sum_{i=1}^{N} \theta_i \left(x^2 - 2x_ix + x_i^2 + y^2 - 2y_iy + y_i{}^2\right) \tag{3.33}$$

$$= A(x^2+y^2) + Bx + Cy + D \tag{3.34}$$

where the parameters in the bottom equation are defined by $A = \sum_{i=1}^N \theta_i$, $B = -2\sum_{i=1}^N \theta_i x_i$, $C = -2\sum_{i=1}^N \theta_i y_i$, and $D = \sum_{i=1}^N \theta_i \left(x_i^2 + y_i{}^2\right)$. For general n, the interpolation matrix will be singular if $N > -1 + \frac{1}{2}(n+1)(n+2)$. Therefore, $g(r) = r^2$ with the Euclidean norm is not a suitable choice of radial basis function when the objective is to interpolate an arbitrary data set using a RBF with centers defined by the data locations [219]. △

Let A be defined as the matrix with elements given by

$$A_{ij} = \|x_j - x_i\|_2,\ i = 1,\ldots,N,\ j = 1,\ldots,m. \tag{3.35}$$

Note that A is symmetric with zero diagonal elements and positive off-diagonal elements that satisfy the triangle inequality (i.e., $A_{ij} \leq A_{il} + A_{lj}$) and $\Phi_{ij} = g(A_{ij})$. It can be shown [167, 219] that if the points $\{x_i\}$ are distinct in $\Re^n$, then the matrix A is positive definite. Examples of singularity for other norms (e.g., the infinity norm) are presented in [219].

The results of Micchelli [167] (reviewed in [79, 219]) give sufficient conditions on $g(\cdot)$ and k such that the RBF LIP interpolation problem is solvable. In particular, the Gaussian, multi-quadratic, and inverse multi-quadratic RBF LIP interpolation problems are solvable independent of n and N. For the linear nodal function the only additional constraint is that $N > 1$. For the cubic nodal function, Φ is nonsingular if $n = 1$, but can be singular if $n > 1$.

The relation of RBFs to splines is investigated in [204, 219].

3.5 CEREBELLAR MODEL ARTICULATION CONTROLLER

The original presentation of the Cerebellar Model Articulation Controller (CMAC) [1, 2] discussed various issues that have been discussed elsewhere in this text. In addition, the original presentation focused on constant, locally supported basis elements. This resulted in piecewise constant approximations. A main contribution of the CMAC approach is the reduction of the amount of memory required to store the coefficient vector denoted herein by θ. Subsequent articles [4, 142, 195] generalized the CMAC approach to generate smooth mappings while retaining the reduced address space of the original CMAC. The following presentation of the CMAC ideas is distinct from that of the original articles to both incorporate the new ideas of the subsequent articles and to conform to the style of this text. The flow of the analysis and some of the examples still follow the presentation in [2]. Applications involving the CMAC have been presented, for example, in [170, 171, 269, 270].

3.5.1 Description

For linear in the parameter approximators, the approximation can be represented in the form

$$\hat{f}(x) = \theta^T \phi(x)$$

where $\theta \in \Re^N$ and $\phi : \mathcal{D} \mapsto \Re^N$. When ϕ is a vector of local basis functions defined on a lattice, then as shown in Section 2.4.8.4, it is possible to define a function $I(x) : \mathcal{D} \mapsto \mathcal{Z}_N^m$ where $\mathcal{Z}_N = \{1, \ldots, N\}$ and $\mathcal{Z}_N^m$ is a set of m elements of $\mathcal{Z}_N$. The set $I(x)$ are the indices (or addresses) of the nonzero elements of $\phi(x)$. Throughout the discussion of the CMAC, the parameter m is a constant. This implies that at any $x \in \mathcal{D}$ there is the same number m of nonzero basis elements. The motivation of this assumption will become clear in the following. Therefore, the approximation of f at x can be calculated exactly and efficiently by

$$\hat{f}(x) = \sum_{k \in I(x)} \theta_k \phi_k(x). \tag{3.36}$$

At this point, an example is useful to ensure that the notation is clear.

■ EXAMPLE 3.7

Let $x \in \mathcal{D} \subset \Re^d$ with $d = 2$. Define $\mathcal{D} = [-1, 1] \times [-1, 1]$. Define the lattice so that there are $r = 201$ basis elements per input dimension with centers defined by $(x_i, y_j) = \left(\frac{i-101}{100}, \frac{j-101}{100}\right)$ for $i, j \in [1, 201]$. Let the basis elements for each input dimension be defined by

$$\phi_i(x) = \chi(x - x_i) \quad \text{and} \quad \phi_j(y) = \chi(y - y_j)$$

where

$$\chi(z) = \begin{cases} 1 & \text{if } -0.01 \leq z < 0.01 \\ 0 & \text{otherwise.} \end{cases}$$

The approximator basis functions for the region $\mathcal{D}$ are defined as

$$\phi_k(x, y) = \chi(x - x_i)\chi(y - y_j)$$

where $k(i, j) = i + 201(j - 1)$. Note that the function $k(i, j)$ maps each integer pair (i, j) to a unique integer in $\mathcal{Z}_N = [1, 40401]$. For any point $(x, y) \in [-1, 1) \times [-1, 1)$, the indices of the $m = r^d = 4$ nonzero elements of the vector ϕ can be directly computed by

$$\begin{aligned} i(x) &= floor(100x + 100) + 1 \\ j(y) &= floor(100y + 100) + 1 \\ I(x, y) &= \{k(i, j), k(i + 1, j), k(i, j + 1), k(i + 1, j + 1)\}, \end{aligned}$$

where $I(x, y)$ is a four element subset of $\mathcal{Z}_N$. △

Since the elements of the vector $\phi(x)$ with indices not in $I(x)$ are all zero and the indices $I(x)$ are simple to calculate, locally supported lattice approximators allow a significant reduction in computation per function evaluation. However, implementation of the approximator still requires memory[3] for the N parameters in θ. The objective of the CMAC approach is to reduce the dimension of the parameter vector θ from N to M where $M \ll N$ without losing the ability to accurately approximate continuous functions over $\mathcal{D}$.

■ **EXAMPLE 3.8**

Assume that $x \in \mathcal{D} \subset \Re^d$ and that the lattice specifies r basis functions per input dimension. In this case, $N = r^d$. The exponential growth implies that computational and memory reduction techniques become increasing important as the dimension of the input space increases. △

The CMAC separates the address space of the parameter vector from the indices of the regressor vector through the introduction of a (deterministic) embedding function $E(i) : \mathcal{Z}_N \mapsto \mathcal{Z}_M$ where $M \ll N$. This results in the approximator being calculated as

$$\hat{f}(x) = \sum_{k \in I(x)} \theta_{E(k)} \phi_k(x). \tag{3.37}$$

Note that the integers $E(k)$ for $k \in I(x)$ are not guaranteed to be unique. The advantage of this representation is that the required physical memory to store $\theta_{E(k)}$ requires only M locations. In the discussion that follows $E(I(x^i))$ will be used to denote the set $\{E(j) | j \in I(x^i)\}$ where $x^i \in \Re^d$ is the i-th evaluation point.

The embedding function E can be implemented, for example, by a hashing function [160]. The embedding function is a deterministic function that maps each integer in $[1, N]$ onto an integer in $[1, M]$. Since $M < N$ the mapping E is not one-to-one. In fact, since it is typical for $M \ll N$ the mapping E is many-to-one. Example embedding functions are $k = mod(j, M)$ and $k =$ ceil $(M$ rand $(j))$ where $j \in [1, N]$. In the latter example, "rand" is a uniform pseudorandom number generator with seed j and with range $[0, 1] \subset \Re^1$.

3.5.2 Properties

Let x^1 and x^2 denote two evaluation points. Using the index sets $I(x^1)$ and $I(x^2)$ it is straightforward to see that adjustment of the parameters affecting $\hat{f}(x)|_{x^1}$ as calculated by eqn. (3.37) will also affect (or *generalize* to) $\hat{f}(x)|_{x^2}$ as calculated by eqn. (3.37) when $I(x^1) \cap I(x^2) \neq \emptyset$, where $\emptyset$ denotes the empty set. When the approximator is computed by eqn. (3.36) and $I(x^1) \cap I(x^2) = \emptyset$, the lattice structure of the approximator is said to *dichotomize* x^1 from x^2, in the sense that changing the parameters to adjust $\hat{f}(x)|_{x^1}$ does not affect $\hat{f}(x)|_{x^2}$. When the function to be approximated is assumed to be continuous, it is desirable to have generalization between nearby points and to dichotomize widely separated points. The term *learning interference* is used to define the possibly negative effects of training at x^1 that affects the value of $\hat{f}(x)|_{x^2}$.

Introduction of the CMAC embedding function in eqn. (3.37 results in increased learning interference. This is true since even if $I(x^1) \cap I(x^2) = \emptyset$ it may be the case that $E\left(I(x^1)\right) \cap$

[3] Due to the lattice definition, the basis function parameters (x_i, y_j) need not be explicitly stored. Therefore, the memory required to store the parameters used to calculate ϕ is much less than N. Throughout this discussion the memory required for the parameters necessary to compute ϕ will be neglected.

d	Number of input dimensions (i.e., $x \in \mathcal{D} \subset \Re^d$)
m	Number of nonzero basis elements at any $x \in \mathcal{D}$
M	Number of physical memory elements
N	Number of basis elements (i.e., $N = r^d$)
r	Number of basis elements per input dimension.
n	Number of elements in $E(I(x^1)) \cap E(I(x^2))$

Table 3.1: Symbols used in the CMAC discussion and their definitions

$E\left(I(x^2)\right) \neq \emptyset$ due to the many to one nature of the embedding function. Although the CMAC approach increases the effects of learning interference, the amount of increase can be designed to be small by increasing the parameter m and by designing the approximator so that the number of elements in the set $E\left(I(x^1)\right) \cap E\left(I(x^2)\right)$ is expected to be small when the number of elements in $I(x^1) \cap I(x^2)$ is small. Increasing m decreases learning interference, since each parameter contributes on the order of $\frac{1}{m}$ to $\hat{f}(x)|_{x^1}$.

To design the CMAC approximator so that the overlap between $E\left(I(x^1)\right) \cap E\left(I(x^2)\right)$ is small when $I(x^1) \cap I(x^2) = \emptyset$ requires some analysis so that the designer can understand the influence of the various design variables. To facilitate the following discussion, the various symbols of this section and their definitions have been summarized in Table 3.1. For any $x \in \mathcal{D}$, $I(x)$ contains m elements of Z_N. Since there exist r^d different cells over the region $\mathcal{D}$, the function $I(x)$ evaluated over $\mathcal{D}$ defines r^d different sets of m-elements of Z_N. Each of these sets maps through $E\left(I(x)\right)$ to a set of m elements selected from Z_M. The number of such distinct sets (ways of selecting m elements from M choices) is

$$\begin{aligned} \binom{M}{m} &= \frac{M!}{m!(M-m)!} \\ &> \frac{(M-m)^m}{m!} > \frac{(M-m)^m}{m^m} = \left(\frac{M}{m} - 1\right)^m. \end{aligned}$$

Therefore, each $I(x)$ can map to a unique $E\left(I(x)\right)$ if

$$\left(\frac{M}{m} - 1\right)^m > r^d. \tag{3.38}$$

This is an existence result. Whether each $I(x)$ actually maps to a unique $E\left(I(x)\right)$ depends on the embedding function that the designer chooses.

■ **EXAMPLE 3.9**

To determine a useful design rule, consider the expression of eqn. (3.38). Taking the $\log_{10}$ of both sides and solving for m yields

$$m > \frac{d \log(r)}{log(\frac{M}{m} - 1)}.$$

The following table displays a few typical values for r and $\frac{M}{m}$ with the corresponding minimum value of m

r	$\frac{M}{m}$	m
100	100	$m > d$
100	1000	$m > \frac{2}{3}d$
1000	100	$m > \frac{3}{2}d$
1000	1000	$m > d.$

All of these lower bounds on m are quite reasonable and easily satisfied. △

■ EXAMPLE 3.10

The purpose of this example is to illustrate that the choice of the embedding function can have serious negative consequences for the capabilities of the approximator.

Assume that a function is to be approximated over the domain $\mathcal{D} = [0, 1] \times [0, 1]$. Let (x, y) denote the two independent variables and define a lattice by

$$\begin{array}{lll} dx = 0.01, & x_i = (i-2)dx, & i = 1, \ldots, 103, \\ dy = 0.01, & y_j = (j-2)dx, & j = 1, \ldots, 103, \end{array}$$

so that $N = 103^2 = 10609$. Define the address of each node by

$$k(i, j) = (i-1)103 + j$$

which given the constraints on i and j has the inverse mapping

$$j = mod(k-1, 103) + 1 \tag{3.39}$$

$$i = \frac{k-j}{103} + 1 \tag{3.40}$$

for $k \in [1, 10609]$ where $mod(m, n) : I \mapsto [0, n-1]$ is the modulus function that returns the remainder of m divided by n. Given any $(x, y) \in \mathcal{D}$, the nodal indices (i.e., indices for the nearest lattice point) can be directly calculated without search as

$$i(x) = 2 + round(100x) \tag{3.41}$$

$$j(y) = 2 + round(100y) \tag{3.42}$$

which allows calculation of $k(i, j)$ as a function of position (x, y). Let the approximator use the basis functions for the nine nearest cells of $\mathcal{D}$, then

$$\begin{array}{rccc} I(x,y) = \{k(i-1, j-1), & k(i-1, j), & k(i-1, j+1), \\ k(i, j-1), & k(i, j), & k(i, j+1), \\ k(i+1, j-1), & k(i+1, j), & k(i+1, j+1)\}, \end{array}$$

where i and j are computed from eqns. (3.41)–(3.42). Define the embedding function to be

$$E(k) = mod(k-1, M) + 1, \text{ where } M < N.$$

Although the conclusions of the example hold for almost any $M < N$, assume in the following that $M = 1000$ so that the discussion can be explicit.

With the above design, $(x, y) \in (0, .005) \times (0, .005)$ corresponding to $i = 2, j = 2, k = 105$ maps to the nodal and physical addresses

$$\begin{array}{rrrr} I(x,y) = E\left(I(x,y)\right) = \{1, & 2, & 3, \\ 104, & 105, & 106, \\ 207, & 208, & 209\}. \end{array}$$

In addition, $(x, y) \in (0.085, 0.095) \times (0.725, 0.735)$ corresponding to $i = 11, j = 75, k = 1105$ with nodal addresses

$$I(x,y) = \{1001, \quad 1002, \quad 1003, \\ 1104, \quad 1105, \quad 1106, \\ 1207, \quad 1208, \quad 1209\}.$$

For $(x, y) \in (0.085, 0.095) \times (0.725, 0.735)$, $E\left(I(x, y)\right)$ maps to exactly the same set of physical addresses as resulted for $(x, y) \in (0, .01) \times (0, .01)$. Therefore, the values of the function approximation at corresponding points on these two regions are identical. In the following discussion, this mapping of two sets of unique nodal addresses to identical sets of physical addresses will be referred to as an *m-element collision*. In fact, in this example each set of nodal addresses corresponding to $k \geq 1105$ will result in an m-element collision with a set previously assigned to another region.

Given the design parameters of this example, eqn. (3.38) shows that there are at least 2×10^{18} combinations of 1000 addresses taken 9 at a time. Since only 10609 combinations of addresses occur in this design, there do exist embedding functions that map each of the 10609 sets of nodal addresses to a unique set of physical addresses. Unfortunately, the selected embedding function is not one of them.

Note that the smoothness of the embedding function assumed in this example allowed the analysis to show that there were many nodal addresses mapping to identical physical addresses. Good embedding functions are typically very discontinuous. When the embedding function is discontinuous the only method for detecting the existence of m-element collisions may be through exhaustive search over all possible nodal addresses. Due to the size of the nodal address space such an exhaustive search is usually not feasible. This is unfortunate since m-element collisions greatly affect the capabilities of the approximator and may result in online performance that is difficult to interpret and debug. △

By introducing the embedding function to decrease the size of the required physical memory the designer is accepting the fact that even though $I(x^1) \cap I(x^2) = \emptyset$ the number of elements n in $E\left(I(x^1)\right) \cap E\left(I(x^2)\right)$ may not be zero. The previous example demonstrated that depending on E, there may exists situations where $n = m$. An objective of the designer is to select E so that n should be significantly less that m. Two separate issues are of interest: repetition of an element of $E(I(x))$ when there is no repetition in $I(x)$; and, $E\left(I(x^1)\right) \cap E\left(I(x^2)\right)$ containing n elements when $I(x^1) \cap I(x^2) = \emptyset$. In both cases, a probabilistic analysis is used; however, once the designer selects the embedding function the mapping is deterministic.

Assuming the $I(x)$ is a set of m distinct nodal addresses, the probability that $E(I(x))$ duplicates at least one address is

$$\sum_{i=1}^{m-1} \frac{i}{M} = \frac{m(m-1)}{2M},$$

which assumes that E uniformly distributes the nodal addresses over the physical address space with probability $\frac{1}{M}$. When $E(I(x))$ duplicates an address, the corresponding parameter receives increased weighting in the calculation of $f(x)$, but it is not too serious of a problem.

Alternatively, when $I(x^1) \cap I(x^2) = \emptyset$, how can the designer determine the probability that number of elements in $E\left(I(x^1)\right) \cap E\left(I(x^2)\right)$ is a particular value of n? Assuming the the elements of $E\left(I(x^1)\right)$ are unique and that the mapping E is uniform in the sense of the previous paragraph, the probability that a single element of $E\left(I(x^2)\right)$ is in $E\left(I(x^1)\right)$ is $q = \frac{m}{M}$. The probability that the same single element of $E\left(I(x^2)\right)$ is not in $E\left(I(x^1)\right)$ is $p = 1 - q$. The probability that n of the m elements of $E\left(I(x^2)\right)$ are in $E\left(I(x^1)\right)$ is, by the binomial distribution,

$$\frac{m!}{n!(m-n)!} q^n p^{m-n}. \tag{3.43}$$

The results of evaluating this expression for various values of m, M, and n are displayed in Table 3.2. The probability decreases rapidly with both n and M. Note that there are tradeoffs involved in the selection of both m and M. Making m large decreases the average contribution of each coefficient (data stored at the physical address) and increases the extent of local generalization, but making m small decreases the amount of computation required and decreases the probability of collisions between non-overlapping sets of nodal addresses (i.e., interference). Selecting M small decreases the physical memory requirements, but increasing M decreases the probability of collisions between non-overlapping sets of nodal addresses.

$m =$	4	9	16	25	4	9	16	25
$M =$	2000	2000	2000	2000	4000	4000	4000	4000
n=0	9.92e-1	9.60e-1	8.79e-1	7.30e-1	9.96e-1	9.80e-1	9.38e-1	8.55e-1
n=1	7.95e-3	3.91e-2	1.13e-1	2.31e-1	3.99e-3	1.99e-2	6.03e-2	1.34e-1
n=2	2.39e-5	7.06e-4	6.87e-3	3.51e-2	5.99e-6	1.79e-4	1.82e-3	1.01e-2
n=3	3.19e-8	7.45e-6	2.58e-4	3.41e-3	3.40e-9	9.44e-7	3.40e-5	4.90e-4
n=4	1.60e-11	5.05e-8	6.77e-6	2.37e-4	1.00e-12	3.19e-9	4.44e-7	1.69e-5

Table 3.2: Probability of n collisions for an physical memory of size M where each input point maps to m addresses.

3.6 MULTILAYER PERCEPTRON

Perceptrons [223] and multilayer perceptron networks [226] have a long history and an extensive literature [170, 296]. Examples of the use of multilayer perceptrons in control applications are contained in [36, 40, 41, 42, 45, 65, 101, 111, 116, 123, 148, 149, 172, 181, 209, 211, 224, 229, 244, 296].

3.6.1 Description

The left image in Figure 3.5 illustrates a perceptron [223]. The output of the perceptron denoted by v_i is

$$v_i = g\left(b_i + \sum_{j=1}^{n} w_{ij} x_j\right). \tag{3.44}$$

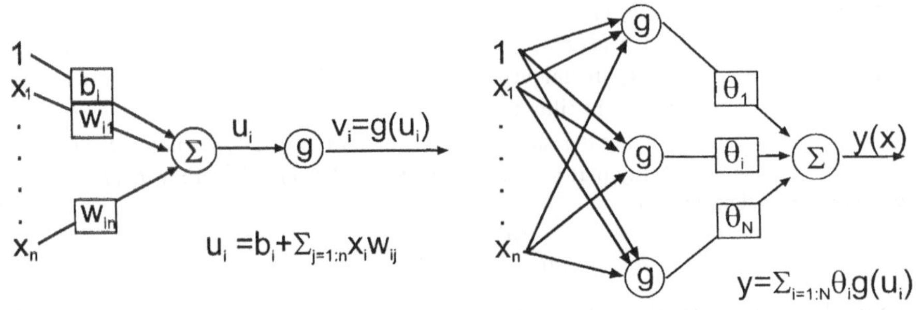

Figure 3.5: *Left*–Single node perceptron. *Right*–Single layer perceptron network. The bold lines in the right figure represents the dot product operation (weighting and summing) performed by the connection and nodal processor.

Often for convenience of notation, this will be written as

$$v_i = g(W_i x)$$

where $W_i = [b_i, w_{i1}, \ldots, w_{in}]$ and $x = [1, x_1, \ldots, x_n]$. The function $g : \Re^1 \mapsto \Re^1$ is a squashing function such as $g(x) = atan(x)$ or $g(x) = \frac{1}{1+e^{-x}}$. Note that the perceptron has multiple inputs and a single output. If $g(x)$ is the signum function, then a perceptron divides its input space into two halves using the hyperplane $u_i = W_i x$. If $u_i < 0$, then $v_i = -1$. If $u_i > 0$, then $v_i = 1$. If $u_i = 0$, then $v_i = 0$. This hyperplane is referred to as a linear discriminant function.

The image on the right side of Figure 3.5 shows a network that forms a linear combination of perceptron outputs. The network output is

$$y = \theta V$$

where $V^\top = [v_1, \ldots, v_N]$ is the vector of outputs from each perceptron defined in eqn. (3.44) and $\theta \in \Re^{q \times N}$ is a parameter matrix. This approximator is referred to as a single hidden layer perceptron network. The parameters in W_i are the hidden layer parameters. The parameters in θ are the output layer parameters. By Theorem 2.4.5, single hidden layer perceptron networks are universal approximators. In the case that y is a scalar (i.e., $q = 1$), the function $g(y)$ with g being a signum function defines a general discriminator function that can be used for classification tasks [152].

If desired, networks with multiple hidden layers can be constructed. This is accomplished by defining θ to be a matrix so that y is a vector. If we define $z = \Lambda g(y)$, then the network has two hidden layers defined by the weights in W and θ.

The perceptron networks defined above are *feedforward* networks. This means that the information flow through the network is unidirectional (from left to right in Figure 3.5). There is not feedback of information either from internal variables or from the network output to the network input. In the case where some of the internal network variables or outputs are fed back to serve as a portion of the input, we would have a *recurrent* perceptron network. In this case, the network is a dynamics system with its own state vector. When such recurrent networks are used, the designer must be concerned with the stability of this network state.

Perceptron networks are sometimes referred to as supervised learning or backpropagation networks, but neither of these names are accurate. Supervised learning refers to the approach of training (i.e., adjusting the parameters) a function approximator $y = \hat{f}(x, \theta, \sigma)$ so that the approximator matches, as closely as possible, a given set of training data described as $\{(y^i, x^i)\}_{i=1}^{P}$. In this scenario a batch of training samples is available for which the desired output y^i is known for each input x^i. Many early applications of perceptron networks were formulated within the supervised learning approach; however, any function approximator can be trained using such a supervised learning scenario. Therefore, referring to a perceptron network as a supervised learning network is not a clear description.

The backpropagation algorithm is described in Section 4.4.2.3. Although the algorithm referred to as backpropagation was derived for perceptron networks, see e.g. [226], that algorithm is based on the idea of gradient descent. Gradient descent parameter adaptation can be derived for any feed forward network that uses a continuous nodal processor, see e.g. [291]. Therefore, referring to a perceptron network as a backpropagation network is again not a clear description of the network. In addition, the fact that a multilayer perceptron network can be trained using backpropagation is not a motivation for using these networks, since gradient descent training is a general procedure that can be used for many families of approximators.

3.6.2 Properties

The literature on neural networks contains several standard phrases that are often used to motivate the use of perceptron networks. For any particular applications, the applicability of these standard phrases should be carefully evaluated.

A typically stated motivation is that perceptron networks are universal approximators. As discussed in Section 2.4.5, numerous families of approximators have this or related properties. Therefore, the fact that perceptron networks are universal approximators is not, by itself, a motivation for using them instead of any other approximator with this property.

A perceptron network with adjustable hidden layer parameters is nonlinearly parameterized. Therefore, another stated motivation is that perceptron networks have certain beneficial "order of approximation" properties, as discussed in Section 2.4.1. On the other hand, perceptron networks are not transparent, see Section 2.4.9. There are no engineering procedures available for defining a suitable network structure (i.e., number of hidden layers, number of nodes per layer, etc.) even in situations where the function f to be approximated is known. Also, since the network is nonlinearly parameterized, the initial choice of parameters may not be in the basin of attraction of the optimal parameters.

Early in the history of neural networks, it was noticed that perceptron networks "offer the inherent potential for parallel computation." However, any approximation structure that can be written in vector product form is suitable for parallel implementation on suitable hardware. Interesting questions are whether any particular application is worth special hardware, or more generally, is any particular approximation structure worth additional research funding to develop special purpose implementation hardware, when hardware optimized for performing matrix vector products already exists.

Another frequently stated motivation is the idea that perceptron networks are models of the nervous systems of biological entities. Since these biological entities can learn to perform complex control actions (balancing, walking, dancing, etc.), perceptron networks should be similarly trainable. There are several directions from which such statements should be considered. First, is a perceptron network a sufficiently accurate model of a nervous system that such an analogy is justified? Second, is the implemented perceptron network comparable in size to a realistic nervous system? Even if those questions could be answered affirmatively, do we understand and can we accurately replicate the feedback and training process that occur in the biological exemplars? Also, the biological nervous system may be optimized for the biochemical environment in which it operates. The optimal implementation approach on an electronic processing unit could be significantly different.

Another frequent motivation for perceptron networks is by analogy to biological control systems. It is stated that biological control systems are semi-fault tolerant because they rely on large numbers of redundant and highly interconnected nonlinear nodal processors and communication pathways. This is referred to as distributed information processing. To motivate perceptron networks, it is argued that highly interconnected perceptron networks have similar properties to biological control systems, since for perceptron networks the approximator information is stored across a large number of "connection" parameters. The idea being that if a few "connections" were damaged, then some information would be retained via the undamaged parameters and these undamaged parameters could be adapted to reattain the prior level of performance. However, the perceptron networks that are typically implemented are much smaller and simpler than such biological systems, resulting in a weak analogy. In addition, this line of reasoning neglects the fact that perceptron networks are typically implemented with a standard CPU and RAM, where there is no "distributed network implementation" since these standard items fail as a unit. Therefore, the CPU and RAM implementation is not currently analogous to a biochemical network implementation.

3.7 FUZZY APPROXIMATION

This section presents the basic concepts necessary for the reader to be able to construct a fuzzy logic controller. The presentation is self-contained, yet succinct. Readers interested in a detailed presentation of the motivation and theory of fuzzy logic should consult, for example, [21, 303, 304]. Detailed discussion of the use of fuzzy logic in fixed and adaptive controllers is presented, for example, in [15, 32, 63, 65, 67, 125, 131, 150, 151, 182, 184, 189, 198, 230, 261, 266, 267, 283, 284, 286, 302]. The main references for this section are [67, 284, 304].

3.7.1 Description

The four basic components of a fuzzy controller are shown in Figure 3.6. In this figure, over-lined quantities represent fuzzy variables and sets while crisp (real valued) variables and sets have no over-lining. This notation will be used throughout this section, unless otherwise specified.

3.7.1.1 Fuzzy Sets and Fuzzy Logic Given a real valued vector variable $x = [x_1, \ldots, x_n]^T$ that is an element of a domain $X = X_1 \times X_2 \times \ldots \times X_n$, the region X_i is referred to as the *universe of discourse* of x_i and X as the universe of discourse of x. The linguistic variable $\bar{x}_i$ can assume the linguistic values defined by $\bar{X}_i = \left\{\bar{X}_i^1, \ldots, \bar{X}_i^{N_i}\right\}$. The degree to which the linguistic variable $\bar{x}_i$ is described by the linguistic value $\bar{X}_i^j$ is defined by a *membership function* $\mu_{\bar{X}_i^j}(x) : X_i \mapsto [0, 1]$. Common membership functions include triangular and Gaussian functions. The fuzzy set $\tilde{X}_i^j$ associated with linguistic variable $\bar{x}_i$, universe of discourse X_i, linguistic value $\bar{X}_i^j$, and membership function $\mu_{\bar{X}_i^j}(x)$

is

$$\tilde{X}_i^j = \left\{ \left(x_i, \mu_{\tilde{X}_i^j}(x_i) \right) \middle| x_i \in X_i \right\}. \tag{3.45}$$

Note that fuzzy sets have members and degrees of membership. The degree of membership is the main feature that distinguishes fuzzy logic from Boolean logic. The support of a fuzzy set $\tilde{F}$ on universe of discourse X is defined as $Supp(\tilde{F}) = \{x \in X | \mu_{\tilde{F}}(x) \neq 0\}$. If the $Supp(\tilde{F})$ is a single point x_s and $\mu_{\tilde{F}}(x_s) = 1$, then x_s is called a *fuzzy singleton.*

EXAMPLE 3.11

To illustrate the concepts of the previous paragraph, consider a vehicle cruise control application. Let the physical variables be $x = [v_e, a]^T$, where $v_e = v - v_c$, v denotes speed, v_c denotes the commanded speed, and a denotes acceleration. The linguistic variables are defined as $\bar{x} = [speederror, acceleration]^T$. The linguistic values for

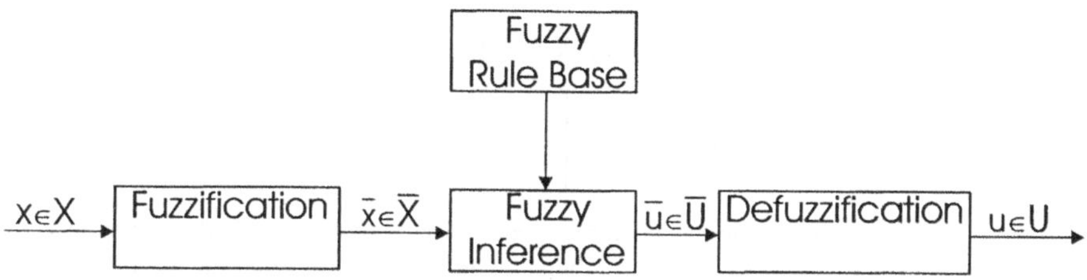

Figure 3.6: Components of a Fuzzy Logic Controller.

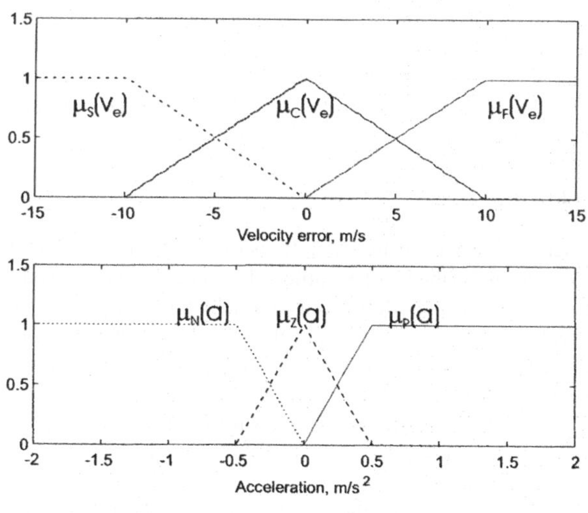

Figure 3.7: Membership functions for speed error and acceleration for the cruise control example.

OR	$\mu_{\bar{A}\cup\bar{B}}(x) = \mu_{\bar{A}}(x) \oplus \mu_{\bar{B}}(x)$	AND	$\mu_{\bar{A}\cap\bar{B}}(x) = \mu_{\bar{A}}(x) \star \mu_{\bar{B}}(x)$
maximum	$\max(\mu_{\bar{A}}(x), \mu_{\bar{B}}(x))$	minimum	$\min(\mu_{\bar{A}}(x), \mu_{\bar{B}}(x))$
algebraic sum	$\mu_{\bar{A}}(x) + \mu_{\bar{B}}(x) - \mu_{\bar{A}}(x)\mu_{\bar{B}}(x)$	algebraic product	$\mu_{\bar{A}}(x)\mu_{\bar{B}}(x)$
bounded sum	$\min(1, \mu_{\bar{A}}(x) + \mu_{\bar{B}}(x))$	bounded product	$\max(0, \mu_{\bar{A}}(x) + \mu_{\bar{B}}(x) - 1)$
drastic sum	$\begin{cases} \mu_{\bar{A}}(x) & \text{if } \mu_{\bar{B}}(x) = 0 \\ \mu_{\bar{B}}(x) & \text{if } \mu_{\bar{A}}(x) = 0 \\ 1 & \text{otherwise} \end{cases}$	drastic product	$\begin{cases} \mu_{\bar{A}}(x) & \text{if } \mu_{\bar{B}}(x) = 1 \\ \mu_{\bar{B}}(x) & \text{if } \mu_{\bar{A}}(x) = 1 \\ 0 & \text{otherwise} \end{cases}$

Table 3.3: Example implementations of fuzzy logic (left) s-norm operations for $\bar{A} \bigcup \bar{B}$ and (right) t-norm operations for $\bar{A} \bigcap \bar{B}$.

each linguistic variable could be defined as

$$\begin{aligned} \bar{X}_1 &= \{Slow, Correct, Fast\} \\ \bar{X}_2 &= \{Negative, Zero, Positive\} \end{aligned}$$

so that $N_1 = N_2 = 3$. Then, the space $\bar{X}$ is defined as

$$\bar{X} = \bar{X}_1 \times \bar{X}_2 = \left\{ \begin{array}{ccc} SN & CN & FN \\ SZ & CZ & FZ \\ SP & CP & FP \end{array} \right\}$$

where each linguistic value has been represented by its first letter. If the universe of discourse is $X = [-15, 15] \times [-2, 2]$, then one possible definition of the membership functions for $\bar{X}_1$ and $\bar{X}_2$ are shown in Figure 3.7. △

In fuzzy logic, the "$\bar{A}$ or $\bar{B}$" operation is represented as "$\bar{A} \bigcup \bar{B}$." The membership function for the fuzzy set $\bar{A} \bigcup \bar{B}$ is calculate by a s-norm operation [284] denoted by $\oplus$, $\mu_{\bar{A}\cup\bar{B}}(x) = \mu_{\bar{A}}(x) \oplus \mu_{\bar{B}}(x)$. In fuzzy logic, the "$\bar{A}$ and $\bar{B}$" operation is represented as "$\bar{A} \bigcap \bar{B}$." The membership function for the fuzzy set $\bar{A} \bigcap \bar{B}$ is calculate by a t-norm operation [284] denoted by $\star$, $\mu_{\bar{A}\cap\bar{B}}(x) = \mu_{\bar{A}}(x) \star \mu_{\bar{B}}(x)$. Table 3.3 contains several of the possible implementations of the $\star$ and $\oplus$ operations. The membership function for the complement of fuzzy set $\bar{A}$ is $\mu_{\bar{A}^c}(x) = 1 - \mu_{\bar{A}}(x)$. The fuzzy complement is used to implement the "not" operation

■ **EXAMPLE 3.12**

Figure 3.8 presents examples of the operations discussed in the previous paragraph for the fuzzy system described in Example 3.11. The algebraic product is used to implement the $\star$ operator. The left mesh plot shows the membership function for the fuzzy set "velocity error is fast and acceleration is positive" (i.e., $\mu_{\bar{F}\cap\bar{P}}(v, a) = \mu_{\bar{F}}(v)\mu_{\bar{P}}(a)$). The center plot shows the membership function for the fuzzy set "acceleration is negative and acceleration is zero" (i.e., $\mu_{\bar{Z}\cap\bar{N}}(a, a) = \mu_{\bar{Z}}(a)\mu_{\bar{N}}(a)$). The right plot shows the membership function for the fuzzy set "acceleration is not negative" (i.e., $\mu_{N^c}(a) = 1 - \mu_N(a)$). △

A *fuzzy relation* $\bar{Q}(U, V)$ between the the universe of discourses U and V is a fuzzy set defined on $U \times V$:

$$\bar{Q} = \{(u, v, \mu_Q(u, v)) | u \in U,\ v \in V,\ \mu_Q(u, v) : U \times V \mapsto [0, 1]\}. \tag{3.46}$$

Logical statements are fuzzy relations with membership function defined by the $\star$, $\oplus$, and complement operators. For example, "(x is small)AND(y is large)" is a relation with the membership function $\mu_{\bar{S}\cap\bar{L}}(x,y) = \mu_{\bar{S}}(x) \star \mu_{\bar{L}}(y)$ where S denotes small and L denotes large.

■ EXAMPLE 3.13

The fuzzy relation for the product xy being small could be defined as

$$\bar{Q}_{xy} \text{ small} = \{(x, y, \exp(-|xy|))\}.$$

△

Fuzzy relations defined for variables with a finite, discrete universe of discourse can be conveniently represented in matrix form.

■ EXAMPLE 3.14

Let $\tilde{A} = \{(1,1),(2,.5)\}$ be a fuzzy set defined over universe of discourse $U = \{1,2\}$. Let $\tilde{B} = \{(1,.9),(2,.7),(3,.5),(4,.1)\}$ be a fuzzy set defined over universe of discourse $V = \{1,2,3,4\}$. The fuzzy relation corresponding to "$\tilde{A}$ OR $\tilde{B}$," using the maximum function to implement the $\oplus$ operation, can be represented as

$$\begin{array}{cc} & \begin{array}{cccc} 1 & 2 & 3 & 4 \end{array} \\ \begin{array}{c} 1 \\ 2 \end{array} & \left[\begin{array}{cccc} 1 & 1 & 1 & 1 \\ .9 & .7 & .5 & .5 \end{array}\right]. \end{array}$$

△

If $\bar{P}(U,V)$ and $\bar{Q}(V,W)$ are fuzzy relations, their composition is a relation on $U \times W$ defined as

$$\bar{P} \circ \bar{Q} = \{ (u, w, \mu_{\bar{P}\circ\bar{Q}}(u,w)) \,|\, u \in U,\ w \in W\} \tag{3.47}$$

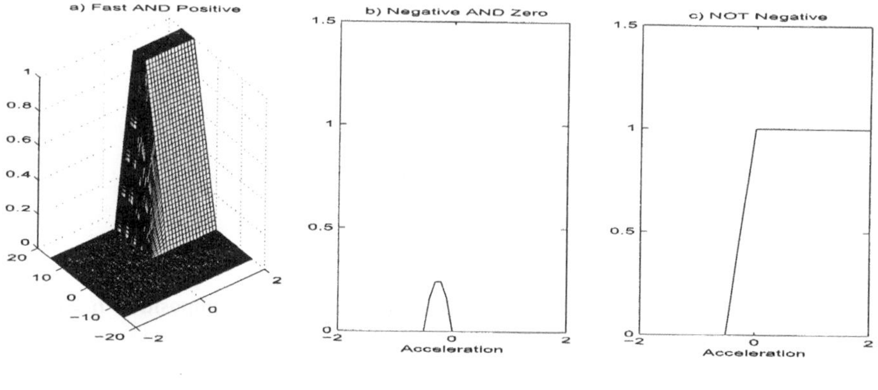

Figure 3.8: Examples of membership functions produced by operations on fuzzy sets. a) Velocity error is fast AND acceleration is positive. b) Acceleration is negative AND acceleration is zero. c) Acceleration is not negative.

where

$$\mu_{\bar{P}\circ\bar{Q}}(u, w) = \max_{v \in V} \left[t\left(\mu_{\bar{P}}(u, v), \mu_{\bar{Q}}(v, w)\right)\right] \tag{3.48}$$

and t represents a $t-$norm (see Table 3.3). Computation of the membership functions for compositions of fuzzy relations can be difficult when the universes of discourse involve continuous variable. When the universes of discourse involve a finite number of discrete variables, computation can be efficiently organized through algebra similar to matrix multiplication.

■ EXAMPLE 3.15

Let $\bar{P}(x, y)$ be the fuzzy relation $x < y$ for $x,\ y \in \Re^1$ described by the membership function

$$\mu_{\bar{P}}(x, y) = \frac{1}{1 + e^{x-y}}.$$

Let $\bar{Q}(y, z)$ be the fuzzy relation $y < z$ for $y,\ z \in \Re^1$ described by the membership function

$$\mu_{\bar{Q}}(y, z) = \frac{1}{1 + e^{y-z}}.$$

Then the membership function for the composition $\bar{P} \circ \bar{Q}$, using the algebraic product for the t-norm, is

$$\begin{aligned} \mu_{\bar{P}\circ\bar{Q}}(x, z) &= \max_{y \in \Re^1} \left[\frac{1}{(1 + e^{x-y})(1 + e^{y-z})} \right] \\ &= \frac{1}{\left(1 + e^{\frac{x-z}{2}}\right)^2}. \end{aligned} \tag{3.49}$$

Derivation of eqn. (3.49) is requested in Exercise 3.9. Examples such as this, where $\mu_{\bar{P}\circ\bar{Q}}(x, z)$ can be explicitly solved, are the exception. △

■ EXAMPLE 3.16

Let the relation $\bar{R}(U, V)$ be represented by the matrix [304]

$$\begin{bmatrix} 0.3 & 0.8 \\ 0.6 & 0.9 \end{bmatrix}$$

and let the relation $\bar{S}(V, W)$ be represented by the matrix

$$\begin{bmatrix} 0.5 & 0.9 \\ 0.4 & 1.0 \end{bmatrix}.$$

If the t-norm is implemented by the min operation, then the composition $\bar{R} \circ \bar{S}$ is represented as

$$\begin{aligned} \bar{R} \circ \bar{S} &= \begin{bmatrix} 0.3 & 0.8 \\ 0.6 & 0.9 \end{bmatrix} \circ \begin{bmatrix} 0.5 & 0.9 \\ 0.4 & 1.0 \end{bmatrix} \\ &= \begin{bmatrix} \max(\min(0.3, 0.5), \min(0.8, 0.4)) & \max(\min(0.3, 0.9), \min(0.8, 1.0)) \\ \max(\min(0.6, 0.5), \min(0.9, 0.4)) & \max(\min(0.6, 0.9), \min(0.9, 1.0)) \end{bmatrix} \end{aligned}$$

$$= \begin{bmatrix} 0.4 & 0.8 \\ 0.5 & 0.9 \end{bmatrix}.$$

△

With these basic tools of fuzzy systems available to us, we are now ready to consider the components of the fuzzy controller shown in Figure 3.6.

3.7.1.2 *Fuzzification* The previous subsection has introduce various aspects of fuzzy logic as operations on fuzzy sets. Since control systems do not directly involve fuzzy sets, a fuzzification interface is used to convert the crisp plant state or output measurements into fuzzy sets, so that fuzzy reasoning can be applied.

Given a measurement x^* of variable x in universe of discourse X, the corresponding fuzzy set is $\bar{X} = \{(x, \mu(x : x^*)\}$. A few common choices are singleton, triangular, and Gaussian fuzzification. For singleton fuzzification,

$$\mu(x : x^*) = \begin{cases} 1 & \text{if } x = x^* \\ 0 & \text{otherwise.} \end{cases}$$

For triangular fuzzification,

$$\mu(x : x^*) = \begin{cases} 1 - \frac{|x - x^*|}{\lambda} & \text{if } |x - x^*| < \lambda \\ 0 & \text{otherwise.} \end{cases}$$

For Gaussian fuzzification,

$$\mu(x : x^*) = \exp\left(\frac{-(x - x^*)^2}{\lambda^2}\right).$$

In each of the above cases, the parameter λ can either be selected by the designer or adapted online.

The *fuzzification* process converts each input variable x^* into a fuzzy set $\bar{X}$. Singleton fuzzification is often used as it greatly simplifies subsequent computations. Other forms of fuzzification may be more appropriate for representing uncertainty (or fuzziness) of the control system inputs due, for example, to measurement noise.

3.7.1.3 *Fuzzy Implication* The *fuzzy rule base* will contain a set of rules $\{R^l, l \in [1, \ldots, N]\}$ of the form

$$R^l\text{: IF } \left(\bar{x}_1 \text{ is } \bar{X}_1^{l_1}\right) \text{ and } \ldots \text{ and } \left(\bar{x}_n \text{ is } \bar{X}_n^{l_n}\right) \text{ THEN } \left(\bar{u} \text{ is } \bar{U}^{l_u}\right) \tag{3.50}$$

where $l_i \in [1, \ldots, N_i]$ and $\bar{U}$ is the set of linguistic values defined for the fuzzy control signal $\bar{u}$. Each term in parenthesis is an *atomic fuzzy proposition.* The antecedent is the compound fuzzy proposition:

$$\bar{A}^l = \left(\bar{x}_1 \text{ is } \bar{X}_1^{l_1}\right) \text{ and } \ldots \text{ and } \left(\bar{x}_n \text{ is } \bar{X}_n^{l_n}\right). \tag{3.51}$$

Each antecedent defines a fuzzy set in $\bar{X} = \bar{X}_1 \times \ldots \times \bar{X}_n$. The antecedent may contain multiple atomic fuzzy propositions using the same variable and need not include all fuzzy variables. The membership function for $\bar{A}^l$ is completely specified once the t-norm and s-norm representation of the "and" and "or" operations are selected. Therefore, the applicability or confidence of rule R^l is calculated by the antecedent as

$$\mu_{\bar{A}^l}(x) = \mu_{\bar{X}^{l_1} \cap \ldots \cap \bar{X}^{l_n}}(\bar{x}_1, \ldots, \bar{x}_n) = \mu_{\bar{X}^{l_1}}(\bar{x}_1) \star \ldots \star \mu_{\bar{X}^{l_n}}(\bar{x}_n). \tag{3.52}$$

Note that when $\star$ is implemented as the algebraic product, then this membership function can have the form of a tensor product. If $\bar{x}_i$ is not a fuzzy singleton, then evaluation of each atomic fuzzy proposition can become computationally difficult.

A rule (implication) of the form

$$R\text{: IF } \left(\bar{x} \text{ is } \bar{A}\right) \text{ THEN } \left(\bar{u} \text{ is } \bar{B}\right) \tag{3.53}$$

for $x \in X$ and $u \in U$ can be interpreted as a relation in $X \times U$. The membership function for this implication may have various forms depending on the interpretation of the implication operation. Four possibilities are displayed in Table 3.4. The first two interpretations are motivated by the fact that $A \Rightarrow B$ has the same truth table as $((\sim A)$ or $B)$. The third row is motivated by the fact that $A \Rightarrow B$ also has the same truth table as $((A$ and $B)$ or $(\sim A))$. Such direct truth table equivalence approaches are not always the most appropriate interpretations of the implication. In some situations, a more causal situation is desired where the implication is interpreted as

IF A
THEN B
ELSE
Nothing.

Such an interpretation of the implication is equivalent (in the truth table sense) to $(A$ and $B)$. The fourth row indicates the membership function corresponding to this interpretation. Such Mamdani implications are widely used in fuzzy control approaches.

Name	Interpretation of $A \Rightarrow B$	Membership Function
Dienes-Reschler	$(\sim A)$ or B	$\mu_R(x,u) = \max(1 - \mu_A(x), \mu_B(u))$
Lukasiewicz	$(\sim A)$ or B	$\mu_R(x,u) = \min(1, (1 - \mu_A(x)) + \mu_B(u))$
Zadeh	$(A$ and $B)$ or $(\sim A)$	$\mu_R(x,u) = \max(\min(\mu_A(x), \mu_B(u)), 1 - \mu_A(x))$
Mamdani	A and B	$\mu_R(x,u) = t(\mu_A(x), \mu_B(u))$

Table 3.4: Interpretations of Fuzzy Implication. The $\sim$ notation denotes logical negation.

■ EXAMPLE 3.17

Consider the fuzzy rule

$$\bar{R}\text{: IF } (\bar{x}_1 \text{ is small) AND } (\bar{x}_2 \text{ is large) THEN } (u \text{ is large}).$$

Let the fuzzy sets for "small" and "large" be defined as

$$\begin{aligned} \mu_{small}(x) &= \exp\left(-x^2\right) \\ \mu_{large}(x) &= \exp\left(-(x-10)^2\right) \\ \mu_{large}(u) &= \exp\left(-(u-10)^2\right). \end{aligned}$$

Using Mamdani implication with the algebraic product for the t-norm representation of the "AND" operation, the membership function relation that corresponds to this rule is

$$\mu_{\bar{R}}(x_1, x_2, u) = \exp\left(-(x_1)^2\right) \exp\left(-(x_2 - 10)^2\right) \exp\left(-(u - 10)^2\right).$$

△

3.7.1.4 *Fuzzy Inference* Given the results of the two previous subsections, from a control system point of view, the inputs to the control system have been converted to fuzzy sets and each rule has been translated into a fuzzy relation. Pertaining to the issue of *inference* there are two related questions. How can the fuzzy set in U that results from a single rule be determined? How can the fuzzy set in U from a set of rules be determined?

According to the *compositional rule of inference* [284, 304], given a rule of the form of eqn. (3.53) and a fuzzy set $\bar{X}$ with membership function $\mu_{\bar{X}}(x)$, then the membership function of the resultant fuzzy set in U can be found by the composition

$$\mu_R(u) = \sup_{x \in X} t\left(\mu_{\bar{X}}(x), \mu_{\bar{R}}(x, u)\right). \tag{3.54}$$

■ **EXAMPLE 3.18**

Let the relevant fuzzy sets corresponding to the (Gaussian) fuzzified control inputs be defined by

$$\begin{aligned} \bar{X}_1 &= \left\{\left(x_1, \exp\left(-9(x_1 - x_1^*)^2\right)\right)\right\} \\ \bar{X}_2 &= \left\{\left(x_2, \exp\left(-9(x_2 - x_2^*)^2\right)\right)\right\} \end{aligned}$$

where (x_1^*, x_2^*) represent the crisp control input variables. Continuing from Example 3.17, let the algebraic product be the t-norm representation of the "AND" operation, with Mamdani implication, then

$$\begin{aligned} \mu_{\bar{R}}(u) = \sup_{(x_1, x_2)} & \left[\exp\left(-9(x_1 - x_1^*)^2\right) \exp\left(-9(x_2 - x_2^*)^2\right)\right. \\ & \left.\exp\left(-(x_1)^2\right) \exp\left(-(x_2 - 10)^2\right) \exp\left(-(u - 10)^2\right)\right]. \end{aligned}$$

Alternatively, let the fuzzy sets corresponding to the fuzzified control inputs be defined by singleton fuzzification. In this case,

$$\mu_{\bar{R}}(u) = \exp\left(-(x_1^*)^2\right) \exp\left(-(x_2^* - 10)^2\right) \exp\left(-(u - 10)^2\right)$$

Note that significant simplification results from singleton fuzzification, since the optimization (sup) over x is effectively eliminated. △

The above text has discussed the method for inferring the fuzzy set output corresponding to a single rule. The remainder of this section will be concerned with the problem of inferring the output fuzzy set that results from a set of rules called the *rule base*. A fuzzy rule base is called *complete* if for any $x \in X$ there exists at least one rule with a nonzero membership function (i.e., $\forall x \in X, \exists l \ni \mu_{\bar{A}^l}(x) \neq 0$). Note that completeness of a fuzzy rule base is similar to the idea of coverage discussed in Section 2.4.8.1. Two methods of inferring the output of a rule base are possible: *compositional inference* and *individual rule inference*.

In compositional inference (see Section 7.2.1 in [284]), the relations corresponding to each rule are combined (through appropriately selected logical operations) into one relation representing the entire rule base. Then, composition with the input fuzzy sets is used to define the output fuzzy set. The composition of all the rules into a single relation can become cumbersome.

In individual rule inference, the output fuzzy set $U_i = \{(u, \mu_{\bar{R}^i}(u))\}$ corresponding to each individual rule is determined according to eqn. (3.54). The output of the inference engine, based on the entire (l rule) rule base, then has membership function described by either

$$\begin{aligned} \mu_{RB}(u) &= \mu_{\bar{R}^1}(u) \oplus \ldots \oplus \mu_{\bar{R}^l}(u) & (3.55)\\ \text{or } \mu_{RB}(u) &= \mu_{\bar{R}^1}(u) \star \ldots \star \mu_{\bar{R}^l}(u). & (3.56) \end{aligned}$$

Eqn. (3.55) is used when the individual rules are interpreted as independent conditional statements intended to cover all possible operational situations. Eqn. (3.56) is used when the rule base is interpreted as a strongly coupled set of conditional statements that all should apply to the given situation. For example, given Mamdani product implication, eqn. (3.55), with the "or" operation implemented as max, the output membership function is

$$\mu_{RB}(u) = \max_{i=1}^{L} \left[\sup_{x \in X} \left(\mu_{\bar{X}}(x : x^*) \mu_{\bar{A}^i}(x) \mu_{\bar{B}^i}(u) \right) \right]. \tag{3.57}$$

Note that the resulting rule base membership function may be multimodal or have disconnected support.

3.7.1.5 Defuzzification The purpose of the *defuzzifier* is to map a fuzzy set, such as $\bar{U} = (u, \mu_{RB}(u))$ for $u \in U$, to a crisp point u^* in U. The point u^* should be in some sense "most representative" of $\bar{U}$. Since there are many interpretations of "most representative," there are also many means to implement the defuzzification process.

Table 3.5 summarizes three methods for performing defuzzification. The first method computes an *indexed center of gravity*. This method is often computationally difficult since the rule base membership function is typically not simple to describe. The middle row of the table describes the *center average* defuzzification process. The function "center" could, for example, select the midpoint of the set $\{u \in U \,|\mu_{\bar{R}^i}(u) > 0\}$. The center average is computationally easier than the indexed center of gravity approach. The final row of the table describes the *maximum* defuzzification process. The set $\text{hgt}_{RB}(U)$ contains all values of u that achieve the maximum value of $\mu_{RB}(u)$ over U. The function g processes $\text{hgt}_{RB}(U)$ to produce a unique value for u^*. The function g could for example select the minimum, center, or maximum of $\text{hgt}_{RB}(U)$.

3.7.2 Takagi–Sugeno Fuzzy Systems

This subsection presents the Takagi–Sugeno fuzzy system. The reasons for presenting this special case are (1) it is commonly used, (2) it is rather straightforward to understand, (3) its parametric form is amenable to stability analysis, and (4) it highlights the parallels between fuzzy approximators and the other approximators discussed in this chapter.

The Takagi–Sugeno fuzzy system uses rules of the form

$$R^l\!: \text{IF } \left(\bar{x}_1 \text{ is } \bar{X}_1^{l_1}\right) \text{ and } \ldots \text{ and } \left(\bar{x}_n \text{ is } \bar{X}_n^{l_n}\right) \text{ THEN } \left(\bar{u} = f_l(x)\right). \tag{3.58}$$

Indexed Center of Gravity	$u^* = \frac{\int_{U_\alpha} \mu_{RB}(u)\, u\, du}{\int_{U_\alpha} \mu_{RB}(u) du}$ $U_\alpha = \{u \in U \vert \mu_{RB}(u) \geq \alpha\}$
Center Average	$c_i = \text{center}\left(\{u \in U \,\vert \mu_{\bar{R}^i}(u) > 0\}\right)$ $h_i = sup_{u \in U}(\mu_{\bar{R}^i}(u)$ $u^* = \frac{\sum_{i=1}^{l} c_i h_i}{\sum_{i=1}^{l} h_i}$
Maximum	$\text{hgt}_{RB}(U) = \{u \in U \vert \mu_{RB}(u) = \sup_{u \in U}(\mu_{RB}(u))\}$ $u^* = g(\text{hgt}_{RB}(U))$

Table 3.5: Example methods of defuzzification.

For the fuzzy logic controllers that are of interest in this book, $f_l(x)$ is a parameterized function (e.g., $f_l(x : \theta_l)$) where the parameters are identified based on experimental data. Typically,

$$f_l(x : \theta_l) = \theta_{l0} + \sum_{i=1}^{n} \theta_{li} x_i,$$

but nonlinear functions in either x or θ can be used. The membership function for the antecedent is formed as in eqn. (3.52). The Takagi–Sugeno approach then calculates the output control action as

$$u = \frac{\sum_i \mu_{\bar{A}^i}(x) f_i(x : \theta_l)}{\sum_j \mu_{\bar{A}^j}(x)} = \sum_i \Gamma_i(x) f_i(x : \theta_l) \text{ where } \Gamma_i(x) = \frac{\mu_{\bar{A}^i}(x)}{\sum_j \mu_{\bar{A}^j}(x)}. \tag{3.59}$$

Note that this approximator has the form of a basis-influence function with basis set $\{f_i(x : \theta_i)\}$ and influence functions $\{\Gamma_i(x)\}$. If the fuzzy rule set is complete and each $\mu_{\bar{A}^i}(x)$ is finite, then this set of influence functions $\{\Gamma_i(x)\}$ will be finite, vanish nowhere, and form a partition of unity.

Eqn. (3.59) has a variety of interesting interpretations. The $f_i(x)$ can be previously existing operating point controllers or local controllers defined by human "experts." Alternatively, this expression can be interpreted as a "gain scheduled" controller. In all these cases, it is of interest to analyze the stability of the *nonlinear* closed-loop control loop that results.

3.7.3 Properties

One of the early motivations for fuzzy systems was there transparency, in the sense that users can (linguistically) read, describe, and understand the rule base. Similarly, a fuzzy system such as the Takagi-Sugeno type is similar to a smoothly interpolated gain scheduled controller, where each control law f_i is applicable over the support of Γ_i.

Fuzzy systems are capable of universal approximation, see for example Chapter 9 in [284]. Adaptation of fuzzy systems, as with any approximator, must be approached with caution. If for example the antecedents of the rule base are adapted, this is a nonlinear estimation process. Adaptation of the antecedents could lead to loss of completeness of the fuzzy rule base.

3.8 WAVELETS

Efficient allocation of approximator resources motivates the tuning of approximator basis functions to the local curvature properties of the function. Similar motivations arise in various application fields. For example, in signal (and image) processing it has proven useful to decompose signals (and images) using a space of basis functions that have local support both in the time and frequency domains. Such motivations across various fields has lead to the development of wavelets, which means small waves. The main references for this section are [51, 60, 260, 262, 306]. A very understandable review of wavelets is maintained on the website of R. Polikar [206]. Example articles discussing the use of wavelets in control applications include [22, 37, 217, 262].

Wavelet algorithms are defined to process data at different scales of resolution in both the time and frequency domains. For our function approximation purposes, we are dealing with a variable x instead of time. Therefore, we will refer to the *space* and *spatial frequency* domains. For a function $f(x)$ in the spatial domain, we will use the notation $\mathcal{F}f(\xi)$ to denote the Fourier transform of f where ξ is the spatial frequency variable. The spatial wavelength is $\sigma = \frac{1}{\xi}$.

The continuous wavelet transform is defined as

$$\Psi_f^{\psi}(\tau, \sigma) = \frac{1}{\sqrt{|\sigma|}} \int f(x) \psi\left(\frac{x-\tau}{\sigma}\right) dx \tag{3.60}$$

where ψ is a real valued *mother wavelet*, τ is the translation parameter, and σ is the scale parameter. Eqn. (3.60) is an inner product between the function f and the scaled and translated mother wavelet. For fixed values of τ and σ, the wavelet transform $\Psi_f^{\psi}(\tau, \sigma)$ quantifies the similarity between f and the wavelet at that scale and translation. The variable τ shifts the mother wavelet along the x-axis. The mother wavelet ψ is selected to have localized support which allows characteristics of f to be accurately resolved along the x-axis when σ is small. The variable σ allows analysis of f at different scales. As σ is increased, the inner product considers a wider range of x which includes lower spatial frequencies. Similarly, as σ is decreased the inner product considers a narrower range of x and higher spatial frequencies.

The continuous wavelet transform is invertible by

$$f(x) = \frac{1}{c_{\psi}^2} \int\int \Psi_f^{\psi}(\tau, \sigma) \frac{1}{\sigma^2} \psi\left(\frac{x-\tau}{\sigma}\right) d\tau d\sigma$$

if the admissibility constant c_{ψ} satisfies

$$c_{\psi} = \left(2\pi \int_{-\infty}^{\infty} \frac{|\hat{\psi}(\xi)|^2}{|\xi|} d\xi\right) < \infty \tag{3.61}$$

where $\hat{\psi} = \mathcal{F}\psi$ is the Fourier transform of $\psi(x)$. For the condition of eqn. (3.61) to be true, it is necessary that $\hat{\psi}(0) = 0$ which is equivalent to

$$\int f(x) dx = 0. \tag{3.62}$$

Examples of two real-valued wavelets are the Mexican hat (or Maar function) described as

$$w_{mh}(x) = A\left(1 - x^2\right) e^{-\frac{1}{2}x^2},$$

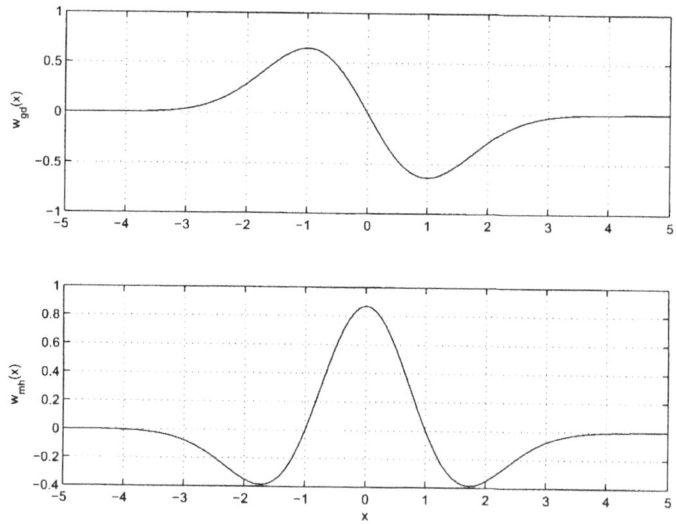

Figure 3.9: Examples of nonorthonormal mother wavelets. *Top* – Gaussian derivative. *Bottom* – Mexican hat.

and the Gaussian derivative described as

$$w_{gd}(x) = -Axe^{-\frac{1}{2}x^2}.$$

These two wavelet functions are illustrated in Figure 3.9. In this figure, the coefficient A of each wavelet is selected so that the $\mathcal{L}_2$ norm of the wavelet is equal to one. Note in particular that each of these wavelets is localized, oscillatory, and satisfies eqn. (3.62). Wavelets defined as higher order derivatives of the function $Ae^{-\frac{1}{2}x^2}$ are often considered.

For function approximation a discretized wavelet basis is selected:

$$\begin{aligned} \hat{f}(x) &= A \sum_{k=-\infty}^{\infty} \sum_{j=-\infty}^{\infty} \theta_{j,k} 2^{j/2} \psi\left(2^j x - k\right) \\ &= \sum_{k=-\infty}^{\infty} \sum_{j=-\infty}^{\infty} \theta_{j,k} \psi_{j,k}(x) \end{aligned}$$

where $\psi_{j,k}(x) = A2^{j/2}\psi\left(2^j x - k\right)$ and A is selected so that the $\mathcal{L}_2$ norm of each $\psi_{j,k}$ is one. This approximation uses an infinite basis set in the same sense as the Fourier or Taylor series involve an infinite basis set. In an application, maximum and minimum values of j are selected to define the minimum and maximum scales of resolution that are of interest. Since the region of approximation $\mathcal{D}$ is compact, at each scale of resolution, a finite range of k can be selected to cover $\mathcal{D}$. Therefore, each application involves a finite basis set,

$$\hat{f}(x) = \sum_{j=j_1}^{j_0-1} \left[\sum_{k=L_j}^{U_j} \theta_{j,k} \psi_{j,k}(x) \right]. \tag{3.63}$$

The properties of the wavelet $\psi_{j,k}$ are of obvious interest. There exist wavelet basis that are orthogonal, biorthogonal, or that form a frame. The following subsections discuss the concept of a multiresolution analysis. Readers interested in frames and biorthogonal wavelets should consult, e.g. [51, 60].

3.8.1 Multiresolution Analysis (MRA)

Consider a function $\xi \in L_2$ (called the scaling function). Dyadic dilations and translations of the scaling function are defined by

$$\xi_{j,k}(x) = 2^{j/2}\xi(2^j x - k) \tag{3.64}$$

with $j,\ k \in Z$. For any $j \in Z$, we can define a space of functions

$$\mathcal{V}_j = \left\{ f \in L_2 \,\middle|\, f(x) = \sum_{k=-\infty}^{\infty} \theta_{j,k}\xi_{j,k}(x) \right\}. \tag{3.65}$$

For certain scaling functions it is possible to define a multiresolution analysis.

Definition 3.8.1 *A multiresolution analysis with scaling function ξ consists of a sequence of successive approximation closed subspaces $\mathcal{V}_j$*

$$\ldots \subset \mathcal{V}_{-1} \subset \mathcal{V}_0 \subset \mathcal{V}_1 \subset \ldots \tag{3.66}$$

with the following properties:

Density.

$$\bigcup_{j\in Z} \mathcal{V}_j \text{ is dense in } L_2(R) \tag{3.67}$$

Separation.

$$\bigcap_{j\in Z} \mathcal{V}_j = \{0\} \tag{3.68}$$

Orthonormality.

$$\{\xi_{0,n}; n \in Z\} \text{ is an orthonormal basis for } \mathcal{V}_0 \tag{3.69}$$

Scaling.

$$f(x) \in \mathcal{V}_j \Leftrightarrow f(2x) \in \mathcal{V}_{j+1}, \quad j \in Z. \tag{3.70}$$

The density property implies that any $f \in L_2$ can be approximated to any specified accuracy $\epsilon > 0$ if j is sufficiently large. The separation property states that the function that is identically zero is the only function common to all the spaces $\mathcal{V}_j$. This property is necessary for functions to have a unique representation under the direct summation operator $\oplus$. The orthonormality property requires that the scaling function be orthonormal to each of its integer translations:

$$\langle \xi_{0,0}, \xi_{0,k} \rangle = \int \xi(x)\xi(x-k)dx = \delta_{0,k}, \text{ for all } k \in Z \tag{3.71}$$

where

$$\delta_{n,m} = \begin{cases} 1 & \text{if } n = m \\ 0 & \text{otherwise.} \end{cases}$$

When this orthonormality condition is satisfied, then for a function $g \in L_2$ if we wish to minimize $\|f - g\|$ for $f \in \mathcal{V}_j$, the optimal value for the parameter $\theta_{j,k}$ in eqn. (3.65) is defined uniquely by the Fourier coefficient:

$$\theta_{j,k} = \int g(x)\xi_{j,k}(x)dx.$$

The main advantage of orthonormality is that the computation of the k-th coefficient of the expansion of a function g in $\mathcal{V}_j$ is independent of the the i-th coefficient or basis function of that space. This greatly simplifies computations.

EXAMPLE 3.19

The simplest scaling function to satisfy these conditions is the characteristic function on the unit interval

$$\xi_H(x) = \chi_{[0,1)}(x) = \begin{cases} 1 & \text{if } x \in [0,1) \\ 0 & \text{otherwise.} \end{cases}$$

With this scaling function, $\mathcal{V}_0$ is the set of functions that are piecewise constant between integers. The space $\mathcal{V}_j$ is the set of functions that are piecewise constant on each interval $[\frac{n}{2^j}, \frac{n+1}{2^j})$. Since the functions that are piecewise constant on the half integer intervals includes the set of functions that are piecewise constant on the integer intervals, it is clear that $\mathcal{V}_0 \subset \mathcal{V}_1$. Repetition of this reasoning can verify the nesting condition of eqn. (3.66). Direct integration of

$$\int_{-\infty}^{-\infty} \xi_H(x)\xi_H(x-j) = \int_{-\infty}^{-\infty} \chi_{[0,1)}(x)\chi_{[0,1)}(x-j) = \chi_{[0,1)}(x)\chi_{[j,j+1)}(x) = 0$$

shows that the orthonormality condition is satisfied. △

The MRA definition shows that $\{\xi_{j,n}\}_{n\in Z}$ is an orthonormal basis for $\mathcal{V}_j$. The fact that the $\mathcal{V}_j$ are dense in L_2 means that $\{\xi_{j,n}\}_{j,n\in Z}$ is a basis for L_2; however, the elements of $\{\xi_{j,n}\}_{n\in Z}$ are not necessarily orthonormal to the elements of $\{\xi_{k,n}\}_{n\in Z}$ for $k \neq j$. Therefore, the dilations and translations of the scaling function do not provide an orthonormal basis for L_2.

If we define $\mathcal{W}_j$ to be the orthogonal complement of $\mathcal{V}_j$ in $\mathcal{V}_{j+1}$, then

$$\mathcal{V}_{j+1} = \mathcal{V}_j \oplus \mathcal{W}_j. \tag{3.72}$$

In particular, we will call the function ψ the wavelet generated by the scaling function ξ, if its translates are mutually orthonormal (i.e., $\int \psi(x-k)\psi(x-m)dx = \delta_{k,m}$ for all $k, m \in Z$), are orthogonal to $\xi_{0,n}$ (i.e., $\int \psi(x-k)\xi(x-n)dx = 0$ for all $k, n \in Z$), and form basis for $\mathcal{W}_0$. The wavelets of resolution j are then defined as

$$\psi_{j,k}(w) = 2^{j/2}\psi(2^j w - k), \quad j,k \in Z. \tag{3.73}$$

The set $\{\psi_{j,k}\}_{k\in Z}$ is an orthonormal basis for $\mathcal{W}_j$. Fortunately, whenever a MRA exists, there exist a wavelet that can be constructed from the scaling function. This construction process is not straightforward. For details on the construction of the wavelets, the reader is referred to [60].

3.8.2 MRA Properties

It follows from the above that

$$L_2(R) = \cdots \oplus \mathcal{W}_{-1} \oplus \mathcal{W}_0 \oplus \mathcal{W}_1 \oplus \cdots$$

That is, the orthonormal wavelet basis generates an orthogonal decomposition of the L_2 space. The following uniform approximation property can be easily verified from the above discussion. It states that any L_2 function can be uniformly approximated using a orthogonal wavelet series.

Theorem 3.8.1 *Any function $f \in L_2(R)$ has the following* unique *series representation*

$$f(w) = \sum_{j=-\infty}^{\infty} \sum_{k=-\infty}^{\infty} \langle f, \psi_{j,k} \rangle \psi_{j,k}(w). \tag{3.74}$$

The above doubly bi-infinite series converges with respect to L_2 norm, that is,

$$\lim_{L_1, L_2, U_1, U_2 \to \infty} \left\| f(w) - \sum_{j=-L_1}^{U_1} \sum_{k=-L_2}^{U_2} \langle f, \psi_{j,k} \rangle \psi_{j,k}(w) \right\| \to 0.$$

The series representation in (3.74) is called a wavelet series, and the coefficients $< f, \psi_{j,k} >$ of the series expansion are called the wavelet coefficients. Note that the (optimal) wavelet coefficients are Fourier coefficients. For the applications of interest in this book, the Fourier coefficients cannot be calculated directly from the inner product since the function f is not known.

The above properties indicate that any function $f(w) \in L_2$ can be written as a unique linear combination of orthogonal wavelets of different resolutions. That is, we can write

$$f(w) = \cdots + g_{-1}(w) + g_0(w) + g_1(w) + \cdots \tag{3.75}$$

where $g_j \in \mathcal{W}_j$ is unique, and $\langle g_i, g_j \rangle \propto \delta_{i,j}$. While many other functional approximators, have the universal approximation property, only wavelets have both the multi-resolution and orthogonal decomposition properties.

■ **EXAMPLE 3.20**

The *Haar wavelet* generated from the Haar scaling function is

$$\psi_H(w) = \begin{cases} 1 & \text{for} \quad 0 \le w < \frac{1}{2}; \\ -1 & \text{for} \quad \frac{1}{2} \le w < 1; \\ 0 & \text{otherwise}. \end{cases} \tag{3.76}$$

To see this, note first that $\psi_{H_{j,k}}(w) = 2^{j/2}\psi_H(2^j w - k); j, k \in Z$. Next, for any $n, k \in Z$ we have $\langle \psi_{H_{0,k}}, \xi_H(x-n) \rangle = 0$. In addition, $\langle \psi_{H_{j,k}}, \psi_{H_{l,m}} \rangle = \delta_{j,k}\delta_{l,m}$. Finally, $\psi_{H_{0,k}}$ are a basis for $\mathcal{W}_0$.

At this point, it is of interest to compare approximation of a function using wavelets with approximation by other methods, for example, with splines. Consider the Haar basis and zeroth order splines. In fact, the Haar scaling function is a first order spline. In the spline expansion, a function is approximated using a series of translates of a rectangular box of a given width, and the coefficients of the expansion are the averages the function taken over the support of the boxes. In the wavelet approach, the scaling function and its translates will be used in conjunction with the wavelets that it generates. The scaling function for the Haar basis captures the average (low frequency) behavior, while the wavelets capture the variation (higher frequency) behavior of the function. An advantage of the wavelet approach is that due to the orthogonality and local support of the basis functions, new basis elements can be added locally as needed without affecting the values of the preexisting basis functions. △

The Haar basis wavelet function is used throughout this section as it is straightforward to understand and is useful for illustration of wavelet concepts. Applications where there are constraints, such as smoothness of the approximation, motivate the use of other orthogonal wavelets with compact support. Alternative wavelets have been formulated in literature. For example, a class of orthogonal wavelets called Daubechies wavelets also are compactly supported. Further, there are classes of orthogonal wavelets such as Meyer wavelets, and Battle-Lemarie wavelets that vanish rapidly outside a compact support. For a detailed study of orthogonal wavelets, the reader is referred to [60].

Let us consider the problem off approximating the function f over the compact set $\mathcal{D}$. Let ψ be a compactly supported orthogonal wavelet and ξ, the associated scaling function as defined in Section 3.8.1. Let $\mathcal{V}_j$ be an MRA and $\mathcal{W}_j$ be as defined in eqn. (3.72). If $\xi(w)$ is the scaling function that generates the space V_j with $\xi_{j,k}(w)$ defined as in (3.64), then from properties (3.66) and (3.67) of the MRA, we get that given any $\epsilon > 0$, there exist an integer j_0, and a function $\hat{f}(w;\mu)$ given by

$$\hat{f}(w;\mu) = \sum_{k=-\infty}^{\infty} \mu_k \xi_{j_0,k}(w)$$

such that

$$\|f(w) - \hat{f}(w;\mu)\| < \epsilon$$

with $\mu^T = (\cdots, \mu_{-1}, \mu_0, \mu_1, \cdots)$. Since $\mathcal{D}$ is a compact set, if the scaling function has a compact support, we can write

$$\hat{f}(w;\mu) = \sum_{k=L}^{U} \mu_k \xi_{j_0,k}$$

for some $L, U \in Z$.

From (3.66)–(3.69) and (3.72) we have

$$\begin{aligned} \mathcal{V}_{j_0} &= \cdots \oplus \mathcal{W}_{j_0-2} \oplus \mathcal{W}_{j_0-1} \\ &= \mathcal{V}_{j_1} \oplus \mathcal{W}_{j_1} \oplus \mathcal{W}_{j_1+1} \oplus \cdots \oplus \mathcal{W}_{j_0-1}, \quad j_1 < j_0 - 1. \end{aligned} \tag{3.77}$$

Hence we can write the approximation for f uniquely as

$$\hat{f}(w;\theta) = \sum_{j=j_1}^{j_0-1} \left[\sum_{k=L_j}^{U_j} \theta_{j,k} \psi_{j,k} \right] + \sum_{k=L_{j_1-1}}^{U_{j_1-1}} \theta_{j_1-1,k} \xi_{j_1,k}. \tag{3.78}$$

The summation inside the square brackets is carried out over orthogonal wavelet translates of a particular resolution. The left summation is carried out over resolutions higher than j_1.

The summation involving $\xi_{j_1,k}$ is carried over orthogonal translates of the scaling function, at the lower resolution level, j_1. Thus (3.78) can be seen as reflecting the fact that any function in L_2 can be decomposed into a scaling function of resolution j_1 and wavelets of higher resolution, with the highest resolution depending upon the desired accuracy of approximation. The analysis leading to (3.78) can be carried out for any wavelet with compact support, with the unique decomposition being a direct sum rather than an orthogonal sum. However, an explicit use of orthogonality is needed for the next step.

The accuracy of approximation can be improved by increasing j_0. Due to the orthogonality of the scaling and wavelet functions, the new approximation is obtained from the existing approximation, simply by adding more basis functions, and evaluating the coefficients corresponding to the new basis elements. The coefficients of the existing basis functions remain the same. New basis elements need to be added only where the function varies rapidly. Care must also be taken, since, as the resolution increases it may be difficult to obtain enough samples to accurately estimate the parameters corresponding to the high resolution wavelets.

3.9 FURTHER READING

This chapter has briefly introduced various approximation structures. Several of these structures have entire books or journals devoted to their study. Therefore, we have only touched the surface in this chapter. Sample references, in addition to those included directly in the text, to publications providing additional information about specific approximator structures are: polynomials and splines [52, 53, 55, 56, 57, 59, 62, 71, 238, 240], CMAC [1, 2, 4, 125, 142, 170, 187, 195], fuzzy logic [15, 21, 63, 67, 131, 150, 182, 184, 198, 201, 285, 283, 284, 302, 303, 304], radial basis functions [30, 31, 35, 79, 193, 204, 205, 219, 232, 290], neural networks [88, 108, 109, 110, 117, 152, 172, 186, 188, 190, 203, 211, 223, 226, 280, 287, 298], and wavelets [22, 37, 51, 60, 260, 281, 306].

3.10 EXERCISES AND DESIGN PROBLEMS

Exercise 3.1 The purpose of this exercise is to exhibit the effect that spatially localized training samples can have on an approximator composed of basis elements with global support.

Consider the approximation of the function $f(x) = \sin(\pi x)$ over the interval $\mathcal{D} = [-1, 1]$ by a third order polynomial. The approximator is $\hat{f}(x) = \sum_{i=0}^{3} \theta_i \phi_i(x)$ where the basis functions are the first four Legendre polynomials defined in eqn. (3.10). The parameter vector $\theta_o = [\theta_0, \ldots, \theta_3] = [0.0000, 0.9549, 0.0000, -1.1582]$ is the least squares optimal set of parameters over $\mathcal{D}$ after truncation to four decimal places. This parameter vector results in the $\mathcal{L}_2$ approximation error

$$\left[\int_{-1}^{1} \left(f(x) - \hat{f}(x)\right)^2 dx\right]^{-\frac{1}{2}} = 0.0937.$$

The $\mathcal{L}_\infty$ approximation error over $\mathcal{D}$ is about 0.2.

1. Numerically computed the $\mathcal{L}_2$ approximation error over $[-1.0, 1.0]$ and over $[0.5, 1.0]$.

2. In control applications, the system may operate in the vicinity of any given operating point for an extended period of time. This results in training samples arriving from a

small subset of the domain $\mathcal{D}$ for that period of time. In this exercise, we will simulate this by selecting training samples only from the region $\mathcal{D}_1 = [0.5, 1.0]$. Randomly generate 1000 training points x_i in $\mathcal{D}_1$. At each x_i, compute the (noise free) value of $f(x_i) = \sin(\pi x_i)$. Update the approximation parameter vector using recursive least squares as defined by eqns. (2.23) and (2.24). Initialize the parameter vector as θ_o and $P_k = A_k^{-1} = I$. Use uniform weights $w_k = 1$. Save the sequence $\hat{\theta}_i$ for $i = 100, 200, \ldots, 1000$.

3. Using $\hat{\theta}_i$ for $i = 100, 200, \ldots, 1000$, compute the $\mathcal{L}_2$ approximation error over $[-1.0, 1.0]$ and over $[0.5, 1.0]$. Plot these values versus the training iteration i.

You should see the $\mathcal{L}_2$ error over $[-1.0, 1.0]$ increasing (not monotonically) and the $\mathcal{L}_2$ error over $[0.5, 1.0]$ decreasing. Why? How would an approximator with locally supported basis elements perform differently?

Exercise 3.2 Select a low order polynomial function such as $f(x) = 1 + x$. Although this is a polynomial function, assume that its functional form is not known and that this unknown function is to be approximated based on noise corrupted measured data. Let m be an integer value varying from the order of $f(x)$ to approximately 10. For each value of m:

1. Generate $m+1$ noise corrupted "measurements" at $x_i = i * \frac{1}{m}$ for $i = 0, \ldots, m$ by evaluating $f(x_i)$ and adding a small amount of random noise (e.g., Gaussian random noise with standard deviation $\sigma = 0.1$). Denote the vector of these measurements by $\tilde{y}$.

2. Fit an m-th order polynomial to the measured data $\{(x_i, \tilde{y}_i)\}_{i=0}^{m}$. Note that this is an interpolation problem. Use the natural polynomial basis $\phi_m(x) = [1, x, \ldots, x^m]$. Let θ_m denote the resulting set of parameters such that $\tilde{y}_i = \phi_m(x_i)\theta_m$.

3. Generate a new set of evaluation points (e.g., $x = [0, 0.01, \ldots, 0.99, 1]$. Evaluate both the original polynomial $f(x) = 1 + x$ and the approximated polynomial $p_m(x) = \phi_m(x)\theta_m$ at each of these evaluation points. Plot x versus both f and p_m. What happens as the order of the interpolating polynomial m increases?

Exercise 3.3 Repeat Exercise 3.2, but use alternative choices of basis functions. Include at least one choice of basis functions that are defined to form a partition of unity.

Exercise 3.4 Select a low order polynomial function such as $f(x) = 1 + x$. For each $n = 11, \ldots, 100$:

1. Generate a set of evaluation points defined as $x_i = i * \frac{1}{n}$ for $i = 0, \ldots, n$.

2. Generate a set of noise corrupted "measurement" data $\tilde{y}_i = f(x_i) + \nu_i$ where ν_i is Gaussian random noise with standard deviation $\sigma = 0.1$.

3. Find $\theta_{10}(n)$ to result in a least squares fit of a tenth order polynomial $p_{10}(x) = \phi_{10}(x_i)\theta_{10}(n)$ to the measurement data $\{(x_i, \tilde{y}_i)\}_{i=0}^{n}$, where $\phi_{10}(x)$ is a basis for the space of 10-th order polynomials defined on $[0, 1]$. Note that this is an approximation, not an interpolation, problem.

4. Evaluate the approximation accuracy defined by the $\mathcal{L}_2$ norm of the approximation error

$$e(n) = \int_0^1 (f(x) - \phi_{10}(x_i)\theta_{10}(n))^2 dx.$$

5. Evaluate the sample variance of the approximation error

$$v(n) = \sqrt{\frac{1}{n}\sum_{i=1}^{n}\left(f(x_i) - \phi_{10}(x_i)\theta_{10}(n)\right)^2}$$

as a function of n.

Plot $e(n)$ and $v(n)$ versus n.

Exercise 3.5 Repeat Exercise 3.4, but use alternative choices of basis functions. Use at least one choice of basis functions that are defined to form a partition of unity. Keep the dimension of the basis vector fixed at 11.

Exercise 3.6 Write a program to interpolate the function

$$f(x) = \frac{1}{1+x^2},$$

at the points $x_i = -5 + 10\left(\frac{i}{m}\right)$ for $i = 0, \ldots, m$, using an m-th order polynomial. For each value of m, denote the interpolating polynomial by $p_m(x)$. Use odd values of $m \in [3, 21]$. For each value of m and for $x \in [-5, 5]$: (1) plot f and $p_m(x)$ versus x; and, (2) plot the error $E(x) = f(x) - p_m(x)$ versus x. Be certain that each plot includes several evaluation points between each pair of interpolation points. Numerically compute

$$e(m) = \int_{-5}^{5} |f(x) - p_m(x)| dx.$$

Plot $e(m)$ versus m. (See Section 3.6 in [240] for a discussion of issues related to this exercise.)

Exercise 3.7 Repeat Exercise 3.6, but use an alternative choice of basis functions (e.g., splines or radial basis functions) that are defined to form a partition of unity.

Exercise 3.8 Consider the problem of estimating the vector θ to minimize the two-norm of the error between Y and $\Phi^\top\theta$ subject to the constraint that $G\theta = b$, where $Y \in \Re^M$ is known, $\Phi \in \Re^{N\times M}$ is known, $\theta \in \Re^N$ is unknown, $G \in \Re^{J\times N}$ is known, and $b \in \Re^J$ is known. This is the restricted least squares problem [82, 164]. Use the method of Lagrange multipliers to show that the the optimal constrained parameter estimate is

$$\hat{\theta} = \theta_{LS} + \left(\Phi\Phi^\top\right)^{-1} G^\top \left(G\left(\Phi\Phi^\top\right)^{-1} G^\top\right)^{-1} (b - G\theta_{LS}).$$

Exercise 3.9 In Example 3.15 , confirm eqn. (3.49).

CHAPTER 4

PARAMETER ESTIMATION METHODS

This chapter has three objectives: the formulation of parametric models for the approximation problem; the design of online learning schemes; and the derivation of parameter estimation algorithms with certain stability and robustness properties. The perspective of this chapter is motivated in Section 4.1, where we use examples to develop some intuition into the adaptive approximation problem for unknown nonlinear functions that appear in the state equation model of a dynamical system. This section includes a formal definition of the adaptive approximation problem and a discussion of various key issues in parametric estimation. In the subsequent sections of this chapter, we describe in detail the procedure for designing online learning algorithms, which consists of three steps: (i) derivation of parametric models; (ii) design of online learning scheme; and (iii) derivation of parameter estimation algorithms. The overall learning approach is developed in a continuous-time framework, where it is assumed that the original dynamical system as well as the adaptive law evolve in continuous-time. The focus on this chapter is parameter estimation methods for adaptive function approximation, not adaptive approximation based control. The methods that are developed here will provide a foundation for the adaptive approximation based control approaches that are developed in Chapters 6 and 7.

Section 4.2 considers the derivation of parametric models. The objective in deriving a suitable *parametric model* is to rewrite the nonlinear differential equation model in a structured way such that the uncertainty appears in a desired fashion. Specifically, any unknown functions in the state variable model are replaced by approximators (potentially, of any form described in Chapter 3), such that the uncertainty is now converted into two components that will be treated differently:

Adaptive Approximation Based Control: Unifying Neural, Fuzzy and Traditional Adaptive Approximation Approaches. By Jay A. Farrell and Marios M. Polycarpou

- parameter uncertainty – unknown "optimal" weights of the approximator;
- functional approximation error – due to the approximator not being able to represent exactly the unknown function.

Based on the derived parametric model, in Section 4.3 we consider the design of *online learning schemes*. This step constructs an architecture for adaptive approximation. The architecture is tightly related to the parametric model derived in Section 4.2. Two types of online learning schemes will be investigated: the *error filtering online learning* scheme, and the *regressor filtering online learning* scheme. The final step of the design procedure, described in Section 4.4, deals with deriving adaptive laws for updating the parameter estimates (weights) that reside in the function approximator.

The stability and convergence properties of the learning architecture (under certain conditions) are formally analyzed in Section 4.5. In Section 4.6, we examine the case where the functional approximation error is nonzero, or there are external time-varying disturbances and/or measurement noise terms that cannot be approximated by the adaptive approximation scheme. In this situation, we consider the modification of the learning algorithms, leading to so called *robust learning algorithms*, and consider the stability and convergence properties of robust learning schemes. Finally, Section 4.7 provides some concluding remarks.

4.1 FORMULATION FOR ADAPTIVE APPROXIMATION

This section describes the general problem of adaptive approximation. The section begins with an example, intended to illustrate the elements that must be defined in any adaptive approximation problem. Next, a series of simple examples illustrate the motivation for parameter estimation within the framework of adaptive approximation. The general adaptive approximation problem is then formulated, and the section concludes with a discussion of key issues that arise in adaptive approximation. These issues will be revisited throughout the chapter, as well as in subsequent chapters dealing with feedback control.

4.1.1 Illustrative Example

As discussed above, the design of the adaptive function approximation schemes consists of three steps: (i) the formulation of a parametric model; (ii) the design of the learning scheme; and (iii) the derivation of parameter estimation algorithms. Next, we consider an example which is intended to illustrate the three steps in the design of adaptive function approximation schemes, and also to illustrate the idea of incorporating a priori information. To avoid (at this stage) some of the complexities associated with dynamical systems, we consider the simple case of a static (memoryless) input-output system of the form

$$y(t) = f^*(u(t)), \tag{4.1}$$

where $u \in \Re^1$ and $y \in \Re^1$ are the input and output signals respectively, and $f^* : \Re^1 \mapsto \Re^1$ is an unknown function. It is assumed that $u(t)$ and $y(t)$ are available for measurement.

One method to make the problem tractable is to replace the unknown function $f^*(u(t))$ by a function approximator $\hat{f}(u(t); \theta^*, \sigma^*)$ with known structure. As discussed in Section 3.1.3, we assume that the structure of $\hat{f}$ has been selected so that there exists (unknown) parameters $\theta^* \in \Re^{q_\theta}$ and $\sigma^* \in \Re^{q_\sigma}$ such that the Minimum Functional Approximation Error (MFAE)

$$\delta(t) = f^*(u(t)) - \hat{f}(u(t); \theta^*, \sigma^*)$$

is small (in some norm sense) on a compact region $\mathcal{D} \subset \Re^1$ that is of interest. Therefore, by rewriting (4.1) we can derive a parametric model written in the form

$$\chi(t) = \hat{f}(u(t); \theta^*, \sigma^*) + \delta(t), \tag{4.2}$$

where $\chi(t) = y(t)$ can be computed from the measured signals. Note that the first step of formulating the parametric model is basically equivalent to rewriting the unknown input-output system into a function approximation model of known structure but unknown parameters (or weights).

Based on the parametric model (4.2), we design the online learning scheme as follows:

$$\hat{\chi}(t) = \hat{f}(u(t); \hat{\theta}(t), \hat{\sigma}(t)), \tag{4.3}$$

where $\hat{\theta}(t)$, $\hat{\sigma}(t)$ are the adjustable weights of an adaptive approximator. The second step, which deals with the design of the online learning scheme consists of replacing the unknown parameters in the parametric model by adjustable parameters (weights).

The third step of the design procedure deals with the derivation of an adaptive law for updating the adjustable parameters of the adaptive approximator. The adaptive law is based on the output estimation error $e(t) = \hat{\chi}(t) - \chi(t)$. By using the gradient optimization method with a simple quadratic cost function, we obtain the following adaptive laws for $\hat{\theta}(t)$ and $\hat{\sigma}(t)$:

$$\begin{aligned} \dot{\hat{\theta}} &= -\Gamma_\theta \frac{\partial \hat{f}}{\partial \hat{\theta}} e(t) \\ \dot{\hat{\sigma}} &= -\Gamma_\sigma \frac{\partial \hat{f}}{\partial \hat{\sigma}} e(t) \end{aligned}$$

where Γ_θ, Γ_σ are positive definite matrices representing the adaptive gain for the update of $\hat{\theta}$ and $\hat{\sigma}$ respectively.

The details of the design procedure, as well as the derivation of the analytical properties, are not discussed in this simple illustrative example. The objective of this chapter is to develop a systematic approach for the design and analysis of parameter estimation methods.

In the above formulation, as illustrated in Figure 4.1, $\chi(t)$ is simply equal to $y(t)$, and therefore the online learning model consists only of the adaptive approximator $\hat{f}$. As we will see, in a general setting of dynamic systems, the online learning model will also contain stable filters.

Now, consider the case where the input-output static system is partially known; i.e.,

$$y(t) = f_0(u(t)) + f^*(u(t))$$

with $f_0(u(t))$ a known function. In this case, the system can be written in the same parametric model form as (4.2):

$$\chi(t) = \hat{f}(u(t); \theta^*, \sigma^*) + \delta(t);$$

however, the measurable variable $\chi(t)$ is given by $\chi(t) = y(t) - f_0(u(t))$. Therefore, the online learning model consists of the adaptive approximator and an identifier structure containing the known component of the input-output system, as shown in Figure 4.2.

To summarize, in the design of an adaptive approximation system, the designer must specify a parametric model for the application, an online learning scheme including a signal χ that is computable from the measured variables and directly affected by the parametric

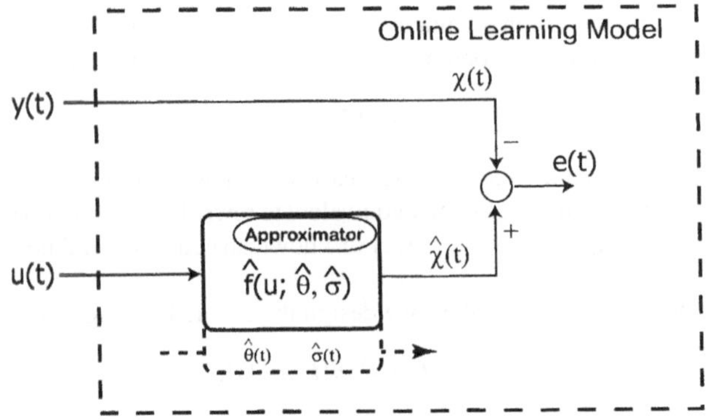

Figure 4.1: Block diagram of online learning model for the unknown static system (4.1). The dashed box underneath the approximator will contain the dynamics associated with estimation of the parameters $\hat{\theta}$ and $\hat{\sigma}$.

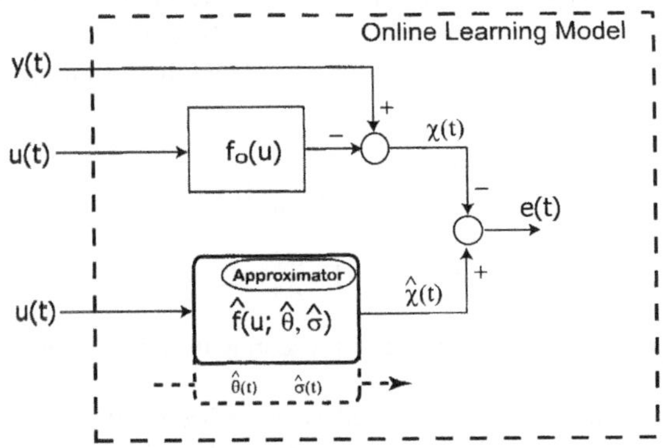

Figure 4.2: Block diagram of online learning model for a partially known static system. The dashed box underneath the approximator will contain the dynamics associated with estimation of the parameters $\hat{\theta}$ and $\hat{\sigma}$.

error, and a parameter adaptation law. One item that sometimes causes confusion and that is easily clarified at this point is that the design will typically include two equations for the signal χ. One of the equations shows the dependence of χ on the parametric error. The other equation shows the method of computation of χ using measured signals in the system.

4.1.2 Motivating Simulation Examples

In this section we consider three simple scalar examples to motivate the use of adaptive approximation. In the first example, the system is a linear model with two unknown parameters. The second example deals with a nonlinear system with an unknown parameter, while the nonlinearity is known. Finally, in the third example we consider a scalar system with an unknown nonlinearity, which is approximated online using a radial basis function network. In these examples we do not include the details of the design and analysis procedure for adaptive approximation, which are presented later in the chapter.

EXAMPLE 4.1

Consider the linear model

$$\dot{y} = ay + bu$$

where $u(t)$ is the input, $y(t)$ is the output and a, b are unknown parameters to be estimated online. In the parameter estimation and adaptive control literature there are various parametric models that have been proposed. We consider the following parametric model and online learning scheme:

$$\begin{aligned} y &= \frac{1}{s+\lambda}[(a+\lambda)y + bu] \\ \hat{y} &= \frac{1}{s+\lambda}[(\hat{a}+\lambda)y + \hat{b}u], \end{aligned}$$

where $\lambda > 0$ is a design constant. In the above formulation, we use the notation $y = H(s)[x]$, where $y(t)$ is the output of a linear system represented by the transfer function $H(s)$ with $x(t)$ as input (see Figure 4.3). Although this notation mixes the time signals $x(t)$, $y(t)$ with the Laplace based transfer function $H(s)$, it turns out to be quite convenient in describing filtering schemes and therefore is used extensively in the adaptive control literature of continuous-time systems [119, 179, 235] and in the remainder of this book. If the initial conditions of the filter are non-zero then there will be an additional term for the initial conditions, however, for simplicity here we assume that the initial conditions are set to zero.

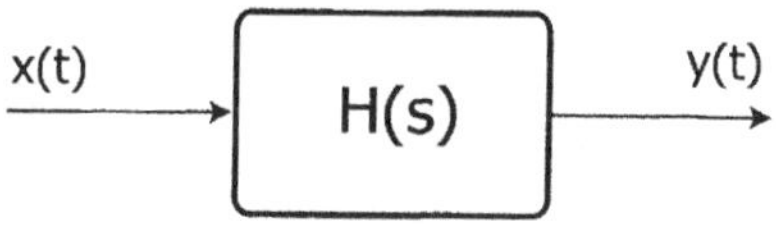

Figure 4.3: Block diagram of the notation $y = H(s)[x]$.

Let $e = \hat{y} - y$ be the output estimation error. The update laws for $\hat{a}$, $\hat{b}$ are generated as follows, based on the so-called Lyapunov synthesis method, which will be described later in the chapter:

$$\begin{aligned} \dot{\hat{a}} &= -\gamma_1 e y \\ \dot{\hat{b}} &= -\gamma_2 e u \end{aligned}$$

where γ_1, γ_2 are positive design constants representing the adaptive gain for the update algorithm of $\hat{a}(t)$ and $\hat{b}(t)$

A simulation example using the above identification scheme is shown in Figure 4.4. We consider two input scenarios. In the first case, $u(t) = \sin(2\pi t)$ and in the second case $u(t) = 3\exp(-t/100)$. For simulation purposes, the unknown parameters are assumed to be $a = -1$, $b = 1$, while the design constants are set to: $\lambda = 2$, $\gamma_1 = \gamma_2 = 10$. The top two plots of Figure 4.4 show the results for $u(t) = \sin(2\pi t)$, while the bottom two are for the second case of $u(t) = 3\exp(-t/100)$. As seen from the plots, in both cases the output estimation error converges to zero. In fact, in the second case the output estimation error converges to zero faster, as compared to the first case. However, the parameter estimates converge to their true values (-1 and 1, respectively), only in the first case, where $u(t) = \sin(2\pi t)$. This is related to the

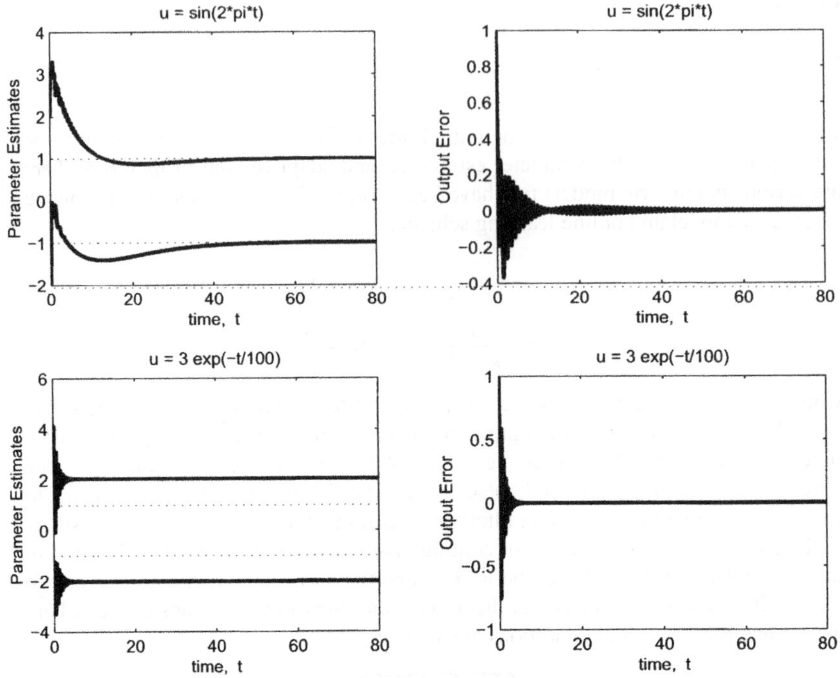

Figure 4.4: Simulation results for Example 4.1. The top two plots show the results for $u(t) = \sin(2\pi t)$, while the bottom two are for the case of $u(t) = 3\exp(-t/100)$. The left plots show the parameters estimates $\hat{a}(t), \hat{b}(t)$ and the right plots show the output estimation error $e = \hat{y} - y$.

fact that, for this problem, the input $u(t) = \sin(2\pi t)$ is a persistently exciting signal, while the signal $u(t) = 3\exp(-t/100)$ is not persistently exciting. The concept of persistency of excitation will be discussed in Section 4.5.4. This example, illustrates the fact that convergence of the output estimation error to zero does not necessarily imply that the parameter estimation error will also converge to zero. It is important to note, however, that convergence of the parameter estimates to their true values is often not a required property of parameter estimation and adaptive approximation tasks.

■ EXAMPLE 4.2

Consider the scalar nonlinear model

$$\dot{y} = af(y) + u$$

where $f(y)$ is a known function, and a is an unknown parameter to be estimated online. We consider the following parametric model and online learning scheme, respectively:

$$
\begin{aligned}
y &= \frac{1}{s+\lambda}[af(y) + \lambda y + u] \\
\hat{y} &= \frac{1}{s+\lambda}[\hat{a}f(y) + \lambda y + u]
\end{aligned}
$$

where $\lambda > 0$ is a design constant. It is noted that the above parametric model and online learning scheme can also be expressed in state-space form as

$$
\begin{aligned}
\dot{y} &= -\lambda y + af(y) + \lambda y + u \\
\dot{\hat{y}} &= -\lambda \hat{y} + \hat{a}f(y) + \lambda y + u.
\end{aligned}
$$

Later in this chapter, we will discuss in more detail the derivation of parametric models and online learning schemes in both an input-output form as well as in state-space form. Based on the above formulation, a stable update law (or adaptive law) for $\hat{a}$ is given by

$$\dot{\hat{a}} = \gamma\,(y - \hat{y})\,f(y)$$

where $\gamma > 0$ is the adaptive gain.

A simulation example using the above identification scheme is shown in Figure 4.5. Again, we consider two input scenarios. In the first case, $u(t) = 10\sin(2\pi t)$ and in the second case $u(t) = 0.2e^{-2t}$. The unknown parameter a is set to $a = 1$, while $f(y)$ is assumed to be $f(y) = e^{-y} - 1$. The design constants are set to: $\lambda = 2$, $\gamma = 10$. The top two plots of Figure 4.5 show the parameter estimate and the output estimation error for the case of $u(t) = 10\sin(2\pi t)$, while the bottom plots show the corresponding results for $u(t) = 0.2e^{-2t}$. As seen from the plots, in both cases the output estimation error converges to zero, while the parameter estimation error converge to zero only for the first case. Again, this is related to the fact that the first input is continuing to change over time (persistently exciting), thus allowing the accurate estimation of the unknown parameter.

EXAMPLE 4.3

Consider the nonlinear model

$$\dot{y} = h(y) + u$$

where $h(y)$ is an unknown function to be estimated online. In this example, we build upon the parameter estimation method of the previous two examples to develop a simple adaptive approximation scheme. The parametric model is chosen as follows:

$$\dot{y} = -\lambda y + \hat{h}(y;\theta^*) + \lambda y + u + \delta(y),$$

where $\hat{h}(y;\theta^*)$ is an adaptive approximator (potentially, any of the approximation models described in Chapter 3), θ^* is a vector of (unknown) optimal parameters (weights), λ is a positive design constant, and $\delta(y) = h(y) - \hat{h}(y,\theta^*)$ is the minimum functional approximation error (MFAE). For simplicity, we assume the use of a linearly parameterized approximator; therefore, $\hat{h}$ is of the form

$$\hat{h}(y,\theta) = \sum_{i=1}^{q_\theta} \theta_i \phi_i(y),$$

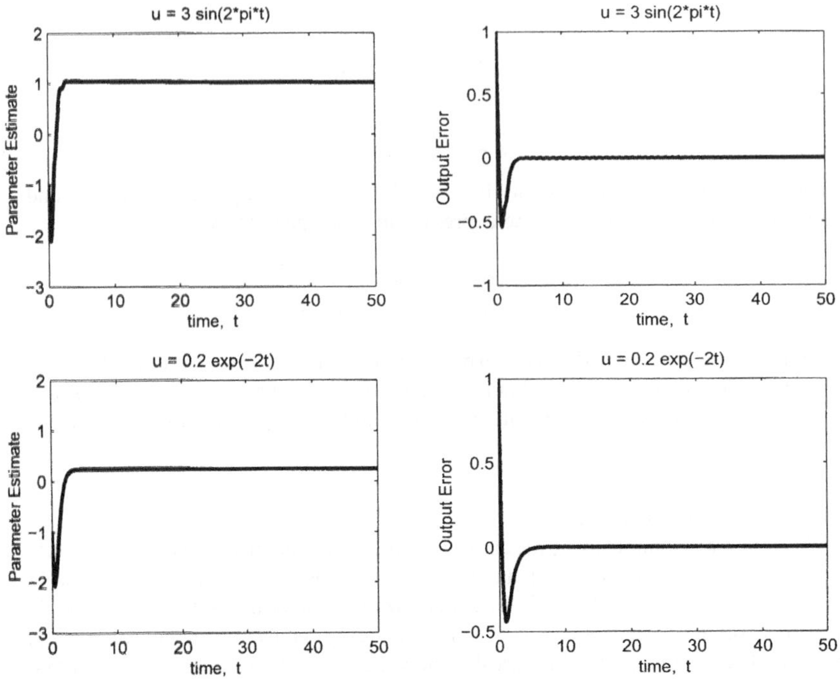

Figure 4.5: Simulation results for Example 4.2. The top two plots show the results for $u(t) = 10\sin(2\pi t)$, while the bottom two are for the case $u(t) = 0.2e^{-2t}$. The left plots show the parameters estimate $\hat{a}(t)$, while the right plots show the output estimation error $e(t) = \hat{y}(t) - y(t)$.

where θ_i is the i-th estimated parameter, ϕ_i is the i-th basis function, and q_θ is the number of basis functions. Therefore, the parametric model can be rewritten as

$$\dot{y} = -\lambda y + \sum_{i=1}^{q_\theta}(\theta_i^* \phi_i(y)) + \lambda y + u + \delta(y).$$

Based on this parametric model, the online learning scheme is given by

$$\dot{\hat{y}} = -\lambda \hat{y} + \sum_{i=1}^{q_\theta}(\hat{\theta}_i \phi_i(y)) + \lambda y + u,$$

where $\hat{\theta}$ is the estimated parameter vector and $\hat{y}$ is used to generate the output estimation error $e(t) = \hat{y}(t) - y(t)$. Using the Lyapunov synthesis method, the update laws for $\hat{\theta}_i$ are given by

$$\dot{\hat{\theta}}_i = -\gamma_i e \phi_i(y), \qquad i = 1, \ldots q_\theta,$$

where $\gamma_i > 0$ is the adaptive gain.

A simulation example using the above adaptive approximation scheme is shown in Figure 4.6. The unknown nonlinearity h is assumed (for simulation purposes) to

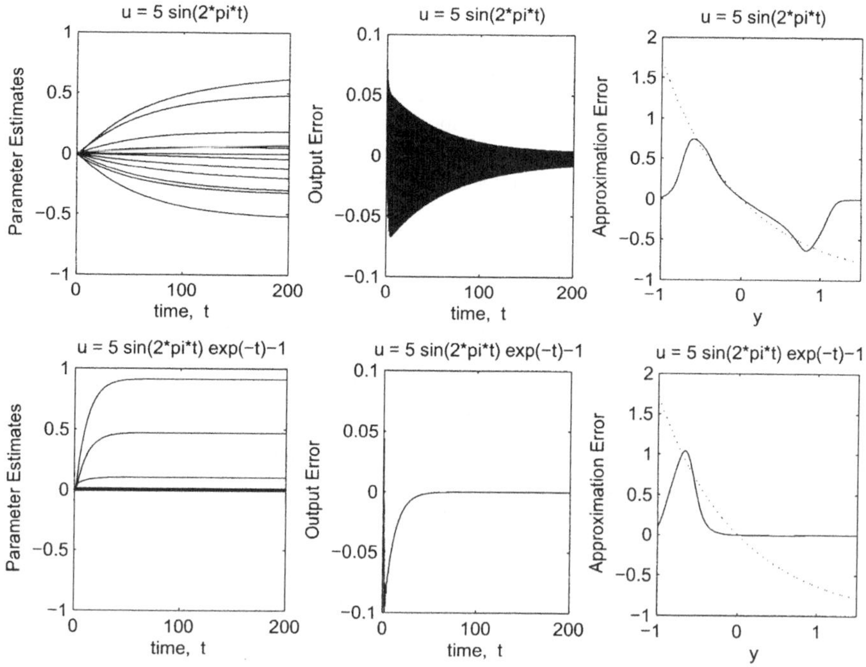

Figure 4.6: Simulation results for Example 4.3. The top three plots show the results for $u(t) = 5\sin(2\pi t)$, while the bottom three are for the case $u(t) = 5e^{-t}\sin(2\pi t)$-1. The left plots show the parameter estimates $\hat{\theta}_i(t)$ for $1 \leq i \leq 12$, while the middle plots show the output estimation error $e(t) = \hat{y}(t) - y(t)$. The right plots show the approximation error by depicting $h(y)$ (dotted line) and the approximation $\hat{h}(y; \hat{\theta}(t))$, evaluated at $t = 200$ (solid line).

be $h(y) = e^{-y} - 1$. The adaptive approximator is a Radial Basis Function (RBF) network with 12 basis functions, where each basis function is a Gaussian function of the form

$$\phi_i(y) = e^{-(y-c_i)^2/\sigma^2},$$

where c_i is the center of the basis function and σ is the width. We assume that $\sigma = \sqrt{2}/10$ and the centers are fixed and uniformly distributed between $[-1\ 1]$. Again, we consider two input scenarios. In the first case, $u(t) = 5\sin(2\pi t)$ and in the second case $u(t) = 5e^{-t}\sin(2\pi t) - 1$. In the second case, the input signal is similar to the first signal with the exception that its variation decays to zero over time. The final value of u is 1 and the corresponding final value of y is -0.693. The design constants are set to: $\lambda = 10$, $\gamma_i = 1$ for all $1 \leq i \leq 12$. The top three plots of Figure 4.6 show the parameter estimates, the output estimation error and the approximation error at the end of the simulation for the case of $u(t) = 5\sin(2\pi t)$, while the bottom three plots show the corresponding results for $u(t) = 5e^{-t}\sin(2\pi t) - 1$. The approximation plots (last plots on the right) show the function $h(y)$ (dotted line) and its adaptive approximator $\hat{h}(y; \hat{\theta}(200))$ (solid line), which denotes the approximation function at time $t = 200$. It is noted that $t = 200$ also coincides with the end of the simulation run, by which time the parameter estimates have pretty much converged to their final values (see the plots on the left). As seen from the plots, in both cases the output estimation error (middle plots) converges toward zero. On the other hand, the approximation error for the first input case becomes close to zero within the range $-0.5 \leq y \leq 0.8$; while for the second input case the approximation error is zero at $y = -0.693$, but remains close to its initial values for $y > -0.1$. Basically, for the second input case there is little learning, even though the output estimation error goes to zero at a specific point and the parameter estimates converge to certain values. In reality, the system does learn, albeit only the single point $y == -0.693$, which is to which the output variable $y(t)$ converges. In the first input case, the output variable $y(t)$ ends up oscillating in a sinusoidal fashion between approximately -0.5 and 0.8, which is the reason that the approximation error is very small in this region. On the other hand, since the learning scheme does not experience any any values of y outside the range $-0.5 \leq y \leq 0.8$, it does not learn anything outside this range and, in fact, the approximator remains close to its initial value.

The above three simulation examples, although quite simple, illustrate nicely some of the properties and issues encountered in adaptive approximation. For example, we note that even though the output estimation error goes to zero, this does not necessarily imply that the parameter estimates converge to their optimal values. We also saw that the approximation error becomes small only in the region in which the input to the approximator varies. This is related to the issue of persistency of excitation, which is discussed Section 4.5.4.

4.1.3 Problem Statement

The adaptive approximation problem can be summarized as follows.

Adaptive Approximation Problem. Given an input/output system containing unknown nonlinear functions, the adaptive approximation problem deals with the design of online learning schemes and parameter adaptive laws for approximating the unknown nonlinearities.

The overall design procedure for solving the adaptive approximation problem consists of the following three steps:

1. Derive a *parametric model* by rewriting the dynamical system in the form

$$\chi(t) = W(s)[\hat{f}(z(t), \theta^*, \sigma^*)] + \delta(t) \tag{4.4}$$

   where $\chi(t) \in \Re^n$ is a vector that can be computed from available signals, $W(s)$ is a known transfer function (in the Laplace s-domain) of dimension $n \times p$, the vector function $\hat{f} : \Re^m \times \Re^{q_\theta} \times \Re^{q_\sigma} \mapsto \Re^p$ represents an adaptive approximator, $z(t) \in \Re^m$ is the input to the adaptive approximator, $\theta^* \in \Re^{q_\theta}$ and $\sigma^* \in \Re^{q_\sigma}$ are unknown "optimal" weights for the adaptive approximator, and $\delta(t) \in \Re^n$ is a possibly filtered version of the unknown Minimum Functional Approximation Error (MFAE) $e_f(x(t))$.

2. Design a *learning scheme* of the form

$$\begin{aligned}\hat{\chi}(t) &= \mathcal{L}(z(t), \hat{\theta}(t), \hat{\sigma}(t)) \\ e(t) &= \hat{\chi}(t) - \chi(t)\end{aligned}$$

where $\hat{\theta}(t), \hat{\sigma}(t)$ are adjustable weights of the adaptive approximator, $\mathcal{L}$ is the structure of the learning scheme, and $\hat{\chi}(t)$ is an estimate of $\chi(t)$ which is used to generate the output estimation error $e(t)$. The output estimation error $e(t)$ provides a measure of how well the estimator approximates the unknown nonlinearities, and therefore is utilized in updating the parameter adaptive laws.

3. Design a *parameter adaptive law* for updating $\hat{\theta}(t)$ and $\hat{\sigma}(t)$, of the form

$$\dot{\hat{\theta}}(t) = \mathcal{A}_\theta(z(t), \chi(t), \hat{\chi}(t), \hat{\theta}(t))$$
$$\dot{\hat{\sigma}}(t) = \mathcal{A}_\sigma(z(t), \chi(t), \hat{\chi}(t), \hat{\sigma}(t))$$

where $\mathcal{A}_\theta$ and $\mathcal{A}_\sigma$ represent the right-hand side of the adaptive law for $\hat{\theta}(t)$ and $\hat{\sigma}(t)$), respectively.

The design of parametric models is discussed in Section 4.2. Design of online learning schemes is discussed in Section 4.3. The design of parameter adaptation schemes is discussed in Section 4.4 for the ideal case, and in 4.6 for the case of the presence of uncertainty. The role of the filter $W(s)$ will become clear in the subsequent presentation. For some applications, the form of the filter $W(s)$ is imposed by the structure of the problem. In other applications, the structure of the problem may purposefully be manipulated to insert the filter in order to take advantage of its beneficial noise reduction properties.

The analysis of the learning scheme consists of proving (under reasonable assumptions) the following properties:

Stable Adaptation Property. In the case of zero mismatch error (i.e., $\delta(t) = 0$), the estimation error $e(t) = \hat{\chi}(t) - \chi(t)$ remains bounded and asymptotically approaches zero (or a small neighborhood of zero).

Stable Learning Property. In the case of zero mismatch error (i.e., $\delta(t) = 0$), the function approximation error $\hat{f}(z(t), \hat{\theta}(t), \hat{\sigma}(t)) - \hat{f}((z(t), \theta^*, \sigma^*)$ remains bounded for all z in some domain of interest $\mathcal{D}$ and asymptotically approaches zero (or is asymptotically less than some threshold ϵ over $\mathcal{D}$).

Robust Adaptive and Learning Properties. In the case of non-zero mismatch error (i.e., $\delta(t) \neq 0$), the function approximation error $\hat{f}(z(t), \hat{\theta}(t), \hat{\sigma}(t)) - \hat{f}((z(t), \theta^*, \sigma^*)$, and the estimation error $e(t) = \hat{\chi}(t) - \chi(t)$ remain bounded for all z in some domain of interest $\mathcal{D}$ and satisfy a small-in-the-mean-square property with respect to the magnitude of the mismatch error.

4.1.4 Discussion of Issues in Parametric Estimation

The parameter estimation methods presented in this text are based on standard estimation techniques but with special emphasis on the adaptive approximation problem. It is important for the reader to note that the methodologies developed in this chapter are not in a research vacuum but the extension of a large number of parameter estimation results. Parametric estimation is a well-established field in science and engineering since it is one of the key components in developing models from observations. Several books are available for parameter estimation in the context of system identification [127, 153, 163, 251], adaptive control [119, 179, 235] and time series analysis [28, 103].

A significant number of results have been developed for offline parameter estimation, where all the data is first collected and then processed to fit an assumed model. Both frequency and time domain approaches can be used, depending on the nature of the input-output data. Moreover, stochastic techniques have been extensively used to deal with measurement noise and other types of uncertainty. A key component in offline parameter estimation is the selection of the norm, which determines the objective function to be minimized.

Most of the parameter estimation methods developed so far in the literature are for linear models. As expected, in the special case of linear models there are more well-established design and analysis tools. However, there is also a large amount of research work that has been developed for nonlinear systems [102, 109, 153]. As illustrated in Examples 4.2 and 4.3, there is a key difference between nonlinear systems where the nonlinearities are known but are multiplied with unknown parameters (Example 4.2), and nonlinear systems where there are unknown nonlinearities that need to be approximated (Example 4.3). The emphasis of the techniques developed in this chapter deals with the latter case of unknown nonlinearities. In this framework, as we saw in Chapter 3, there are several adaptive approximation models that can be used to estimate the unknown nonlinearities.

Next, we discuss some fundamental issues that arise in parameter estimation, as they relate to the contents of this chapter.

Recursive estimation — no data storage. This chapter deals exclusively with online parameter estimation methods; that is, techniques that are based on the idea of first choosing an initial estimate for the unknown parameter, then recursively updating the estimate based on the current set of measurements. This is in contrast to offline parameter estimation methods where a set of data is first collected and then fit to a model. One of they key characteristics of online parameter estimation methods that the reader should keep in mind is that as streaming data becomes available in real-time, it is processed, via updating the of parameter estimates, and then thrown away. Therefore, the presented techniques require no data storage during real-time processing applications, except possibly for some buffering window that can be used to filter measurement noise. In general, the information presented by the past history of measurements (in time and/or space) is encapsulated by the current value of the parameter estimate. Adaptive parameter estimation methods are used extensively in various applications, especially those dealing with time-varying systems or unstable open-loop systems. It is also used as a way of avoiding long delays and high costs that result from offline system identification methods.

Linearly versus nonlinearly parameterized approximators. As discussed in Chapter 2, adaptive approximators can be classified into two categories of interest: linearly parameterized and nonlinearly parameterized. In the case of linearly parameterized approximators, the parameters denoted by σ are selected *a priori* and remain fixed. Therefore, the remaining adaptable weights $\hat{\theta}$ appear linearly. For nonlinearly parameterized approximators, both θ and σ weights are updated online. As we will see, the case of linearly parameterized approximators provides alternative approaches for designing online learning schemes and allows the derivation of stronger analytical results for stability and convergence. It is important for the reader to note the difference between linear models and linearly parameterized approximators. In linear models, the entire structure of the system is assumed to be linear, as in Example 4.1. In linearly parameterized approximators, the unknown nonlinearities are estimated by nonlinear approximators, where the weights (parameter estimates) appear linearly with respect to some basis functions, as in Example 4.3.

Continuous-time versus discrete-time. The adaptive parameter estimation problem can be formulated both in a continuous-time as well as a discrete-time framework. In practical applications, the actual plant typically evolves in continuous-time, while data processing (parameter estimation, monitoring, etc.) and feedback control is implemented in discrete-time using computing devices. Therefore, real-time applications yield so called *hybrid systems*, where both continuous-time and discrete-time signals are intertwined [9, 265]. Unfortunately, the theory of hybrid systems is still at an early stage, and the analysis of parameter estimation techniques for such systems is difficult to achieve. The approach followed in this chapter is to describe the relevant formulation and results in continuous-time. Naturally, the continuous-time framework is in line with the rest of the book. The discrete-time framework is briefly illustrated with some example and exercises.

Parameter convergence and persistency of excitation. It is important to keep in mind that different applications may have different objectives relevant to parameter convergence. In most control applications that focus on accurate tracking of reference input signals, the main objective is not necessarily to make the parameter estimates $\hat{\theta}(t)$ and $\hat{\sigma}(t)$ converge to the optimal values θ^* and σ^*, respectively, since accurate tracking performance can be achieved without convergence of the parameters. Of course, if parameter convergence occurs, then the designer should be ecstatic! Parameter convergence is a strong requirement. In applications where parameter convergence is desired, the input to the approximator, denoted by $z(t)$, must also satisfy a so-called *persistency of excitation* condition. The structure of the persistence of excitation condition can be strongly affected by the choice of function approximator. The issue of persistency of excitation and parameter convergence is further discussed in Section 4.5.4.

4.2 DERIVATION OF PARAMETRIC MODELS

From a mathematical viewpoint the selection of a function approximator provides a way for parameterizing an unknown function. As discussed in Chapter 2, several approximator properties such as localization, generalization and parametric linearity need to be considered.

In this section we present a procedure for creating parametric models suitable for developing adaptive parameter estimation algorithms. The procedure for deriving parametric models basically consists of rewriting the nonlinear differential equation model that describes the system in such a way that unknown parameters appear in a desired fashion. There are two key steps to pay attention to:

- in replacing the unknown nonlinearities by approximators and unknown parameters by their estimates, we make sure that we use, as much as possible, any available plant knowledge;

- to avoid the use of differentiators and to facilitate the derivation of convenient parametric models, we employ a number of filtering techniques, where certain signals are passed through a stable (usually low-pass) filter.

As we will see, the objective is to define a signal χ that is computable from measured signals and is affected by the parametric error.

4.2.1 Problem Formulation for Full-State Measurement

To further examine the construction of parametric models let us focus on the nonlinear system represented by

$$\dot{x}(t) = f(x(t), u(t)) \tag{4.5}$$
$$y(t) = x(t) \tag{4.6}$$

where $u(t) \in \Re^m$ is the control input vector, $x(t) \in \Re^n$ is the state variable vector, $y(t)$ is the measured output and $f : \Re^n \times \Re^m \mapsto \Re^n$ is a vector field representing the dynamics of the system. Therefore, in this problem the full state vector $x(t)$ is assumed to be available for measurement. In most applications the vector field f is partially known. The known part of f, usually referred to as the *nominal* model, is derived either by analytical methods using first principles or by offline identification methods. Therefore, it is assumed that f can be decomposed as

$$f(x, u) = f_0(x, u) + f^*(x, u), \tag{4.7}$$

where f_0 represents the known system dynamics and f^* represents the discrepancy between the actual dynamics f and the nominal dynamics f_0. The above decomposition is crucial because it allows the control designer to incorporate any prior information; therefore, the function approximator is needed to approximate only the uncertainty f^*, whose magnitude is typically small, instead of the overall function f. If there is no prior information, then f_0 is simply set to zero.

The nonlinear system (4.5) can be rewritten as

$$\dot{x} = f_0(x, u) + \hat{f}(x, u; \theta^*, \sigma^*) + e_f(x, u), \tag{4.8}$$

where $\hat{f}$ is an approximating function of the type described in Chapter 3, and (θ^*, σ^*) is a set of "optimal" parameters that minimize a suitable cost function between f^* and $\hat{f}$ for all (x, u) belonging to a compact set $\mathcal{D} \subset (\Re^n \times \Re^m)$. The error term e_f, defined as

$$e_f(x, u) = f^*(x, u) - \hat{f}(x, u; \theta^*, \sigma^*), \tag{4.9}$$

represents the *minimum functional approximation error* (MFAE), which is the minimum possible deviation between the unknown function f^* and the adaptive approximator $\hat{f}$ in the ∞-norm sense over the compact set $\mathcal{D}$:

$$(\theta^*,\ \sigma^*) = \arg \min_{(\theta,\sigma)\in\Re^{q_\theta}\times\Re^{q_\sigma}} \left\{ \sup_{(x,u)\in\mathcal{D}} \left\| f^*(x,u) - \hat{f}(x,u;\theta,\sigma) \right\| \right\}.$$

In general, increasing the number of adjustable parameters in the adaptive approximator reduces the MFAE. Universal approximation results (discussed in Chapter 2) indicate that as the number of adjustable parameters becomes sufficiently large, the MFAE, e_f, can be made arbitrarily small (over a compact domain). However, in most practical cases the number of adjustable parameters is not extremely high and therefore the designer has to deal with non-zero MFAE.

If $\dot{x}$ is available for measurement then from (4.8) the parameter estimation problem becomes a *static* nonlinear approximation problem of the general form

$$\chi = \hat{f}(x, u; \theta^*, \sigma^*) + e_f(x, u), \tag{4.10}$$

where $\chi = \dot{x} - f_0(x, u)$ is a measurable variable, e_f is the minimum functional approximation error (or noise term) and (θ^*, σ^*) are the unknown parameter vectors to be estimated.

4.2.2 Filtering Techniques

Frequently in applications only x is available for measurement. The use of differentiation to obtain $\dot{x}$ is not desirable. Therefore, the assumption of $\dot{x}$ being available should be avoided. One way to avoid the use of differentiators is to use filtering techniques. By filtering each side of (4.8) with a stable first-order filter $\frac{\lambda}{s+\lambda}$, where $\lambda > 0$, we obtain

$$\chi(t) = \frac{\lambda}{s+\lambda}\left[\hat{f}(z(t);\theta^*,\sigma^*)\right] + \delta(t) \tag{4.11}$$

where $z(t) = (x(t), u(t))$ is the input vector to the adaptive approximator $\hat{f}$, $\chi(t)$ is a measurable variable computed as

$$\chi(t) = \frac{\lambda s}{s+\lambda}[x(t)] - \frac{\lambda}{s+\lambda}\left[f_0(x(t), u(t))\right], \tag{4.12}$$

and $\delta(t)$ is the filtered MFAE:

$$\delta(t) = \frac{\lambda}{s+\lambda}\left[e_f(x(t), u(t))\right]. \tag{4.13}$$

It is noted that in deriving (4.12) we use the fact that $\dot{x}(t) = s[x(t)]$.

The reader is reminded that the parametric model described by (4.11) is of the general form (4.4) described in Section 4.1.3, where the filter $W(s)$ is given by

$$W(s) = \frac{\lambda}{s+\lambda} I_{n\times n},$$

and $I_{n\times n}$ is the $n \times n$ identity matrix. Therefore, in this case, the matrix transfer function consists of n identical first-order filters. The parameter $\lambda > 0$ is a design parameter that could influence the convergence rate of the adaptive scheme.

A reader may ask: what's the use of rewriting (4.5) as (4.11), since the functional uncertainty in f^* is still present in the form of δ? The answer to this question is that the magnitude of the uncertainty f^* can be significantly larger than the magnitude of the filtered MFAE δ. Moreover, the magnitude of δ can be further reduced, if desired, by increasing the dimension of the basis vector $\phi(z)$ in the adaptive approximator. In the limit, as this dimension increases toward infinity, the MFAE e_f converges to zero (over a compact domain), as shown by universal approximation results. Since δ is small, it can be more easily accommodated in the nonlinear identification and control design. The "price" paid for reducing the uncertainty from f^* to δ is the presence of unknown parameters θ^* and σ^*, which need to be estimated online. This cost becomes a design tradeoff, in the sense that the smaller the dimension of $\phi(z)$ (or the number of adjustable parameters) that are used the smaller the difference between f^* and δ.

■ **EXAMPLE 4.4**

Consider the second-order system

$$\begin{aligned}\dot{x}_1 &= x_2 - g_1(x_2)\\ \dot{x}_2 &= x_1 + g_2(x_1, x_2) + 2u\end{aligned}$$

where g_1 and g_2 are the unknown functions. In this example, f_0 and f^* are given by

$$f_0(x_1, x_2, u) = \begin{bmatrix} x_2 \\ x_1 + 2u \end{bmatrix}, \qquad f^*(x_1, x_2) = \begin{bmatrix} -g_1(x_2) \\ g_2(x_1, x_2) \end{bmatrix}.$$

If we let $\hat{f}_1(x_2;\theta_1,\sigma_1)$ and $\hat{f}_2(x_1,x_2;\theta_2,\sigma_2)$ be the adaptive approximators for $-g_1(x_2)$ and $g_2(x_1,x_2)$ respectively then the parametric model (4.11) becomes

$$\begin{bmatrix} \chi_1(t) \\ \chi_2(t) \end{bmatrix} = \frac{\lambda}{s+\lambda} \begin{bmatrix} \hat{f}_1(x_2;\theta_1^*,\sigma_1^*) \\ \hat{f}_2(x_1,x_2;\theta_2^*,\sigma_2^*) \end{bmatrix} + \begin{bmatrix} \delta_1(t) \\ \delta_2(t) \end{bmatrix},$$

where δ_1 and δ_2 are the filtered MFAEs associated with each approximator, and χ_1, χ_2 are measurable variables generated by (see eqn. (4.12))

$$\begin{aligned} \begin{bmatrix} \chi_1(t) \\ \chi_2(t) \end{bmatrix} &= \frac{\lambda s}{s+\lambda} \begin{bmatrix} x_1(t) \\ x_2(t) \end{bmatrix} - \frac{\lambda}{s+\lambda} \begin{bmatrix} x_2(t) \\ x_1(t) + 2u(t) \end{bmatrix} \\ &= \lambda \begin{bmatrix} x_1(t) \\ x_2(t) \end{bmatrix} - \frac{\lambda}{s+\lambda} \begin{bmatrix} \lambda x_1(t) + x_2(t) \\ \lambda x_2(t) + x_1(t) + 2u(t) \end{bmatrix}. \end{aligned}$$

△

EXAMPLE 4.5

Consider the second-order system

$$\ddot{y} - g(y,\dot{y},u) = 0,$$

which can be written in state-space form as

$$\begin{aligned} \dot{x}_1 &= x_2 \\ \dot{x}_2 &= g(x_1,x_2,u) \end{aligned}$$

where $x_1 = y$, $x_2 = \dot{y}$ and g is an unknown function. Now, we have

$$f_0(x_1,x_2,u) = \begin{bmatrix} x_2 \\ 0 \end{bmatrix}, \qquad f^*(x_1,x_2,u) = \begin{bmatrix} 0 \\ g(x_1,x_2,u) \end{bmatrix}.$$

It is clear that in this example, $\chi_1(t) = 0$, and therefore it does not require any further consideration. Hence, we can proceed to derive a parametric model only for the second state equation—since the first does not contain any uncertainty. If we let $\hat{f}_2(x_1,x_2,u;\theta_2,\sigma_2)$ be the adaptive approximator of $g(x_1,x_2,u)$ then the parametric model for χ_2 becomes

$$\chi_2(t) = \frac{\lambda}{s+\lambda}[\hat{f}_2(x_1,x_2,u;\theta_2^*,\sigma_2^*)] + \delta_2(t),$$

where δ_2 is the filtered MFAEs associated with $\hat{f}_2$, and χ_2 is generated as follows:

$$\chi_2(t) = \frac{\lambda s}{s+\lambda}[x_2(t)].$$

This examples shows that often the parametric model can be simplified, thereby leading to a simpler estimation scheme. △

4.2.3 SPR Filtering

Instead of the simple filter $\frac{\lambda}{s+\lambda}$, the designer can select to use a more complicated filter $W(s)$. In this case, by filtering each side of (4.8) with an appropriate stable filter $W(s)$, we obtain

$$\chi(t) = W(s)\left[\hat{f}(z(t);\theta^*,\sigma^*)\right] + \delta(t), \tag{4.14}$$

where $\delta(t)$ and $\chi(t)$ are given by

$$\delta(t) = W(s)\left[e_f(x(t),u(t))\right] \tag{4.15}$$
$$\chi(t) = sW(s)[x(t)] - W(s)\left[f_0(x(t),u(t))\right]. \tag{4.16}$$

For reasons that will become apparent in the subsequent analysis of the adaptive approximation scheme using the general filter $W(s)$, we assume that $W(s)$ is a strictly positive real (SPR) filter. A detailed presentation of SPR function and their properties is beyond the scope of this book. A thorough treatment of the SPR condition and it's use in parameter estimation problems is given for example in [119, 235]. Some of the key features of SPR functions that will be used subsequently are summarized in Section A.2.3 of Appendix A.

4.2.4 Linearly Parameterized Approximators

The nonlinear system (4.5)–(4.6) can be rewritten as a parametric model of the form (4.11) whether the adaptive approximator used is linearly or nonlinearly parameterized. However, in the special case of a linearly parameterized approximator, a different type of parametric model can be derived.

It is recalled that for linearly parameterized approximators, σ is selected a priori, and therefore the approximation function $\hat{f}$ can be written as $\hat{f}(z;\theta^*,\sigma^*) = \theta^{*\top}\phi(z)$. Therefore, the parametric model (4.11) becomes

$$\chi(t) = \frac{\lambda}{s+\lambda}\left[\theta^{*\top}\phi(z(t))\right] + \delta(t). \tag{4.17}$$

Since θ^* is a constant vector, it can be pulled in front of the linear filter, resulting in

$$\chi(t) = \theta^{*\top}\zeta(t) + \delta(t), \tag{4.18}$$

where $\zeta(t)$ is a vector of filtered basis functions; i.e.,

$$\zeta(t) = \frac{\lambda}{s+\lambda}\left[\phi(z(t))\right].$$

Of course, the extension also works when (4.14) is used with the more general filter $W(s)$.

It is interesting to note that the parametric model (4.18) is an algebraic equation with the unknown coefficient vector θ^* appearing linearly. As we will see in the next two sections, this type of parametric model allows the application of powerful and well-understood optimization algorithms, such as the gradient algorithm and the recursive least squares algorithm.

Incorporating Partial *A Priori* Knowledge. From a mathematical perspective, any nonlinear function f in (4.5) can be broken in up in two components f_0 and f^*, as in (4.7), where f^* contains all the uncertain and unknown terms, and f_0 contains the remaining (known) terms. However, in many practical applications the system under consideration may have a partially known structure with unknown nonlinearities multiplying known functions. As discussed earlier, in general the designer is interested in taking advantage of any known structure. Therefore, instead of collapsing all the nonlinearities together into one "big" nonlinearity f^*, sometimes it is better to leave the underlying known structure intact, and proceed to approximate each nonlinearity separately.

To formulate such a scenario, let the unknown function f be written as

$$f(x,u) = f_0(x,u) + \sum_{i=1}^{M} f_i^*(x,u)\rho_i(x,u), \tag{4.19}$$

where $f_0 : \Re^n \times \Re^m \mapsto \Re^n$ is a known function, $f_i^* : \Re^n \times \Re^m \mapsto \Re^n$ are unknown functions, and $\rho_i : \Re^n \times \Re^m \mapsto \Re^1$ are known functions. The integer M simply represents the number of ρ_i terms that are multiplied by unknown nonlinearities f_i^*. In this case, the derivation of parametric models can proceed in a similar fashion as presented above. It can readily be verified (see Exercise 4.1) that the parametric model is of the form

$$\chi(t) = \frac{\lambda}{s+\lambda}\left[\sum_{i=1}^{M} \hat{f}_i(x(t),u(t);\theta_i^*,\sigma_i^*)\rho_i(x(t),u(t))\right] + \delta(t) \tag{4.20}$$

where $\chi(t)$ is given by (4.12), and the filtered MFAE is given by

$$\delta(t) = \frac{\lambda}{s+\lambda}\left[\sum_{i=1}^{M} \left(f_i^*(x(t),u(t)) - \hat{f}_i(x(t),u(t);\theta_i^*,\sigma_i^*)\right)\rho_i(x(t),u(t))\right]. \tag{4.21}$$

Each approximating function $\hat{f}_i$ has a corresponding set of (unknown) "optimal parameters" (θ_i^*, σ_i^*) that minimize the $\max_{(x,u)\in\mathcal{D}} \|f_i^* - \hat{f}_i\|$. The presence of the known multiplier terms ρ_i, in general, does not present any additional challenges for adaptive approximation.

In the case that $\hat{f}_i$ is linearly parameterized, the parametric model can be further simplified. If each $\hat{f}_i$ is parameterized as $\hat{f}_i(z;\theta_i^*) = \theta_i^{*\top}\phi_i(z)$, then the resulting filtered regressor form is

$$\chi(t) = \sum_{i=1}^{M} \theta_i^{*\top} \underbrace{\frac{\lambda}{s+\lambda}\left[\phi_i\left(z(t)\right)\rho_i\left(z(t)\right)\right]}_{\zeta_i(t)} + \delta(t).$$

■ **EXAMPLE 4.6**

Consider the second-order system

$$\begin{aligned}\dot{x}_1 &= x_2 - x_1 g_1(x_2)\\ \dot{x}_2 &= x_1 g_2(x_1,x_2) - x_2 g_3(u)\end{aligned}$$

where the above structure of the system is known, but the functions g_1, g_2, g_3 are unknown. One approach to deriving parametric models, is to follow the direct breaking up of the known and unknown components, as described by (4.7). In this case, f_0 and f^* are given by

$$f_0(x_1,x_2,u) = \begin{bmatrix} x_2 \\ 0 \end{bmatrix}, \qquad f^*(x_1,x_2,u) = \begin{bmatrix} -x_1 g_1(x_2) \\ x_1 g_2(x_1,x_2) - x_2 g_3(u) \end{bmatrix}.$$

In this case, two functions are approximated. One function has two arguments and the other has three arguments.

Alternatively, the designer can choose to incorporate the known structure of the system into the formulation of the adaptive approximation problem, as described by (4.19). In this case $f = f_0 + f_1^* \rho_1 + f_2^* \rho_2$, where $\rho_1(x_1) = x_1$, $\rho_2(x_2) = -x_2$ and

$$f_0(x_2) = \begin{bmatrix} x_2 \\ 0 \end{bmatrix}, \quad f_1^*(x_1, x_2) = \begin{bmatrix} -g_1(x_2) \\ g_2(x_1, x_2) \end{bmatrix}, \quad f_2^*(u) = \begin{bmatrix} 0 \\ g_3(u) \end{bmatrix}.$$

In this case, three functions would be approximated; however, two have a single argument and the third has two arguments. In addition, each approximated function is simpler than in the former case. △

Choosing the most suitable formulation is not usually obvious. Sometimes it is preferable to collapse all the nonlinearities together, while at other times it is more convenient to leave them separate. In general, it is wise to collapse nonlinear functions together only if they are not needed at a later time, for example, to design feedback control laws. For the readers familiar with elementary circuit theory, the decision is analogous to simplifying electrical circuits: if the voltages and currents through a part of the network are not needed, then that part of the network can be collapsed into a simpler network containing only a voltage source and an impedance (Thevenin's and Norton's equivalent circuits). A similar dilemma occurs in parameter estimation problems for simple linear systems: sometimes, it is more convenient to collapse several parameters together and estimate only one parameter, in other cases the physical significance of a certain parameter necessitates that it be estimated separately.

Another motivation for not collapsing the nonlinearities to a single function with several inputs is that the memory requirements grow exponentially with input dimensions, but only linearly with the number of approximated functions.

4.2.5 Parametric Models in State Space Form

The filtering techniques developed above have conveniently been described in terms of time signals (or functions of time signals) passed through a transfer function. In this section we present the same results in state-space form. The rationale for considering this parallel formulation in state-space is two-fold. First, it provides a way to view the parametric modeling derivation that may be more suitable to readers that are more comfortable with the state-space domain for representing dynamical systems. Second, it provides an alternative approach for parametric modeling that is more convenient for time-varying and nonlinear systems.

In the case of nonlinearly parameterized approximators, (4.11) can be written in state-space form as

$$\dot{\chi}(t) = -\lambda\chi(t) + \lambda\hat{f}(z(t); \theta^*, \sigma^*) + \lambda e_f(z(t)) \tag{4.22}$$

utilizing the definition of (4.13). This equation shows the dependence of χ on θ^* and σ^*, but is not directly computable since θ^*, σ^*, e_f are unknown. The value of the variable $\chi(t)$ is computed by (4.12), which can be rewritten as

$$\chi(t) = \lambda x(t) - \frac{\lambda^2}{s+\lambda}[x(t)] - \frac{\lambda}{s+\lambda}\left[f_0(x(t), u(t))\right].$$

Therefore, $\chi(t)$ is generated in state-space form as follows:

$$\dot{\xi}(t) = -\lambda\xi(t) - \lambda x(t) - f_0(x(t), u(t)) \tag{4.23}$$
$$\chi(t) = \lambda\xi(t) + \lambda x(t) \tag{4.24}$$

where

$$\xi = -\frac{\lambda}{s+\lambda}[x(t)] - \frac{1}{s+\lambda}\left[f_0(x(t), u(t))\right]$$

is an intermediate state-variable. It is important to note that the state-space representation (4.23)-(4.24) is not unique. Using a change of variables, it is possible to use a different state-space form to represent the input-output system characterized by (4.12).

In the case of linearly parameterized approximators, the parametric model can be written in the form of (4.18), where ζ and δ are generated as follows:

$$\dot{\zeta}(t) = -\lambda\zeta(t) + \lambda\phi(x(t), u(t))$$
$$\dot{\delta}(t) = -\lambda\delta(t) + \lambda e_f(x(t), u(t)).$$

4.2.6 Parametric Models of Discrete-Time Systems

In the case of discrete-time systems with full state measurement, the equations corresponding to (4.5) and (4.6) are given by

$$x(k+1) = f\left(x\left(k\right), u\left(k\right)\right) \tag{4.25}$$
$$y(k) = x(k) \tag{4.26}$$

where $u(k) \in \Re^m$ is the control input vector at sampled time $t = kT_s$ (T_s is the sampling time), $x(k) \in \Re^n$ is the state variable vector, $y(k)$ is the measured output and $f : \Re^n \times \Re^m \mapsto \Re^n$ is a vector field representing the dynamics of the discrete-time system. Again, it is assumed that f can be broken up into two components, f_0 and f^*, where f_0 represents the known part and f^* represents the unknown part, which is to be approximated online.

Similar to the formulation developed in Section 4.2.1, the state difference equation (4.25) can be rewritten as

$$x(k) = f_0(x(k-1), u(k-1)) + \hat{f}(x(k-1), u(k-1); \theta^*, \sigma^*) + e_f(x(k-1), u(k-1)), \tag{4.27}$$

where $\hat{f}$ is an approximating function and e_f is the minimum functional approximation error (MFAE):

$$e_f(x(k), u(k)) = f^*(x(k), u(k)) - \hat{f}(x(k), u(k); \theta^*, \sigma^*).$$

Therefore, the discrete-time parametric model is of the form

$$\chi(k) = \hat{f}(x(k-1), u(k-1); \theta^*, \sigma^*) + \delta(k) \tag{4.28}$$

where the discrete-time measurement model is $\chi(k) = x(k) - f_0(x(k-1), u(k-1))$ with the filtered error $\delta(k) = e_f(x(k-1), u(k-1))$.

In comparing the continuous-time parametric model (4.11) and discrete-time parametric model (4.28) we notice that the two models are almost identical, with the filter $\frac{\lambda}{s+\lambda}$ being replace by the delay function z^{-1}, where z is defined based on the z-transform variable. In a more general setting, the discrete-time parametric model can be described by

$$\chi(k) = W(z)[\hat{f}(x(k), u(k); \theta^*, \sigma^*)] + \delta(k) \tag{4.29}$$

where the discrete-time measurement is

$$\chi(k) = zW(z)[x(k)] - W(z)[f_0(x(k), u(k))]$$

with the filtered model error

$$\delta(k) = W(z)[e_f(x(k), u(k))].$$

The matrix $W(z)$ is a stable discrete-time filter, whose denominator degree is at least one higher than the degree of the numerator (in order for $zW(z)$ to be a proper transfer function).

In many applications, the discrete-time system is represented in terms of tapped delays of the input/output instead of the full-state model described by (4.25) and 4.26). This is sometimes referred to as a nonlinear auto-regressive moving average (NARMA) model [153]. In this case, the output $y(k)$ is described by

$$y(k) = f(y(k-1), y(k-2), \ldots, y(k-n_y), u(k-1), u(k-2), \ldots, u(k-n_u)), \quad (4.30)$$

where n_y and n_u are the maximum delays in the output and input variables respectively that influence the current output. By letting

$$z(k) = [y(k),\ y(k-1),\ \ldots\ , y(k+1-n_y),\ u(k),\ u(k-1),\ \ldots\ , u(k+1-n_u]^\top,$$

and rewriting the difference equation, we obtain a similar discrete-time parametric model as in (4.28); i.e.,

$$\chi(k) = \hat{f}(z(k-1); \theta^*, \sigma^*) + \delta(k), \quad (4.31)$$

where $\chi(k) = y(k)$ and $\delta(k) = f(z(k-1)) - \hat{f}(z(k-1); \theta^*, \sigma^*)$.

In summary, we see that a class of nonlinear discrete-time systems can be represented by a parametric model of the general form (4.29), which is quite similar to the corresponding continuous-time formulation. As with continuous-time systems, in the special case of linearly parameterized approximators, $\hat{f}$ can be written as $\hat{f}(z; \theta^*, \sigma^*) = \theta^{*\top}\phi(z)$, and therefore the discrete-time parametric model (4.31) can be written as

$$\chi(k) = \theta^{*\top}\zeta(k-1) + \delta(k), \quad (4.32)$$

where $\zeta(k) = \phi(z(k))$.

■ **EXAMPLE 4.7**

Let us now consider the following discrete-time time nonlinear system

$$\begin{aligned} y(k) &= \frac{1}{2}y(k-1) - \frac{2}{3}y(k-2)^2 + f(y(k-1), y(k-2)) \\ &\quad + g(u(k-1), u(k-2)), \end{aligned} \quad (4.33)$$

where f and g are unknown nonlinear functions. It is assumed that the above general structure of the dynamic system is known by the designer, including the fact that f and g are functions of $y(k-1), y(k-2)$ and $u(k-1), u(k-2)$ respectively; however, the functions f and g are not known.

The systems described by (4.33) can be rewritten as

$$\begin{aligned} y(k) - \frac{1}{2}y(k-1) + \frac{2}{3}y(k-2)^2 &= \hat{f}(y(k-1), y(k-2); \theta_f^*, \sigma_f^*) \\ &\quad + \hat{g}(u(k-1), u(k-2); \theta_g^*, \sigma_g^*) \\ &\quad + \delta(k), \end{aligned} \quad (4.34)$$

where

$$\begin{aligned}\delta(k) &= f(y(k-1),y(k-2)) - \hat{f}(y(k-1),y(k-2);\theta_f^*,\sigma_f^*)\\ &\quad + g(u(k-1),u(k-2)) - \hat{g}(u(k-1),u(k-2);\theta_g^*,\sigma_g^*).\end{aligned}$$

Therefore (4.34) can be written in the form

$$\chi(k) = \hat{f}(y(k-1),y(k-2);\theta_f^*,\sigma_f^*) + \hat{g}(u(k-1),u(k-2);\theta_g^*,\sigma_g^*) + \delta(k),$$

where

$$\chi(k) = y(k) - \frac{1}{2}y(k-1) + \frac{2}{3}y(k-2)^2.$$

It is noted that $\hat{f} + \hat{g}$ can also be represented with only one adaptive approximator $\hat{h}$ which has four inputs ($y(k-1)$, $y(k-2)$, $u(k-1)$, $u(k-2)$), instead of two approximators each with two inputs. However, in general this is not beneficial since adaptive approximation is more difficult with one network having four inputs, as compared to two networks having two inputs each. The former will require on the order of d^4 parameters whereas the latter will require on the order of $2d^2$ parameters where $d \gg 1$ is the number of basis functions per input dimension. This is related to the "curse of dimensionality" issue, which was discussed in Chapter 2. △

4.2.7 Parametric Models of Input-Output Systems

So far (with the exception of the discrete-time NARMA model (4.30)) the derivation of a parametric models has assumed that the full-state is available for measurement. In this section, we show that a similar procedure also works for a class of input-output systems. A key requirement for input-output systems is that any unknown nonlinearity $f^*(y,u)$ is a function of measurable variables.

■ **EXAMPLE 4.8**

Consider a second order system of the form

$$\begin{aligned}\dot{x}_1 &= x_2\\ \dot{x}_2 &= -x_1 + 2x_2 + f^*(x_1) + u\\ y &= x_1\end{aligned}$$

where u is the input, y is the measurable output and f^* is an unknown function of x_1. The system can be rewritten as

$$\ddot{y} - 2\dot{y} + y - f^*(y) = u.$$

In this example, y is measurable, but $\dot{y}$ and $\ddot{y}$ are not. By introducing (i.e., adding and subtracting) an adaptive approximator $\hat{f}(y;\theta^*,\sigma^*)$ and then filtering both sides by a transfer function of the form $\frac{\lambda^2}{(s+\lambda)^2}$, we obtain

$$\begin{aligned}\frac{\lambda^2(s^2-2s+1)}{(s+\lambda)^2}[y(t)] - \frac{\lambda^2}{(s+\lambda)^2}[u(t)] &= \frac{\lambda^2}{(s+\lambda)^2}\left[\hat{f}(y(t);\theta^*,\sigma^*)\right] +\\ &\quad \frac{\lambda^2}{(s+\lambda)^2}\left[f(y(t)) - \hat{f}(y(t);\theta^*,\sigma^*)\right].\end{aligned}$$

This can be written (similar to the general parametric form (4.11)) as

$$\chi(t) = \frac{\lambda^2}{(s+\lambda)^2}\left[\hat{f}(y(t);\theta^*,\sigma^*)\right] + \delta(t),$$

where $\chi(t)$ and $\delta(t)$ are defined as

$$\begin{aligned}
\chi(t) &= \frac{\lambda^2(s^2-2s+1)}{(s+\lambda)^2}[y(t)] - \frac{\lambda^2}{(s+\lambda)^2}[u(t)] \\
\delta(t) &= \frac{\lambda^2}{(s+\lambda)^2}\left[f(y(t)) - \hat{f}(y(t);\theta^*,\sigma^*)\right]
\end{aligned}$$

which is of the same parametric modeling structure as that derived for the full-state measurement. △

It is clear from this simple example that for the procedure to work out it is crucial that the unknown nonlinearity f be a function of the measurable variable y (and not $\dot{y}$). Otherwise, it would have been necessary for $\dot{y}$ to be an input to the adaptive approximator.

The above procedure for deriving a parametric model for input-output systems can be extended to a more general class of systems described by

$$y^{(n)} + \alpha_{n-1}y^{(n-1)} + \cdots + \alpha_2\ddot{y} + \alpha_1\dot{y} + \alpha_0 y + g_0(u) + f^*(y,u) = 0, \tag{4.35}$$

where the coefficients $\{\alpha_0,\ \alpha_1,\ \ldots\ \alpha_{n-1}\}$ and the function g_0 are known, while f^* is an unknown nonlinear function which is to be approximated online.

Let the n-th order filter $W(s)$ be of the form

$$W(s) = \frac{\lambda^n}{(s+\lambda)^n}.$$

Then by filtering both sides of (4.35) we obtain

$$\chi(t) = W(s)[\hat{f}(y(t),u(t);\theta^*,\sigma^*)] + \delta(t),$$

where $\chi(t)$ and $\delta(t)$ are defined as follows:

$$\begin{aligned}
\chi(t) &= \frac{\lambda^n\left(s^n + \alpha_{n-1}s^{(n-1)} + \cdots + \alpha_2 s^2 + \alpha_1 s + \alpha_0\right)}{(s+\lambda)^n}[y(t)] \\
&\quad + \frac{\lambda^n}{(s+\lambda)^n}[g_0(u(t))] \\
&= W(s)\left(s^n + \alpha_{n-1}s^{(n-1)} + \cdots + \alpha_2 s^2 + \alpha_1 s + \alpha_0\right)[y(t)] \\
&\quad + W(s)[g_0(u(t))] \\
\delta(t) &= \frac{\lambda^n}{(s+\lambda)^n}\left[f^*(y(t),u(t)) - \hat{f}(y(t),u(t);\theta^*,\sigma^*)\right] \\
&= W(s)\left[f^*(y(t),u(t)) - \hat{f}(y(t),u(t);\theta^*,\sigma^*)\right].
\end{aligned}$$

Therefore, again we obtain a similar parametric modeling structure.

4.3 DESIGN OF ONLINE LEARNING SCHEMES

The previous section has dealt with rewriting the nonlinear system, in particular the functional uncertainty f^*, into a form that is convenient for designing online learning models and parameter adaptive laws. In that section we defined the utility variable χ. For each type of system, we presented two equations: the parametric model equation shows the dependence of χ on the parametric function approximator; and, the measurement equation shows how χ can be computed from measured signals. In this section, we consider the design of online learning models for nonlinear function approximation, based on the parametric forms derived in the previous section. The online learning model will generate a training signal $e(t)$ that will be used to approximate the unknown nonlinearities in the system. The online learning model consists of the adaptive approximator augmented by identifier dynamics. The identifier dynamics are used to incorporate any *a priori* knowledge into the identification design and to filter some of the signals to avoid the use of differentiators and decrease the effects of noise.

We now proceed to the design of online learning schemes for dynamic systems. We will consider the design of two approaches: (i) the Error Filtering Online Learning (EFOL) scheme, and (ii) the Regressor Filtering Online Learning (RFOL) scheme.

4.3.1 Error Filtering Online Learning (EFOL) Scheme

Based on the general parametric model (4.11), the EFOL model is described by

$$\hat{\chi}(t) = \frac{\lambda}{s+\lambda}\left[\hat{f}(z(t);\hat{\theta}(t),\hat{\sigma}(t))\right]. \tag{4.36}$$

Therefore, the estimator is obtained by replacing the unknown "optimal" weights θ^* and σ^*, by their parameter estimates $\hat{\theta}(t)$ and $\hat{\sigma}(t)$, respectively. The output estimation error $e(t)$, which will be used in the update of the parameter estimates, is given by

$$e(t) = \hat{\chi}(t) - \chi(t), \tag{4.37}$$

where $\chi(t)$ generated by (4.12) is a measurable variable. The architecture of the EFOL scheme is depicted as a block diagram in Figure 4.7. As can be seen from the diagram, the inputs to the EFOL scheme are the plant input vector $u(t)$ and measurable state vector $x(t)$. The output estimation error $e(t)$, used in the update of the parameter estimates $\hat{\theta}(t)$ and $\hat{\sigma}(t)$, can be regarded as the output of the EFOL model.

Alternatively, one may consider the EFOL model as consisting of two components: (1) the adaptive approximator, which is selected based on the considerations outlined in Chapters 2 and 3; and (2) the rest of the parts, referred to as the *estimator*, which contains the filters and *a priori* known nonlinearities f_0. The block diagram of this configuration is depicted in Figure 4.8. As seen from the diagram, this configuration for viewing the EFOL model isolates the approximator, which is usually a convenient way for implementing the online learning design, as it requires fewer filters.

To extract some intuition behind this online learning scheme and to understand why it is referred to as "error filtering" scheme, we use (4.36) and (4.11) to rewrite the output estimation error as

$$e(t) = \frac{\lambda}{s+\lambda}\left[\hat{f}(z(t);\hat{\theta}(t),\hat{\sigma}(t)) - \hat{f}(z(t);\theta^*,\sigma^*)\right] - \delta(t)$$

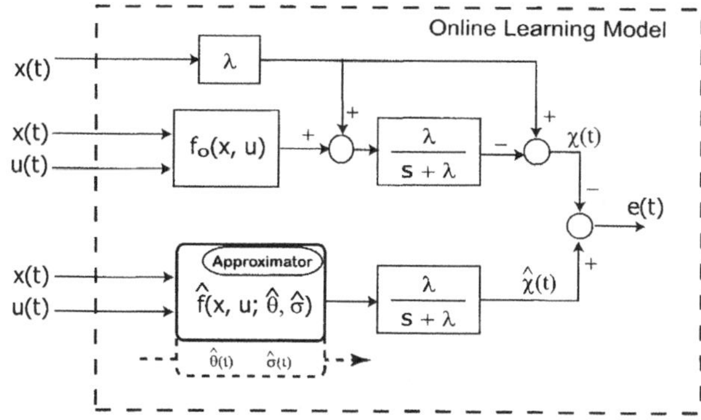

Figure 4.7: Block diagram of error filtered online learning system. The dashed box under the approximator indicates the dynamics of the parameter estimator.

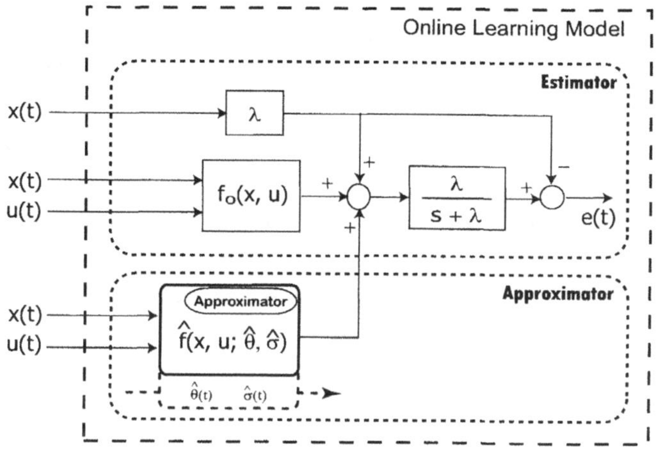

Figure 4.8: Alternative block diagram configuration for EFOL model for dynamical systems.

$$
\begin{aligned}
&= \frac{\lambda}{s+\lambda}\left[\hat{f}(z(t);\hat{\theta}(t),\hat{\sigma}(t)) - \hat{f}(z(t);\theta^*,\sigma^*) - e_f(z(t))\right] \\
&= \frac{\lambda}{s+\lambda}\left[\hat{f}(z(t);\hat{\theta}(t),\hat{\sigma}(t)) - f^*((z(t))\right].
\end{aligned}
$$

Therefore, $e(t)$ is equal to the filtered version of the approximation error $\hat{f}(z(t);\hat{\theta}(t),\hat{\sigma}(t)) - f^*(z(t))$ at time t; thus the term "error filtering."

A key observation is that if at some specific time $t = t_1$, the estimation error $e(t_1) = 0$, this does not necessarily imply that $\hat{f}(z(t_1);\hat{\theta}(t_1),\hat{\sigma}(t_1)) = f^*((z(t_1))$. Moreover, the reverse is also not valid; the fact that $\hat{f}(z(t_1);\hat{\theta}(t_1),\hat{\sigma}(t_1)) = f^*((z(t_1))$ does not imply that $e(t_1) = 0$ (see Exercise 4.2). In general, the estimation error signal $e(t)$ follows the approximation error signal $\hat{f}(z(t);\hat{\theta}(t),\hat{\sigma}(t)) - f^*((z(t))$ with some decay dynamics that depend on the value of λ. It is easy to see that the larger the value of λ the closer the estimation error will follow the approximation error. On the other hand, in the presence of measurement noise, a large value of λ will allow noise to have a greater effect on the approximator parameters. This may also be seen from Figures 4.7 and 4.8, where λ multiplies the state measurement vector $x(t)$.

The EFOL scheme can be applied both to linearly as well nonlinearly parameterized approximators. In the special case of linearly parameterized approximators, the EFOL model described by (4.36) becomes

$$\hat{\chi}(t) = \frac{\lambda}{s+\lambda}\left[\hat{\theta}(t)^{\top}\phi(z(t))\right], \tag{4.38}$$

where $\hat{\theta}(t)$ are the adjustable parameters and $\phi(z(t))$ is a vector of basis functions. The remaining components of the online learning model remain the same. As presented in Figure 4.8, any of the approximators described in Chapter 3 can be inserted as the approximator component of the online learning scheme.

Eqn. (4.38) should be contrasted with eqn. (4.17). In (4.17) θ^* is a constant vector that can be factored through the filter without affecting the validity of the equation. In (4.38), $\hat{\theta}(t)$ cannot be pulled through the filter as it is not a constant vector.

For readers who are more comfortable with state space representations, the EFOL model can be readily described in state space form using the same procedure described in Section 4.2. Specifically, $\hat{\chi}(t)$ is described in state space form as

$$\dot{\hat{\chi}}(t) = -\lambda\hat{\chi}(t) + \lambda\hat{f}(z(t);\hat{\theta}(t),\hat{\sigma}(t)). \tag{4.39}$$

To compute the output estimation error $e(t) = \hat{\chi}(t) - \chi(t)$, the variable $\chi(t)$ is generated according to (4.23)-(4.24). Therefore, the estimation error $e(t)$ is described in state space form as:

$$\dot{\xi}(t) = -\lambda\xi(t) - \lambda x(t) - f_0(x(t),u(t)) \tag{4.40}$$

$$e(t) = \hat{\chi}(t) - \lambda\xi(t) - \lambda x(t). \tag{4.41}$$

Although in this section we have worked only with the filter $\frac{\lambda}{s+\lambda}$, the same design procedure can be applied to any SPR filter $W(s)$. Based on the parametric model (4.17), the EFOL model is of the form

$$\hat{\chi}(t) = W(s)\left[\hat{f}(z(t);\hat{\theta}(t),\hat{\sigma}(t))\right]. \tag{4.42}$$

4.3.2 Regressor Filtering Online Learning (RFOL) Scheme

The second class of learning models that we consider is called Regressor Filtering Online Learning (RFOL) scheme. The way it is introduced here, this learning model can be designed only for linearly parameterized approximators. It is important to reiterate that the RFOL scheme is not based on the EFOL model (4.38).

Based on the linearly parameterized model (4.18), the RFOL model is described by:

$$\hat{\chi}(t) = \hat{\theta}(t)^{\top}\zeta(t), \tag{4.43}$$

where ζ is a vector of filtered basis functions

$$\zeta(t) = \frac{\lambda}{s+\lambda}\left[\phi(z(t))\right]. \tag{4.44}$$

In the more general case of a filter of the form $W(s)$, ζ becomes

$$\zeta(t) = W(s)\left[\phi(z(t))\right]. \tag{4.45}$$

The name "regressor filtering" is due to the filtering $W(s)$ being placed in between the basis functions ϕ (sometimes referred to as regressor) and the adaptable parameters $\hat{\theta}$, as shown in Figure 4.9. As we will see later on, RFOL models allow the use of powerful optimization methods, for deriving parameter adaptive laws, with provable convergence properties.

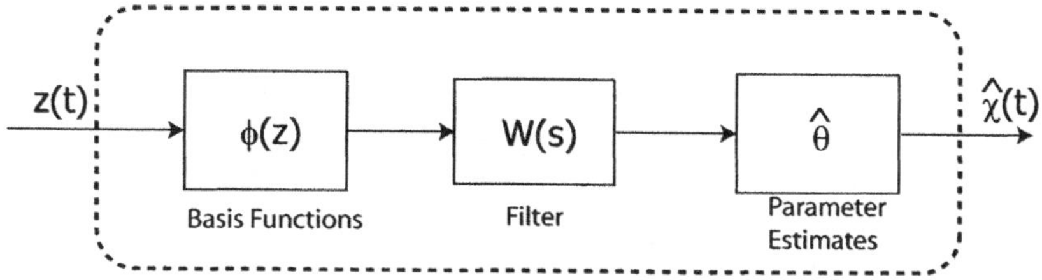

Figure 4.9: Online learning scheme based on regressor filtering.

An important observation from Figure 4.9 is that the adaptive approximator as used in generating $\hat{\chi}(t)$ is no longer a static mapping, since it contains filters in the middle, which have dynamics. At any time instant, a static approximator can still be produced as $\hat{f}(z) = \hat{\theta}(t)^\top \phi(z(t))$, but it is not utilized in the learning scheme.

In the state space representation, the RFOL model is described by

$$\dot{\zeta}(t) = -\lambda\zeta(t) - \lambda\phi(z(t)) \tag{4.46}$$

$$\hat{\chi}(t) = \hat{\theta}^\top(t)\zeta. \tag{4.47}$$

To compute the output estimation error $e(t) = \hat{\chi}(t) - \chi(t)$, the variable $\chi(t)$ is again generated according to (4.23)–(4.24). A key characteristic of the RFOL model is that the output estimation error $e(t)$ satisfies

$$e(t) = \left(\hat{\theta}(t) - \theta^*\right)^\top \zeta(t) - \delta(t). \tag{4.48}$$

Therefore, the relationship between the output estimation error $e(t)$ and the parameter estimation error $\tilde{\theta} = \hat{\theta}(t) - \theta^*$ is a simple linear and static relationship, which allows the direct use of linear regression methods.

A block diagram representation of the overall configuration for the RFOL model is depicted in Figure 4.10. In comparing the EFOL and RFOL configurations, as shown in Figure 4.8 and Figure 4.10, we notice that the EFOL requires only n filters (where n is the number of state variables), whereas the RFOL requires $n+N$ filters, where N is the number of basis functions. In general, the number of basis functions is quite large, especially in cases where the dimension of the input z is large. Therefore, the RFOL scheme is significantly more demanding computationally than the EFOL scheme.

4.4 CONTINUOUS-TIME PARAMETER ESTIMATION

This is a good time to pause momentarily and summarize the overall learning procedure. So far, we have achieved two tasks:

- First, we derived a class of parametric models by rewriting the (partially) unknown differential equation as a parametric model, for example, converting eqn. (4.5) to

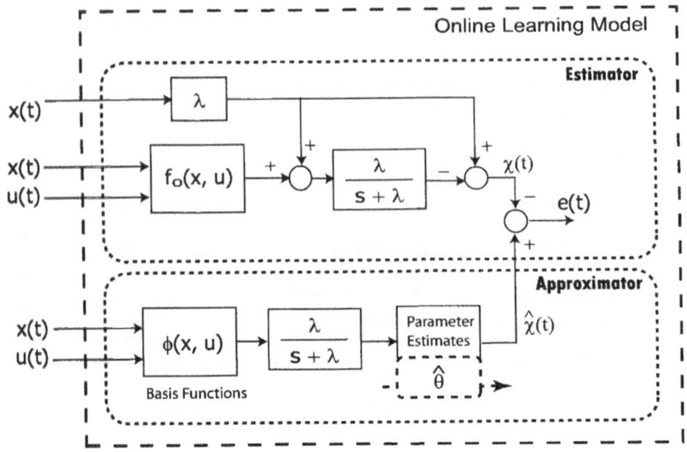

Figure 4.10: Block diagram configuration for RFOL model for dynamical systems. The dashed box below the Parameter Estimates indicates the dynamics of the parameter estimation process.

eqn. (4.11). This parametric model converts the original functional uncertainty (described by $f^*(x,u)$ in eqn. (4.7)) into parametric uncertainty (described by the unknown θ^* and σ^* in eqn. (4.8)) and the filtered MFAE, represented by $\delta(t)$. In addition to the model conversion, the procedure provides a method in eqn. (4.12) to compute χ using available signals.

- Second, based on the parametric model (4.11), we designed online learning schemes by replacing the unknown parameters θ^* and σ^* by their estimates $\hat{\theta}(t)$ and $\hat{\sigma}(t)$ with appropriate filtering, to generate a signal $e(t)$ that will be useful for parameter estimation. We treated linearly parameterized approximators as a special case, which in addition to the design of the EFOL model, allows the design of the so called RFOL model.

The natural next step is the selection of adaptive laws for adjusting the parameter estimates $\hat{\theta}(t)$ and $\hat{\sigma}(t)$. In this section, we study two methods for designing continuous-time parameter estimation algorithms: (i) the *Lyapunov synthesis method* and (ii) the *optimization method*. The Lyapunov synthesis method is applied to the EFOL scheme to derive parameter estimation algorithms with inherent stability properties. On the other hand, the optimization method is applied to the RFOL scheme, and relies on minimizing a suitably chosen cost function by standard optimization methods, such as the gradient (steepest descent) and recursive least-squares methods. It is noted that the pairing of the Lyapunov synthesis method with the EFOL scheme and the optimization method with the RFOL scheme is not coincidental. These specific combinations allow the design of adaptive approximation schemes whose performance can be analyzed and some stability properties can be derived, as shown in Section 4.5. This section will focus on the case where $\delta(t)$ is identically zero. In order to address the presence of the filtered MFAE δ, in Section 4.6 we discuss the use of robust learning algorithms.

Section 4.4.1 presents the Lyapunov synthesis method, while Section 4.4.2 presents various optimization methods for designing parameter estimation algorithms. Section 4.4.3 presents a summary discussion.

4.4.1 Lyapunov-Based Algorithms

Lyapunov stability theory, and in particular Lyapunov's direct method, is one of the most celebrated methods for investigating the stability properties of nonlinear systems [134, 234, 249, 279]. The principal idea is that it enables one to determine whether or not the equilibrium state of a dynamical system is stable without explicitly solving for the solution of the differential equation. The procedure for deriving such stability properties involves finding a suitable scalar function $V(x,t)$, in terms of the state variables x and time t, and investigating its time derivative

$$\frac{d}{dt}V(x,t) = \left(\frac{\partial V}{\partial x}\right)^{\top}\left(\frac{dx}{dt}\right) + \frac{\partial V}{\partial t}$$

along the trajectories of the system. Based on the properties of $V(x,t)$ (known as the Lyapunov function) and its derivative, various conclusions can be made regarding the stability of the system.

In general, there are no well-defined methods for selecting a Lyapunov function. However, in adaptive control problems there is a standard class of Lyapunov function candidates that are known to yield useful results. Furthermore, in some applications, such as mechanical systems, the Lyapunov function can be thought to represent a system's total energy, which provides an intuitive means to select the Lyapunov function. In terms of energy considerations, the intuitive reasoning behind Lyapunov stability theory is that in a purely dissipative system the energy stored in the system is always positive and its time derivative is nonpositive. Lyapunov theory is reviewed in more detail and several useful results are discussed in Appendix A.

The derivation of parameter estimation algorithms using the Lyapunov stability theory is crucial to the design of stable adaptive and learning systems. Historically, Lyapunov-based techniques provided the first algorithms for globally stable adaptive control systems in the early 1960s. In the recent history of neural control and adaptive fuzzy control methods, most of the results that deal with the stability of such schemes are based, to some extent, on Lyapunov synthesis methods. In many nonlinear control problems, Lyapunov synthesis methods are used not only for the derivation of learning algorithms but also for the design of the feedback control law.

According to the Lyapunov synthesis method, the problem of designing an adaptive law is formulated as a stability problem where the differential equation of the adaptive law is chosen such that certain stability properties can be established using Lyapunov theory. Since such algorithms are derived based on stability methods, by design they have some inherent stability and convergence properties.

4.4.1.1 Illustrative Scalar Example of Lyapunov Synthesis Method. To illustrate the Lyapunov synthesis method, we consider a very simple first-order example. Let the parametric model (4.11) be given by

$$\chi(t) = \frac{\lambda}{s+\lambda}\left[\theta^{*}\phi(z)\right], \tag{4.49}$$

where for simplicity we assume that there is a single parameter θ^* to be estimated, and it is linearly parameterized. The filtered MFAE is assumed to be zero. Using the error filtering online learning (EFOL) scheme, the estimator is given by

$$\hat{\chi}(t) = \frac{\lambda}{s+\lambda}\left[\hat{\theta}(t)\phi(z)\right]. \tag{4.50}$$

We let the output estimation error be given by $e(t) = \hat{\chi}(t) - \chi(t)$, and the parameter estimation error is defined as $\tilde{\theta}(t) = \hat{\theta}(t) - \theta^*$. To apply the Lyapunov synthesis method, we select the Lyapunov function

$$V(e,\tilde{\theta}) = \frac{\mu}{2\lambda}e^2 + \frac{1}{2\tilde{\gamma}}\tilde{\theta}^2 \tag{4.51}$$

where μ and $\tilde{\gamma}$ are positive constants to be selected. This is a standard Lyapunov function candidate, which is a quadratic function of the output estimation error e and the parameter estimation error $\tilde{\theta}$. By taking the time derivative of V and using the fact that θ^* is constant (i.e., $\dot{\tilde{\theta}} = \dot{\hat{\theta}}$) we obtain

$$\frac{d}{dt}V(e,\tilde{\theta}) = \dot{V} = \frac{\mu}{\lambda}e\dot{e} + \frac{1}{\tilde{\gamma}}\tilde{\theta}\dot{\hat{\theta}}.$$

From (4.49) and (4.50), the output estimation error satisfies

$$e(t) = \frac{\lambda}{s+\lambda}\left[\tilde{\theta}(t)\phi(z)\right],$$

which implies that $\dot{e} = -\lambda e + \lambda\tilde{\theta}\phi(z)$. Therefore

$$\begin{aligned}\dot{V} &= -\mu e^2 + \mu e\tilde{\theta}\phi(z) + \frac{1}{\tilde{\gamma}}\tilde{\theta}\dot{\hat{\theta}} \\ &= -\mu e^2 + \frac{1}{\tilde{\gamma}}\tilde{\theta}\left(\dot{\hat{\theta}} + \tilde{\gamma}\mu e\phi(z)\right).\end{aligned} \tag{4.52}$$

To obtain desirable stability and convergence properties, we want the derivative of V to be at least negative semidefinite. The first term of (4.52) is negative, while the second term is indefinite; in other words, it can be positive or negative. Furthermore, it is not possible to force the second term to be negative because the sign of the variable $\tilde{\theta}$ is unknown. Therefore, the best we can do is try to force it to zero. This can be done by selecting $\dot{\hat{\theta}} = -\tilde{\gamma}\mu e\phi(z)$, which yields

$$\dot{V}(t) = -\mu e^2. \tag{4.53}$$

From an implementation viewpoint, both $\tilde{\gamma}$ and μ are positive constants that can be collapsed into a single constant γ. Hence, the parameter adaptive law is chosen as

$$\dot{\hat{\theta}} = -\gamma\phi(z)e. \tag{4.54}$$

The main idea behind the Lyapunov synthesis method is that the Lyapunov function candidate has indicated what the parameter adaptive law needs to be in order to obtain some desirable stability properties. Now, let us examine what those properties are.

Uniform Boundedness. By selecting the parameter adaptive law as (4.54), the derivative of the Lyapunov function satisfies $\dot{V} = -\mu e^2$. By Lyapunov Theorem A.2.1, the fact that V is positive definite and $\dot{V}$ is negative semidefinite implies that the equilibrium point $(e, \tilde{\theta}) = (0, 0)$ is uniformly stable. It is also clear that $0 \leq V(t) \leq V(0)$, which shows that $V(t)$ is also uniformly bounded (i.e., $V(t) \in \mathcal{L}_\infty$). Therefore, both $e(t)$ are $\tilde{\theta}(t)$ are uniformly bounded (i.e., $e(t) \in \mathcal{L}_\infty$ and $\tilde{\theta}(t) \in \mathcal{L}_\infty$). Moreover, since θ^* is a finite constant, $\hat{\theta}(t) = \tilde{\theta}(t) + \theta^*$ is also uniformly bounded ($\hat{\theta}(t) \in \mathcal{L}_\infty$).

Convergence of output estimation error. To show convergence to zero of the output estimation error $e(t)$, we will employ a version of Barbălat's Lemma (see Lemma A.2.4 in Appendix A) according to which if $e, \dot{e} \in \mathcal{L}_\infty$ and $e \in \mathcal{L}_2$ then $\lim_{t\to\infty} e(t) = 0$. We start by noting that since $\dot{V}(t) \leq 0$ and by definition $V(t) \geq 0$, it implies that $V(t)$ converges to some value; i.e., $\lim_{t\to\infty} V(t) = V_\infty$ exists and is finite. Integrating both sides of (4.53) for $t \in [0, \infty)$ we obtain

$$\int_0^\infty e^2(t)dt = \frac{1}{\mu}\left(V(0) - V_\infty\right),$$

which implies $e(t)$ is square integrable; i.e., $e(t) \in \mathcal{L}_2$. To show that $\dot{e}(t) \in \mathcal{L}_\infty$, we need to assume that $\phi(z(t)$ is uniformly bounded; in this case, $\dot{e}(t) = -\lambda e(t) + \lambda\tilde{\theta}(t)\phi(z(t))$ is also uniformly bounded. Since the requirements of Barbălat's Lemma are satisfied, we conclude that $\lim_{t\to\infty} e(t) = 0$. Moreover, since

$$\dot{\hat{\theta}}(t) = -\gamma\phi(z(t))e(t),$$

using the uniform boundedness of $\phi(z(t))$ and the convergence of $e(t)$, we obtain that

$$\lim_{t\to\infty} \dot{\hat{\theta}}(t) = \lim_{t\to\infty} \dot{\tilde{\theta}}(t) = 0.$$

Convergence of parameter estimation error. The above analysis showed that the rate of change of the parameter estimate approached zero, but did not show that the parameter error converged to zero. In fact, it did not even show that the parameter error had a limit (see Example A.5). To show convergence of $\hat{\theta}(t)$ to the "true" value θ^* we need additional conditions on $\phi(z(t))$. Specifically, it is required that there exists positive constants α and δ such that for all $t_0 \geq 0$, ϕ satisfies

$$\int_{t_0}^{t_0+\delta} \phi(z(t))^2 dt \geq \alpha.$$

This condition is called *persistency of excitation conditions*, and it is discussed further in Section 4.5.4.

This example is simple enough to illustrate the main ideas behind the Lyapunov synthesis method. Next, we extend this procedure to two more general classes of parametric systems.

The methodology of the proof of this example has several features that are relatively standard to proofs that will follow throughout this book. Therefore, to decrease redundancy, we have included several useful lemmas in Section A.3 that will be called upon in subsequent proofs.

4.4.1.2 Lyapunov Synthesis Method for Linearly Parameterized Systems.

First, we consider the extension of the previous example to the case of a parameter vector. Therefore the parametric model (4.11) is given by

$$\chi(t) = \frac{\lambda}{s+\lambda}\left[\theta^{*\top}\phi(z)\right], \tag{4.55}$$

and the EFOL scheme is described by

$$\hat{\chi}(t) = \frac{\lambda}{s+\lambda}\left[\hat{\theta}(t)^{\top}\phi(z)\right]. \tag{4.56}$$

The same procedure followed earlier for the case of a scalar parameter, can be applied again here. The main difference is that, since $\hat{\theta}(t)$ is now a vector, the Lyapunov function candidate is

$$V(e,\tilde{\theta}) = \frac{\mu}{2\lambda}e^2 + \tilde{\theta}^{\top}\Gamma^{-1}\tilde{\theta} \tag{4.57}$$

where Γ is a positive definite matrix that will ultimately appear in the adaptive law for updating $\hat{\theta}(t)$ as the learning rate or adaptive gain. Using the same procedure as for the scalar case we obtain the following parameter adaptive law:

$$\dot{\hat{\theta}}(t) = -\Gamma\phi(z)e. \tag{4.58}$$

The details are left for the reader, as Exercise 4.7. In general, the adaptive gain Γ is a positive-definite (symmetric) matrix. In many applications, it is simplified to $\Gamma = \gamma I$, which implies that each element $\hat{\theta}_i(t)$ of the parameter estimate vector uses the same adaptive gain. Another useful special case is that of a diagonal adaptive gain

$$\Gamma = \begin{bmatrix} \gamma_1 & & & \\ & \gamma_2 & & \\ & & \ddots & \\ & & & \gamma_q \end{bmatrix}.$$

In this case, each element $\hat{\theta}_i(t)$ of the parameter estimate vector has its own adaptive gain γ_i, but there is no coupling between them.

Next, let us consider the case of a general filter $W(s)$ instead of the first-order filter $\frac{\lambda}{s+\lambda}$. From the parametric model $\chi(t) = W(s)[\theta^{*\top}\phi(z)]$ and it's estimate $\hat{\chi}(t) = W(s)[\hat{\theta}^{\top}\phi(z)]$, we obtain that for $\delta = 0$ the output error $e(t) = \hat{\chi}(t) - \chi(t)$ satisfies

$$e(t) = W(s)[\tilde{\theta}^{\top}\phi(z)].$$

We assume that the filter $W(s) = C(sI - A)^{-1}B$ is SPR where $(A,\ B,\ C)$ is a minimal state-space realization of $W(s)$. The state-space model is

$$\begin{aligned} \dot{e}_0 &= Ae_0 + B\tilde{\theta}^{\top}\phi(z) \\ e &= Ce_0, \end{aligned} \tag{4.59}$$

where e_0 is the state variable of the realization. Note that (4.59) is a theoretical tool that supports the following analysis. The error e is still computed using (4.41).

To apply the Lyapunov synthesis method we select the Lyapunov function

$$V(e_0,\tilde{\theta}) = \frac{\mu}{2}e_0^{\top}Pe_0 + \frac{1}{2}\tilde{\theta}^{\top}\Gamma^{-1}\tilde{\theta},$$

where $P > 0$ is a positive definite matrix. The time derivative of V along the solutions of (4.59) satisfies

$$\dot{V} = \frac{\mu}{2}e_0^{\top}\left(A^{\top}P + PA\right)e_0 + \mu\tilde{\theta}^{\top}\phi(z)B^{\top}Pe_0 + \tilde{\theta}^{\top}\Gamma^{-1}\dot{\tilde{\theta}}.$$

Now, using the Kalman-Yakubovich-Popov Lemma (see page 392), since $W(s)$ is SPR there exist positive definite matrices P, Q such that $A^\top P + PA = -Q$ and $B^\top P = C$. Therefore

$$\begin{aligned}\dot{V} &= -\frac{\mu}{2}e_0^\top Q e_0 + \mu\tilde{\theta}^\top \phi(z) C e_0 + \tilde{\theta}^\top \Gamma^{-1}\dot{\hat{\theta}} \\ &= -\frac{\mu}{2}e_0^\top Q e_0 + \tilde{\theta}^\top \Gamma^{-1}\left(\dot{\hat{\theta}} + \mu\Gamma\phi(z)e\right),\end{aligned}$$

which leads to the parameter adaptive law

$$\dot{\hat{\theta}} = -\mu\Gamma\phi(z)e.$$

The reader will notice that this adaptive law is exactly of the same form as (4.58), even though the filter $W(s)$ is different.

4.4.1.3 Lyapunov Synthesis Method for Nonlinearly Parameterized Systems.

Now, we consider the case of nonlinearly parameterized approximators. The parametric model (4.11) is given by

$$\chi(t) = \frac{\lambda}{s+\lambda}\left[\hat{f}(z(t);\theta^*,\sigma^*)\right], \tag{4.60}$$

and the EFOL scheme is described by

$$\hat{\chi}(t) = \frac{\lambda}{s+\lambda}\left[\hat{f}(z(t);\hat{\theta}(t),\hat{\sigma}(t))\right]. \tag{4.61}$$

We attempt to follow a similar procedure as for the case of linearly parameterized approximators. In this case, the output estimation error $e(t) = \hat{\chi}(t) - \chi(t)$ satisfies

$$e(t) = \frac{\lambda}{s+\lambda}[\hat{f}(z(t);\hat{\theta},\hat{\sigma}) - \hat{f}(z(t);\theta^*,\sigma^*)],$$

which can also be written in state-space form as follows:

$$\dot{e} = -\lambda e + \lambda\left(\hat{f}(z(t);\hat{\theta},\hat{\sigma}) - \hat{f}(z(t);\theta^*,\sigma^*)\right).$$

Following the formulation of Chapter 2, $\hat{f}$ is assumed to be of the form

$$\hat{f}(z;\theta,\sigma) = \phi(z,\sigma)^\top\theta.$$

Using the Taylor series expansion

$$\hat{f}(z(t);\hat{\theta},\hat{\sigma}) - \hat{f}(z(t);\theta^*,\sigma^*) = \phi(z,\hat{\sigma})^\top\tilde{\theta} + \hat{\theta}^\top\frac{\partial\phi}{\partial\sigma}(z,\hat{\sigma})\tilde{\sigma} + \mathcal{F}(z(t);\tilde{\theta},\tilde{\sigma})),$$

where $\tilde{\theta} = \hat{\theta} - \theta^*$, $\tilde{\sigma} = \hat{\sigma} - \sigma^*$ are the parameter estimation errors and $\mathcal{F}$ is a term that contains the higher-order components of the Taylor series expansion. If these higher-order terms are ignored for the purpose of deriving adaptive laws for $\hat{\theta}(t)$ and $\hat{\sigma}(t)$, we obtain

$$\dot{\hat{\theta}}(t) = \Gamma_\theta\phi(z,\hat{\sigma})e(t) \tag{4.62}$$

$$\dot{\hat{\sigma}}(t) = \Gamma_\sigma\frac{\partial\phi}{\partial\sigma}(z(t),\hat{\sigma})^\top\hat{\theta}e(t) \tag{4.63}$$

where Γ_θ, Γ_σ are positive definite matrices representing the adaptive gains for the corresponding update laws for $\hat{\theta}(t)$ and $\hat{\sigma}(t)$, respectively.

We note that the adaptive laws (4.62), (4.63) are of similar form as the adaptive algorithm (4.58) obtained for linearly parameterized networks. Key differences include the presence of the higher-order terms $\mathcal{F}$, which can cause convergence problems; the presence of the argument $\hat{\sigma}$ in $\phi(z, \hat{\sigma})$ that can cause $\hat{\theta}$ to adapt in different directions at the same location z depending on the value of $\hat{\sigma}$; and the quantity $\frac{\partial \phi}{\partial \sigma}$ is a matrix that may have poor numeric properties for particular ranges of $(z, \hat{\sigma})$.

4.4.2 Optimization Methods

In this subsection, we present a methodology for applying optimization approaches to RFOL schemes. Even though, in principle, optimization methods can also be applied to the EFOL scheme, the combination of the error filtering formulation with optimization techniques is not suitable for deriving stable adaptive schemes since the filtering of the error function creates problems in the stability analysis. The presented optimization schemes are based on solid analytical properties, which are presented in Section 4.5. Since we restrict ourselves to the RFOL scheme, the optimization methodology is developed for linearly parameterized approximators. In Subsection 4.4.2.1 we present the gradient method, which is based on the principle of steepest descent. Then, in Subsection 4.4.2.2, we present the recursive least squares (RLS) method. In Subsection 4.4.2.3, we describe the backpropagation algorithm for supervised learning in static systems, which is an algorithm that has been extensively studied in the neural network literature.

4.4.2.1 Gradient Method. One of the most straightforward and widely used approaches for parameter estimation is the gradient (or steepest descent) method. The main idea behind the gradient method is to start with an initial estimate $\hat{\theta}(0) \in \Re^{M_1}$ of the unknown parameter θ^* and at each time t update the parameter estimate $\hat{\theta}(t)$ in the direction that yields the greatest rate of decrease in a certain suitable cost function $J(\hat{\theta})$. Several variations of the standard gradient algorithm have also been used in the parameter estimation literature. For example, the stochastic gradient approach leads to the well known *Least-Mean-Square* (LMS) algorithm, first developed by Widrow and Hoff [297, 299]. Another useful modification of the gradient algorithm is the *gradient projection* algorithm, which restricts the parameter estimates to be within a specified region [119].

In this section we focus on the deterministic, continuous-time version of the gradient learning algorithm. For continuous-time adaptive algorithms, infinitesimally small step lengths yield the following update law with respect to a specified cost function:

$$\dot{\hat{\theta}}(t) = -\nabla J(\hat{\theta}(t)),$$

where $\nabla J(\hat{\theta})$ denotes the gradient of the cost function J with respect to $\hat{\theta}$. A key consideration is the selection of the cost function $J(\hat{\theta})$, which needs to be selected such that the resulting update law is in terms of measurable quantities. For example, one might attempt to minimize the following desirable cost function: $J(\hat{\theta}) = |\tilde{\theta}|^2 = |\hat{\theta} - \theta^*|^2$; however, such a cost function leads to an update law which is in terms of the unknown parameter θ^* and cannot therefore be implemented.

To derive an implementable update law, based on eqns. (4.18) and (4.43), consider the cost function

$$J(\hat{\theta}) \quad = \quad \frac{\gamma}{2} e^2(t) \tag{4.64}$$

$$= \frac{\gamma}{2}\left(\hat{\chi}(t) - \chi(t)\right)^2$$
$$= \frac{\gamma}{2}\left(\hat{\theta}(t)^\top \zeta(t) - \chi(t)\right)^2$$

where $\gamma > 0$ is a positive design constant and the filtered MFAE, $\delta(t)$, is assumed to be zero for the time being. If we minimize this cost function using the gradient method we obtain the following adaptive law:

$$\dot{\hat{\theta}}(t) = -\gamma\zeta(t)e(t) \tag{4.65}$$

which is computable as discussed relative to (4.47).

We note that the adaptive law (4.65) is of the same general form as the adaptive law (4.54) which was derived using the Lyapunov synthesis method. Specifically, notice that both adaptive laws have three terms:

- The positive constant γ represents the adaptive gain, or, in the context of optimization theory, the step size. In discrete-time update laws the step size cannot be too large or otherwise it may cause divergence. In the case of continuous-time adaptation, the adaptive gain can be allowed to be any positive number. However, this is only in theory; in practice, there are some key trade-offs in the selection of the adaptive gain, even for continuous-time adaptation. Intuitively, if the adaptive gain is small then the adaptation and learning are slow. On the other hand, if the adaptive gain is large then adaptation is faster, however, in the presence of noise the approximator may over-react to random effects. This may lead the parameter estimate to become unbounded. Therefore, even though the theory of continuous-time adaptation for the ideal case may indicate that large adaptive gains are acceptable (and may result in faster learning), the designer needs to judiciously select this design variable based on the specific application and any a priori information about the measurement noise levels. As we will see later in the design of robust adaptive schemes (Section 4.6), other type of modeling errors can also play a crucial role in the selection of the adaptive gain.

- The second term, $\zeta(t)$, is the filtered regressor. Recall that there is a close relationship between ζ, which is used here, and the regressor $\phi(z(t))$, which is used in the adaptive law (4.54) derived using the Lyapunov synthesis method. This relationship is described by

$$\zeta(t) = \frac{\lambda}{s+\lambda}\left[\phi(z(t))\right],$$

or in the case of a general filter $W(s)$, the relationship is given by

$$\zeta(t) = W(s)\left[\phi(z(t))\right].$$

From (4.65), it is clear that if the filtered regressor becomes zero then adaptation stops, even if the error $e(t)$ is non-zero. Intuitively, the regressor can be thought of as containing the information used by the learning approach to allocate the error $e(t)$ among the elements of the parameter estimate $\hat{\theta}$. If the filtered regressor is zero (i.e., no allocation information) then the error $e(t)$ is not allocated to any element of the parameter estimate and nothing is learned. Similarly, if the regressor is non-zero, but contains the same allocation information repeatedly, then the learning scheme is able to learn that specific information but nothing else. This is closely related to the issue of persistency of excitation (see Section 4.5.4), which requires the regressor to change sufficiently over any time interval in order for the parameter estimate vector $\hat{\theta}(t)$ to converge to its true vector value θ^*.

- The third term, $e(t)$, is the measurable output estimation error. This can be viewed as the feedback information for the learning scheme. If the error $e(t)$ is non-zero, it provides two key pieces of information to the learning system: (i) the sign of $e(t)$ indicates to the learning scheme the direction in which the parameter estimate vector should be changed to enhance learning; (ii) the magnitude of $e(t)$ indicates to the learning scheme by how much to update: large errors require larger modifications, while small errors require only small modifications in the weights of the approximator. If for some period of time $t \in [t_0, \ t_1]$ the error $e(t) \approx 0$, where $(t_1 - t_0) > \lambda$ this implies that the learning system already knows (or has already learned) this information (contained in the parameter subspace spanned by the regressor $\zeta(t)$ for $t \in [t_0, \ t_1]$) and therefore there is no need to make any modifications to the value of its parameter estimate vector during this time period. If we use the analogy of classroom teaching, if the professor lectures on material that the students are already familiar with, there is no learning taking place (surprise, surprise!).

The adaptive law described by (4.65) can be generalized to the case where the scalar adaptive gain γ is replaced by a positive definite matrix Γ of dimension q_θ-by-q_θ, where q_θ is the dimension of $\hat{\theta}(t)$. This is achieved by re-scaling the optimization problem [157]. In this case, the adaptive law becomes

$$\dot{\hat{\theta}}(t) = -\Gamma\zeta(t)e(t). \tag{4.66}$$

The *normalized gradient algorithm* is a variation of the gradient algorithm, which is sometimes used to improve the stability and convergence properties of the algorithm. The normalized gradient algorithm is described by

$$\dot{\hat{\theta}}(t) = -\Gamma\zeta(t)\frac{e(t)}{1 + \beta\zeta^\top(t)\zeta(t)} \tag{4.67}$$

where $\beta \geq 0$ is a design constant. If β is set to zero, the we obtain the standard (non-normalized) gradient adaptive law.

The stability properties of the gradient algorithm are discussed in Section 4.5.2, while the non-ideal case of $\delta(t) \neq 0$ and the derivation of robust learning algorithms are examined in more detail in Section 4.6.

In this section we focused on an instantaneous cost function of a simple quadratic form. The parameter estimation literature also contains some more advanced gradient algorithms which are based on more complex cost functions. One such cost function that has attracted some attention is the integral cost function of the form

$$J(\hat{\theta}) = \frac{1}{2}\int_0^t e^{-\beta(t-\tau)}\left(\hat{\theta}(t)^\top\zeta(\tau) - \chi(\tau)\right)^2 d\tau.$$

The application of the gradient method to this cost function yields a new adaptive law whose stability properties have been investigated in [119, 138].

4.4.2.2 Least Squares Algorithms. Least squares methods have been widely used in parameter estimation both in batch (nonrecursive) and in recursive form [11, 119]. The basic idea behind the least squares method is to fit a mathematical model to a sequence of observed data by minimizing the sum of the squares of the difference between the observed and computed data. To illustrate the least squares method, consider the problem of computing the parameter vector $\hat{\theta}$ at time t that minimizes the cost function

$$J(\hat{\theta}) = \int_0^t \left(\hat{\theta}(t)^\top \zeta(\tau) - \chi(\tau)\right)^\top \left(\hat{\theta}(t)^\top \zeta(\tau) - \chi(\tau)\right) d\tau, \tag{4.68}$$

where $\chi(\tau)$ is the measured data at time τ, and $\zeta(\tau)$ is the filtered regressor vector at time τ. The above cost function penalizes all the past errors $\zeta(\tau)^\top \hat{\theta}(t) - \chi(\tau)$ for $\tau \in [0, t]$. By setting the gradient (with respect to $\hat{\theta}$) of the cost function to zero ($\nabla J(\hat{\theta}) = 0$), we obtain the least squares estimate for $\hat{\theta}(t)$:

$$\hat{\theta}(t) = \left[\int_0^t \zeta(\tau)\zeta(\tau)^\top d\tau\right]^{-1} \int_0^t \zeta(\tau)\chi(\tau)^\top d\tau, \tag{4.69}$$

provided that the inverse exists. The validity of this assumption is determined by the level of regressor excitation. In the above formulation, we have considered the general case where $\chi(t)$ is a vector (say of dimension m), which implies that $\hat{\theta}(t)$ is a matrix of dimension q_θ-by-m.

The least squares estimate given by (4.69) is derived for batch processing; in other words, all the data in the time interval $[0,\ t]$ is gathered before it is processed. In adaptive approximation, the estimated parameter vector $\hat{\theta}(t)$ needs to be computed in real-time, as new data becomes available. The recursive version of the least squares algorithm for the vector $\hat{\theta}$ is given by

$$\dot{\hat{\theta}}(t) = -P(t)\zeta(t)e(t), \qquad \hat{\theta}(0) = \hat{\theta}_0 \tag{4.70}$$

$$\dot{P}(t) = -P(t)\zeta(t)\zeta(t)^\top P(t), \qquad P(0) = P_0 \tag{4.71}$$

where $P(t)$ is a square matrix of the same dimension as the parameter estimate $\hat{\theta}(t)$. The initial condition P_0 of the P matrix is chosen to be positive-definite. In applications where the measurements are corrupted by noise, the least squares algorithm can be derived within a stochastic framework. In such a derivation, the matrix P represents the covariance of the parameter estimation error. In deterministic analysis, even though this interpretation is not applicable, P is often referred to as the *covariance matrix*.

It is interesting to note that the update law for $\hat{\theta}$, described by (4.70), is similar to the gradient learning algorithm (4.66), with $P(t)$ representing a time-varying learning rate. In practice, recursive least squares can converge considerably faster than the gradient algorithm at the expense of the increased computation required to compute P. However, in its "pure" form the recursive least squares may result in the covariance matrix $P(t)$ becoming arbitrarily small. This problem, which is referred to as the *covariance wind-up* problem, can slow down adaptation in some directions and, as a result, critically dampen the ability of the algorithm to track time-varying parameters.

Several modifications to the "pure" least squares algorithm have been considered. One such modification is *covariance resetting* according to which the covariance matrix is reset to $P(t_r) = P_0$ at time t_r if the minimum eigenvalue of $P(t_r)$ is less then a predefined small positive constant. This modification helps in preventing the covariance matrix from becoming too small, but may result in large estimation transients immediately following $t = t_r$. A second commonly used modification to the least squares algorithm leads to the *least squares with forgetting factor*, which is given by

$$\dot{\hat{\theta}}(t) = -P(t)\zeta(t)e(t) \quad (4.72)$$
$$\dot{P}(t) = -P(t)\zeta(t)\zeta(t)^{\mathsf{T}}P(t) + \mu P(t) \quad (4.73)$$

where $\mu > 0$ is typically a small positive constant, referred to as the forgetting factor. The extra term $\mu P(t)$ in (4.73) prevents the covariance matrix from becoming too small, but it may, on the other hand, cause it to become too large. To avoid this complication, $P(t)$ is either reset to P_0 or adaptation is disabled (i.e., $\dot{P}(t) = 0$) in the case that $P(t)$ becomes too large. The literature on parameter estimation and adaptive control has several rules of thumb on how to choose the design variables that appear in the least squares algorithm and its various modified versions [119]. The stability and convergence properties of the least squares algorithm are presented in Section 4.5.3.

4.4.2.3 Error Backpropagation Algorithm The error backpropagation algorithm (or simply backpropagation algorithm) is a learning method that has been studied and applied extensively in the neural networks literature. It appears that the term *backpropagation* was first used around 1985 [227] and became popular with the publication of the seminal edited book by Rumelhart and McClelland [226]. However, the backpropagation algorithm was discovered independently by two other researchers at about the same time [145, 194]. After the error backpropagation algorithm became popular, it was found out that the algorithm had also been described earlier by Werbos in his doctoral thesis in 1974 [291]. Moreover, the basic idea behind the backpropagation algorithm can be traced even further back to the control theory literature, and specifically the book by Bryson and Ho [33].

In hindsight, it is not surprising that the error backpropagation algorithm was independently discovered by so many researchers over the years, since it is based on the well-known steepest descent method, as it applies to the multi-layer perceptron. In this subsection, we will describe briefly the error backpropagation algorithm for the training of multi-layer perceptrons and we will relate it to the other learning algorithms that we developed in this chapter.

The error backpropagation algorithm is derived by using the steepest descent optimization method on the multi-layer perceptron. It provides a computationally efficient method for training multi-layer perceptrons due to the fact that it can be implemented in a distributed manner. Moreover, the derivation of the local gradient for each network weight (parameter) can be computed by propagating the error through the network in reverse; i.e., in the opposite direction of processing the input signal. This is the reason for being called *error backpropagation algorithm*. In contrast to the other learning algorithms developed in this chapter for adaptive approximation of dynamic systems, the backpropagation development herein is based on *supervised learning* for static systems.

The multi-layer perceptron was described in Section 3.6. The input-output ($z \mapsto y$) relationship of a multi-layer perceptron with n inputs, a single output, and one hidden layer with q_θ nodes is given by

$$v_i = g\left(b_i + \sum_{j=1}^{n} w_{ij} z_j\right) \qquad i = 1, \ldots q_\theta$$

$$y = \sum_{i=1}^{q_\theta} \theta_i v_i$$

where z_j is the j-th input, y is the output, θ_i, b_i, w_{ij} (for $i = 1, \ldots q_\theta$ and $j = 1, \ldots n$) are the adjustable weights and $g : \Re^1 \mapsto \Re^1$ is the activation function. As discussed in Section 3.6, the activation function is typically a squashing function, where the output is constrained within a bounded interval. Two examples of squashing functions are:

$$\begin{aligned} g(x) &= \tanh(x) & g &: \Re^1 \mapsto [-1,\ 1] \\ g(x) &= \frac{1}{1+e^x} & g &: \Re^1 \mapsto [0,\ 1]. \end{aligned}$$

Let us consider the problem of discrete-time supervised learning by minimizing the quadratic error function

$$J(\theta_i, b_i, w_{ij}) = \frac{1}{2}e^2(k) = \frac{1}{2}\left(y(k) - y^*(k)\right)^2$$

where $y^*(k) = f(z(k))$ is the target output at sample time k. Let ϑ denote one of the adjustable weights of the multi-layer perceptron. Then, according to the steepest descent optimization method, the update law for $\vartheta(k)$ is given by

$$\begin{aligned} \vartheta(k+1) &= \vartheta(k) - \gamma\frac{\partial J}{\partial \vartheta} \\ &= \vartheta(k) - \gamma e(k)\frac{\partial y}{\partial \vartheta}. \end{aligned}$$

If ϑ is one of the output weights θ_i then

$$\frac{\partial y}{\partial \vartheta} = v_i.$$

If ϑ is one of the input weights b_i or w_{ij} then by the chain rule

$$\frac{\partial y}{\partial \vartheta} = \frac{\partial y}{\partial v_i}\cdot\frac{\partial v_i}{\partial u_i}\cdot\frac{\partial u_i}{\partial \vartheta}$$

where $v_i = g(u_i)$ and $u_i = b_i + \sum_{j=1}^{n} w_{ij}z_j$. We note that:

- $\frac{\partial y}{\partial v_i}$ corresponds to the output weight θ_i;
- $\frac{\partial v_i}{\partial u_i}$ is the derivative of g evaluated at u_i, which is denoted by $g'(u_i)$;
- $\frac{\partial u_i}{\partial \vartheta}$ is equal to 1 for the offset weights b, and corresponds to z_j for the weight parameter w_{ij}.

These partial derivatives illustrate how the error is propagating backwards in the network as the gradient for each weight is located closer to the input layer. Using the chain rule, this idea can be easily further extended to multi-layer perceptrons with more than one hidden layer. The same ideas can also be extended to apply to the nonlinear parameters of any other network type, for example the centers of radial basis function networks.

The error backpropagation algorithm development above is for static systems. When the unknown nonlinearity is a portion of a differential equation, especially in control applications, the target output of the function approximator y^* may not be available for measurement; therefore, the error measure needs to use a different output. In the formulation that we derived in this chapter, the measurable output, which is used to generate the so-called output error, is denoted by $\hat{\chi}$. Therefore the standard backpropagation algorithm is not directly applicable to the adaptive approximation problem considered in this chapter, but it can be used indirectly, as a component of the adaptive law, in the computation of the partial derivative if the multi-layer perceptron is used as the adaptive approximator.

Furthermore, it is worth noting that the concept of the error backpropagation algorithm has also been extended to dynamical systems using learning algorithms such as *dynamic backpropagation* [181] and *backpropagation through time* [292], although the stability properties of these algorithms are not established. One of the difficulties associated with dynamic backpropagation type of algorithms is the fact it yields adaptive laws that typically require the sensitivity of the output with respect to variations in the unknown parameters θ^*. Since these sensitivity functions are not available, implementation of such adaptive laws is not possible and instead the designer needs to use an approximation of the sensitivity functions instead of the actual ones. One type of approximation used in dynamic backpropagation is to replace the gradient with respect to the unknown parameters by the gradient with respect to the estimated parameters. Such adaptive laws were used extensively in the early neural control literature, and simulations indicated that they performed well under certain conditions. Unfortunately, with approximate sensitivity functions, it is not possible, in general, to prove stability and convergence. It is interesting to note that approximate sensitivity function approaches also appeared in the early days of adaptive linear control, in the form of the so-called MIT rule [124].

4.4.3 Summary

In the previous sections we have developed a number of learning schemes. At this time, the reader maybe overwhelmed by the different possible combinations. For example, one could employ the error filtering scheme or the regressor filtering scheme; in the derivation of the update law, there is the option of using a the Lyapunov synthesis method or optimization approaches such as the gradient and the recursive least squares. Moreover, there are options in selecting the filter: as we discussed, one could proceed with a first-order filter of the form $\frac{\lambda}{s+\lambda}$ or a more complicated filter $W(s)$. There is also the selection of the approximator, which can be linearly or nonlinearly parameterized. Lastly, within each selection there are a number of design constants that need to be selected. In this subsection, we attempt to put some order in the design of learning schemes by tabulating some of the different schemes. The reader can obtain a better understanding of the issues by simulating the learning schemes and varying some of the design variables.

Table 4.1 summarizes the design options for the Error Filtering Online Learning (EFOL) scheme. The stability properties of this approach are summarized in Theorem 4.5.1.

Table 4.2 summarizes the design options for the Regressor Filtering Online Learning (RFOL) scheme, which is only applicable for LIP approximators. The stability properties of this approach are summarized in Theorems 4.5.2 and 4.5.3.

4.5 ONLINE LEARNING: ANALYSIS

The previous three sections have introduced the idea of designing parametric models, learning schemes, and parameter estimation algorithms; the overall adaptive approximation scheme was presented with a minimum of formal analysis. In this section, we examine the stability and convergence properties of the developed learning schemes. In addition to obtaining guarantees about the performance of the learning scheme, this stability analysis provides valuable intuition about the underlying properties of the online learning methods and in the selection of the design variables. The formal analysis of this section only consid-

Table 4.1: Error Filtering Online Learning (EFOL) scheme.

Plant	$\dot{x} = f_0(x,u) + f^*(x,u)$
Online Learning Model	$\dot{\xi} = -\lambda\xi + \lambda^2 x + \lambda f_0(x,u) + \lambda \underbrace{\hat{f}(x,u;\hat{\theta},\hat{\sigma})}_{\text{Approximator}}$ $e = \xi - \lambda x$
Adaptive Law	$\dot{\hat{\theta}} = -\Gamma\phi(x,u;\hat{\sigma})e$ $\dot{\hat{\sigma}} = \begin{cases} 0 & \text{if approximator is LIP} \\ -\Gamma\frac{\partial\phi(x,u;\hat{\sigma})}{\partial\sigma}\hat{\theta}e & \text{if approximator is NLIP} \end{cases}$
Design Variables	λ: filtering constant Γ: adaptive gain matrix $\hat{\theta}(0)$: initial parameter estimate $\hat{f}(\cdot)$: Adaptive Approximator

ers the case where $\delta = 0$. This section will informally discuss the $\delta \neq 0$ case to motivate the formal analysis of that case which is presented in Section 4.6.

4.5.1 Analysis of LIP EFOL Scheme with Lyapunov Synthesis Method

First, we consider the EFOL scheme with the adaptive law derived using the Lyapunov synthesis method. The following theorem, describes the properties of this learning scheme, with a linearly parameterized approximator and a first-order filter.

Theorem 4.5.1 *The learning scheme described in Table 4.1 with a linear parametric model (and $\delta = 0$) has the following properties:*

- $e(t) \in \mathcal{L}_2 \cap \mathcal{L}_\infty$, $\quad\hat{\theta}(t) \in \mathcal{L}_\infty$, $\quad\hat{\chi}(t) \in \mathcal{L}_\infty$.

If, in addition, the regressor vector ϕ is uniformly bounded (i.e., $\phi(z(t)) \in \mathcal{L}_\infty$) then the following properties also hold:

- $\lim_{t\to\infty} e(t) = 0$, $\quad\lim_{t\to\infty} \dot{\hat{\theta}}(t) = \lim_{t\to\infty} \dot{\tilde{\theta}}(t) = 0$.

Table 4.2: Regressor Filtering Online Learning (RFOL) scheme.

Plant	$\dot{x} = f_0(x,u) + f^*(x,u)$	
Online Learning Model	$\dot{\xi} = -\lambda\xi + \lambda^2 x + \lambda f_0(x,u)$ $\dot{\zeta} = -\lambda\zeta + \lambda\phi(x,u;\hat{\sigma})$ $e = \xi - \lambda x + \zeta^\top \hat{\theta}$	
Adaptive Laws	$\dot{\hat{\theta}} = -\Gamma\zeta e$	Gradient Algorithm
	$\dot{\hat{\theta}} = -\frac{\Gamma\zeta e}{1+\beta\|\zeta\|^2}$	Normalized Gradient Algorithm
	$\dot{\hat{\theta}} = -P\zeta e$ $\dot{P} = -P\zeta\zeta^\top P$	Recursive Least Squares Algorithm
	$\dot{\hat{\theta}} = -P\zeta e$ $\dot{P} = -P\zeta\zeta^\top P + \mu P$	Recursive Least Squares Algorithm with Forgetting Factor
Design Variables	λ: filtering constant Γ: adaptive gain matrix β: normalizing constant μ: forgetting factor $\hat{\theta}(0)$: initial parameter estimate $P(0)$: initial covariance matrix $\phi(\cdot)$: Basis Function of Adaptive Approximator	

Proof: Based on (4.55) and (4.56), the output estimation error $e(t) = \hat{\chi}(t) - \chi(t)$ satisfies the differential equation

$$\dot{e}(t) = -\lambda e(t) + \lambda \tilde{\theta}^\top(t)\phi(z(t)). \tag{4.74}$$

Consider the Lyapunov function candidate

$$V(e, \tilde{\theta}) = \frac{\mu}{2\lambda}e^2 + \frac{\mu}{2}\tilde{\theta}^\top \Gamma^{-1}\tilde{\theta} \tag{4.75}$$

where μ is a positive constant. By taking the time derivative of V along the differential equations (4.74) and (4.58), and using the fact that θ^* is constant we obtain

$$\begin{aligned} \dot{V} &= \frac{\mu}{\lambda}e\dot{e} + \mu\tilde{\theta}^\top\Gamma^{-1}\dot{\tilde{\theta}} \\ &= -\mu e^2. \end{aligned} \tag{4.76}$$

We are now in a position to utilize Lemma A.3.1 to show that $e(t) \in \mathcal{L}_2$, $\lim_{t\to\infty} e(t) = 0$, $e(t) \in \mathcal{L}_\infty$ and $\tilde{\theta}(t) \in \mathcal{L}_\infty$. Moreover, since θ^* is a finite constant, $\hat{\theta}(t) = \tilde{\theta}(t) + \theta^*$ is also uniformly bounded (i.e., $\hat{\theta}(t) \in \mathcal{L}_\infty$). Finally, since $\dot{\hat{\theta}}(t) = \Gamma\phi e$, with $\phi \in \mathcal{L}_\infty$ and $e(t) \to 0$, it can be readily seen that $\lim_{t\to\infty} \dot{\hat{\theta}}(t) = \lim_{t\to\infty} \dot{\tilde{\theta}}(t) = 0$. ■

If the first-order filter $\frac{\lambda}{s+\lambda}$ is replaced by a general filter $W(s)$, which is Strictly Positive Real (SPR), then it is possible to obtain similar results. The details of the proof for an SPR filter is left as an exercise (see Exercise 4.7).

Effect of model error. In the case where $\delta \neq 0$, the error dynamics of eqn. (4.74) become

$$\dot{e}(t) = -\lambda e(t) + \lambda\tilde{\theta}^\top(t)\phi(z(t) - \lambda e_f, \tag{4.77}$$

where the relation between δ and e_f is given by

$$\delta(t) = \frac{\lambda}{s+\lambda}[e_f(x(t), u(t))].$$

Therefore, the derivative of the same Lyapunov function becomes

$$\dot{V} = -\mu e^2 - \mu e_f e \tag{4.78}$$

which is not negative definite. Note that

$$\dot{V} \leq -\mu|e|\left(|e| - |e_f|\right). \tag{4.79}$$

Therefore, $\dot{V}$ is only guaranteed to be negative semidefinite when $|e| \geq |e_f|$. When $|e| < |e_f|$, the Lyapunov function may increase. In fact, there is no bound on $\|\tilde{\theta}\|$ while $|e| < |e_f|$. Let (t_1, t_2) denote a time period for which $|e| < |e_f|$. In this time period, it is possible for $\|\tilde{\theta}\|$ to grow large, while maintaining $\tilde{\theta}(t)^\top\phi(z(t)) \approx 0$. If at $t = t_2$, the vector $\phi(z(t))$ changes significantly due to changes in $z(t)$, then $\tilde{\theta}(t_2)^\top\phi(z(t_2))$ can become large which causes $|e|$ to become large. Therefore, even if it is known that $|e_f(t)| \leq \bar{\epsilon}$ for all $t > 0$, where $\bar{\epsilon}$ is a small positive constant, it is *not* valid to state that $|e(t)|$ is ultimately bounded by $|\bar{\epsilon}|$.

Therefore, in the presence of noise, disturbances, or modeling errors that can be represented by e_f, there are no guaranteed stability or performance properties. Appropriate robust methods to recover these properties will be discussed in Section 4.6.

4.5.2 Analysis of LIP RFOL Scheme with the Gradient Algorithm

Here we consider the RFOL scheme with the adaptive law derived using the gradient optimization method. The following theorem, describes the properties of this learning scheme. As we will see, these properties are similar to the corresponding stability properties obtained for the EFOL scheme with the Lyapunov synthesis method.

Theorem 4.5.2 *The normalized gradient algorithm (4.67) with the RFOL scheme (with $\delta(t) = 0$) has the following properties:*

- $\hat{\theta}(t) \in \mathcal{L}_\infty$, $\quad \dot{\hat{\theta}}(t) \in \mathcal{L}_\infty$, $\quad \frac{e(t)}{\sqrt{1+\beta\|\zeta(t)\|^2}} \in \mathcal{L}_2 \cap \mathcal{L}_\infty$.

If, in addition, the regressor vector $\zeta(t)$ is uniformly bounded then the following properties also hold:

- $\hat{\chi}(t) \in \mathcal{L}_\infty$, $\quad e(t) \in \mathcal{L}_2 \cap \mathcal{L}_\infty$
- $\lim_{t\to\infty} e(t) = 0$, $\quad \lim_{t\to\infty} \dot{\hat{\theta}}(t) = \lim_{t\to\infty} \dot{\tilde{\theta}}(t) = 0$.

Proof: Since it is assumed that $\delta(t) = 0$, from (4.48) we have that the output estimation error satisfies

$$e(t) = \tilde{\theta}(t)^\top \zeta(t). \tag{4.80}$$

Consider the Lyapunov function candidate

$$V(\tilde{\theta}) = \frac{1}{2}\tilde{\theta}^\top \Gamma^{-1}\tilde{\theta}.$$

By taking the time derivative of V along the solution of the differential equation (4.67) we obtain

$$\dot{V} = \frac{-\tilde{\theta}(t)^\top \zeta(t) e(t)}{1+\beta\|\zeta(t)\|^2} \tag{4.81}$$

$$= \frac{-|e(t)|^2}{1+\beta\|\zeta(t)\|^2} \leq 0. \tag{4.82}$$

Since $\dot{V}$ is negative semidefinite, $V, \tilde{\theta} \in \mathcal{L}_\infty$. This implies that $\hat{\theta}(t) \in \mathcal{L}_\infty$.

Furthermore, $\dot{V}(t) \leq 0$ and $V(t) \geq 0$ implies that $V(t)$ converges to some value; i.e., $\lim_{t\to\infty} V(t) = V_\infty$ exists and is finite. By taking the integral of (4.82) for $t \in [0,\infty)$ we obtain that

$$\int_0^\infty \frac{|e(t)|^2}{1+\beta\|\zeta(t)\|^2} dt < \infty.$$

Therefore,

$$\frac{e(t)}{\sqrt{1+\beta\|\zeta(t)\|^2}} \in \mathcal{L}_2.$$

Note that for any $\zeta(t)$,

$$\frac{\|\zeta(t)\|^2}{1+\beta\|\zeta(t)\|^2} \in \mathcal{L}_\infty;$$

therefore, since $\tilde{\theta} \in \mathcal{L}_\infty$ we obtain

$$\frac{|e(t)|^2}{1+\beta\|\zeta(t)\|^2} = \frac{|\tilde{\theta}(t)^\top \zeta(t)|^2}{1+\beta\|\zeta(t)\|^2} \in \mathcal{L}_\infty.$$

This implies

$$\frac{e(t)}{\sqrt{1+\beta\|\zeta(t)\|^2}} \in \mathcal{L}_2 \cap \mathcal{L}_\infty.$$

Moreover,

$$\dot{\hat{\theta}} = \frac{-\Gamma e \zeta}{1+\beta\|\zeta(t)\|^2} = \frac{-\Gamma \zeta^\top \tilde{\theta} \zeta}{1+\beta\|\zeta(t)\|^2}.$$

Therefore, we obtain that $\dot{\hat{\theta}} \in \mathcal{L}_\infty$.

Now, if we assume that $\zeta(t)$ is uniformly bounded, we can easily obtain that $\hat{\chi}(t) = \hat{\theta}(t)^\top \zeta(t) \in \mathcal{L}_\infty$ and $e(t) \in \mathcal{L}_2 \cap \mathcal{L}_\infty$. Next, consider the error derivative

$$\dot{e}(t) = \dot{\tilde{\theta}}(t)^\top \zeta(t) + \tilde{\theta}(t)^\top \dot{\zeta}(t).$$

Using the normalized adaptive law for $\hat{\theta}(t)$ and the fact that $e(t), \zeta(t) \in \mathcal{L}_\infty$, we obtain $\dot{\tilde{\theta}} \in \mathcal{L}_\infty$. Moreover, since $\zeta(t) = W(s)[\phi(z(t))]$ is the output of a stable SPR filter $W(s)$ with a bounded input ϕ, we obtain that $\dot{\zeta} \in \mathcal{L}_\infty$. Therefore, $\dot{e} \in \mathcal{L}_\infty$. Since $e \in \mathcal{L}_2 \cap \mathcal{L}_\infty$ and $\dot{e} \in \mathcal{L}_\infty$, using Barbălat's Lemma we conclude that $\lim_{t\to\infty} e(t) = 0$. Moreover, it can be readily seen that $\lim_{t\to\infty} \dot{\hat{\theta}}(t) = \lim_{t\to\infty} \dot{\tilde{\theta}}(t) = 0$. ■

It is important to note that even in the restrictive case of no approximation errors and a linearly parameterized approximator, it cannot be established that the parameter estimate vector $\hat{\theta}(t)$ will converge to the optimal vector θ^*. To guarantee that $\hat{\theta}(t)$ will converge to θ^*, the regressor vector $\zeta(t)$ needs to satisfy a so-called *persistency of excitation* condition. Intuitively, this implies that there should be sufficient variation in $\zeta(t)$ to allow the parameter estimates to converge to their optimal values. The concept of persistency of excitation is discussed in Section 4.5.4.

Effect of model error. In the presence of approximation errors (i.e., $\delta(t) \neq 0$), the eqn. (4.80) becomes

$$e(t) = \tilde{\theta}(t)^\top \zeta(t) - \delta(t).$$

Therefore, the derivative of the Lyapunov function becomes

$$\dot{V} = \frac{-e(t)\left(e(t) + \delta(t)\right)}{1+\beta\|\zeta(t)\|^2}. \tag{4.83}$$

This is only negative semidefinite if $e(t)^2 > -\delta(t)e(t)$ for all t. Even if $\delta(t)$ is known to be upper bounded, this condition cannot be guaranteed for small $e(t)$. Therefore, the stability of the gradient algorithm (4.66) cannot be guaranteed. In fact, it is known from adaptive parameter estimation of linear systems that even if $\delta(t)$ is a small signal it can be sufficient to make the adaptive system unstable. The instability is typically caused by drift of the adaptive parameter estimates. To address this problem, the standard update law described by (4.66) needs to be modified. Several modifications exist in the literature for enhancing the robustness of adaptive schemes. These modifications are discussed in Section 4.6.

4.5.3 Analysis of LIP RFOL Scheme with RLS Algorithm

The recursive least squares (RLS) algorithm described by (4.70)–(4.71) has similar stability properties as the gradient algorithm.

Theorem 4.5.3 *The Recursive Least Squares algorithm (4.70)–(4.71) with the RFOL scheme (with $\delta = 0$) has the following properties:*

- $\hat{\theta}(t) \in \mathcal{L}_\infty$, $\quad P(t) \in \mathcal{L}_\infty$, $\quad e(t) \in \mathcal{L}_2$
- $\lim_{t\to\infty} \hat{\theta}(t) = \bar{\theta}$, $\quad \lim_{t\to\infty} P(t) = P_\infty$ $\quad$ *(where $\bar{\theta}$, P_∞ are constants).*

If, in addition, the regressor vector $\zeta(t)$ is uniformly bounded, then the following properties also hold:

- $e(t) \in \mathcal{L}_\infty$, $\quad \hat{\chi}(t) \in \mathcal{L}_\infty$
- $\lim_{t\to\infty} e(t) = 0$, $\quad \lim_{t\to\infty} \dot{\hat{\theta}}(t) = \lim_{t\to\infty} \dot{\tilde{\theta}}(t) = 0$.

Proof: From (4.71) we note that $P(t)$ is symmetric for all $t \geq 0$. Moreover, $P(t) \geq 0$ and bounded from below; therefore, $P(t)$ has a limit: $\lim_{t\to\infty} P(t) = P_\infty$, where P_∞ is a constant positive definite matrix.

Using the fact $P^{-1}P = I$, we obtain the identity

$$\frac{d}{dt}(P^{-1}) = \dot{P}^{-1} = -P^{-1}\dot{P}P^{-1}. \tag{4.84}$$

Now, consider the time derivative of $P(t)^{-1}\tilde{\theta}(t)$. Using the RLS algorithm (4.70)–(4.71) and the identity (4.84) we obtain

$$\begin{aligned}
\frac{d}{dt}(P(t)^{-1}\tilde{\theta}(t)) &= \dot{P}^{-1}\tilde{\theta} + P^{-1}\dot{\tilde{\theta}} \\
&= -P^{-1}\dot{P}P^{-1}\tilde{\theta} + P^{-1}\dot{\tilde{\theta}} \\
&= \zeta\zeta^\top\tilde{\theta} - \zeta e \\
&= \zeta e - \zeta e \;=\; 0.
\end{aligned}$$

Therefore, $P(t)^{-1}\tilde{\theta}(t) = P(0)^{-1}\tilde{\theta}(0)$, which implies

$$\begin{aligned}
\lim_{t\to\infty} \tilde{\theta}(t) &= \lim_{t\to\infty} P(t)P(0)\tilde{\theta}(0) \\
&= P_\infty P(0)\tilde{\theta}(0) = \bar{\theta}.
\end{aligned}$$

So far we have established that $\tilde{\theta}$, $\hat{\theta} \in \mathcal{L}_\infty$ and that $\bar{\theta} = \lim_{t\to\infty} \tilde{\theta}(t)$ and $P_\infty = \lim_{t\to\infty} P(t)$ exist.

Now consider the Lyapunov function candidate

$$V(\tilde{\theta}, P) = \frac{1}{2}\tilde{\theta}(t)^\top P(t)^{-1}\tilde{\theta}(t).$$

The time derivative of V along (4.70), (4.71) satisfies

$$\dot{V} \;=\; \tilde{\theta}^\top P^{-1}\dot{\tilde{\theta}} + \frac{1}{2}\tilde{\theta}^\top \dot{P}^{-1}\tilde{\theta}$$

$$\begin{aligned}
&= -\tilde{\theta}^\top \zeta e - \frac{1}{2}\tilde{\theta}^\top P^{-1}\dot{P}P^{-1}\tilde{\theta} \\
&= -e^2 + \frac{1}{2}\tilde{\theta}^\top \zeta\zeta^\top \tilde{\theta} \\
&= -e^2 + \frac{1}{2}e^2 \\
&= -\frac{1}{2}e^2.
\end{aligned}$$

This implies $V \in \mathcal{L}_\infty$ and $e \in \mathcal{L}_2$. If $\zeta(t)$ is uniformly bounded then $e \in \mathcal{L}_\infty$. Using a similar procedure as in the stability proof of the gradient algorithm, we obtain that $\dot{e} \in \mathcal{L}_\infty$. Therefore, using Barbălat's Lemma we conclude that $\lim_{t\to\infty} e(t) = 0$. ∎

In comparing the stability properties of the gradient and least squares algorithms we notice that in addition to the other boundedness and convergence properties, the recursive least squares also guarantees that the parameter estimate $\hat{\theta}(t)$ converges to a constant vector $\bar{\theta}$. If the regressor vector ζ satisfies the persistency of excitation condition then $\hat{\theta}(t)$ converges to the optimal parameter vector θ^*.

Despite its fast convergence properties, the recursive least squares algorithm has not been widely used in problems involving large function approximation structures, mainly due to its heavy computational demands. Specifically, if the number of adjustable parameters is N, then updating of the covariance matrix $P(t)$ requires adaptation of N^2 parameters. Issues related to least-squares-based learning and its computational requirements are discussed in some detail in Exercise 4.4. An alternative *locally weighted learning* approach that can have considerably smaller computational requirements, referred to as *receptive field weighted regression* [13, 236, 237], is discussed in Exercise 4.5.

Effect of model error. When $\delta(t) \neq 0$, then $e(t) = \tilde{\theta}(t)^\top \zeta(t) - \delta(t)$. Therefore, the derivative of the Lyapunov function becomes

$$\begin{aligned}
\dot{V} &= -\tilde{\theta}^\top \zeta e + \frac{1}{2}\tilde{\theta}^\top \zeta\zeta^\top \tilde{\theta} \\
&= -(e+\delta)e + \frac{1}{2}(e+\delta)^2 \\
&= -\frac{1}{2}e^2 + \frac{1}{2}\delta^2.
\end{aligned}$$

Therefore, $\dot{V}$ is negative semidefinite only if $|e(t)| > |\delta(t)|$ for all t. Once $|e(t)|$ becomes smaller than $|\delta(t)|$ then the derivative becomes positive.

4.5.4 Persistency of Excitation and Parameter Convergence

In Section 4.5 it was established that under certain conditions the parameter estimates remain bounded and the output estimation error converges to zero asymptotically. We also saw that the various adaptive approximation schemes presented in this section could not establish that the parameter estimation vector $\hat{\theta}(t)$ will converge to the optimal parameter vector θ^*, even in the special case of linearly parameterized approximators with no approximation error ($\delta(t) = 0$).

The observation that it is possible for the output error $e = y - \hat{y}$ to be zero while the parameter estimation error $\tilde{\theta} = \hat{\theta} - \theta^*$ is non-zero was also made in Example 4.1 for the linear case and in Example 4.2 for the nonlinear case, where for certain inputs the output estimation error $e(t) \to 0$, while the parameter estimate $\hat{\theta}(t) \to \hat{\theta}_\infty \neq \theta^*$.

In this subsection, we consider the issue of parameter convergence, and present conditions under which the parameter estimation error $\tilde{\theta}(t) = \hat{\theta} - \theta^*$ converges to zero. Convergence conditions are related to the issue of persistency of excitation, which is an important topic when the objective is to achieve parameter convergence. In adaptive approximation based control the objective typically is to track a desired signal, not to achieve convergence of the parameter estimation error.

To extract some intuition behind persistency of excitation and parameter convergence, let us consider the gradient algorithm within the RFOL scheme. In this case, the parameter update law and output estimation error $e(t)$ satisfy

$$\dot{\tilde{\theta}} = \dot{\hat{\theta}} = -\Gamma\zeta(t)e(t), \text{ and} \tag{4.85}$$

$$e(t) = \zeta(t)^\top \tilde{\theta}. \tag{4.86}$$

From (4.85)–(4.86) we obtain

$$\dot{\tilde{\theta}} = -\Gamma\zeta(t)\zeta(t)^\top \tilde{\theta}. \tag{4.87}$$

As long as the adaptive gain matrix Γ is positive definite, it does not play a role in whether the parameter estimation error converges to zero or not, but it does influence (significantly) the rate of convergence. Therefore, we note that the convergence of the parameter estimation error $\tilde{\theta}(t)$ depends on the matrix $\zeta(t)\zeta(t)^\top$.

In general, for parameter convergence it is desired that $\zeta(t)\zeta(t)^\top$ stays away from zero in some sense — this is exactly the concept that the persistency of excitation condition formalizes.

Definition 4.5.1 *A bounded vector signal $\zeta \in \Re^{q_\theta}$ is persistently exciting (PE) if there exists $\alpha > 0$ and $\delta > 0$ such that*

$$\int_t^{t+\delta} \zeta(\tau)\zeta(\tau)^\top d\tau \geq \alpha I, \qquad \textit{for all } t \geq 0.$$

We note that at any time instance t the $q_\theta \times q_\theta$ matrix $\zeta(t)\zeta(t)^\top$ has rank 1. Therefore, the PE condition is not expected to hold instantaneously, but the idea is that over every time period $[t,\ t+\delta]$ the integral of $\zeta(t)\zeta(t)^\top$ retains a rank equal to q_θ.

It can be shown [138, 235] that if $\zeta(t)$ is PE and piecewise continuous then the equilibrium $\tilde{\theta} = 0$ of the differential equation (4.87) is globally exponentially stable.

It is recalled that the filtered regressor vector $\zeta(t)$ is obtained by filtering the regressor ϕ; that is,

$$\zeta(t) = W(s)\left[\phi\left(x(t), u(t)\right)\right].$$

Therefore, the condition of PE on ζ is influenced, in general, by the signals $u(t)$, $x(t)$, and also possibly by the filter $W(s)$. Since $x(t)$ is the output of the system with $u(t)$ as input, we see that the unknown system also influences the PE condition on $\zeta(t)$. For the special case of a linear system, with the unknown parameters θ^* being the coefficients of the numerator and denominator polynomial of the transfer function, it can be shown that the persistency of excitation condition on $\zeta(t)$ can be converted to a "richness" condition [119] on the input $u(t)$. Specifically, under such conditions, $\zeta(t)$ is PE if $u(t)$ has at least $2q_\theta$ frequencies [235]. In this case, $u(t)$ is said to be *sufficiently rich*.

The above results show the relationship between the PE condition and parameter convergence. Although the above formulation has considered the RFOL scheme with the gradient algorithm, similar results can be obtained for the RLS algorithm as well as the error filtering scheme. For a detailed treatment of parameter convergence in various linear identification schemes, the interested reader is referred to [119, 179, 235].

Subsequent chapters will focus on the problem of designing approximation based tracking controllers for nonlinear systems. In such tracking control applications, a goal is to force the system state vector $x(t)$ to converge to a desired state vector $x_c(t)$. The control input $u(t)$ is determined by the history of $x(t)$, $x_c(t)$, and the error between them. Assuming that the controller is able to achieve its goal of forcing $\tilde{x} = x - x_c$ toward zero, then the reference trajectory x_c plays a very significant role in determining whether ϕ and hence ζ satisfy the persistence of excitation condition.

For local basis elements, especially radial basis functions, various authors have considered the issue of persistence of excitation in adaptive approximation types of applications, e.g., [74, 75, 100, 141, 233]. The problem is particularly interesting with locally supported basis elements. For example, the results in [100, 141] demonstrate that persistence of excitation of the vector ϕ is achieved if for a specified $\epsilon > 0$ there exists $T > \mu > 0$ such that in every time interval of length T the state x spends at least μ seconds within an ϵ neighborhood of each radial basis function center. Note that since the centers are distributed across the operating region $\mathcal{D}$, this type of condition would require the state (and the commanded trajectory x_c) to fully explore the operating region in each time interval of length T. This is impractical in many control applications, but is required if the objective is to achieve convergence of the parameters over the entire region $\mathcal{D}$.

If S_k denotes the support of the k-th element of ϕ and each S_k is small relative to $\mathcal{D}$ (e.g., splines that become zero instead of Gaussian RBFs that approach zero asymptotically), then the results in [74, 75] present local persistence of excitation results that ensure convergence of the approximator parameters associated with ϕ_k while $x \in S_k$. These local persistence of excitation results are very reasonable to achieve in applications, but approximator convergence is only obtained in those regions S_k that lie along the state trajectory corresponding to x_c.

4.6 ROBUST LEARNING ALGORITHMS

The learning algorithms designed by the procedure described in Sections 4.2)–(4.4, and analyzed in Section 4.5 are based on the assumption that $\delta = 0$. In other words, it was assumed that the only uncertainty in the dynamical system is due to the unknown $f^*(x,u)$, which can be represented *exactly* by an adaptive approximation function $\hat{f}(x,u;\theta^*,\sigma^*)$ for some unknown parameter vectors θ^* and σ^*. In practice, the adaptive approximation function $\hat{f}(x,u;\theta^*,\sigma^*)$ may not be able to match exactly the modeling uncertainty $f^*(x,u)$, even if it was possible to select the parameter vectors $\hat{\theta}$ and $\hat{\sigma}$ optimally. This discrepancy is what we defined as "minimum functional approximation error" (MFAE) in Section 3.1.3 and Section 4.2.

In addition to the MFAE, there are other types of modeling errors that may occur:

- **Unmodeled dynamics.** The dimension of the state space model described by (4.5) may be less than the dimension of the real system. It is quite typical in practice to utilize reduced order models. This may be done either purposefully, in order to reduce the complexity of the model, or due to unknown dynamics of the full-order model. Indeed, in some applications (such as in flexible structures) the full-order model may be of infinite dimension.

- **Measurement noise.** The measured input and output variables may be corrupted by random noise. Therefore, there may be some discrepancy between the actual values of $u(t)$ and $y(t)$ and the corresponding values that are used in the learning scheme.

- **External disturbances.** In some applications, the measured output $y(t)$ is influenced not only by the measurable input $u(t)$ — usually referred to as "controlled" input — but also by other, "uncontrolled" inputs. Such inputs create disturbances, which may influence the plant in unpredictable ways. External disturbances are, in general, time-varying functions, which may appear only for a limited time, or they may influence the measured output persistently. In special cases, disturbances may have known time-varying characteristics (e.g., they may be periodic with known frequency, but unknown magnitude).

- **Time variations.** It has been implicitly assumed that the unknown function $f^*(x, u)$ is not an explicit function of time; in other words, the modeling uncertainty is not varying with time. In cases where f^* is time varying then the optimal parameters θ^*, σ^* are also time varying. In general, and especially when the time variations are fast and of significant magnitude, it creates additional problems for online learning schemes.

In this section, we consider modifications to the standard learning algorithms in order to provide stability and improve performance in the presence of modeling errors. These modifications lead to what are known as *robust learning algorithms*. The term "robust" is used to indicate that the learning algorithm retains some stability properties in the presence of modeling errors within the specifications for which the algorithm was designed. It is well known from the adaptive control literature of linear systems [119] that in the presence of even small modeling errors such as the ones itemized above, the standard adaptive laws in Tables 4.1 and 4.2 may exhibit *parameter drift*—a phenomenon in which the parameter vectors $\hat{\theta}(t)$, $\hat{\sigma}$ drift further from their optimal values and possibly to infinity.

Intuitively, parameter drift occurs as a result of the learning algorithm attempting to adjust the parameters in order to match a function for which an exact match does not exist for any value of the parameters (either due to MFAE or other modeling errors such as external disturbances and measurement noise). There are two categories of approaches for preventing parameter drift. In the first category of approaches, the learning algorithm is modified such that it directly restricts the parameter estimates from drifting to infinity. The so-called σ-modification, ϵ-modification, and projection algorithms belong to this category. In the second category of approach, the parameter estimates are prevented from drifting to infinity indirectly by not performing parameter adaptation when the training error is too small. The dead-zone approach has this characteristic.

To illustrate the various options for robustifying the adaptive laws summarized in Tables 4.1 and 4.2 we consider a generic adaptive law

$$\dot{\hat{\theta}}(t) = -\Gamma\xi(t)\varepsilon(t), \tag{4.88}$$

where Γ is the learning rate matrix, $\xi(t)$ is the regressor vector, and $\varepsilon(t)$ is the training error. In the case of the gradient algorithm (4.66) based on the RFOL scheme, the regressor is $\xi(t) = \zeta(t)$, while for the EFOL scheme, the regressor is $\xi(t) = \phi(x(t), u(t))$. Based on (4.88), four different modifications for enhancing robustness are described.

4.6.1 Projection Modification

One of the most straightforward and effective ways to prevent parameter drift is to restrain the parameter estimates within a predefined bounded and convex region $\mathcal{S}$, which is designed to ensure that $\theta^* \in \mathcal{S}$. In addition, the initial conditions $\hat{\theta}(0)$ are chosen such that $\hat{\theta}(0) \in \mathcal{S}$. The projection modification implements this idea as follows: if the parameter estimate $\hat{\theta}(t)$ is inside the desired region $\mathcal{S}$, or is on the boundary (denoted by $\delta\mathcal{S}$) with its direction of change toward the inside the region $\mathcal{S}$, then the standard adaptive law (4.88) is implemented. In the case that $\hat{\theta}(t)$ is on the boundary $\delta\mathcal{S}$ and its derivative is directed outside the region, then the derivative is projected onto the hyperplane tangent to $\delta\mathcal{S}$. Therefore, the projection modification keeps the parameter estimation vector within the desired convex region $\mathcal{S}$ for all time.

Next, we make the projection modification more precise. Let the desirable region $\mathcal{S}$ be a closed convex set with a smooth boundary defined by

$$\mathcal{S} = \{\theta \in \Re^{q_\theta} \mid \kappa(\theta) \leq 0\}$$

where $\kappa : \Re^{q_\theta} \mapsto \Re$ is a smooth function. According to the projection algorithm, the standard adaptive law (4.88) is modified as follows:

$$\dot{\hat{\theta}}(t) = \mathcal{P}[-\Gamma\xi\varepsilon] = \begin{cases} -\Gamma\xi\varepsilon & \text{if } \hat{\theta} \in \mathcal{S}^0 \text{or if} \\ & \hat{\theta} \in \delta\mathcal{S} \text{ and } \nabla\kappa^\top\Gamma\xi\varepsilon \geq 0 \\ -\Gamma\xi\varepsilon + \Gamma\frac{\nabla\kappa\nabla\kappa^\top}{\nabla\kappa^\top\Gamma\nabla\kappa}\Gamma\xi\varepsilon & \text{otherwise} \end{cases} \tag{4.89}$$

where $\mathcal{S}^0$ is the interior of $\mathcal{S}$, $\delta\mathcal{S}$ is the boundary of $\mathcal{S}$, and $\nabla\kappa = \frac{d\kappa}{d\theta}$.

To illustrate the use of the projection algorithm, we now consider some examples.

■ **EXAMPLE 4.9**

Consider a desirable region $\mathcal{S}$ defined by all the values of $\hat{\theta} \in \Re^q$ that satisfy $\hat{\theta}^\top\hat{\theta} \leq M^2$ where M is a positive constant. In this case, the parameter estimates are prevented from becoming too large by restricting them within the region $\hat{\theta}^\top\hat{\theta} \leq M^2$. By defining $\kappa(\hat{\theta}) = \hat{\theta}^\top\hat{\theta} - M^2$, we obtain the column vector $\nabla\kappa = 2\hat{\theta}$. Therefore, the projection algorithm (4.89) becomes

$$\dot{\hat{\theta}}(t) = \begin{cases} -\Gamma\xi\varepsilon & \text{if } \|\hat{\theta}\|_2 < M \\ & \text{or if } \|\hat{\theta}\|_2 = M \text{ and } \hat{\theta}^\top\Gamma\xi\varepsilon \geq 0 \\ -\Gamma\xi\varepsilon + \Gamma\frac{\hat{\theta}\hat{\theta}^\top}{\hat{\theta}^\top\Gamma\hat{\theta}}\Gamma\xi\varepsilon & \text{otherwise.} \end{cases} \tag{4.90}$$

The above modification guarantees that $\|\hat{\theta}(t)\|_2 \leq M$ for all $t \geq 0$ as long as $\|\hat{\theta}(0)\|_2 \leq M$. △

■ **EXAMPLE 4.10**

Now consider a two-dimensional parameter estimate $\hat{\theta} = [\hat{\theta}_1 \;\; \hat{\theta}_2]^\top$, where it is known that $\underline{\theta}_1 \leq \hat{\theta}_1 \leq \overline{\theta}_1$ and $\underline{\theta}_2 \leq \hat{\theta}_2 \leq \overline{\theta}_2$. The lower and upper limits $\underline{\theta}_1, \underline{\theta}_2$ and $\overline{\theta}_1, \overline{\theta}_2$ are assumed to be known. Therefore, in this case the desirable region $\mathcal{S}$ is a rectangle. For simplicity, let us choose the learning rate matrix to be diagonal; i.e., $\Gamma = \text{diag}(\gamma_1, \gamma_2)$.

The regressor ξ is defined by $\xi = [\xi_1 \;\; \xi_2]^\top$. By using simple algebraic computations, it can be easily shown that in this case the projection algorithm (4.89) for updating $\hat{\theta}_1, \hat{\theta}_2$, becomes

$$\dot{\hat{\theta}}_1(t) = \begin{cases} -\gamma_1\xi_1\varepsilon & \text{if } \underline{\theta}_1 < \hat{\theta}_1 < \overline{\theta}_1 \\ & \text{or if } \hat{\theta}_1 = \underline{\theta}_1 \text{ and } \gamma_1\xi_1\varepsilon \leq 0 \\ & \text{or if } \hat{\theta}_1 = \overline{\theta}_1 \text{ and } \gamma_1\xi_1\varepsilon \geq 0 \\ 0 & \text{otherwise;} \end{cases} \tag{4.91}$$

$$\dot{\hat{\theta}}_2(t) = \begin{cases} -\gamma_2\xi_2\varepsilon & \text{if } \underline{\theta}_2 < \hat{\theta}_2 < \overline{\theta}_2 \\ & \text{or if } \hat{\theta}_2 = \underline{\theta}_2 \text{ and } \gamma_2\xi_2\varepsilon \leq 0 \\ & \text{or if } \hat{\theta}_2 = \overline{\theta}_2 \text{ and } \gamma_2\xi_2\varepsilon \geq 0 \\ 0 & \text{otherwise.} \end{cases} \tag{4.92}$$

The initial conditions need to be chosen such that $\underline{\theta}_1 \leq \hat{\theta}_1(0) \leq \overline{\theta}_1$ and $\underline{\theta}_2 \leq \hat{\theta}_2(0) \leq \overline{\theta}_2$. △

One of the key properties of the projection modification is that it does not destroy the stability properties obtained using the standard adaptive laws in the case where $\delta = 0$. As the following theorem shows, in addition to guaranteeing that $\hat{\theta}(t) \in \mathcal{S}$ for all $t \geq 0$, the projection algorithm retains the stability properties obtained without the projection modification.

Theorem 4.6.1 *Suppose $\hat{\theta}(0) \in \mathcal{S}$ and $\theta^* \in \mathcal{S}$. In the case where $\delta = 0$, the projection modification algorithm given by (4.89) retains the stability properties of the EFOL and RFOL schemes established in the absence of the projection and, in addition, guarantees that $\hat{\theta}(t) \in \mathcal{S}$ for all $t \geq 0$.*

Proof: First, we prove that $\hat{\theta}(t) \in \mathcal{S}$ for all $t \geq 0$. If $\hat{\theta}(t) \in \delta\mathcal{S}$ then it follows from (4.89) that if $\nabla\kappa^\top \Gamma\xi\varepsilon \geq 0$ then no modification is employed; therefore $\dot{\hat{\theta}}^\top \nabla\kappa \leq 0$. On the other hand, if the projection modification is used (i.e., $\nabla\kappa^\top \Gamma\xi\varepsilon < 0$) then it can be easily seen that the modified projection algorithm satisfies $\dot{\hat{\theta}}^\top \nabla\kappa = 0$. Therefore, if $\hat{\theta}$ is on the boundary $\delta\mathcal{S}$, then we have $\dot{\hat{\theta}}^\top \nabla\kappa \leq 0$. This implies that the vector $\dot{\hat{\theta}}$ points either inside $\mathcal{S}$ or along the tangent plane of $\delta\mathcal{S}$ at point $\hat{\theta}$. This implies that $\hat{\theta}(t)$ will never leave $\mathcal{S}$.

The projection algorithm has the same form as the standard algorithm except for the additional term

$$Q = \Gamma \frac{\nabla\kappa\nabla\kappa^\top}{\nabla\kappa^\top \Gamma \nabla\kappa} \Gamma\xi\varepsilon, \tag{4.93}$$

which goes into effect if $\hat{\theta} \in \delta\mathcal{S}$ and $\nabla\kappa^\top \Gamma\xi\varepsilon < 0$. If we use the same Lyapunov function candidate V as with the standard adaptive algorithm, then the time derivative $\dot{V}$ will have an additional term due to Q. This additional term is given by

$$\tilde{\theta}^\top \Gamma^{-1} Q = \tilde{\theta}^\top \frac{\nabla\kappa\nabla\kappa^\top}{\nabla\kappa^\top \Gamma \nabla\kappa} \Gamma\xi\varepsilon.$$

Since $\mathcal{S}$ is convex, and by assumption $\theta^* \in \mathcal{S}$, we have that $\tilde{\theta}^\top \nabla\kappa = (\hat{\theta} - \theta^*)^\top \nabla\kappa \geq 0$ when $\hat{\theta} \in \mathcal{S}$. Moreover, by definition, $\nabla\kappa^\top \Gamma\xi\varepsilon < 0$. Hence, the extra term in the derivative of the Lyapunov function satisfies $\tilde{\theta}^\top \Gamma^{-1} Q \leq 0$. Since the projection modification can only make the Lyapunov function derivative more negative, the stability properties derived for the standard algorithm still hold. ■

Remark: In the above proof, we use the following standard result from vector calculus: for two vectors $a, b \in \Re^n$, if $a^\top b > 0$ then the angle between the two vectors is less than $90°$. If $a^\top b = 0$ then the two vectors are orthonormal, and the angle between them is $90°$.

Effect of Model Error. Note that the projection operator has no effect on the parameter estimation as long as $\hat{\theta} \in \mathcal{S}$. Therefore, in the case where $\delta \neq 0$, the projection method does not prevent an increase in the Lyapunov function (i.e., an increase in $\|\tilde{\theta}\|_\Gamma$) when e is small relative to δ. The projection operator only prevents $\hat{\theta}$ from leaving $\mathcal{S}$. Therefore, use of the projection method does not guarantee a small ultimate bound on $e(t)$ in the case where $\delta \neq 0$. Consider the following example which extends the EOFL analysis that is on page 144.

■ **EXAMPLE 4.11**

In this example, we will let $\delta(t) = \frac{\lambda}{s+\lambda}[\epsilon]$ where $\epsilon(t)$ is the combined modeling errors due to disturbances, MFAE, etc. that add into the $\dot{x}$ equation. The EOFL learning system variables are defined as

$$\chi = \frac{\lambda}{s+\lambda}\left[(\theta^*)^\top \phi\right] + \delta, \qquad \hat{\chi} = \frac{\lambda}{s+\lambda}\left[\hat{\theta}^\top \phi\right] \qquad \dot{\hat{\theta}}(t) = \mathcal{P}[-\Gamma\xi\varepsilon]$$
$$e = \hat{\chi} - \chi, \qquad \tilde{\theta} = \hat{\theta} - \theta^*.$$

Therefore,

$$\dot{e} = -\lambda e + \lambda\tilde{\theta}^\top\phi - \lambda\epsilon.$$

Consider the Lyapunov function V of eqn. (4.75). The time derivative of V using the projection form of parameter adaptation is

$$\begin{aligned}\dot{V} &= \frac{\mu}{\lambda}e\dot{e} + \mu\tilde{\theta}^\top\Gamma^{-1}\dot{\tilde{\theta}} \\ &= \mu e\left(-e + \tilde{\theta}^\top\phi - \epsilon\right) + \mu\tilde{\theta}^\top\Gamma^{-1}\left(-\Gamma\phi e + Q\right) \\ &= -\mu e^2 - \mu e\epsilon + \mu\tilde{\theta}^\top\Gamma^{-1}Q\end{aligned}$$

where Q is defined in (4.93). On the interior of $\mathcal{S}$, $Q = 0$; therefore,

$$\dot{V} = -\mu e^2 - \mu e\epsilon.$$

Even if ϵ is bounded as $|\epsilon| < \bar{\epsilon}$, the term $e\epsilon$ is sign indefinite. Using the upper bound,

$$\dot{V} < -\mu|e|\left(|e| - \bar{\epsilon}\right)$$

we can show that that V will decrease for $|e| > \bar{\epsilon}$; however, when $|e| < \bar{\epsilon}$ it is possible that V will increase until $\theta \in \delta\mathcal{S}$. Figure 4.11 shows the type of trajectory that could occur. In this figure, the parameter error and e decrease until $|e| < \bar{\epsilon}$. Once that inequality is satisfied, the parameter error diverges until $\theta \in \delta\mathcal{S}$. Eventually e increases. Once $|e| > \bar{\epsilon}$, the Lyapunov function again decreases. Note that such behavior could occur repetitively. △

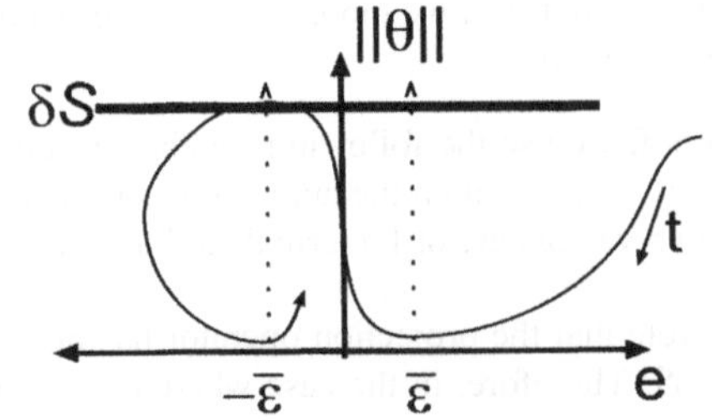

Figure 4.11: Depiction of possible projection-based parameter adaptation in the presence of model error.

4.6.2 σ-Modification

In this approach, the adaptive law (4.88) is modified to

$$\dot{\hat{\theta}}(t) = -\Gamma\xi(t)\varepsilon(t) - \Gamma\sigma\left(\hat{\theta}(t) - \theta_o\right) \tag{4.94}$$

where σ is a small positive constant and θ_o is a vector design parameter that is often selected to be the zero vector, unless there is better prior information about the value of θ^*. When $\delta \neq 0$, the additional term $-\Gamma\sigma\left(\hat{\theta}(t) - \theta_o\right)$ prevents $\hat{\theta}(t)$ from drifting to ∞ by pulling it toward θ_o. For example, if due to nonzero δ the parameter estimate $\hat{\theta}(t)$ starts drifting to large positive values, then $-\Gamma\sigma\left(\hat{\theta}(t) - \theta_o\right)$ becomes large and negative, thus forcing the parameter estimate to decrease.

■ **EXAMPLE 4.12**

In this example, we consider the same problem as Example 4.11, but using the σ-modification adaptation specified in (4.94). As in Example 4.11 we use the Lyapunov function of Equation (4.75). In this case, for simplicity we set $\mu = 1$. We do not make any assumptions regarding the size of ϵ other than its being in $\mathcal{L}_\infty$. The time derivative of the Lyapunov function using the σ-modification form of parameter adaptation is

$$\begin{aligned}
\dot{V} &= \frac{1}{\lambda}e\dot{e} + \tilde{\theta}^\top\Gamma^{-1}\dot{\tilde{\theta}} \\
&= e\left(-e + \tilde{\theta}^\top\phi - \epsilon\right) + \tilde{\theta}^\top\Gamma^{-1}\left(-\Gamma\phi e - \Gamma\sigma\left(\hat{\theta} - \theta_o\right)\right) \\
&= -e^2 - e\epsilon - \sigma\tilde{\theta}^\top\left(\hat{\theta} - \theta_o\right) \\
&= -e^2 - e\epsilon - \frac{\sigma}{2}\left(\tilde{\theta}^\top\tilde{\theta} + \left(\hat{\theta} - \theta_o\right)^\top\left(\hat{\theta} - \theta_o\right) - (\theta^* - \theta_o)^\top(\theta^* - \theta_o)\right) \\
&\leq -e^2 - e\epsilon - \frac{\sigma}{2}\tilde{\theta}^\top\tilde{\theta} + \frac{\sigma}{2}(\theta^* - \theta_o)^\top(\theta^* - \theta_o) \\
&= -\frac{1}{2}e^2 - \frac{1}{2}(e - \epsilon)^2 + \frac{\epsilon^2}{2} - \frac{\sigma}{2}\tilde{\theta}^\top\tilde{\theta} + \frac{\sigma}{2}(\theta^* - \theta_o)^\top(\theta^* - \theta_o) \\
&\leq -\frac{1}{2}e^2 + \frac{\epsilon^2}{2} - \frac{\sigma}{2}\tilde{\theta}^\top\tilde{\theta} + \frac{\sigma}{2}(\theta^* - \theta_o)^\top(\theta^* - \theta_o) \\
\dot{V} &\leq -cV + \rho
\end{aligned}$$

where

$$
\begin{aligned}
c &= \frac{\min(\sigma,1)\min(\lambda,1)}{2\max(\gamma,1)} \\
\rho &= \frac{\epsilon^2}{2}+\frac{\sigma}{2}\left(\theta^*-\theta_o\right)^{\mathrm{T}}\left(\theta^*-\theta_o\right).
\end{aligned}
$$

Therefore, the function V converges exponentially until $V(e(t),\tilde{\theta}(t)) \leq \frac{\rho}{c}$. Theoretically this bound and exponential convergence look great, but it is important to note that at least when the basis vectors ϕ form a partition of unity over $\mathcal{D}$, then $\|\theta^*\|_\infty$ is the same order of magnitude as $\sup_{x\in\mathcal{D}}(f^*(x))$. Since θ^* is unknown, θ_o is often set to zero. Also, c is typically much less than one. Therefore, the ultimate bound $\frac{\rho}{c}$ is not necessarily small. In addition, the ultimate bound is not directly related to the MFAE, so enhancing the approximator structure does not necessarily decrease the bound. △

Although the σ-modification does not require *apriori* information such as an upper bound on δ, the robustness is achieved at the expense of destroying some of the convergence properties of the ideal case ($\delta = 0$). For example, parameter estimation using the σ-modification no longer has an equilibrium at $(\varepsilon,\tilde{\theta}) = (0,0)$, since $\varepsilon = 0$ causes $\hat{\theta}$ to converge to θ_o. Therefore, several modifications have been suggested for addressing this issue, including the so-called switching σ-modification [119].

4.6.3 ϵ-Modification

The ϵ-modification was motivated as an attempt to eliminate some of the drawbacks associated with the σ-modification. It is given by

$$
\dot{\hat{\theta}}(t) = -\Gamma\xi(t)\varepsilon(t) - \Gamma|\varepsilon(t)|\nu\left(\hat{\theta}(t)-\theta_o\right) \tag{4.95}
$$

where $\nu > 0$ and θ_o are design constants. The idea behind this approach is to retain the equilibrium at $(\varepsilon,\tilde{\theta}) = (0,0)$ by forcing the additional term $-\Gamma|\varepsilon|\nu\left(\hat{\theta}(t)-\theta_o\right)$ to be zero in the case that $\varepsilon(t)$ is zero. In the case that the parameter estimate vector $\hat{\theta}(t)$ starts drifting to large values then the ϵ-modification term again acts as a stabilizing force if $\varepsilon \neq 0$. Note that without such modifications it is possible for the parameter estimate to diverge to ∞ while maintaining ε near zero, since without persistence of excitation it is very possible that $\tilde{\theta}$ lies in the subspace defined by $\varepsilon = \tilde{\theta}^T\xi(t) = 0$.

Now let us consider the same formulation as Example 4.12, where instead of the σ-modification we use the ϵ-modification. In this case, the time derivative of the Lyapunov function is given by

$$
\begin{aligned}
\dot{V} &= \frac{1}{\lambda}e\dot{e} + \tilde{\theta}^{\mathrm{T}}\Gamma^{-1}\dot{\tilde{\theta}} \\
&= e\left(-e+\tilde{\theta}^{\mathrm{T}}\phi-\epsilon\right) + \tilde{\theta}^{\mathrm{T}}\Gamma^{-1}\left(-\Gamma\phi e - \Gamma|\varepsilon|\nu\left(\hat{\theta}(t)-\theta_o\right)\right) \\
&= -e^2 - e\epsilon - \nu\tilde{\theta}^{\mathrm{T}}\left(\hat{\theta}-\theta_o\right)|\varepsilon| \\
&\leq -e^2 - e\epsilon - \frac{\nu}{2}\tilde{\theta}^{\mathrm{T}}\tilde{\theta} + \frac{\nu}{2}\left\|\tilde{\theta}-(\hat{\theta}-\theta_o)|\varepsilon|\right\|^2
\end{aligned}
$$

$$
\begin{aligned}
&\leq -\frac{1}{2}e^2 + \frac{\epsilon^2}{2} - \frac{\nu}{2}\left\|\tilde{\theta}\right\|^2 + \frac{\nu}{2}\left\|\tilde{\theta} - (\hat{\theta} - \theta_o)|\varepsilon|\right\|^2 \\
\dot{V} &\leq -cV + \rho
\end{aligned}
$$

where

$$
\begin{aligned}
c &= \frac{\min(\nu, 1)\min(\lambda, 1)}{2\max(\gamma, 1)} \\
\rho &= \frac{\epsilon^2}{2} + \frac{\nu}{2}\left\|\tilde{\theta} - (\hat{\theta} - \theta_o)|\varepsilon|\right\|^2.
\end{aligned}
$$

Therefore, we obtain similar results as with the σ-modification.

4.6.4 Dead-Zone Modification

When $\delta = \frac{\lambda}{s+\lambda}[\epsilon] \neq 0$ (e.g., in the presence of approximation errors), the adaptive law (4.88) tries to drive the estimation error ε to zero, sometimes at the expense of increasing the magnitude of the parameter estimates. The idea behind the dead-zone modification is to enhance robustness by turning off adaptation when the estimation error becomes relatively small compared to ϵ. Note, for example, that in eqn. (4.79) the time derivative of the Lyapunov function is negative semidefinite for $|\varepsilon| > |\epsilon|$. Therefore, for $|\varepsilon| > |\epsilon|$ the Lyapunov function is decreasing. When $|\varepsilon| < |\epsilon|$, then the parameter estimates may diverge and the Lyapunov function may increase. The apparently simple solution is to stop parameter estimation when $|\varepsilon| < |\epsilon|$.

The dead-zone modification is given by

$$
\dot{\hat{\theta}}(t) = \begin{cases} -\Gamma\xi(t)\varepsilon(t) & \text{if } |\varepsilon| \geq \epsilon_0 \\ 0 & \text{if } |\varepsilon| < \epsilon_0 \end{cases} \tag{4.96}
$$

where ϵ_0 is a positive design constant intended to be an upper bound on $\epsilon(t)$. One of the drawbacks of the dead-zone modification is that the designer needs an upper bound on the model error, which is usually not available. Therefore, ϵ_0 must be selected conservatively to ensure that it overbounds $\epsilon(t)$. A second drawback of the dead-zone approach is that even in the case where $\epsilon(t) = 0$, asymptotic stability of the origin cannot be proved; instead, uniform ultimate boundedness of the origin is attained with the size of the bound determined by ϵ_0 and the control parameters.

If $\epsilon(t) > \epsilon_0$ for any interval of time for which $|\varepsilon| < \epsilon$, then the Lyapunov function may increase. Note that the dead-zone approach can be combined with the other approaches (e.g., projection). Such combined approaches are considered further in the example at the end of this section.

■ EXAMPLE 4.13

In this example, we consider the same problem as Example 4.11, but using the dead-zone adaptation specified in eqn. (4.96). As in Example 4.11, we use the Lyapunov function of eqn. (4.75) (with $\mu = 1$) and we assume that $\epsilon < \bar{\epsilon}$. The time derivative of the Lyapunov function using the dead-zone form of parameter adaptation for $|e| > \epsilon_0$ is

$$
\dot{V} = \frac{1}{\lambda}e\dot{e} + \tilde{\theta}^\top\Gamma^{-1}\dot{\tilde{\theta}}
$$

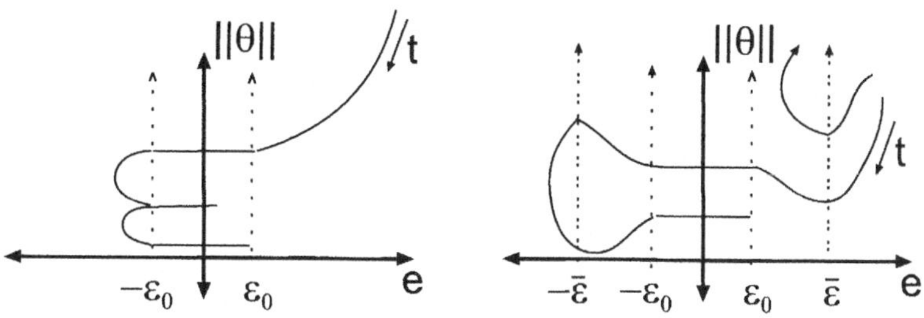

Figure 4.12: Depiction of possible dead-zone based parameter adaptation in the presence of model error.

$$
\begin{aligned}
&= e\left(-e+\tilde{\theta}^{\mathrm{T}}\phi-\epsilon\right)-\tilde{\theta}^{\mathrm{T}}\Gamma^{-1}\left(\Gamma\phi e\right)\\
&= -e^2-e\epsilon\\
&< -|e|\left(|e|-\bar{\epsilon}\right)
\end{aligned}
$$

There are now two cases to consider $\epsilon_0 > \bar{\epsilon}$ and $\epsilon_0 < \bar{\epsilon}$. The designer of course will try to select $\epsilon_0 > \bar{\epsilon}$, but since $\bar{\epsilon}$ may not be know it is important to understand the consequences of having $\epsilon_0 < \bar{\epsilon}$.

If $\epsilon_0 > \bar{\epsilon}$, then $\dot{V} < 0$ whenever parameter adaptation is active (i.e., $|e| \geq \epsilon_0$). When $|e| < \epsilon_0$, parameter adaptation stops. Note that if the trajectory enters the dead-zone at time t_1 and leaves the dead-zone at time t_2, then $|e(t_1)| = |e(t_2)| = \epsilon_0$ and $\tilde{\theta}(t_1) = \tilde{\theta}(t_2)$; therefore, $V(e(t_1), \tilde{\theta}(t_1)) = V(e(t_2), \tilde{\theta}(t_2))$. If odd subscripted times (i.e., t_{2i+1} for $i = 1, 2, \ldots$) denote times at which the trajectory leaves the dead-zone and even subscripted times (i.e., t_{2i} for $i = 1, 2, \ldots$) denote times at which the trajectory enters the dead-zone, then extension of the above argument shows that $V(e(t_{2i-1}), \tilde{\theta}(t_{2i-1})) = V(e(t_{2i}), \tilde{\theta}(t_{2i}))$ and $V(e(t_{2i+1}), \tilde{\theta}(t_{2i+1})) \leq V(e(t_{2i}), \tilde{\theta}(t_{2i}))$. In fact, if we denote $\alpha = \epsilon_0 - \bar{\epsilon} > 0$ then outside the dead-zone

$$
\dot{V} < -\epsilon_0\alpha < 0;
$$

therefore, $V(e(t_{2i+1}), \tilde{\theta}(t_{2i+1})) \leq V(e(t_{2i}), \tilde{\theta}(t_{2i})) - \epsilon_0\alpha(t_{2i+1} - t_{2i})$. This shows that the total time outside the dead-zone must satisfy the following inequality

$$
\sum_{i=0}^{q}(t_{2i+1}-t_{2i}) < \frac{1}{\epsilon_0\alpha}V(e(0),\tilde{\theta}(0)) < \infty
$$

where q may be finite or infinite, but the cumulative time outside the dead-zone is finite [87]. Therefore, in this example, $|e(t)|$ is ultimately bounded by ϵ_0. Such a possible trajectory is depicted in the left image of Figure 4.12.

If $\epsilon_0 < \bar{\epsilon}$, then, even though parameter adaptation will stop for $|e| < \epsilon_0$, the Lyapunov function and in particular the parameter estimation error may increase for $\bar{\epsilon} > |e| > \epsilon_0$. Two possible trajectories are depicted in the image in the right half of Figure 4.12. Note that while $\bar{\epsilon} > |e| > \epsilon_0$ the Lyapunov function can increase without bound. △

4.6.5 Discussion and Comparison

In the presence of model errors (i.e., $\delta \neq 0$), the above robust adaptive laws guarantee, under certain conditions, that the parameter estimates $\hat{\theta}(t)$ and the estimation error $\varepsilon(t)$ remain bounded. We have included several examples in the previous subsections to clarify and allow comparison between the bounds available from the alternative approaches. To be useful as design tools, the designer should be able to clearly understand how to make the bound smaller as a function of the approximation structure or the control and estimation design parameters. Although, in the presence of approximation error, it cannot be established that $\varepsilon(t)$ will converge to zero, it can be shown that the estimation error is *small-in-the-mean-squared* sense [119], in the sense that integral square error over a finite interval is proportional to the integral square approximation error (see Section A.2.2.4).

In the introduction to this section, we stated that there were two categories of approaches for increasing the robustness of parameter adaptation methods to model error. As the discussion of this section has pointed out, the first category of methods (i.e., σ-modification, ϵ-modification, and projection) do not require any assumptions about upper bounds on the model error and do prevent the parameter estimates from diverging to infinity, but also are not guaranteed to maintain the accuracy of the parameter estimates when the training error is small relative to the model error. The second category of methods (i.e., dead-zones) require an assumption of a known bound on the model error. If this assumption is valid, then the dead-zone maintains the accuracy of the parameter estimate when the training error is small relative to the modeling error. If the assumed size of the bound is invalid, then there are no guarantees. Note that the best of both approaches is easily achievable by implementing one of the approaches from each category.

■ **EXAMPLE 4.14**

In this example, we consider the same problem as Example 4.11, but using the projection and dead-zone adaptation:

$$\dot{\hat{\theta}}(t) = \mathcal{P}[\Gamma \xi d(\varepsilon)] \tag{4.97}$$

where the projection operator is defined in eqn. (4.89) and the dead-zone is implemented as

$$d(\varepsilon) = \begin{cases} \varepsilon & \text{if } \varepsilon \geq \epsilon_0 \\ 0 & \text{otherwise.} \end{cases}$$

The analysis for this approach must consider a few cases. If the assumption that $\bar{\epsilon} < \epsilon_0$ is valid, then projection maintains $\hat{\theta} \in \mathcal{S}$ while the dead-zone maintains the accuracy of the parameter estimate when the training error is small; thus preventing the possible divergence of the parameter estimate depicted in Figure 4.11. Alternatively, if the assumption that $\bar{\epsilon} < \epsilon_0$ is not valid then projection would prevent divergence to infinity as depicted in the right image of Figure 4.12 when $\bar{\epsilon} > |e| > \epsilon_0$. In both of these cases, performance of parameter estimation using both projection and a dead-zone is better than using either approach alone. △

Implementation of the dead-zone or projection methods as written would involved discontinuous differential equations. Therefore, implementations usually involve smoothing of the discontinuities.

4.7 CONCLUDING SUMMARY

One of the key components of adaptive approximation based control is the design of estimation schemes for approximating, online, the unknown nonlinearities. In this chapter, the emphasis was on adaptive approximation without regard to the feedback control problem, which will be discussed in the next three chapters. Invariably, the problem of adaptive approximation is closely related to parameter estimation. Once a certain approximation structure is selected, based on the options presented in Chapter 3 and following the properties described in Chapter 2, then the approximation problem to a large extent reduces to the estimation of unknown parameters.

The literature has a large number of formulations and parameter estimation techniques. For example, there are techniques based on optimization methods, there are techniques that are based on Lyapunov design methods, and there are also methods for modifying the standard update laws so that they are made robust to certain types of modeling errors. This chapter has provided a structured formulation for parameter estimation in the context of adaptive approximation of dynamical systems. First, we considered the derivation of parametric models, which basically amounts to rewriting the system equation so that the uncertainty appears in a suitable way for designing estimation schemes. Then, we considered the design of online learning schemes. The last part of the design procedure was the derivation of adaptive laws for updating the parameter estimates. The stability and convergence properties of the designed adaptive schemes were analyzed under certain ideal conditions. Finally, we investigated the design and analysis of robust learning algorithms, which are able to address the case of modeling errors.

4.8 EXERCISES AND DESIGN PROBLEMS

Exercise 4.1 For the case where the unknown nonlinearities are of the form described by eqn. (4.19), work out the details in deriving the parametric model eqn. (4.20).

Exercise 4.2 Consider the filtering scheme

$$e(t) = \frac{\lambda}{s+\lambda}[q(t)]$$

where $q(t)$ is the input to the filter and $e(t)$is the filter output. Simulate this and plot $q(t)$ and $e(t)$ on the same figure for these scenarios:

(a) $\lambda = 1$, $q(t) = e^{-t}(\sin(2\pi t) + 0.4(\cos(20\pi t)))$;

(a) $\lambda = 10$, $q(t) = e^{-t}(\sin(2\pi t) + 0.4(\cos(20\pi t)))$;

(a) $\lambda = 1$, $q(t) = e^{-0.1t^2}\cos(2\pi t)$ for $t \leq 3$ and $g(t) = -0.1$ for $t > 3$;

(a) $\lambda = 10$, $q(t) = e^{-0.1t^2}\cos(2\pi t)$ for $t \leq 3$ and $g(t) = -0.1$ for $t > 3$.

Assume zero initial condition for the filter, and in your simulations consider the time interval $t \in [0,\ 6]$.

Exercise 4.3 Consider the following methodology that is phrased in terms of state estimation. Let

$$\begin{aligned} \dot{x} &= f(x) \\ y &= x \end{aligned}$$

where $f(x) = \theta^\top \phi(x) + e_f(x)$ with $|e_f(x)| < \epsilon$ on $\mathcal{D}$. Define

$$\begin{aligned}\dot{\hat{x}} &= \hat{f}(x) + L(y - \hat{y}) \\ \hat{y} &= \hat{x}\end{aligned}$$

where $\hat{f}(x) = \hat{\theta}^\top \phi(x)$. Also, define $e = x - \hat{x}$ and $\tilde{\theta} = \hat{\theta} - \theta$. The above defines a parametric model and learning scheme with training signal e that can be computed from available signals.

1. Find the differential equation for e.

2. Use the Lyapunov candidate function $\mathcal{V} = \frac{1}{2}\left(e^2 + \tilde{\theta}^T \Gamma^{-1} \tilde{\theta}\right)$ to derive a stable parameter update law for the case that $\epsilon = 0$. What constraint is required for L?

3. For the case that $\epsilon = 0$, prove the properties of e and $\tilde{\theta}$.

4. For the case that $\epsilon \neq 0$, but an upper bound is known, what is the appropriate dead-zone size to ensure uniform boundedness of the solution. What is the uniform bound on $|e(t)|$?

Exercise 4.4 In this exercise, we consider the second order case where the system model is

$$\begin{bmatrix} \dot{x}_1 \\ \dot{x}_2 \end{bmatrix} = \begin{bmatrix} x_2 \\ f(y) + g(y)u \end{bmatrix} \tag{4.98}$$

$$y = x_1 \tag{4.99}$$

where y and u are available signals. In particular, the derivative of the output x_2 is not directly measured. The functions f and g are not known and will be approximated. An important aspect of this problem is that unknown nonlinearities f and g depend only on the directly measurable signal y. If this approach is understood, then generalization to the n-th order case is straightforward.

1. Assuming that $f(y) = \theta_f^\top \phi_f(y)$ and $g(y) = \theta_g^\top \phi_g(y)$, show that the state differential equation can be written as

$$\begin{bmatrix} \dot{x}_1 \\ \dot{x}_2 \end{bmatrix} = \begin{bmatrix} x_2 \\ \Theta^\top \Phi(y, u) \end{bmatrix} \tag{4.100}$$

   where $\Theta^\top = [\theta_f^\top, \theta_g^\top] \in \Re^{2N}$ and $\Phi(y, u) = [\phi_f^\top, u\phi_g^\top]$.

2. Add and subtract $a_1\dot{x}_1 + a_2 x_1$ to both sides of the equation

$$\ddot{x}_1 = f(y) + g(y)u$$

   to show that

$$y = \Theta^\top \Phi_F + y_{F_1} + y_{F_2}$$

   where

$$\begin{aligned}\Phi_F &= \frac{1}{s^2 + a_1 s + a_2}[\Phi(y, u)] \\ y_{F_1} &= \frac{a_1 s}{s^2 + a_1 s + a_2}[y] \\ y_{F_2} &= \frac{a_2}{s^2 + a_1 s + a_2}[y].\end{aligned}$$

Note that $(s^2 + a_1 s + a_2) = 0$ must be a Hurwitz polynomial.

3. Let

$$\hat{y} = \hat{\Theta}^\top \Phi_F + y_{F_1} + y_{F_2}$$

and $e = \hat{y} - y$. Show that $e = \tilde{\Theta}^\top \Phi_F$ where $\tilde{\Theta} = \hat{\Theta} - \Theta$.

4. Relative to the cost function

$$J(\Theta(t)) = \int_0^t e^{-\mu(t-\tau)} \left(y(\tau) - \hat{y}\left(\hat{\Theta}(t), \Phi_F(\tau), y_{F_1}(\tau), y_{F_2}(\tau)\right)\right)^2 d\tau \quad (4.101)$$

the adaptation algorithm for least squares with forgetting is:

$$\begin{aligned} \dot{P} &= -P\Phi_F\Phi_F^\top P + \mu P \quad \text{with } P(0) \text{ positive definite} \\ \dot{\hat{\Theta}} &= -P\Phi_F e. \end{aligned}$$

(a) Show that the time derivative of the Lyapunov function $V = \tilde{\Theta}^\top P^{-1} \tilde{\Theta}$ is $\dot{V} = -\left(\mu V + e^2\right)$. Show that $V \in \mathcal{L}_\infty$, $V \in \mathcal{L}_2$, and $e \in \mathcal{L}_2$.

(b) Show that implementation of the this least squares approach requires implementation of $2N + 2$ second-order filters plus solution of $(4N^2 + 2N)$ ordinary differential equations.

5. Implement a simulation using $f = 0$, $g = 2 + \sin(y^2)$, and

$$u = \frac{1}{2}\left(-K_1 \left(y - 2\sin(\pi t)\right) - K_2 \left(\frac{a_2}{a_1} y_{F_1} - 2\pi \cos(\pi t) \right) + 2\pi^2 \sin(\pi t) \right).$$

The choice of parameters $K_1 = 1$, $K_2 = 1$, $a_1 = 3.5$, $a_2 = 49$, and $\mu = 0.01$ works reasonably well. Let $\hat{f} = 0$ and $\hat{g} = \theta_g^\top \phi_g(y)$ where ϕ_g is composed of Gaussian radial basis functions defined by eqn. (3.23) with centers separated by $\gamma = 0.3$ and uniformly covering $y \in \mathcal{D} = [-6, 6]$. Let the simulation run for at least 100 s.

(a) Since g is available in simulation, you can compute θ_g. Use this known vector to plot the norm of the approximator parameter error as a function of time. Discuss.

(b) Use the known value of θ_g to plot the plot the value of the Lyapunov function versus time. Discuss.

(c) Plot g and $\hat{g}$ (at least) at the beginning and end of the simulation. Discuss why it is more accurate in some regions of $\mathcal{D}$ than others.

(d) Repeat the above simulation using alternative definitions of the approximator. Be certain to try some approximator with globally supported basis functions. Compare such items as the number of filters needed to compute Φ_F and the approximation accuracy.

6. In the above controller, $\frac{a_2}{a_1} y_{F_1}$ is used as an estimate for the unmeasured quantity $\dot{y} = x_2$. Use Laplace analysis to show that this approximation is reasonable at low frequencies (i.e., s near zero).

Note that in this approach P has row and column dimensions equal to $q = dim(\theta) = N$ which can be quite large, especially when the dimension of $\mathcal{D}$ is larger than one. A similarly large number of filters is required to compute Φ_F. The computations required for this least squares implementation can become impractical in some applications. For comparison, see the approach of Exercise 4.5.

Exercise 4.5 This exercise considers an alternative estimation approach that is referred to in the literature as receptive field weighted regression (RFWR) [236, 237]. For convenience, we will consider the same application as Exercise 4.4, only the approximator and estimation algorithms will change.

1. Assume that

$$f(x,\theta) = \frac{\sum_{k=1}^{M}\omega_k(x)f_k(x)}{\sum_{k=1}^{M}\omega_k(x)} = \sum_{k=1}^{M} f_k(x)\bar{\omega}_k(x), \tag{4.102}$$

$$g(x,\theta) = \frac{\sum_{k=1}^{M}\omega_k(x)g_k(x)}{\sum_{k=1}^{M}\omega_k(x)} = \sum_{k=1}^{M} g_k(x)\bar{\omega}_k(x). \tag{4.103}$$

The following items clarify the constraints assumed in this decomposition.

(a) $\{\omega_k(x)\}_{k=1}^{M}$ defines a set of continuous, positive, locally supported weighting functions.

(b) $S_k = \{x \in \mathcal{D}\,|\omega_k(x) \neq 0\}$ denotes the support of $\omega_k(x)$. The weighting functions ω_k are defined so that each set S_k is convex and connected and $\mathcal{D} = \bigcup_{k=1}^{M} S_k$. An example of a weighting function satisfying the above conditions is the biquadratic kernel defined as

$$\omega_k(x) = \begin{cases} \left(1 - \left(\frac{||x-c_k||}{\mu_k}\right)^2\right)^2, & \text{if } ||x-c_k|| < \mu_k \\ 0, & \text{otherwise.} \end{cases} \tag{4.104}$$

where c_k is the center location of the k-th weighting function and μ_k is a constant which represents the radius of the region of support.

(c) To simplify expressions used later, define

$$\bar{\omega}_k(x) = \frac{\omega_k(x)}{\sum_k \omega_k(x)}.$$

The set of non-negative functions $\{\bar{\omega}_k(x)\}_{k=1}^{M}$ forms a *partition of unity* on $\mathcal{D}$: $\sum_{k=1}^{M}\bar{\omega}_k(x) = 1$, for all $x \in \mathcal{D}$. Note that the support of $\omega_k(x)$ is exactly the same as the support of $\bar{\omega}_k(x)$.

(d) On each region S_k,

$$f_k(x,\theta_{f_k}) = \psi_{f_k}^\top(x)\theta_{f_k} \quad \text{and} \quad g_k(x,\theta_{g_k}) = \psi_{g_k}^\top(x)\theta_{g_k}$$

are local estimates to f and g. Since each region S_k is small, the local approximations can be quite simple. For example, an affine approximation such as $\hat{f}_k(x,\theta_{f_k}) = \theta_{f_{k1}} + \theta_{f_{k2}}(x - c_k)$ would yield $\psi_{f_k} = [1, (x-c_k)]^\top$ and $\theta_{f_k} = [\theta_{f_{k1}}, \theta_{f_{k2}}]^\top$.

Under the assumption that, for $y \in S_k$ it is true that $f(y) = f_k(y)$ and $g(y) = g_k(y)$, show that the state differential equation of (4.98) can be written as

$$\begin{bmatrix} \dot{x}_1 \\ \dot{x}_2 \end{bmatrix} = \begin{bmatrix} x_2 \\ \Theta_k^\top \Psi_k(y, u) \end{bmatrix} \tag{4.105}$$

where $\Theta_k^\top = [\theta_{f_k}^\top, \theta_{g_k}^\top]$ and $\Psi_k(y, u) = [\psi_{f_k}^\top, u\psi_{g_k}^\top]$.

2. Add and subtract $a_1\dot{x}_1 + a_2 x_1$ to both sides of the equation

$$\ddot{x}_1 = \Theta_k^\top \Psi_k(y, u)$$

to show that

$$y = \Theta_k^\top \Psi_{F_k} + y_{F_1} + y_{F_2}$$

where

$$\Psi_{F_k} = \frac{1}{s^2 + a_1 s + a_2} [\Psi_k(y, u)]$$

and y_{F_1}, y_{F_2}, and $(s^2 + a_1 s + a_2)$ are as defined in Exercise 4.4.

3. Let

$$\hat{y}_k = \hat{\Theta}_k^\top \Psi_{F_k} + y_{F_1} + y_{F_2}$$

and $e_k = \hat{y}_k - y$. Show that $e_k = \tilde{\Theta}_k^\top \Psi_{F_k}$ where $\tilde{\Theta}_k = \hat{\Theta}_k - \Theta_k$.

4. In contrast to the least squares cost function of eqn. (4.101), which allows cooperation between all the elements of Θ in fitting the data over $\mathcal{D}$, RFWR uses the *locally weighted* error criterion:

$$J_k(\Theta_k) = \int_0^t e^{-\mu(t-\tau)} \bar{\omega}_k(y(.)) \left(y(\cdot) - \hat{y}_k\left(\hat{\Theta}_k(t), \Psi_{F_k}(\cdot), y_{F_1}(\cdot), y_{F_2}(\cdot)\right)\right)^2 d\tau \tag{4.106}$$

where all arguments indicated by $(\cdot)$ are (τ)'s that were eliminated to make the equation fit on the line. In this approach, each $\hat{\Theta}_k$ is optimized independently over S_k. The RFWR adaptation algorithm is:

$$\begin{aligned} \dot{P}_k &= -\bar{\omega}_k P_k \Psi_F \Psi_F^\top P_k + \bar{\omega}_k \mu P_k \quad \text{with } P_k(0) \text{ positive definite} \\ \dot{\hat{\Theta}}_k &= -\bar{\omega}_k P_k \Psi_F e_k. \end{aligned}$$

Note that both differential equations automatically turn off when $y \notin S_k$. Note also that both forgetting and learning are localized to the regions S_k corresponding to active weighting functions.

(a) Show that the time derivative of the Lyapunov function $V = \sum_{k=1}^M V_k$ with $V_k = \tilde{\Theta}_k^\top P_k^{-1} \tilde{\Theta}_k$ is

$$\dot{V} = -\sum_{k=1}^{M} \left[\bar{\omega}_k \left(\mu V_k + e_k^2\right)\right]$$

where the term in square brackets is the $\dot{V}_k$. Show that V, $V_k \in \mathcal{L}_\infty$, V_k, V, $e \in \mathcal{L}_2$.

(b) Let $q_k = dim(\theta_{f_k})$. Show that implementation of the RFWR requires implementation of $2Mq_k + 2$ second-order filters plus solution of $M(4q_k^2 + 2q_k)$ ordinary differential equations. Compare these computations with those for the least squares approach of Exercise 4.4. First consider $N = M$ and $q_k = 1$. This is a direct comparison. The difference in computational requirements is due to the relative sizes of P and P_k. The difference is significant for large N. Next consider the situation where you increase to $q_k = 2$ in the RFWR, i.e., using an affine local approximator. Show that if $N > 4$ then the RFWR approach is still computationally less expensive than the least squares approach.

5. Repeat the simulation exercise of Exercise 4.4 using the RFWR approach to parameter adaptation.

Exercise 4.6 Given the Lyapunov function of eqn. (4.57) complete the analysis required to derive eqn. (4.58). What properties can be derived for the signals e, $\tilde{\theta}$, $\hat{\theta}$, and $V(t)$.

Exercise 4.7 Prove a similar stability result as in Theorem 4.5.1, with the first-order filter $\frac{\lambda}{s+\lambda}$ being replaced by an SPR filter $W(s)$. (Hint: see the SPR discussion in Appendix A.)

CHAPTER 5

NONLINEAR CONTROL ARCHITECTURES

This chapter presents an introduction to some of the dominant methods that have been developed for nonlinear control design. The objective of this chapter is to introduce the methods, analysis tools, and key issues of nonlinear control. In this chapter, we set the foundation, but do not yet discuss the use of adaptive approximation to improve the performance of nonlinear controller operation in the presence of nonlinear model uncertainty. Chapters 6 and 7 will discuss the methods, objectives, and outcomes of augmenting nonlinear control with approximation capabilities assuming that the reader is familiar with the material in this chapter.

This chapter begins with a discussion of the traditional and still commonly used approaches of small-signal linearization and gain scheduling. These approaches are based on the principle of linearizing the system around a certain operation point, or around multiple operating points, as in gain scheduling. The method of feedback linearization is presented in Section 5.2. This is one of the most commonly used nonlinear control design tools. In Section 7.2, feedback linearization is extended to include adaptive approximation. The method of backstepping is discussed in Section 5.3 and its extension using adaptive approximation is discussed in Section 7.3. A modification to the standard backstepping approach that simplifies the algebraic manipulations and online computations, especially in adaptive approaches, is presented in Section 5.3.3. Section 5.4 presents a set of robust nonlinear control design techniques, which are based on the principle of assuming that the unknown component of the nonlinearities is bounded by a known function. The methods include bounding control, sliding mode control, Lyapunov redesign, nonlinear damping, and adaptive bounding. These techniques rely on the design of a nonlinear controller that is able to

Adaptive Approximation Based Control: Unifying Neural, Fuzzy and Traditional Adaptive Approximation Approaches. By Jay A. Farrell and Marios M. Polycarpou

handle all nonlinearities within the assumed bound. As a result, they may result in high-gain control algorithms. As we will see, one of the key motivations of adaptive approximation is to reduce the need for such conservative control design. Finally, Section 5.5 briefly presents the adaptive nonlinear control methodology, which is based on the estimation of unknown parameters in nonlinear systems.

Naturally, it is impossible to cover in a single chapter all nonlinear control design and analysis methods. By necessity, many of the technical details have been omitted. An excellent treatment of nonlinear systems and control methods is given in [134]. The intent of the present chapter is to introduce selected nonlinear control methods, highlight some methods that are robust to nonlinear model errors, and to motivate the use of adaptive approximation in certain situations. Throughout this chapter, the main focus is on tracking control problems, even though where convenient we also consider the regulation problem. Also, the presentation focuses on systems where the full state is measured; output feedback methods are not discussed.

5.1 SMALL-SIGNAL LINEARIZATION

Consider the nonlinear system

$$\dot{x} = f(x, u) \tag{5.1}$$

where $f(x,u)$ is continuously differentiable in a domain $\mathcal{D}_x \times \mathcal{D}_u \subset \Re^n \times \Re^m$. First, we consider the linearization around an equilibrium point x_e, which for notational simplicity is assumed to be the origin; i.e., $x = 0$, $u = 0$. Then we consider the linearization around a nominal trajectory $x^*(t)$. Finally, we describe the concept of gain scheduling, which is a feedback control technique based on linearization around multiple operating points.

The main idea behind linearization is to approximate the nonlinear system in (5.1) by a linear model of the form

$$\dot{x} = Ax + Bu$$

where A, B are matrices of dimension $n \times n$ and $n \times m$, respectively. Typically, the linear model is an accurate approximation to the nonlinear system only in a neighborhood of the point around which the linearization took place. This is illustrated in Figure 5.1, which depicts linearization around $x = 0$. As shown in the diagram, the linearized model Ax is a good approximation of $f(x)$ for x close to zero; however, if $x(t)$ moves significantly away from the equilibrium point $x = 0$, then the linear approximation is inaccurate. As a consequence, a linear control law that was designed based on the linear approximation may very well be unsuitable once the trajectory moves away from the equilibrium, possibly due to modeling errors or disturbances.

The term *small-signal linearization* is used to characterize the fact that the linear model is close to the real nonlinear system if the system trajectory $x(t)$ remains close to the equilibrium point x_e or to the nominal trajectory $x^*(t)$. Therefore, for sufficiently small signals $x(t) - x_e$, the linearized system is an accurate approximation of the nonlinear system. The term "small-signal" linearization also distinguishes this type of linearization from *feedback linearization*, which will be studied in the next section.

In general, feedback control techniques based on the linear model work well when applied to the nonlinear system if the uncertainty of the system is small, thus allowing the feedback controller to keep the trajectory close to the equilibrium point x_e. Obviously, linear controllers derived based on small-signal linearization have good closed-loop performance in cases where the system nonlinearities are not dominant or they do not have a destabilizing

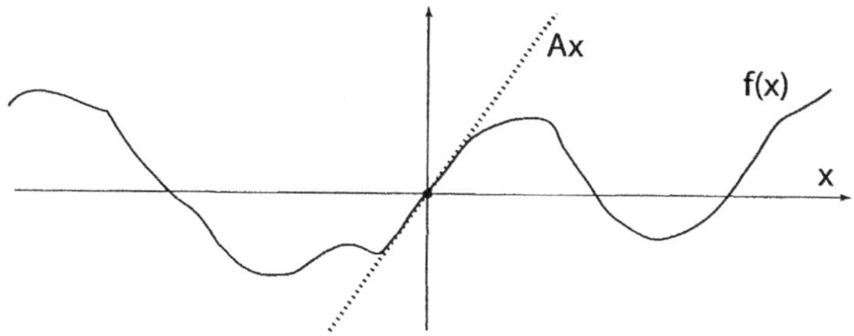

Figure 5.1: Diagram to illustrate small-signal linearization around $x = 0$.

effect. For example, stabilization of the origin, the nonlinearity of the scalar system

$$\dot{x} = x - x^3 + u$$

has a stabilizing effect, thus if the control law $u = -2x$ is used then for the resulting closed-loop system $\dot{x} = -x - x^3$ the origin is asymptotically stable even though the nonlinear term $-x^3$ has not been removed by the control law.

5.1.1 Linearizing Around an Equilibrium Point

If the nonlinear system of (5.1) is linearized around $(x,\ u) = (0,\ 0)$ then the linear model is described by

$$\dot{x} = Ax + Bu \tag{5.2}$$

where the matrices $A \in \Re^{n\times n}$ and $B \in \Re^{n\times m}$ are given by

$$A = \frac{\partial f}{\partial x}(x,u)\bigg|_{x=0,\ u=0}; \qquad B = \frac{\partial f}{\partial u}(x,u)\bigg|_{x=0,\ u=0}. \tag{5.3}$$

If we assume that the pair $(A,\ B)$ is stabilizable [10, 19, 39], then there exists a matrix $K \in \Re^{n\times n}$ such that the eigenvalues of $A+BK$ are located strictly in the left-half complex plane. Therefore, if the control law $u = Kx$ is selected then the closed-loop linear model is given by

$$\dot{x} = (A + BK)x.$$

Since all the eigenvalues of $A + BK$ are in the left-half complex plane, $x(t)$ will converge to zero asymptotically (exponentially fast).

Now, if the control law $u = Kx$ is applied to the nonlinear system (5.1) then the closed-loop dynamics are

$$\dot{x} = f(x,\ Kx). \tag{5.4}$$

Linearization of (5.4) around $x = 0$ yields

$$\begin{aligned} \dot{x} &= \left[\frac{\partial f}{\partial x}(x,Kx) + \frac{\partial f}{\partial u}(x,u)K\right]_{x=0,\ u=0} x \\ &= (A + BK)x. \end{aligned}$$

Therefore, the linear control law $u = Kx$ not only makes the linear model asymptotically stable but also makes the equilibrium point $x = 0$ of the nonlinear system asymptotically stable. Unfortunately, in the case of the nonlinear system, the asymptotic stability is only *local*. This implies that if the initial condition $x(0)$ is sufficiently close to $x = 0$ then there is asymptotic convergence of $x(t)$ to zero; if not, then the trajectory may not converge to zero. In fact, it may also become unbounded.

If the nonlinear system has an output function then we can proceed to obtain the C and D matrices as well. Specifically, consider the system

$$\dot{x} = f(x, u) \tag{5.5}$$

$$y = h(x, u) \tag{5.6}$$

where $h(x, u)$ is continuously differentiable in the domain $\mathcal{D}_x \times \mathcal{D}_u \subset \Re^n \times \Re^m$. Linearization about $x = 0$, $u = 0$ yields the linear model

$$\begin{aligned} \dot{x} &= Ax + Bu \\ y &= Cx + Du \end{aligned}$$

where A, B are given by (5.3), while $C \in \Re^{p\times n}$ and $D \in \Re^{p\times m}$ are given by

$$C = \frac{\partial h}{\partial x}(x, u)\bigg|_{x=0,\ u=0} ; \qquad D = \frac{\partial h}{\partial u}(x, u)\bigg|_{x=0,\ u=0} .$$

Assuming $(A,\ B)$, is stabilizable and $(A,\ C)$ is detectable, then based on the linear model one can design a linear dynamic output feedback controller to achieve regulation. An observer-based controller is an example of such an approach [134, 159, 279].

It is interesting to note that, similar to adaptive approximation based control, linear control is also based on an approximation, albeit a very simple one: a linear function, which is accurate only in a small neighborhood of an operating point. The basic idea behind approximation based control using nonlinear models is to expand the region where the approximation is valid from a small neighborhood around the linearizing point (in the case of linear models) to an expanded region $\mathcal{D}$, where $\mathcal{D}$ can be relatively large (i.e., defining the state space region of possible operation). It should be noted, however, that similar to linear control methods, if the state trajectories move outside the approximating region $\mathcal{D}$, then the approximation-based controller may not be effective in achieving the desired control objectives. Methods to ensure that the state trajectory remains in the region $\mathcal{D}$ will be an important topic in Chapters 6 and 7.

■ EXAMPLE 5.1

Consider the third-order nonlinear system

$$\begin{aligned} \dot{x}_1 &= x_2 + x_2 x_3 \\ \dot{x}_2 &= x_3 + x_1 x_3 - x_2 u \\ \dot{x}_3 &= x_1 + u + x_3 u \\ y &= x_1. \end{aligned}$$

It can be readily verified that $x^* = [0\ \ 0\ \ 0]^\top$, $u^* = 0$ is an equilibrium point of the nonlinear system. Linearizing the system around the equilibrium point $x = x^*$, $u = u^*$ gives

$$\begin{bmatrix} \dot{x}_1 \\ \dot{x}_2 \\ \dot{x}_3 \end{bmatrix} = \begin{bmatrix} 0 & 1 & 0 \\ 0 & 0 & 1 \\ 1 & 0 & 0 \end{bmatrix} \begin{bmatrix} x_1 \\ x_2 \\ x_3 \end{bmatrix} + \begin{bmatrix} 0 \\ 0 \\ 1 \end{bmatrix} u$$

$$y = \begin{bmatrix} 1 & 0 & 0 \end{bmatrix} \begin{bmatrix} x_1 \\ x_2 \\ x_3 \end{bmatrix}.$$

Suppose the control objective is to achieve regulation of y with the closed-loop poles located at $s = -1 \pm j$ and $s = -2$. Hence the desired characteristic equation is

$$s^3 + 4s^2 + 6s + 4 = 0.$$

This can be achieved by selecting the control law as

$$u = -5x_1 - 6x_2 - 4x_3.$$

If the same linear control law is applied to the nonlinear system then we obtain the following closed-loop nonlinear dynamics:

$$\dot{x}_1 = x_2 + x_2 x_3 \tag{5.7}$$

$$\dot{x}_2 = x_3 + x_1 x_3 + 5x_1 x_2 + 6x_2^2 + 4x_2 x_3 \tag{5.8}$$

$$\dot{x}_3 = -4x_1 - 6x_2 - 4x_3 - 5x_1 x_3 - 6x_2 x_3 - 4x_3^2. \tag{5.9}$$

Linearization of the above closed-loop system (5.7)–(5.9) yields

$$\widetilde{A} = A + BK = \begin{bmatrix} 0 & 1 & 0 \\ 0 & 0 & 1 \\ -4 & -6 & -4 \end{bmatrix}.$$

As expected, the eigenvalues of $\widetilde{A}$ are $\lambda_{1,2} = -1 \pm j$ and $\lambda_3 = -2$. △

5.1.2 Linearizing Around a Trajectory

Consider the nonlinear system system (5.1), where in this case the control objective is to design a control law such that the state $x(t)$ tracks a desired vector signal $x_d(t)$. Let the tracking error be denoted by $e(t) = x(t) - x_d(t)$. If a tracking controller is designed based on a linearization valid at some operating point $x_e \in \Re^n$, then as $x_d(t)$ moves away from the equilibrium point, the state $x(t)$ will try to follow it. However, as the distance between $x(t)$ and x_e increases, the linear approximation may become increasingly inaccurate. As the accuracy of the linear approximation decreases, the designed linear controller may become unsuitable, thus possibly forcing $x(t)$ even further away from the equilibrium x_e.

The tracking objective is in general more suitably addressed by a control law that is designed based on linearization about the desired trajectory $x_d(t)$. Obviously, linearization around $x_d(t)$ assumes that this signal is available *a priori*. If $x_d(t)$ is not available but needs to be generated online possibly by an outer-loop controller, then small-signal linearization can be performed around a nominal trajectory $x^*(t)$, which is available *a priori*. Associated with a nominal trajectory $x^*(t)$ is a nominal control signal $u^*(t)$ and initial conditions $x^*(0) = x_0^*$ such that $x^*(t)$ satisfies

$$\dot{x}^*(t) = f(x^*(t), u^*(t)), \qquad x^*(0) = x_0^*.$$

Let $\tilde{x}(t) = x(t) - x^*(t)$ and $\tilde{u}(t) = u(t) - u^*(t)$. Then

$$\begin{aligned} \dot{\tilde{x}} &= f(x, u) - f(x^*, u^*) \\ &= f(\tilde{x} + x^*, \; \tilde{u} + u^*) - f(x^*, u^*). \end{aligned} \tag{5.10}$$

Using the Taylor series expansion of $f(\tilde{x}+x^*,\ \tilde{u}+u^*)$ around (x^*, u^*) we obtain

$$f(\tilde{x}+x^*,\ \tilde{u}+u^*) = f(x^*,u^*) + \frac{\partial f}{\partial x}(x^*,u^*)\tilde{x} + \frac{\partial f}{\partial u}(x^*,u^*)\tilde{u} + \mathcal{F}(t,\tilde{x},\tilde{u})$$

where $\mathcal{F}$ represents the higher-order terms of the Taylor series expansion. Since $\mathcal{F}$ contains the higher-order terms, it satisfies

$$\lim_{\|(\tilde{x},\tilde{u})\|\to 0}\left(\sup_{t\geq 0}\frac{\|\mathcal{F}(t,\tilde{x},\tilde{u})\|}{\|(\tilde{x},\tilde{u})\|}\right) = 0.$$

In other words, as $\tilde{x}$ and $\tilde{u}$ become small, $\mathcal{F}$ goes to zero faster than $\|(\tilde{x},\tilde{u})\|$. Using the Taylor series expansion, (5.10) can be rewritten

$$\dot{\tilde{x}} = \frac{\partial f}{\partial x}(x^*,u^*)\tilde{x} + \frac{\partial f}{\partial u}(x^*,u^*)\tilde{u} + \mathcal{F}(t,\tilde{x},\tilde{u}).$$

In a linear approximation the higher-order terms are ignored. Hence, the small-signal linearization of (5.1) around the nominal trajectory $x^*(t)$ is given by

$$\dot{z} = A(t)z + B(t)\tilde{u},$$

where z is the state of the linear model and the matrices $A(t) : [0,\ \infty) \mapsto \Re^{n\times n}$ and $B(t) : [0,\ \infty) \mapsto \Re^{n\times m}$ are given by

$$A(t) = \left.\frac{\partial f}{\partial x}(x,u)\right|_{x=x^*,\ u=u^*} = \begin{bmatrix} \frac{\partial f_1}{\partial x_1}(x^*,u^*) & \cdots & \frac{\partial f_1}{\partial x_n}(x^*,u^*) \\ \vdots & & \vdots \\ \frac{\partial f_n}{\partial x_1}(x^*,u^*) & \cdots & \frac{\partial f_n}{\partial x_n}(x^*,u^*) \end{bmatrix} \tag{5.11}$$

$$B(t) = \left.\frac{\partial f}{\partial u}(x,u)\right|_{x=x^*,\ u=u^*} = \begin{bmatrix} \frac{\partial f_1}{\partial u_1}(x^*,u^*) & \cdots & \frac{\partial f_1}{\partial u_m}(x^*,u^*) \\ \vdots & & \vdots \\ \frac{\partial f_n}{\partial u_1}(x^*,u^*) & \cdots & \frac{\partial f_n}{\partial u_m}(x^*,u^*) \end{bmatrix}. \tag{5.12}$$

Now, suppose we select the control law as $\tilde{u} = K(t)z(t)$. The closed-loop dynamics for the linear system system are given by

$$\dot{z} = [A(t) + B(t)K(t)]\, z(t). \tag{5.13}$$

If the pair $(A(t),\ B(t))$ is uniformly completely controllable, then there exists $K(t)$ such that the closed-loop system (5.13) is asymptotically stable; therefore, $z(t) \to 0$, which implies that $x(t) \to x^*(t)$. If the nominal trajectory $x^*(t)$ coincides with the desired vector signal $x_d(t)$, then we achieve asymptotic convergence of the tracking error to zero.

Clearly, the above stability arguments were based on the linear model. Applying the same control law to the nonlinear system we have

$$u(t) = K(t)\,(x(t) - x_d(t)) + u^*(t),$$

which implies

$$\dot{x}(t) = f(x(t),\ K(t)(x(t) - x_d(t)) + u^*(t)).$$

Again, linearizing the closed-loop system around $x = x^* = x_d$, $u = u^*$ yields

$$\dot{e}(t) = (A(t) + B(t)K(t))\, e(t).$$

Therefore, applying the linear control law to the nonlinear system yields a locally asymptotically stable closed-loop system. In this case, locality is defined relative to the nominal trajectory (i.e., $\|x(t) - x^*(t)\|$ sufficiently small for all $t > 0$).

If the nonlinear system has an output function then, again, we can proceed to obtain the $C(t)$ and $D(t)$ matrices. Linearization of the nonlinear system (5.5)–(5.6) around a nominal trajectory $x^*(t)$ produces a linear model of the form

$$\begin{aligned}\dot{z} &= A(t)z + B(t)\tilde{u} \\ \tilde{y} &= C(t)z + D(t)\tilde{u},\end{aligned}$$

where $A(t)$, $B(t)$ are given by (5.11)–(5.12), while $C(t) \in \Re^{p\times n}$ and $D(t) \in \Re^{p\times m}$ are given by

$$C(t) = \frac{\partial h}{\partial x}(x,u)\bigg|_{x=x^*(t),\ u=u^*(t)} \quad ; \qquad D = \frac{\partial h}{\partial u}(x,u)\bigg|_{x=x^*(t),\ u=u^*(t)} .$$

Therefore, we see that linearizing around a trajectory yields similar results as linearizing around an equilibrium point, with the key difference that in the former case the linear model is time-varying.

Next we present an example of linearizing around a nominal trajectory to illustrate the concepts introduced in this subsection.

■ EXAMPLE 5.2

A simple model of a satellite of unit mass moving in a plane can be described by the following equations of motion in polar coordinates [225]:

$$\begin{aligned}\ddot{r}(t) &= r(t)\dot{\theta}^2(t) - \frac{\beta}{r^2(t)} + u_1(t) \\ \ddot{\theta}(t) &= -\frac{2\dot{r}(t)\dot{\theta}(t)}{r(t)} + \frac{u_2(t)}{r(t)}\end{aligned}$$

where, as shown in Figure 5.2, $r(t)$ is the radius from the origin to the mass, $\theta(t)$ is the angle from a reference axis, $u_1(t)$ is the thrust force applied in the radial direction, $u_2(t)$ is the thrust force applied in the tangential direction, and β is a constant parameter. With zero thrust forces (i.e., $u_1(t) = 0$ and $u_2(t) = 0$), the resulting solution can take various forms (ellipses, parabolas, or hyperbolas) depending on the initial conditions. In this example, we consider a simple circular trajectory with constant angular velocity (i.e., $r(t)$ and $\dot{\theta}(t)$ are both constant). It is easy to verify that with zero thrusts forces and the initial conditions $r(0) = r_0$, $\dot{r}(0) = 0$, $\theta(0) = \theta_0$, $\dot{\theta}(0) = \omega_0 := \sqrt{\beta/r_0^3}$ the resulting nominal trajectory is $r^*(t) = r_0$ and $\theta^*(t) = \omega_0 t + \theta_0$. The objective is to linearize the model around this nominal trajectory.

To construct the state equation representation, let $x_1(t) = r(t)$, $x_2(t) = \dot{r}(t)$, $x_3(t) = \theta(t)$, $x_4(t) = \dot{\theta}(t)$. The equations of motion in the state coordinates are given by

$$\begin{bmatrix}\dot{x}_1\\ \dot{x}_2\\ \dot{x}_3\\ \dot{x}_4\end{bmatrix} = f(x,u) = \begin{bmatrix} x_2 \\ x_1x_4^2 - \frac{\beta}{x_1^2} + u_1 \\ x_4 \\ \frac{-2x_2x_4}{x_1} + \frac{1}{x_1}u_2\end{bmatrix}.$$

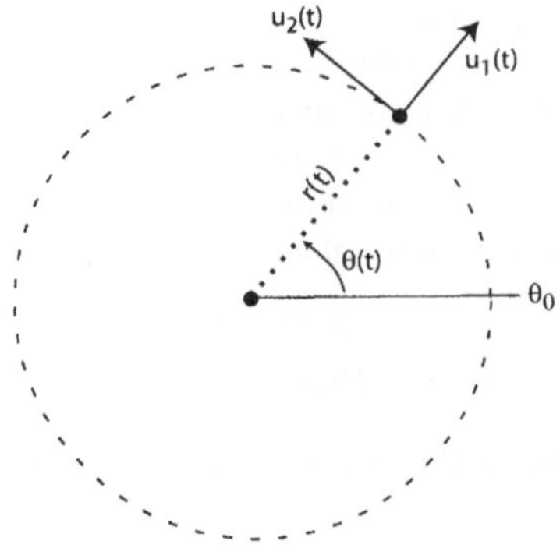

Figure 5.2: Point mass satellite moving in a planar gravitational orbit.

The nominal trajectory is described by

$$u^* = \begin{bmatrix} u_1^* \\ u_2^* \end{bmatrix} = \begin{bmatrix} 0 \\ 0 \end{bmatrix}, \qquad x^* = \begin{bmatrix} x_1^* \\ x_2^* \\ x_3^* \\ x_4^* \end{bmatrix} = \begin{bmatrix} r_0 \\ 0 \\ \omega_0 t + \theta_0 \\ \omega_0 \end{bmatrix}, \qquad \omega_0 = \sqrt{\frac{\beta}{r_0^3}}.$$

If we define $\tilde{x}(t) = x(t) - x^*(t)$, $\tilde{u}(t) = u(t) - u^*(t)$, then the small-signal linearized system (around the nominal trajectory) is given by

$$\begin{aligned} \begin{bmatrix} \dot{z}_1 \\ \dot{z}_2 \\ \dot{z}_3 \\ \dot{z}_4 \end{bmatrix} &= \begin{bmatrix} 0 & 1 & 0 & 0 \\ 3\omega_0^2 & 0 & 0 & 2r_0\omega_0 \\ 0 & 0 & 0 & 1 \\ 0 & \frac{-2\omega_0}{r_0} & 0 & 0 \end{bmatrix} \begin{bmatrix} z_1 \\ z_2 \\ z_3 \\ z_4 \end{bmatrix} + \begin{bmatrix} 0 & 0 \\ 1 & 0 \\ 0 & 0 \\ 0 & \frac{1}{r_0} \end{bmatrix} \begin{bmatrix} u_1 \\ u_2 \end{bmatrix} \\ \dot{z} &= A(t)z + B(t)u. \end{aligned}$$

We notice that in this special case of a circular orbit, the matrices $A(t)$ and $B(t)$ happen to be time-invariant. This is a coincidence — in general, the matrices will be time-varying. △

5.1.3 Gain Scheduling

In the previous two subsections we have described the procedure for linearizing around an operating point x_e or around a trajectory $x^*(t)$. As discussed, a key limitation of the small-signal linearization approach is the fact that the linear model is accurate only in a neighborhood around the operating point x_e or the nominal trajectory x^*. Consequently, the linear control law that is designed based on the linear model is, in general, effective only if the system state remains in that same neighborhood. In this subsection we introduce the gain scheduling control approach, which is based on small-signal linearization around *multiple* operating points. For each linear model we design a feedback controller, thus creating a family of feedback control laws, each applicable in the neighborhood of a specific operating point. The family of feedback controllers can be combined into a single control whose parameters are changed by a scheduling scheme based on the trajectory or some other scheduling variables.

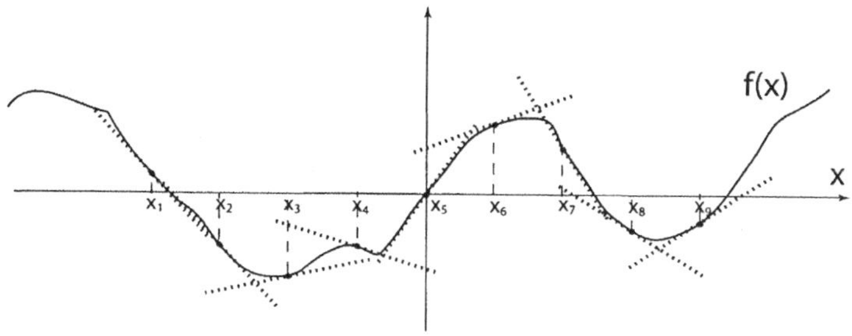

Figure 5.3: Diagram to illustrate the gain scheduling approach, which is based on the linearization around multiple operating points $\{x_1,\ x_2, \ldots\ x_9\}$. The multiple linear models constitute an example of approximating a nonlinear system.

Consider again the nonlinear system (5.1), where in this case we linearize the system into N linear models

$$\dot{z} = A_i z + B_i u, \qquad i = 1,\ 2,\ \ldots\ N$$

where $z = x - x_i$ for each i. Each linear model parameterized by $(A_i,\ B_i)$ is valid around an operating point point x_i. This is illustrated in Figure 5.3, where the nonlinear function $f(x)$ is linearized around nine operating points $\{x_1,\ x_2, \ldots\ x_9\}$. For each linear model we design a control law based on the control objective associated with a particular operating point. Suppose that the control law $u = K_i z$ corresponds to the linear model $\dot{z} = A_i z + B_i u$. A key element of the gain scheduling approach is the design of a scheduler for switching between the various control laws parameterized by $\{K_1,\ K_2, \ldots\ K_N\}$. Typically, transitions between different operating points are handled by interpolation methods. The gain scheduler can be viewed as a look-up table with the appropriate logic for selecting a suitable controller gain K_i based on identifying the corresponding operating point x_i.

Intuitively, we note that if the region of attraction associated with each linearization is larger than the scheduled operating region corresponding to each operating point, then the resulting gain scheduling control scheme will be stable. However, special care needs to be taken since the resulting controller is time-varying, hence the stability analysis needs to be treated as a time-varying system. A formal stability analysis for gain scheduling is beyond the scope of this text (see the work of Shamma and Athans [242, 243]).

Despite the derivation of some stability results under certain conditions, gain scheduling is still considered to some degree an ad hoc control method. However, it has been used in several application examples, especially in flight control [118, 161, 183, 256, 257, 258]. The gain scheduling approach has also been utilized in other applications such as in process control and automotive applications [11, 113, 126].

One of the key limitations of gain scheduling is that the controller parameters are precomputed offline for each operating condition. Hence, during operation the controller is fixed, even though the linear control gains are changing as the operating conditions change. In the presence of modeling errors or changes in the system dynamics, the gain scheduling controller may result in deterioration of the performance since the method does not provide any learning capability to correct—during operation—any inaccurate schedules. Another possible drawback of the gain scheduling approach is that it requires considerable offline effort to derive a reliable gain schedule for each possible situation that the plant will encounter.

The gain scheduling approach can be conveniently viewed as a special case of the adaptive approximation approach developed in this text. A local linear model is an example case of a local approximation function. For example, the linear functions in Figure 5.3 can be replaced by other approximation functions. Typically, the adaptive approximation based techniques developed in this book consist of approximation functions with at least some overlap between, which intuitively can be viewed as a way to obtain a smoother approximation from one operating position (or from one node) to another.

One of the key differences between the standard gain scheduling technique and the adaptive approximation based control approach is the ability of the latter to adjust certain parameters (weights) during operation. Unlike gain scheduling, adaptive approximation is designed around the principle of "learning" and thus reduces the amount of modeling effort that needs to be applied offline. Moreover, it allows the control scheme to deal with unexpected changes in plant dynamics due to faults or severe disturbances.

5.2 FEEDBACK LINEARIZATION

This section describes the approach of cancelling the nonlinearities by the combined use of feedback and change of coordinates. This approach, referred to as Feedback Linearization, is one of the most powerful and commonly found techniques in nonlinear control. The presentation begins with a simple single variable plant to illustrate the main ideas of the approach and proceeds to generalize the approach to wider classes of systems.

In this section, we restrict ourselves to the case of completely known nonlinearities. In Chapters 6 and 7 we will deal with the case that the nonlinearities are partially or completely unknown. For convenience, it is appropriate to distinguish between input-state linearization methods and input-output linearization methods.

5.2.1 Scalar Input-State Linearization

To illustrate the main intuitive idea behind feedback linearization, we start by considering the simple scalar system

$$\dot{y} = f(y) + g(y)u,$$

where u is the control input, y is the measured output, and the nonlinear functions f, g are assumed to be known a priori. The control objective is to design a control law that generates u such that $u(t)$ and $y(t)$ remain bounded and $y(t)$ tracks a desired function $y_d(t)$. We will assume throughout that $y_d(t)$ and all of its derivatives that are required for computing the control signal are in fact available, continuous, and bounded. Section A.4 of the appendix discusses prefiltering, which is one method to ensure the validity of this assumption. For this scalar system it is straightforward to see that, assuming that $g(y) \neq 0$, the control law

$$u = \frac{1}{g(y)}\left[-a_m(y - y_d) + \dot{y}_d - f(y)\right], \tag{5.14}$$

where $a_m > 0$ is a design constant, achieves the control objective. Specifically, with the above feedback control algorithm, the tracking error $e(t) = y(t) - y_d(t)$ satisfies $\dot{e} = -a_m e$. Hence, the tracking error converges to zero exponentially fast from any initial condition (global stability results).

A key observation for the reader is that implementation of the feedback control algorithm (5.14) is feasible in all scenarios of desired trajectories y_d only if the function $g(y) \neq 0$

for all $y \in \Re$. Otherwise, if $g(y)$ approaches zero then the control effort becomes large, causing saturation of the control input and possibly leading to instability. This problem, which arises due to the lack of controllability at some values of the state-space, is referred to as the *stabilizability* problem.

EXAMPLE 5.3

It is important to note that even if $g(y) = 0$ at a crucial part of the state-space, that does not necessarily imply that the system is uncontrollable. For example, consider the input-output system

$$\dot{y} = yu$$

where the objective is to track the signal $y_d(t) = 0$. Therefore, in this case the singularity point $y = 0$ is actually the desired setpoint. The regulation problem can be solved by simply selecting $u = -1$ (which does not contain any feedback information), or by selecting $u = -y^2$. Therefore, it is not necessary for the control law to cancel $g(y)$ in order to stabilize the closed-loop system. If the control objective is for y to track an arbitrary signal $y_d(t)$ then the problem becomes more difficult, and in fact it becomes necessary to address the stabilizability problem. △

The control law (5.14) illustrates the use of the controller for cancelling nonlinearities. Specifically, as we can see from (5.14), the nonlinearities f and g in the open-loop system are cancelled by the controller. This converts the system into one with linear error dynamics, for which there are known control design and analysis methods. In fact, (5.14), can be rewritten as

$$u = \frac{1}{g(y)}\left[-f(y) + v\right] \quad (5.15)$$

$$v = -a_m(y - y_d) + \dot{y}_d, \quad (5.16)$$

where (5.15) is a feedback linearizing operator that causes the closed-loop system to transform to the linear system $\dot{y} = v$, and (5.16) is a linear stabilizing controller for the linearized tracking problem. Many other linear controllers could be selected. Even for this simple system we can extract some key observations:

- The feedback linearizing operator of (5.15) exactly linearizes the model $\dot{y} = f(y) + g(y)u$ over the domain of validity of that model. There are no approximations. This is distinct from the small signal linearization of Section 5.1, which was exact only at a single point.
- The role of the design parameter $a_m > 0$ is to set the time constant of the exponential convergence of the tracking error in response to initial condition errors and disturbances.
- The parameter a_m does *not* determine the bandwidth of the overall control system in the sense of the bandwidth of input signals y_d that can be tracked. Note that the exponential convergence of the tracking error dynamics is independent of the input signal y_d. This is achieved by feeding forward the derivative of the input signal, $\dot{y}_d$. Therefore, from a theoretical perspective, the reference input tracking bandwidth of this controller is infinite. In fact, this bandwidth will be limited by physical constraints, such as the actuators, and must be accounted for in the design of the system that generates y_d and its derivatives.

- The linearization achieved by the feedback operator (5.15) requires exact knowledge of f and g. The effect of model errors requires further analysis.

These comments also apply to feedback linearization when it is applied to higher order systems.

Appended Integrators. One role of integrators in control laws is to force the tracking error to zero in the presence of model error, disturbances, and input type. The required number of integrators as a function of the type of the input to be tracked is discussed in most text books on control system design, e.g., [66, 86, 140]. Integrators can have similar utility in nonlinear control applications. Integrators can be included in the control law and control design analysis by various approaches such as that discussed in Exercise 1.3 and the following.

In addition to the tracking error $e(t) = y(t) - y_d(t)$ define

$$e_F(t) = e(t) + c\int_0^t e(\tau)d\tau, \tag{5.17}$$

where $c > 0$. It is noted that $e_F(t)$ is a linear combination of the tracking error and the integral of the tracking error that can be thought of as providing a PI controller (proportional-integral control). For implementation and analysis, the system state space model will include one appended controller state to compute the integral of the tracking error. From (5.17), we obtain $\dot{e}_F = \dot{e} + ce$; hence, to force $e_F(t)$ to zero, the control law (5.15) is modified to

$$u = \frac{1}{g(y)}\left(-a_m e_F - c(y - y_d) + \dot{y}_d - f(y)\right) \tag{5.18}$$

(see also (6.35)). This control law results in $\dot{e}_F = -a_m e_F$. It is easy to see that if $e_F(t)$ converges to zero then so does $e(t)$ (notice that $e = \frac{s}{c+s}[e_F]$).

5.2.2 Higher-Order Input-State Linearization

Similar ideas can be developed for n-th order systems in the so called companion form:

$$\left.\begin{aligned} \dot{x}_1 &= x_2 \\ \dot{x}_2 &= x_3 \\ &\vdots \\ \dot{x}_n &= f(x) + g(x)u. \end{aligned}\right\} \tag{5.19}$$

The nonlinearities can be cancelled by using a feedback linearizing control law of the form

$$u = \frac{1}{g(x)}\left[-f(x) + v\right].$$

This results in a simple linear relation (n integrators in series) between v and x, given by

$$\begin{aligned} \dot{x}_1 &= x_2 \\ \dot{x}_2 &= x_3 \\ &\vdots \\ \dot{x}_n &= v. \end{aligned}$$

Therefore, we can choose v as

$$v = y_d^{(n)} - \lambda_0 e - \lambda_1 \dot{e} - \cdots - \lambda_{n-1} e^{(n-1)},$$

where $e(t) = x_1(t) - y_d(t)$ is the tracking error. In this case, the characteristic equation for the tracking error dynamics of the closed-loop system is

$$s^n + \lambda_{n-1} s^{n-1} + \cdots + \lambda_1 s + \lambda_0 = 0.$$

Choosing the design coefficients $\{\lambda_0,\ \lambda_1,\ \ldots\ \lambda_{n-1}\}$ so that this characteristic equation is a Hurwitz polynomial (i.e., the roots of the polynomial are all in the left-half complex plane) implies that the closed-loop system is exponentially stable and $e(t)$ converges to zero exponentially fast, with a rate that depends on the choice of the design coefficients.

A important question is: Can all functions in nonlinear systems be cancelled by such feedback methods?

Clearly, the extent of the designer's ability to cancel nonlinearities depends on the structural and physical limitations that are applicable. For example, if control actuators could be placed to allow control of every state independently (an unrealistic assumption in almost all practical applications), then under some conditions on the invertibility of the actuator gain $g(x)$, we would be able to use each control signal to cancel the nonlinearities of the corresponding state. In general, however, it is not possible to cancel all nonlinearities by feedback linearization methods. To achieve such nonlinearity cancellation, certain structural properties in the nonlinear system must be satisfied.

A first cut at the class of feedback linearizable systems is nonlinear systems described by

$$\dot{x} = Ax + B\beta^{-1}(x)\left[u - \alpha(x)\right], \tag{5.20}$$

where u is a m-dimensional control input, x is an n-dimensional state vector, A is an $n \times n$ matrix, B is an $n \times m$ matrix, and the pair (A, B) is controllable. The nonlinearities are contained in the functions $\alpha : \Re^n \mapsto \Re^m$ and $\beta : \Re^n \mapsto \Re^{m \times m}$, which are defined on an appropriate domain of interest, with the matrix $\beta(x)$ assumed to be nonsingular for every x in the domain of interest and the symbol β^{-1} denotes an inverse matrix. Systems described by (5.20) can be linearized by using a state feedback of the form

$$u = \alpha(x) + \beta(x)v,$$

which results in

$$\dot{x} = Ax + Bv.$$

For stabilization, a state feedback $v = Kx$ can be designed such that the closed-loop system $\dot{x} = (A + BK)x$ is asymptotically stable. This is achieved by selecting K such that all the eigenvalues of $A + BK$ are in the left-half complex plane. A similar design procedure, based on linear control designed methods, can be used to select v for tracking problems.

The reader will undoubtedly notice that the class of systems described by (5.20) is significantly more general than the class of nonlinear systems in companion form (5.19). The class of feedback linearizable systems is actually even larger than the systems described by (5.20) since it includes nonlinear systems that can be transformed to (5.20) by a coordinate transformation. This topic is discussed in detail in the next subsection.

If a nonlinear system is not feedback linearizable, it does not imply that it cannot be controlled. There are several classes of nonlinear systems that cannot be put into the standard form for feedback linearizable systems, but they can be controlled by other methods.

Feedback linearization, although a very useful tool with a beautiful mathematical theory for dealing with nonlinear systems, has some serious drawbacks in practical applications. Two of these drawbacks are discussed below:

- Feedback linearization may not be the most efficient way of controlling a nonlinear system. To illustrate this concept consider the (frequently used) simple system

$$\dot{x} = -x^3 + u.$$

For stabilization around $x = 0$, a feedback linearizing controller would cancel the term x^3. However, this is a "stable" term so there is not real need to cancel it. Instead, a simple linear feedback control law of the form $u = -x$, could achieve similar results without a large control effort, as compared to linearizing feedback controller of the form $u = -x + x^3$. The reader will undoubtedly note that if the initial state $x(0)$ is far away from zero, then the feedback linearizing controller will require significantly larger control effort than a linear control law. The bottom line is that, in this case, the controller is working hard to cancel a nonlinearity that is actually a stable term helping the control effort. The concept of cancelling useful nonlinearities is also present in higher dimensional systems, however it becomes less evident due to the complexity of the problem. Note that this issue is less important when the objective is tracking. In the above example, when the objective is to cause x to track y_d, then the x^3 term would have to be addressed, e.g.,

$$u = x^3 + \dot{y}_d - a_m(y - y_d).$$

- Feedback linearization relies heavily on the exact cancellation of nonlinearities. In practice, the nonlinear terms of a dynamical system are not known exactly, therefore exact cancellation may not be possible. By their nature, linearization methods are not "robust" with respect to modeling or other uncertainties. For example, consider a feedback linearizable system of the form

$$\dot{x} = x^2 + \varepsilon x^4 + u$$

where in the actual system, $\varepsilon > 0$. However, because of lack of knowledge about the value of ε, the designer had assumed that $\varepsilon = 0$, thus designing a stabilizing control law of the form $u = -x - x^2$. In this case, the closed-loop system is given by

$$\dot{x} = -x + \varepsilon x^4,$$

which is unstable if

$$x(0) > \varepsilon^{-1/3}.$$

Moreover, if $x(0) > \varepsilon^{-1/3}$, then $x(t) \to \infty$ in finite time — this is called *finite escape time*. For tracking control, the issue of modeling errors can be even more critical, since the signal y_d may cause the state to move into the portion of the state space where the model error is significant (e.g., $y_d > \varepsilon^{-1/3}$ in the example of this paragraph). Methods to accommodate modeling error are presented in Section 5.4 using bounding techniques, in Section 5.5 using adaptive techniques and in Chapters 6 and 7 using adaptive approximation methods.

5.2.3 Coordinate Transformations and Diffeomorphisms

Fortunately, the class of systems described by (5.20) does not include all the possible systems the are feedback linearizable. The reason is that a large number of systems are not immediately in the form described by (5.20), but they can be put into that form by a nonlinear change of coordinates, or as it is sometimes called, *state transformations*. In this section, we attempt to make the concept of coordinate transformation intuitively understandable without going into all the mathematical details that are sometimes associated with it.

Since we are dealing with nonlinear systems, we are interested in nonlinear state transformations. A nonlinear state transformation is a natural extension of the same concept from linear systems. For example, consider the linear input/output system

$$\Sigma_x \quad \begin{cases} \dot{x} &= A_x x + B_x u \\ y &= C_x x + D_x u \end{cases} \tag{5.21}$$

where $u \in \Re^m$ is the input, $y \in \Re^p$ is the output and $x \in \Re^n$ is the state. The above system can be transformed to a new state coordinate system $z = Tx$, where T is an invertible matrix. In the new z-coordinate, the system is described by

$$\Sigma_z \quad \begin{cases} \dot{z} &= A_z z + B_z u \\ y &= C_z z + D_z u \end{cases} \tag{5.22}$$

where

$$\begin{aligned} A_z &= T A_x T^{-1} \\ B_z &= T B_x \\ C_z &= C_x T^{-1} \\ D_z &= D_x. \end{aligned}$$

Clearly, from an input/output ($u \mapsto y$) viewpoint, the two systems Σ_x, Σ_z are exactly the same. As discussed in basic control courses and linear system theory textbooks [10, 19, 39], state transformations can be useful for putting the system into a new coordinate framework which can make the control design and analysis more convenient.

In the case of nonlinear state transformations, we have $z = T(x)$, where $T : \Re^n \mapsto \Re^n$ is a function which is required to be a *diffeomorphism*. This means that T is smooth and its inverse T^{-1} exists and is also smooth. It is important for the reader to distinguish between a local diffeomorphism, where T is defined over a region $\Omega \subset \Re^n$, and a global diffeomorphism, which is defined over the whole space $\Re^n$.

In the special case of a linear transformations, a diffeomorphism is equivalent to the matrix (which represents the linear operator relative to some basis) being invertible (i.e., non-zero determinant). For nonlinear transformations, one can check whether a function is a diffeomorphism by attempting to find a smooth inverse function T^{-1} such that $x = T^{-1}(z)$. In cases of complex multivariable transformations it may be difficult to derive such an inverse function. In these cases, one can show local existence of a diffeomorphism by using Lemma 5.2.1, which follows from the well-known implicit function theorem.

Lemma 5.2.1 *Let $T(x)$ be a smooth function defined in a region $\Omega \subset \Re^n$. If the Jacobian matrix*

$$\nabla T = \frac{\partial T}{\partial x}$$

is nonsingular at a point $x_0 \in \Omega$, then $T(x)$ is a local diffeomorphism in a subregion of Ω.

Once a diffeomorphism $T(x)$ is defined then it is possible to follow a similar procedure as for linear systems to derive the model appropriate relative to the new set of coordinates $z = T(x)$. Consider the following affine nonlinear dynamical system:

$$\Sigma_x \quad \begin{cases} \dot{x} &= f_x(x) + G_x(x)u \\ y &= h_x(x) \end{cases} \tag{5.23}$$

where $f_x : \Omega_x \mapsto \Re^n$, $G_x : \Omega_x \mapsto \Re^{n\times m}$ and $h_x : \Omega_x \mapsto \Re^p$ are smooth functions in a region $\Omega_x \subset \Re^n$. The above system can be transformed to a new state coordinate system $z = T(x)$, where T is a diffeomorphism. In the new z-coordinate, the system is described by

$$\Sigma_z \quad \begin{cases} \dot{z} &= f_z(z) + G_z(z)u \\ y &= h_z(z) \end{cases} \tag{5.24}$$

where

$$\begin{aligned} f_z(z) &= \left.\frac{dT}{dx}(x)f_x(x)\right|_{x=T^{-1}(z)} \\ G_z(z) &= \left.\frac{dT}{dx}(x)G_x(x)\right|_{x=T^{-1}(z)} \\ h_z(z) &= h(x)|_{x=T^{-1}(z)} \,. \end{aligned}$$

It is important to note that, while in linear change of coordinates the transformation is always global (i.e., T it is a global diffeomorphism), for nonlinear change of coordinates it is often the case that the transformation is local.

Following the development of the concept of a diffeomorphism, we can now define the class of feedback linearizable systems. A nonlinear system

$$\dot{x} = f(x) + G(x)u \tag{5.25}$$

is said to be *input-state feedback linearizable* if there exists a diffeomorphism $z = T(x)$, with $T(0) = 0$, such that

$$\dot{z} = Az + B\beta^{-1}(z)\left[u - \alpha(z)\right], \tag{5.26}$$

where $(A,\ B)$ is a controllable pair and $\beta(z)$ is an invertible matrix for all z in a domain of interest $D_z \subset \Re^n$.

Therefore, we see that the class of feedback linearizable systems includes not only systems described by (5.20), but also systems that can be transformed into that form by a nonlinear state transformation. Determining if a given nonlinear system is feedback linearizable and what is an appropriate diffeomorphism are not obvious issues, and in fact they can be extremely difficult since in general they involve solving a set of partial differential equations.

Given a nonlinear system (5.25), consider a diffeomorphism $z = T(x)$. In the z-coordinates we have

$$\dot{z} = \frac{\partial T}{\partial x}f(x) + \frac{\partial T}{\partial x}G(x)u. \tag{5.27}$$

For feedback linearizable systems, (5.27) needs to be of the form

$$\begin{aligned} \dot{z} &= Az + B\beta^{-1}(z)\left[u - \alpha(z)\right] \\ &= AT(x) - B\beta^{-1}(T(x))\alpha(T(x)) + B\beta^{-1}(T(x))u. \end{aligned}$$

Therefore, the diffeomorphism $T(x)$ that we are looking for needs to satisfy

$$\begin{aligned} \frac{\partial T}{\partial x} f(x) &= AT(x) - B\beta^{-1}(T(x))\alpha(T(x)) && (5.28) \\ \frac{\partial T}{\partial x} G(x) &= B\beta^{-1}(T(x)). && (5.29) \end{aligned}$$

Hence, we conclude that that for a diffeomorphism to be able to transform (5.25) into (5.26), it needs to satisfy the partial differential equations (5.28)–(5.29) for some $\alpha(\cdot)$ and $\beta(\cdot)$. Whether a given system belongs to the class of feedback linearizable systems or not can be determined by checking two types of necessary and sufficient conditions: (i) a controllability condition and (ii) an involutivity condition [121, 134, 249]. The derivation of this result, while interesting from a mathematical viewpoint, is beyond the scope of this book.

■ EXAMPLE 5.4

Consider a model of a single-link manipulator with flexible joints, which is described by

$$\begin{aligned} J_1\ddot{q}_1 + MgL\sin q_1 + k(q_1 - q_2) &= 0 \\ J_2\ddot{q}_2 - k(q_1 - q_2) &= u, \end{aligned}$$

where J_1, J_2, M, g, L, k are known constants. The system can be written in state-space form by defining $x_1 = q_1$, $x_2 = \dot{q}_1$, $x_3 = q_2$, $x_4 = \dot{q}_2$. Thus, we obtain

$$\begin{bmatrix} \dot{x}_1 \\ \dot{x}_2 \\ \dot{x}_3 \\ \dot{x}_4 \end{bmatrix} = \begin{bmatrix} x_2 \\ -\frac{MgL}{J_1}\sin x_1 - \frac{k}{J_1}(x_1 - x_3) \\ x_4 \\ \frac{k}{J_2}(x_1 - x_3) \end{bmatrix} + \begin{bmatrix} 0 \\ 0 \\ 0 \\ \frac{1}{J_2} \end{bmatrix} u.$$

Consider the following diffeomorphism $z = T(x)$

$$\begin{aligned} z_1 &= x_1 \\ z_2 &= x_2 \\ z_3 &= -\frac{MgL}{J_1}\sin x_1 - \frac{k}{J_1}(x_1 - x_3) \\ z_4 &= -\frac{MgL}{J_1}x_2\cos x_1 - \frac{k}{J_1}(x_2 - x_4). \end{aligned} \qquad (5.30)$$

Proving that (5.30) is indeed a diffeomorphism is left as an exercise (see Exercise 5.8). The dynamics of the system in the z-coordinates are given by

$$\begin{aligned} \dot{z}_1 &= z_2 \\ \dot{z}_2 &= z_3 \\ \dot{z}_3 &= z_4 \\ \dot{z}_4 &= -z_3\left(\frac{MgL}{J_1}\cos z_1 + \frac{k}{J_1} + \frac{k}{J_2}\right) + \frac{MgL}{J_1}\left(z_2^2 - \frac{k}{J_2}\right)\sin z_1 + \frac{k}{J_1J_2}u. \end{aligned} \qquad (5.31)$$

Therefore, if we choose the control law u as

$$u = \frac{J_1J_2}{k}\left[v + z_3\left(\frac{MgL}{J_1}\cos z_1 + \frac{k}{J_1} + \frac{k}{J_2}\right) - \frac{MgL}{J_1}\left(z_2^2 - \frac{k}{J_2}\right)\sin z_1\right]$$

we obtain the following set of linear equations

$$\begin{aligned}\dot{z}_1 &= z_2\\ \dot{z}_2 &= z_3\\ \dot{z}_3 &= z_4\\ \dot{z}_4 &= v.\end{aligned} \tag{5.32}$$

Finally, the performance of the closed-loop system can be adjusted by selecting the intermediate control function v. Since (5.32) is controllable, by appropriately selecting v it is possible to arbitrarily place the closed-loop poles. △

5.2.4 Input-Output Feedback Linearization

Feedback linearization has been studied extensively in the nonlinear systems literature (see, for example, [121, 134, 159, 185]). In this text, we cover only some of the basic background to help the reader understand some of the techniques that will be used in Chapters 6 and 7 in the context of adaptive approximation based control. In this subsection, we present the concept of input-output linearization.

Consider the single-input single-output (SISO) nonlinear system

$$\dot{x} = f(x) + g(x)u \tag{5.33}$$

$$y = h(x) \tag{5.34}$$

where $u \in \Re^1, y \in \Re^1, x \in \Re^n$ and f, g and h are sufficiently smooth in a domain $D \subset \Re^n$. The time derivative of $y = h(x)$ is given by

$$\dot{y} = \frac{\partial h}{\partial x}(x)f(x) + \frac{\partial h}{\partial x}(x)g(x)u.$$

If $\frac{\partial h}{\partial x}(x)g(x) \neq 0$ for any $x \in D_0$ then the nonlinear system is said to have *relative degree one* on D_0. Intuitively, this implies that the control variable u appears explicitly in the differential equation for the first derivative of the output y; i.e., the input and output are separated by a single integrator. If $\frac{\partial h}{\partial x}(x)g(x) = 0$ (i.e., u does not directly affect $\dot{y}$), then we keep on differentiating the output until u appears explicitly. In order to define the second, third (and so on) derivatives, it is convenient to define the concept of a *Lie derivative*, which is used in advanced calculus.

The notation for the Lie derivative of h with respect to f is defined as

$$L_f h(x) = \frac{\partial h}{\partial x}(x)f(x).$$

This notation is convenient for dealing with repeated derivatives, as shown below:

$$\begin{aligned}L_f^k h(x) &= L_f L_f^{k-1} h(x) = \frac{\partial}{\partial x}\left(L_f^{k-1}h(x)\right) f(x)\\ L_g L_f^k h(x) &= \frac{\partial}{\partial x}\left(L_f^k h(x)\right) g(x).\end{aligned}$$

Based on the definition of the Lie derivative, if

$$L_g h(x) = \frac{\partial h}{\partial x}(x)g(x) = 0,$$

we keep on taking derivatives until $L_g L_f^{r-1} h(x) \neq 0$, which implies that u first appears explicitly in the equation for $y^{(r)}$, the r-th derivative of the output.

The nonlinear system (5.33)-(5.34) is said to have a relative degree r in a region $D_0 \subset D$ if the following conditions are satisfied for any $x \in D_0$:

$$\begin{aligned} L_g L_f^i h(x) &= 0, \qquad i = 0,\ 1,\ 2,\ \ldots,\ r-2 \\ L_g L_f^{r-1} h(x) &\neq 0. \end{aligned}$$

If a system has relative degree r, then

$$y^{(r)} = L_f^r h(x) + L_g L_f^{r-1} h(x) u.$$

Hence, the system is input-output linearizable, since the state feedback control

$$u = \frac{1}{L_g L_f^{r-1} h(x)} \left[-L_f^r h(x) + v \right] \tag{5.35}$$

gives the following linear input-output mapping:

$$y^{(r)} = v.$$

This is simply a chain of r integrators, which can be easily controlled by an appropriate selection of v. However, unless $r = n$, there are more states in the system that are not affected by the control input u. The dynamics of this set of $(n - r)$ states will define the so-called *zero dynamics* of the system, which are discussed below.

Using differential geometric tools, it can be shown that for a system with relative degree r there exists a diffeomorphism $z = T(x)$ that would transform the nonlinear system (5.33)–(5.34) into the input-output linearizable canonical form

$$\dot{\eta} = \phi(\eta, \zeta) \tag{5.36}$$

$$\dot{\zeta} = A_0 \zeta + B_0 \beta_0^{-1}(\eta, \zeta) \left[u - \alpha_0(\eta, \zeta) \right] \tag{5.37}$$

$$y = C_0 \zeta \tag{5.38}$$

where $z = [\eta^\top\ \zeta^\top]^\top$, with $\eta \in \Re^{n-r}$, $\zeta \in \Re^r$, and $(A_0,\ B_0,\ C_0)$ is a canonical form representation of a system consisting of r integrators in series:

$$A_0 = \begin{bmatrix} 0 & 1 & 0 & \cdots & 0 \\ 0 & 0 & 1 & & 0 \\ \vdots & & & \ddots & \vdots \\ 0 & & & & 1 \\ 0 & 0 & & \cdots & 0 \end{bmatrix}, \quad B_0 = \begin{bmatrix} 0 \\ 0 \\ \vdots \\ 0 \\ 1 \end{bmatrix}, \quad C_0 = \begin{bmatrix} 1 & 0 & \cdots & 0 & 0 \end{bmatrix}. \tag{5.39}$$

The diffeomorphism $z = T(x)$, with $T(0) = 0$, which transforms the nonlinear system (5.33)–(5.34) into the canonical form (5.36)–(5.38) is characterized by

$$\begin{aligned} \frac{\partial T_i}{\partial x} g(x) &= 0, \qquad \text{for } i = 1,\ 2,\ \ldots\ n-r \\ T_{n-r+1} &= h(x) \\ T_{n-r+2} &= L_f h(x) \\ &\ \vdots \\ T_n &= L_f^{r-1} h(x). \end{aligned}$$

The transformed system described by (5.36)–(5.38) is said to be in *normal form*. Basically, the nonlinear system is decomposed in two parts, the ζ-dynamics, which can be linearized by feedback, and the η variables, which characterize the internal dynamics of the system. The ζ-dynamics can be linearized and controlled by utilizing a feedback controller of the form

$$u = \alpha_0(\eta, \zeta) + \beta_0(\eta, \zeta)v,$$

where v can be chosen to set the convergence rate of the ζ-dynamics or to achieve reference input tracking. The feedback linearizing control functions α_0 and β_0 are computable based on the Lie derivatives obtained by differentiating the output variable:

$$\begin{aligned} \alpha_0(\eta, \zeta) &= \left. \frac{L_f^r h(x)}{L_g L_f^{r-1} h(x)} \right|_{x=T^{-1}(\eta,\zeta)} \\ \beta_0(\eta, \zeta) &= \left. \frac{1}{L_g L_f^{r-1} h(x)} \right|_{x=T^{-1}(\eta,\zeta)} . \end{aligned}$$

The zero dynamics are obtained by setting $\zeta = 0$ in the η-dynamics:

$$\dot{\eta} = \phi(\eta, 0). \tag{5.40}$$

The nonlinear system is said to be *minimum phase* if the zero dynamics described by (5.40) have an asymptotically stable equilibrium point in D.

The concepts of relative degree, coordinate decomposition into the η and ζ dynamics, minimum phase, zero dynamics, etc. all have their corresponding equivalents for linear systems. Of course, for linear systems we have the concept of a transfer function which characterizes both the stability of the input-output system (by the location of the roots of the denominator polynomial — the poles), as well as the stability of the internal dynamics, which are given by the roots of the numerator polynomial — the zeros.

Consider the n-th order linear system described by the transfer function

$$H(s) = k\frac{s^{n-r} + b_{n-r-1}s^{n-r-1} + \cdots \; b_1 s + b_0}{s^n + a_{n-1}s^{n-1} + \cdots \; a_1 s + a_0}$$

where r is the relative degree of the system; i.e., the difference between the order of the denominator and the order of the numerator. A state model (non-unique) for the system is given by

$$\begin{aligned} \dot{x} &= Ax + Bu \\ y &= Cx, \end{aligned}$$

where we are assuming that $r \geq 1$, thus the D matrix is zero. By taking the first time derivative of the output $y(t)$, we obtain

$$\dot{y} = CAx + CBu.$$

If $r = 1$ (relative degree 1) then $CB \neq 0$. On the other hand, if $CB = 0$, then the relative degree is larger than 1 so we continue to take time derivatives of the output. Following this procedure it can be shown that for linear systems with relative degree r,

$$\begin{aligned} CA^i B &= 0, \qquad \text{for } i = 1,\ 2,\ \ldots\ r-2 \\ CA^{r-1} B &\neq 0 \end{aligned}$$

and the r-th derivative of $y(t)$ satisfies

$$y^{(r)} = CA^r x + CA^{r-1}Bu.$$

Moreover, the dynamics of the linear system can be broken up into two components as follows:

$$\begin{aligned} \dot{\eta} &= P\eta + Q\zeta & (5.41)\\ \dot{\zeta} &= A_0\zeta + B_0\left[C_\zeta^T\zeta + C_\eta^T\eta + ku\right] & (5.42)\\ y &= C_0\zeta & (5.43) \end{aligned}$$

where $\eta \in \Re^{n-r}$, $\zeta \in \Re^r$, the triple $(A_0,\ B_0,\ C_0)$ is a canonical form representation of r integrators, as described by (5.39), P and Q are matrices of appropriate dimension, C_ζ is a vector of dimension r, and C_η is a vector of dimension $(n-r)$.

The reader will note that (5.41)–(5.43) is a linear special case of the normal form described by (5.36)–(5.38). The zero dynamics of the linear system, as defined earlier for the general normal form of nonlinear systems, are obtained by setting $\zeta = 0$ in (5.41). This yields

$$\dot{\eta} = P\eta. \qquad (5.44)$$

The stability of the zero dynamics are determined by the eigenvalues of P. The model is said to be minimum phase if all the eigenvalues are in the left open-half complex plane. It is important to note that the eigenvalues of P turn out to be the same as the roots of the numerator of the transfer function $H(s)$. This justifies the use of the term *zero dynamics* for nonlinear systems.

One question that may be raised in obtaining the normal form for nonlinear systems is whether any system can be put into the canonical normal form. In general, the answer is negative since for some systems the relative degree is undefined. This may happen, for example, if $L_gL_f^ih(x) = k_0x_1$, where k_0 is a scalar constant. This implies that $L_gL_f^ih(x)$ is zero for $x_1 = 0$ but it is nonzero in any neighborhood of $x_1 = 0$.

Next, we consider the tracking control design for input-output feedback linearizable systems. We assume that the control objective is for $y(t)$ to track a desired signal $y_d(t)$. Let $e(t) = y(t) - y_d(t)$ be the tracking error. Starting from the normal form (5.36)–(5.38) we design the feedback control law

$$\begin{aligned} u &= \alpha_0(\eta,\zeta) + \beta_0(\eta,\zeta)v \\ &= -\frac{L_f^r h(x)}{L_gL_f^{r-1}h(x)} + \frac{1}{L_gL_f^{r-1}h(x)}v & (5.45) \end{aligned}$$

where v is selected as follows:

$$\begin{aligned} v &= y_d^{(r)} - k_{r-1}e^{(r-1)} - k_{r-2}e^{(r-2)} - \cdots - k_1\dot{e} - k_0 e \\ &= y_d^{(r)} - \sum_{j=0}^{r-1} k_j\left(L_f^j h(x) - y_d^{(j)}\right). & (5.46) \end{aligned}$$

The control law (5.45)–(5.46) results in the linear error dynamics

$$e^{(r)} + k_{r-1}e^{(r-1)} + k_{r-2}e^{(r-2)} + \cdots + k_1\dot{e} + k_0 e = 0.$$

Therefore, by appropriately selecting the coefficients $\{k_0,\ k_1,\ \ldots\ k_{r-2},\ k_{r-1}\}$ the roots of the characteristic equation

$$s^r + k_{r-1}s^{r-1} + k_{r-2}s^{r-2} + \cdots + k_1 s + k_0 = 0$$

can be arbitrarily assigned. This implies that the tracking error can be made to converge to zero asymptotically (exponentially fast).

From the normal form (5.36)–(5.38), we note that the above control design has taken care only of the ζ variables. The designer also needs to be assured that the internal dynamics, $\dot{\eta} = \phi(\eta, \zeta)$, remain bounded when the control law is designed for the ζ-dynamics. This issue is addressed next.

Let

$$\bar{y}_d(t) = \left[y_d(t)\ \dot{y}_d(t)\ y_d^{(2)}(t)\ \cdots y_d^{(r-1)}(t) \right]^{\mathrm{T}}.$$

As shown by Isidori [121], if we assume that $\bar{y}_d(t)$ is bounded for all $t \geq 0$ and the solution of

$$\dot{\eta} = \phi(\eta, \bar{y}_d(t)), \qquad \eta(0) = 0$$

is well defined, bounded, and uniformly asymptotically stable, then using the control law (5.45)–(5.46) guarantees that the whole state remains bounded and the tracking error $e(t)$ converges to zero exponentially fast.

In the special case of regulation to the origin; i.e., $y_d(t) = 0$ for all $t \geq 0$, then it is required that the zero dynamics

$$\dot{\eta} = \phi(\eta, 0)$$

are asymptotically stable in order to ensure that the overall system states remain bounded and the tracking error converges to zero.

In summary, we note that for input-output linearizable systems there are two components to be taken care of:

- the ζ-dynamics, which can be linearized by the control variable u, and
- the η-dynamics, referred to as internal dynamics, which are rendered unobservable by the u defined in (5.35), but which need to have some stability properties (minimum phase) in order to allow stable control of the overall system.

■ EXAMPLE 5.5

Consider the flexible manipulator model of Example 5.4, whose state representation is given by

$$\begin{bmatrix} \dot{x}_1 \\ \dot{x}_2 \\ \dot{x}_3 \\ \dot{x}_4 \end{bmatrix} = \begin{bmatrix} x_2 \\ -\frac{MgL}{J_1}\sin x_1 - \frac{k}{J_1}(x_1 - x_3) \\ x_4 \\ \frac{k}{J_2}(x_1 - x_3) \end{bmatrix} + \begin{bmatrix} 0 \\ 0 \\ 0 \\ \frac{1}{J_2} \end{bmatrix} u.$$

First consider the case where the output $y = x_1$. In this case the diffeomorphism $z = T(x)$ given by (5.30) transforms the system into the normal form since

$$\begin{aligned} \dot{z}_1 &= z_2 \\ \dot{z}_2 &= z_3 \\ \dot{z}_3 &= z_4 \\ \dot{z}_4 &= -z_3\left(\frac{MgL}{J_1}\cos z_1 + \frac{k}{J_1} + \frac{k}{J_2}\right) + \frac{MgL}{J_1}\left(z_2^2 - \frac{k}{J_2}\right)\sin z_1 + \frac{k}{J_1 J_2}u \end{aligned} \qquad (5.47)$$

is already in the form described by (5.36)–(5.36). The relative degree is 4, which is the same as the order of the nonlinear system; hence, there are no internal dynamics.

Next, consider the case where the output is given by $y = x_3$. By taking time derivatives of $y(t)$, we note that the control input u appears in the second derivative:

$$\begin{aligned} \dot{y} &= x_4 \\ \ddot{y} &= \frac{k}{J_2}(x_1 - x_3) + \frac{1}{J_1}u. \end{aligned}$$

Therefore, in this case the relative degree is 2. The input-output feedback linearizing controller designed for tracking the signal y_d is

$$u = J_1\left(\ddot{y}_d - \frac{k}{J_2}(x_1 - x_3) - \lambda_1(y - y_d) - \lambda_2(\dot{y} - \dot{y}_d\right),$$

for $\lambda_1, \lambda_2 > 0$. This controller renders x_1 and x_2 unobservable from y. The system is already in normal form, without any transformation, since the first two variables x_1, x_2, are the η-dynamics which characterize the internal dynamics of the system. The last two variables x_3, x_4, are the ζ-dynamics, which are in the canonical form. The zero dynamics are obtained from the η variables by setting x_3 and x_4 to zero. Therefore the zero dynamics are given by

$$\begin{aligned} \dot{x}_1 &= x_2 \\ \dot{x}_2 &= -\frac{MgL}{J_1}\sin x_1 - \frac{k}{J_1}x_1 \end{aligned}$$

or equivalently

$$J_1\ddot{q}_1 + MgL\sin q_1 + kq_1 = 0.$$

This example illustrates that, while in the general case, the transformation of a system into normal form can be quite tedious, in practice it may often turn out that the normal form can be obtained quite trivially. △

■ EXAMPLE 5.6

Consider the system

$$\begin{aligned} \dot{x}_1 &= x_2 - \alpha x_1^3 \\ \dot{x}_2 &= 2x_3^2 + u \\ \dot{x}_3 &= x_1 + x_2^3 - 3x_3^3 \\ y &= x_1 \end{aligned}$$

where α is a constant. The objective is to transform the system into normal form and to design a feedback linearizing controller. By taking the first two time derivatives of $y(t)$ we obtain

$$\begin{aligned} \dot{y} &= x_2 - \alpha x_1^3 \\ \ddot{y} &= 2x_3^2 - \alpha x_1^2 x_2 + 3\alpha^2 x_1^5 + u. \end{aligned}$$

Therefore, the relative degree of the system is 2. By using the diffeomorphism

$$\begin{aligned} \eta &= x_3 \\ \zeta_1 &= x_1 \\ \zeta_2 &= x_2 - \alpha x_1^3 \end{aligned}$$

we can convert the system into the normal form:

$$\begin{aligned} \dot{\eta} &= \zeta_1 + \left(\zeta_2 + \alpha\zeta_1^3\right)^3 - 3\eta^3 \\ \dot{\zeta}_1 &= \zeta_2 \\ \dot{\zeta}_2 &= -3\alpha\zeta_1^2\zeta_2 + 2\eta^2 + u. \end{aligned}$$

By selecting the feedback control law

$$u = 3\alpha\zeta_1^2\zeta_2 - 2\eta^2 + v$$

we obtain

$$\begin{aligned} \dot{\eta} &= \zeta_1 + \left(\zeta_2 + \alpha\zeta_1^3\right)^3 - 3\eta^3 \\ \dot{\zeta}_1 &= \zeta_2 \\ \dot{\zeta}_2 &= v. \end{aligned}$$

The zero dynamics are obtained by setting ζ_1, ζ_2 to zero in the η-dynamics, which yields

$$\dot{\eta} = -3\eta^3.$$

Therefore, the zero dynamics are globally asymptotically stable, which implies that the system is minimum phase. This can be seen from the fact that the solutions of the zero dynamics with initial conditions $\eta(t_0) = \eta_0$ are given by

$$\eta(t) = \frac{\eta_0}{\sqrt{6\eta_0^2(t - t_0) + 1}}.$$

It is important to note that both the diffeomorphism as well as the normal form depend on the parameter α. Therefore, if the parameter α is unknown or uncertain, then both the transformation and the normal form will be incorrect. Consequently, the feedback linearizing controller will not cancel all the nonlinearities; i.e., it will not be a true feedback linearizing controller. △

The last example illustrates one of the key drawbacks of feedback linearization: it depends on exact cancellation of nonlinear functions. If one of the functions is uncertain then cancellation is not possible. This is one of the motivations for adaptive approximation based control.

Another possible difficulty with feedback linearization is that not all systems can be transformed to a linearizable form. The next section presents another technique, referred to as backstepping, which can be applied to a class of systems which may not be feedback linearizable.

5.3 BACKSTEPPING

This section describes the backstepping control design procedure. In Section 5.3.1 we consider a second order system with known nonlinearities. In Section 5.3.2 we present a lemma that can be applied recursively to extend the backstepping control design method to higher order systems. One of the drawbacks of the backstepping approach is the complexities involved with the computation of the control signal for higher order systems. Section 5.3.3 presents an alternative formulation of the backstepping approach to address this issue. These methods will be revisited in Chapter 7 for the case of unknown nonlinearities, where adaptive approximation methods will be developed.

5.3.1 Second Order System

To illustrate the concept of backstepping, or integrator backstepping, we start with a simple second-order system:

$$\dot{x}_1 = f(x_1) + g(x_1)x_2 \tag{5.48}$$

$$\dot{x}_2 = u, \tag{5.49}$$

where $(x_1, x_2) \in \Re^2$ is the state, $g(x_1) \neq 0$ for x_1 in some domain $\mathcal{D}$ that defines the operating envelope, and $u \in \Re$ is the control input. The objective is to design a feedback control algorithm to cause $x_1(t)$ to converge to $y_d(t)$. In this section, we assume that both $f(x_1)$ and $g(x_1)$ are known functions.

The key idea behind the backstepping procedure is that the tracking problem would be solved if the control input u could force $x_2(t)$ to satisfy

$$x_2 = \frac{1}{g(x_1)}\left[-f(x_1) - k_1(x_1 - y_d(t)) + \dot{y}_d(t)\right],$$

with $k_1 > 0$. In this case, x_1 satisfies $\dot{x}_1 - \dot{y}_d = -k_1(x_1 - y_d)$, which implies that $x_1(t)$ converges to $y_d(t)$. This is equivalent to treating x_2 as a *virtual* control input for the x_1 subsystem. Therefore, we introduce the virtual control variable $\alpha(x_1, y_d, \dot{y}_d)$, which is defined as

$$\alpha(x_1, y_d, \dot{y}_d) = \frac{1}{g(x_1)}\left[-f(x_1) - k_1(x_1 - y_d(t)) + \dot{y}_d(t)\right].$$

By adding and subtracting $g(x_1)\alpha(x_1, y_d, \dot{y}_d)$ in (5.48) we obtain

$$\dot{x}_1 = f(x_1) + g(x_1)\alpha + g(x_1)\left(x_2 - \alpha\right).$$

If we let $z_1 = x_1 - y_d$, then z_1 satisfies

$$\dot{z}_1 = -k_1 z_1 + g(x_1)\left(x_2 - \alpha\right).$$

Now, consider a coordinate transformation

$$z_2 = x_2 - \alpha(x_1, y_d, \dot{y}_d),$$

whose derivative is given by

$$\begin{aligned} \dot{z}_2 &= u - \frac{\partial \alpha}{\partial x_1}\dot{x}_1 - \frac{\partial \alpha}{\partial y_d}\dot{y}_d - \frac{\partial \alpha}{\partial \dot{y}_d}\ddot{y}_d \\ &= v, \end{aligned}$$

where

$$v = u - \frac{\partial \alpha}{\partial x_1}\dot{x}_1 - \frac{\partial \alpha}{\partial y_d}\dot{y}_d - \frac{\partial \alpha}{\partial \dot{y}_d}\ddot{y}_d \tag{5.50}$$

is referred to as a *modified control input*.

With this change of variables, we have rewritten the original system (5.48)–(5.49) as the tracking error dynamics:

$$\dot{z}_1 = -k_1 z_1 + g(x_1) z_2 \tag{5.51}$$

$$\dot{z}_2 = v. \tag{5.52}$$

The main, and key difference, between the original system (5.48)–(5.49) and the modified system (5.51)–(5.52) is that the modified system has an equilibrium at the origin and the z_1 dynamics of that equilibrium are asymptotically stable when $z_2 = 0$ and $v = 0$.

Now consider the Lyapunov function

$$V(z_1, z_2) = \frac{1}{2} z_1^2 + \frac{1}{2} z_2^2,$$

whose time derivative along the solutions of (5.51)– (5.52) is given by

$$\dot{V} = -k_1 z_1^2 + z_1 g(x_1) z_2 + z_2 v.$$

If we select the modified control input as

$$v = -z_1 g(x_1) - k_2 z_2, \qquad k_2 > 0 \tag{5.53}$$

then

$$\dot{V} = -k_1 z_1^2 - k_2 z_2^2,$$

which shows that the equilibrium point $(z_1, z_2) = (0, 0)$ of the closed-loop tracking error dynamics is globally asymptotically stable.

From the definition of v we conclude (by combining (5.50) and (5.53)) that the feedback control law u given by

$$\begin{aligned} u &= \frac{\partial \alpha}{\partial x_1}\dot{x}_1 + \frac{\partial \alpha}{\partial y_d}\dot{y}_d + \frac{\partial \alpha}{\partial \dot{y}_d}\ddot{y}_d - z_1 g(x_1) - k_2 z_2 \\ &= \frac{\partial \alpha}{\partial x_1}\left[f(x_1) + g(x_1) x_2\right] + \frac{k_1 \dot{y}_d}{g(x_1)} + \frac{\ddot{y}_d}{g(x_1)} - z_1 g(x_1) \\ &\quad - k_2 \left[x_2 + \frac{1}{g(x_1)}\left(f(x_1) + k_1(x_1 - y_d(t)) - \dot{y}_d(t)\right)\right] \end{aligned} \tag{5.54}$$

results in a globally asymptotically stable origin for the (z_1, z_2) system that ensures perfect tracking of y_d by x_1, assuming of course, that $g(x_1)$ is bounded away from zero for all $x_1 \in \Re$.

Some remarks:

- Even with a simple second-order system, the feedback control algorithm (5.54) becomes quite complex. Once the backstepping procedure gets extended to the n-th order case, it becomes considerably more complex. In fact, as we will see, for the n-th order case, the feedback control law is usually not written in a closed form, as in (5.54), but recursively based on a so-called backstepping procedure, which has as many steps as the number of state variables.

- A key assumption in the above backstepping procedure is that both $f(x_1)$ and $g(x_1)$ are known exactly. In the case where they are partially or completely unknown then it may be appropriate that these functions be estimated online, which is the topic of discussion in the next two chapters.

5.3.2 Higher Order Systems

Consider the system model

$$\dot{x}_1 = f_1(x_1) + g_1(x_1)x_2 \tag{5.55}$$

$$\dot{x}_2 = f_2(x) + g_2(x)u \tag{5.56}$$

where $x = [x_1^\top \; x_2^\top]^\top \in \Re^n$ and $x_2 \in \Re^1$. Define $z_1 = x_1 - y_d$ where $y_d(t)$ is the signal vector to be tracked. For this system, we assume that we know scalar virtual control functions $\alpha(x_1, y_d, \dot{y}_d)$ and positive definite $V_1(z_1)$ such that

$$\frac{\partial V_1}{\partial z_1}\left[f_1 + g_1\alpha - \dot{y}_d\right] \le -W_1(z_1) \tag{5.57}$$

where $W_1(z_1)$ is a positive definite function. Our objective is to define u such that the system of equations (5.55)–(5.56) will have x_1 tracking y_d (i.e., z_1 convergent to zero).

We define $z_2 = x_2 - \alpha$. Then the (z_1, z_2) dynamics are described by

$$\dot{z}_1 = f_1(x_1) + g_1(x_1)\alpha + g_1(x_1)z_2 - \dot{y}_d \tag{5.58}$$

$$\dot{z}_2 = f_2(x) + g_2(x)u - \dot{\alpha} \tag{5.59}$$

where

$$\dot{\alpha} = \frac{\partial \alpha}{\partial x_1}\dot{x}_1 + \frac{\partial \alpha}{\partial y_d}\dot{y}_d + \frac{\partial \alpha}{\partial \dot{y}_d}\ddot{y}_d.$$

Consider $V(z_1, z_2) = V_1(z_1) + \frac{1}{2}z_2^2$. The time derivative of V along the solutions of (5.58)–(5.59) is given by

$$\begin{aligned}
\dot{V} &= \frac{\partial V_1}{\partial z_1}\left[f_1(x_1) + g_1(x_1)\alpha - \dot{y}_d + g_1(x_1)z_2\right] + z_2\left(f_2(x) + g_2(x)u - \dot{\alpha}\right) \\
&\le -W_1(z_1) + \frac{\partial V_1}{\partial z_1}g_1(x_1)z_2 + z_2\left(f_2(x) + g_2(x)u - \dot{\alpha}\right).
\end{aligned}$$

Therefore, if $g_2(x) \ne 0$ and the control signal u is selected as

$$u = \frac{1}{g_2(x)}\left(-f_2(x) - k_2 z_2 + \dot{\alpha} - \frac{\partial V_1}{\partial z_1}g_1(x_1)\right) \tag{5.60}$$

with $k_2 > 0$ being a design parameter, then we have

$$\dot{V} \le -W_1(z_1) - k_2 z_2^2$$

which is negative definite.

Therefore, we have proven Lemma 5.3.1. Note that this lemma can be applied recursively to achieve tracking control for higher order systems.

Lemma 5.3.1 *Given a system in the form of (5.55)–(5.56) and known functions $\alpha_1(x_1, y_d, \dot{y}_d)$ and positive definite $V_1(z_1)$ satisfying (5.57), then for u specified according to (5.60), the tracking error dynamics of (5.58)–(5.59) are asymptotically stable. If V_1 is radially unbounded and all assumptions hold globally, then the tracking error dynamics are globally asymptotically stable.*

EXAMPLE 5.7

Consider the third-order system

$$\dot{v}_1 = v_1^2 + (1 + v_1^2)v_2 \tag{5.61}$$
$$\dot{v}_2 = v_1 v_2 + (2 + \cos v_2)v_3 \tag{5.62}$$
$$\dot{v}_3 = v_3^2 + (1 + v_1^2 v_2^2)u. \tag{5.63}$$

The tracking control design problem is solved in three steps, where the second and third steps will utilize Lemma 5.3.1.

Step 1. In this step, we find a control signal α_1 to solve the tracking control problem for the system

$$\dot{v}_1 = v_1^2 + (1 + v_1^2)\alpha_1.$$

If we select

$$\alpha_1 = \frac{-v_1^2 - k_1 z_1 + \dot{y}_d}{(1 + v_1^2)}$$

where $z_1 = v_1 - y_d$ and $k_1 > 0$, then the controlled z_1 dynamics are

$$\dot{z}_1 = -k_1 z_1$$

and the time derivative of $V_1 = \frac{1}{2} z_1^2$ is given by

$$\dot{V} = -k_1 z_1^2 = -W_1(z_1),$$

where $W_1(z_1) = k_1 z_1^2$.

Step 2. We are now in a position to use Lemma 5.3.1 to specify a control signal α_2 to solve the tracking problem for the second order subsystem

$$\dot{v}_1 = v_1^2 + (1 + v_1^2)v_2$$
$$\dot{v}_2 = v_1 v_2 + (2 + \cos v_2)\alpha_2.$$

To utilize the lemma, we let $x_1 = v_1$, $x_2 = v_2$, $f_1 = v_1^2$, $g_1 = (1 + v_1^2)$, $f_2 = v_1 v_2$, $g_2 = (2 + \cos v_2)$, and define $z_2 = v_2 - \alpha_1$. Application of Lemma 5.3.1, specifies that

$$\alpha_2 = \frac{1}{(2 + \cos v_2)} \left(-v_1 v_2 - k_2 z_2 - z_1(1 + v_1^2) + \dot{\alpha}_1 \right)$$

where k_2 is a positive design parameter. The Lyapunov function for the second order tracking error dynamics would be $V_2 = \frac{1}{2}\left(z_1^2 + z_2^2\right)$, which has a time derivative satisfying

$$\dot{V}_2 = -W_2(z_1),$$

where $W_2(z_1, z_2) = k_1 z_1^2 + k_2 z_2^2$.

Step 3. Now, we are in a position to use Lemma 5.3.1 to specify a control signal u to solve the original three state tracking problem To utilize the lemma, we let

$$x_1 = [v_1 \; v_2]^\top$$
$$x_2 = v_3$$

$$\begin{aligned}
f_1 &= \begin{bmatrix} v_1^2 + (1+v_1^2)v_2 \\ v_1 v_2 \end{bmatrix} \\
g_1 &= \begin{bmatrix} 0 \\ 2+\cos v_2 \end{bmatrix} \\
f_2 &= v_3^2 \\
g_2 &= (1+v_1^2 v_2^2)
\end{aligned}$$

and define $z_3 = v_3 - \alpha_2$. Application of Lemma 5.3.1, specifies that

$$u = \frac{1}{(1+v_1^2 v_2^2)}\left(-v_3^2 - k_3 z_3 - z_2(2+\cos v_2) + \dot{\alpha}_2\right) \tag{5.64}$$

where k_3 is a positive design parameter. As a result of the lemma, the control law given by (5.64) results in globally exponentially stable tracking error dynamics.

Implementation of this controller requires analytic computation of $\alpha_1, \dot{\alpha}_1, \alpha_2, \dot{\alpha}_2$, and finally u. These computations will involve y_d, $\dot{y}_d$, and $\ddot{y}_d$. In general, the computation of the quantities $\dot{\alpha}_i$ can be algebraically tedious, especially for systems of order larger than two or three. △

5.3.3 Command Filtering Formulation

Much of the complexity that arises in the backstepping control laws that result from recursive application of Lemma 5.3.1 is due to the computation of the time derivative of the virtual control variables $\alpha_i(x_1, \ldots, x_i, y_d, \ldots, y_d^{(i)})$. The computation of these time derivatives becomes even more complex in applications where the functions f and g are approximated online. This section presents an alternative formulation of the backstepping approach that decreases the algebraic complexity of the backstepping control law from that of eqn. (5.54).

Consider the second-order system

$$\dot{x}_1 = f_1(x_1) + g_1(x_1)x_2 \tag{5.65}$$

$$\dot{x}_2 = f_2(x) + g_2(x)u \tag{5.66}$$

where $x = [x_1\ x_2]^\top \in \Re^2$ is the state, $x_2 \in \Re^1$, and u is the scalar control signal. A region $\mathcal{D}$ is the specified operation region of the system. The functions f_i, g_i for $i = 1, 2$ are known locally Lipschitz functions. The functions g_i are assumed to be nonzero for all $x \in \mathcal{D}$. There is a desired trajectory $x_{1c}(t)$, with derivative $\dot{x}_{1c}(t)$, both of which lie in a region $\mathcal{D}$ for $t \geq 0$ and both signals are assumed known. Define the tracking errors

$$\begin{aligned}
\tilde{x}_1 &= x_1 - x_{1c} \\
\tilde{x}_2 &= x_2 - x_{2c}
\end{aligned}$$

where x_{2c} will be defined by the backstepping controller. Let

$$\alpha_1(x_1, \tilde{x}_1, \dot{x}_{1c}) = \frac{1}{g_1}\left[-f_1 - k_1\tilde{x}_1 + \dot{x}_{1c}\right] \text{ with } k_1 > 0 \tag{5.67}$$

be a smooth feedback control and define the smooth positive definite function $V_1(\tilde{x}_1) = \frac{1}{2}\tilde{x}^\top \tilde{x}$ such that

$$\frac{\partial V_1}{\partial \tilde{x}_1}\left[f_1 + g_1\alpha_1 - \dot{x}_{1c}\right] = -W(\tilde{x}_1) \tag{5.68}$$

where $W(\tilde{x}_1) = k_1 \tilde{x}_1^\top \tilde{x}_1$ is positive definite in $\tilde{x}_1$.

To solve the tracking control problem for the system of eqns. (5.65)–(5.66) we use the following procedure:

1. Define

$$x_{2c}^o = \alpha_1(x_1, \tilde{x}_1, \dot{x}_{1c}) - \xi_2 \tag{5.69}$$

$$\dot{\xi}_1 = -k_1\, \xi_1 + g_1(x_1)\,(x_{2c} - x_{2c}^o), \tag{5.70}$$

where ξ_2 will be defined in step 3. The signal x_{2c}^o is filtered to produce the command signal x_{2c} and its derivative $\dot{x}_{2c}$. Such a filter is defined in Appendix A.4. Note that by the design of this *command filter*, the signal $(x_{2c} - x_{2c}^o)$ is bounded and small. Therefore, as long as $g_1(x_1)$ is bounded, then ξ_1 is bounded because it is the output of a stable linear filter with a bounded input.

2. Define the compensated tracking errors as

$$\bar{x}_i = \tilde{x}_i - \xi_i, \text{ for } i = 1, 2. \tag{5.71}$$

3. Define

$$u_c^o = \frac{1}{g_2}\left(-k_2\tilde{x}_2 + \dot{x}_{2c} - f_2 - \bar{x}_1^\top g_1\right) \text{ with } k_2 > 0 \tag{5.72}$$

$$\dot{\xi}_2 = -k_2\xi_2 + g_2(x)(u_c - u_c^o) \tag{5.73}$$

where u_c^o is filtered to produce u_c and $\dot{u}_c$ where $u = u_c$ is the control signal applied to the actual system. By the design of the command filter, the signal $(u_c - u_c^o)$ is bounded and small; therefore, if $g_2(x)$ is bounded, then ξ_2 is the bounded output of a stable linear filter with a bounded input. If $u_c^o = u_c = u$, then $\xi_2 = 0$.

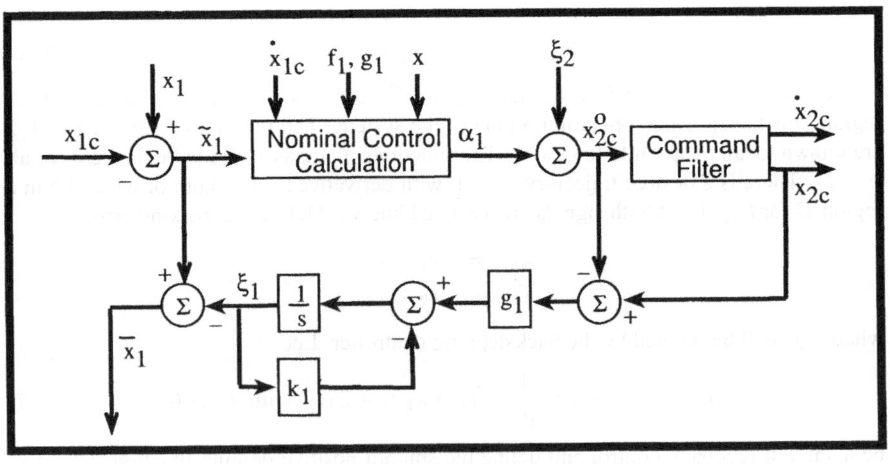

Figure 5.4: Diagram illustrating the command filter computations related to x_1. The nominal control block refers to eqn. (5.67). The diagram for x_2 would be similar.

Figure 5.4 displays a block diagram implementation of the above procedure. Note that u_c^o is computed using $\dot{x}_{2c}$, not $\dot{x}_{2c}^o$. The quantity $\dot{x}_{2c}$ is available as the output of the filter in step 1. The quantity $\dot{x}_{2c}^o$ is not used in the control law. It is not directly available and is tedious to compute for higher order systems.

Given the above procedure, we now analyze the stability of the control law. The tracking error dynamics can be written as

$$\begin{aligned}
\dot{\tilde{x}}_1 &= f_1 + g_1\, x_{2c}^o - \dot{x}_{1c} + g_1(x_{2c} - x_{2c}^o) + (g_1 x_2 - g_1\, x_{2c}) \\
&= f_1 + g_1\alpha_1 - \dot{x}_{1c} - g_1\, \xi_2 + g_1(x_{2c} - x_{2c}^o) + (g_1 x_2 - g_1\, x_{2c}) \\
&= -k_1\tilde{x}_1 - g_1\, \xi_2 + g_1(x_{2c} - x_{2c}^o) + g_1(x_2 - x_{2c}) && (5.74) \\
&= -k_1\tilde{x}_1 + g_1\, \bar{x}_2 + g_1(x_{2c} - x_{2c}^o) + && (5.75) \\
\dot{\tilde{x}}_2 &= f_2 + g_2\, u_c^o - \dot{x}_{2c} + g_2(u_c - u_c^o) \\
&= -k_2\tilde{x}_2 - g_1^{\mathsf{T}}\bar{x}_1 + g_2(u_c - u_c^o). && (5.76)
\end{aligned}$$

As defined in (5.70) and (5.73), the variables ξ_1, ξ_2 represent the filtered effect of the errors $(x_{2c} - x_{2c}^o)$ and $(u_c - u_c^o)$, respectively. The variables $\bar{x}_i$ represent the compensated tracking errors, obtained after removing the corresponding unachieved portion of x_{2c}^o and u_c^o. After some algebraic manipulation, the dynamics of the compensated tracking errors are described by

$$\begin{aligned}
\dot{\bar{x}}_1 &= \dot{\tilde{x}}_1 - \dot{\xi}_1 \\
&= -k_1\bar{x} + g_1\bar{x}_2 && (5.77) \\
\dot{\bar{x}}_2 &= -k_2\bar{x}_2 - g_1^{\mathsf{T}}\bar{x}_1. && (5.78)
\end{aligned}$$

Consider the following Lyapunov function candidate

$$V = \sum_{i=1}^{2} V_i(\bar{x}_i) \ \text{ where } V_i(\bar{x}_i) = \frac{1}{2}\bar{x}_i^{\mathsf{T}}\bar{x}_i. \qquad (5.79)$$

The time derivative of V along the solution of (5.77)–(5.78) is

$$\begin{aligned}
\dot{V}_1 &= \bar{x}_1^{\mathsf{T}}\left(f_1 + g_1\alpha_1 + k_1\tilde{x}_1 - \dot{x}_{1c} - k_1\,\bar{x}_1 + g_1\bar{x}_2\right) \\
&= -k_1\bar{x}_1^{\mathsf{T}}\bar{x}_1 + \bar{x}_1^{\mathsf{T}}g_1\bar{x}_2 \\
\dot{V}_2 &= \bar{x}_2\left(-k_2\bar{x}_2 - g_1^{\mathsf{T}}\bar{x}_1\right) \\
&= -k_2\bar{x}_2^2 - \bar{x}_2 g_1^{\mathsf{T}}\bar{x}_1 \\
\dot{V} &= \dot{V}_1 + \dot{V}_2 = -k_1\bar{x}_1^{\mathsf{T}}\bar{x}_1 - k_2\bar{x}_2^2 \le -\lambda\, V && (5.80)
\end{aligned}$$

where $\lambda = 2\min(k_1, k_2) > 0$. The fact that $\dot{V} \le -\lambda\, V$ shows that the origin of the $(\bar{x}_1, \bar{x}_2)$ system is exponentially stable. Therefore, we can summarize these results in the following theorem.

Lemma 5.3.2 *Let the control law α_1 solve the tracking problem for system*

$$\dot{x}_1 = f_1(x_1) + g_1(x_1)\alpha_1 \ \textit{ with } x_1 \in \Re^{n-1}$$

with Lyapunov function V_1 satisfying (5.68). Then the controller of (5.69)–(5.73) solves the tracking problem (i.e., guarantees that $x_1(t)$ converges to $y_d(t)$) for the system described by (5.65)–(5.66).

Note that this lemma can be applied recursively $n-1$ times to address a system with n states. An example of this will be presented below. Note that the result guarantees desirable properties for the compensated tracking errors $\bar{x}_i$, not the actual tracking errors $\tilde{x}_i$. The difference between these two quantities is ξ_i, which is the output of the stable linear filter

$$\dot{\xi}_i = -k_i\,\xi_i + r_i$$

with input

$$r_i = g_i\left(x_{(i+1)c} - x^o_{(i+1)c}\right).$$

The magnitude of the portion of the input defined by $\left(x_{(i+1)c} - x^o_{(i+1)c}\right)$ is determined by the design of the $(i+1)$st command filter. This portion can be made arbitrarily small by appropriate design of the command filter. If the function g_i is bounded, then ξ_i is bounded. When r_i approaches zero, then $\xi_i \to 0$ and $\tilde{x}_i \to \bar{x}_i$ all i.

The goal of the derivation of this theorem was to avoid tedious algebraic manipulations involved in the computation of the backstepping control signal. Avoiding such computations will become increasingly important in backstepping approaches that include parameter adaptation.

In the following example, we return to the problem of Example 5.7 using Lemma 5.3.2.

■ **EXAMPLE 5.8**

From (5.61)–(5.63) and (5.69), we have that

$$\begin{aligned}
x^o_{2c} &= \frac{-v_1^2 - k_1\bar{x}_1 + \dot{y}_d}{(1+v_1^2)} - \xi_2 \\
x^o_{3c} &= \frac{1}{(2+\cos v_2)}\left(-v_1v_2 - k_2\bar{x}_2 + \dot{x}_{2c} - (1+v_1^2)\bar{x}_1\right) - \xi_3 \\
u^o_c &= \frac{1}{(1+v_1^2v_2^2)}\left(-v_3^2 - k_3\bar{x}_3 + \dot{x}_{3c} - (2+\cos v_2)\bar{x}_2\right),
\end{aligned}$$

where

$$\begin{aligned}
\dot{\xi}_1 &= -k_1\xi_1 + (1+v_1^2)(x_{2c} - x^o_{2c}) \\
\dot{\xi}_2 &= -k_2\xi_2 + (2+\cos v_2)(x_{3c} - x^o_{3c}) \\
\dot{\xi}_3 &= -k_3\xi_3 + (1+v_1^2v_2^2)(u_c - u^o_c)
\end{aligned}$$

and for $i = 1,2,3$ we have $\tilde{x}_i = v_i - x_{ic}$, $\bar{x}_i = \tilde{x}_i - \xi_i$. Each pair $(x_{2c}, \dot{x}_{2c})$ and $(x_{3c}, \dot{x}_{3c})$ is the output of second-order, low-pass, unity-gain filter of Figure A.4 with input x^o_{2c} or x^o_{3c}, respectively. If u^o_c is used as the control signal, then $u_c = u^o_c$ and $\xi_3 = 0$. △

This example should be compared with Example 5.7. For an n-th order system, standard backstepping will require as controller inputs $y_d^{(i)}$ for $i = 0, \ldots, n$ and will analytically compute α_i and $\dot{\alpha}_i$. The command filtered approach will require as controller inputs only y_d and $\dot{y}_d$ and will analytically compute only α_i. The tradeoff is that the command filtered approach will require n scalar filters for the ξ variables and n command filters.

5.4 ROBUST NONLINEAR CONTROL DESIGN METHODS

In the previous three sections of this chapter we have examined three methods for controlling nonlinear systems, namely small-signal linearization, feedback linearization, and backstepping. The methodologies developed were based on the key assumption that the control designer exactly knows the system nonlinearities. In practice, this is not a realistic assumption. Consequently, it is important to consider ways to make these approaches more robust with respect to modeling errors. In this section we introduce a set of nonlinear control design tools that are based on the principle of assuming that the unknown component of the nonlinearities are bounded in some way by a known function. If this assumption is satisfied then it is possible to derive nonlinear control schemes that utilize these known bounding functions instead of the unknown nonlinearities.

Although these techniques have been extensively studied in the nonlinear control literature, they tend to yield conservative control laws, especially in cases where the uncertainty is significant. The term "conservative" is used among control engineers to indicate the fact that due to the uncertainty the control effort applied is more than needed. As a result, the control signal $u(t)$ may be large (high-gain feedback), which may cause several problems, such as saturation of the actuators, large error in the presence of measurement noise, excitation of unmodeled dynamics, and large transient errors. Furthermore, as we will see, these techniques typically involve a switching control function, which may cause chattering.

The robust nonlinear control design methods developed in this section provide an important perspective for the adaptive approximation based control described in Chapters 6 and 7. Specifically, adaptive approximation based control can be viewed as a way of reducing uncertainty during operation such that the need for conservative robust control can be eliminated or reduced. Another reason for studying these techniques in the context of adaptive approximation is their utilization, as we will see, to guarantee closed-loop stability outside of the approximation region $\mathcal{D}$.

This section presents five nonlinear control design tools: (i) bounding control, (ii) sliding mode control, (iii) Lyapunov redesign method, (iv) nonlinear damping, and (v) adaptive bounding. As we will see, these techniques are, in fact, quite similar.

5.4.1 Bounding Control

Bounding control is one of the simplest approaches for dealing with unknown nonlinearities. Here, we consider a simple scalar system with one unknown nonlinearity, which lies within certain known bounds. This approach can be extended to more complex systems. In Chapter 6, we will revisit bounding control as a way of motivating adaptive approximation of the unknown component of nonlinear systems.

Consider the scalar nonlinear system

$$\dot{x} = f(x) + u \tag{5.81}$$

where the objective is to design a control law such that $y(t) = x(t)$ tracks a desired signal $y_d(t)$. Let $e(t) = y(t) - y_d(t)$ be the tracking error. We assume that the function f is unknown but belongs to a certain known range as follows:

$$f_L(x) \leq f(x) \leq f_U(x), \qquad \forall x \in \Re^1$$

where f_L and f_U are known lower and upper bounds, respectively, on the unknown function f. In general, the bounds f_L and f_U may be positive or negative, or their sign may change as x varies.

Consider the following control law:

$$u = \begin{cases} -a_m(x - y_d) + \dot{y}_d - f_U(x) & \text{if } e \geq 0 \\ -a_m(x - y_d) + \dot{y}_d - f_L(x) & \text{if } e < 0 \end{cases} \tag{5.82}$$

where $a_m > 0$. Using the above control, it is easy to see that the tracking error dynamics satisfy

$$\begin{cases} \dot{e} = -a_m e + f(x) - f_U(x) & \text{if } e \geq 0 \\ \dot{e} = -a_m e + f(x) - f_L(x) & \text{if } e < 0. \end{cases}$$

Now, let $V = \frac{1}{2}e^2$ be a Lyapunov function candidate. The time derivative of V satisfies

$$\begin{aligned} \dot{V} &= e\dot{e} \\ &= \begin{cases} -a_m e^2 + (f(x) - f_U(x))e & \text{if } e \geq 0 \\ -a_m e^2 + (f(x) - f_L(x))e & \text{if } e < 0 \end{cases} \\ &\leq -a_m e^2. \end{aligned}$$

Therefore, the tracking error converges to zero exponentially fast. It is noted that, in general, the control law (5.82) is discontinuous at $e = 0$. This may result in the trajectory $x(t)$ going back and forth between y_d^+ and y_d^-, causing the control law to be switching, thus creating chattering problems. By y_d^+ we denote a value of the trajectory $y(t)$ which is slightly larger than $y_d(t)$, thus causing the tracking error e to be slightly positive, and correspondingly, y_d^- denotes a value of the trajectory which is slightly smaller than $y_d(t)$. The chattering can be remedied by using a smooth approximation to the control law of the form

$$u = \begin{cases} -a_m(x - y_d) + \dot{y}_d - f_U(x) & \text{if } e \geq \delta \\ -a_m(x - y_d) + \dot{y}_d - \frac{1}{2\delta}\left[(\delta - e)f_L(y_d - \delta) + (\delta + e)f_U(y_d + \delta)\right] & \text{if } |e| \leq \delta \\ -a_m(x - y_d) + \dot{y}_d - f_L(x) & \text{if } e < -\delta \end{cases}$$

where $\delta > 0$ is a small design constant. Exercise 5.18 asks the reader to prove that the closed-loop system with the above smooth approximation of the discontinuous bounding control achieves convergence to the set $|x| < \delta$ in finite time.

5.4.2 Sliding Mode Control

Sliding mode control is a methodology based on the principle that it is easier to control a first-order system than a n-th order system. Therefore, this approach can be viewed as a way to reduce a higher-order control problem into a simpler one for which there are known feedback control methods. This simplification comes at the expense of using a large control effort, which, as discussed earlier in the chapter, could be the source of other potential problems, especially in the presence of measurement noise or high frequency unmodeled dynamics. The sliding mode control methodology can be applied to several classes of nonlinear systems. Here, we consider its application to a class of feedback linearizable systems.

Consider an n-th order nonlinear system of the form

$$\left.\begin{aligned} \dot{x}_1 &= x_2 \\ \dot{x}_2 &= x_3 \\ &\vdots \\ \dot{x}_{n-1} &= x_n \\ \dot{x}_n &= f(x) + g(x)u, \end{aligned}\right\} \tag{5.83}$$

where it is assumed that f and g are unknown and $g(x) \geq g_0 > 0$ for all $x \in \Re^n$. The control objective is for $y(t) = x_1(t)$ to track a desired signal $y_d(t)$. Let $e = y - y_d$ be the tracking error. The sliding mode surface s is defined as

$$s = e^{(n-1)} + \lambda_{n-1} e^{(n-2)} + \cdots + \lambda_2 \dot{e} + \lambda_1 e = 0, \tag{5.84}$$

where the coefficients $\{\lambda_1,\ \lambda_2,\ \ldots\ \lambda_{n-1}\}$ are selected such that the characteristic polynomial (in p)

$$p^{n-1} + \lambda_{n-1} p^{n-2} + \cdots + \lambda_2 p + \lambda_1 = 0 \tag{5.85}$$

is Hurwitz (i.e., all the roots of the polynomial are in the left-half complex plane). The manifold described by $s = 0$ is referred to as the *sliding manifold* or *sliding surface* and has dimension $(n - 1)$. The objective of sliding mode control is to steer the trajectory onto this sliding manifold. This is achieved by forcing the variable s to zero in finite time. By design of the sliding surface, if x is on the sliding surface defined by $s = 0$, then

$$e^{(n-1)} = -\lambda_{n-1} e^{(n-2)} - \cdots - \lambda_2 \dot{e} - \lambda_1 e.$$

Since the polynomial given by (5.85) is Hurwitz, once on the sliding manifold the tracking error will go to zero with a transient behavior characterized by the selected coefficients $\{\lambda_1,\ \lambda_2,\ \ldots\ \lambda_{n-1}\}$ (i.e., exponentially fast).

The sliding mode control objective can be achieved if the control law u is chosen such that

$$\frac{d}{dt}\frac{1}{2}s^2 \leq -\kappa |s|,$$

where $\kappa > 0$. In this case, the upper right-hand derivative of $|s(t)|$ satisfies the differential inequality

$$\frac{d^+}{dt}|s(t)| \leq -\kappa,$$

which implies that the trajectory reaches the manifold $s = 0$ in finite time.

Following (5.84), the derivative of $s(t)$ satisfies

$$\begin{aligned} \dot{s} &= e^{(n)} + \lambda_{n-1} e^{(n-1)} + \cdots + \lambda_2 \ddot{e} + \lambda_1 \dot{e} \\ &= f(x) + g(x)u - y_d^{(n)} + \lambda_{n-1} e^{(n-1)} + \cdots + \lambda_2 \ddot{e} + \lambda_1 \dot{e}. \end{aligned}$$

If f and g were known function, then we could choose the control law

$$u = \frac{1}{g(x)} \left[-f(x) + y_d^{(n)} - \lambda_{n-1} e^{(n-1)} - \cdots - \lambda_2 \ddot{e} - \lambda_1 \dot{e} - \kappa \operatorname{sgn}(s) \right],$$

where $\kappa > 0$ is a design variable and $\operatorname{sgn}(\cdot)$ denotes the sign function:

$$\operatorname{sgn}(s) = \begin{cases} 1 & \text{if } s > 0 \\ 0 & \text{if } s = 0 \\ -1 & \text{if } s < 0. \end{cases}$$

Based on this control law, the derivative of $s(t)$ satisfies

$$\dot{s} = -\kappa \operatorname{sgn}(s),$$

which implies

$$\begin{aligned}\frac{d}{dt}\frac{1}{2}s^2 &= s\dot{s} \\ &= -s\kappa \operatorname{sgn}(s) \\ &= -\kappa|s|.\end{aligned}$$

Now consider the case where f and g are unknown but the designer has a known upper bound $\eta(x,t)$ such that

$$\left| \frac{f(x) - y_d^{(n)} + \lambda_{n-1}e^{(n-1)} + \cdots + \lambda_2\ddot{e} + \lambda_1\dot{e}}{g(x)} \right| \leq \eta(x,t).$$

Suppose that the control law is selected as

$$u = -(\eta(x,t) + \eta_0) \operatorname{sgn}(s), \tag{5.86}$$

where $\eta_0 > 0$ is a design constant. Now, let

$$V = \frac{1}{2}s^2$$

be the Lyapunov function candidate. The derivative of V is given by

$$\begin{aligned}\dot{V} &= s\dot{s} \\ &= s\left(f(x) - y_d^{(n)} + \lambda_{n-1}e^{(n-1)} + \cdots + \lambda_2\ddot{e} + \lambda_1\dot{e}\right) + sg(x)u \\ &\leq |s|\,\eta(x,t)\,g(x) + s\,g(x)\,u \\ &= -\eta_0\,g_0\,|s|,\end{aligned}$$

where g_0 is defined in (5.83). Therefore, we have achieved the desired objective of forcing the trajectory onto the sliding manifold in finite time. It is interesting to note that this is achieved without specific knowledge of f and g, just the upper bound $\eta(x,t)$.

Despite the resulting stability and convergence properties of the sliding mode control approach, it has two key drawbacks in its standard form. The sliding mode control law given by (5.86) has two components, the gain $\eta(x,t)+\eta_0$ and the switching function $\operatorname{sgn}(s)$, both of which can create problems:

- **(High-Gain)** Note that the gain term is the result of taking an upper bound on the uncertainty. In general, this creates a high-gain feedback control, which can create problem in the presence of measurement noise and high-frequency unmodeled dynamics. Moreover, high-gain feedback may require significant control effort, which can be expensive and/or may cause saturation of the actuators. In practice, high-gain feedback control is to be avoided.

- **(Chattering)** The switching function $\operatorname{sgn}(s)$ causes the control gain to switch from $\eta(x,t)+\eta_0$ to $-(\eta(x,t)+\eta_0)$ every time the trajectory crosses the sliding manifold. Although in theory the trajectory is suppose to "slide" on the sliding manifold, in

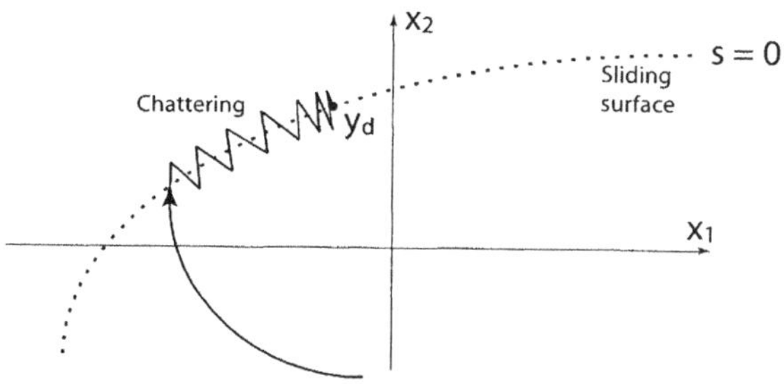

Figure 5.5: Graphical illustration of sliding mode control and chattering as a result of imperfection in the switching.

practice there are imperfections and delays in the switching devices, which lead to *chattering*. This is illustrated in Figure 5.5. Chattering causes significant problems in the feedback control system, especially if it is associated with high gains. For example, chattering may excite high-frequency dynamics which were neglected in the design model, it can cause wear and tear of moving mechanical parts and it can cause high heat losses in electrical power systems.

Research in sliding mode control has developed some techniques for addressing the above two issues. The high gain problem can be reduced by using as much *a priori* information as possible, thus cancelling the known nonlinearities and employing an upper bound only for the unknown portions of the nonlinearities. The chattering problem can also be addressed, partially, by employing a continuous approximation of the sign function. The tradeoff in the use of this approximation is that only uniform boundedness of solutions can be proved. Despite these remedies, the sliding mode methodology is based on the principle of bounding the uncertainty by a larger function, and as a result it is a conservative control approach. In this text, we present a methodology for "learning" or approximating the uncertainty online, instead of using an upper bound for it. However, the approximation will be valid only within a certain compact region $\mathcal{D}$. In order to achieve stability outside this region, we will rely on bounding control techniques such as sliding mode.

5.4.3 Lyapunov Redesign Method

Consider a nonlinear system described by

$$\dot{x} = f(x) + G(x)u, \tag{5.87}$$

where $x \in \Re^n$ is the state and $u \in \Re^m$ is the controlled input. Assume that the vector field $f(x)$ and the matrix $G(x)$ each consist of two components: a known nominal part and an unknown part. Therefore,

$$f(x) = f_0(x) + f^*(x) \tag{5.88}$$

$$G(x) = G_0(x) + G^*(x) \tag{5.89}$$

where f_0 and G_0 characterize the known nominal plant, and f^*, G^* represent the uncertainty. Later we will assume that the unknown portion satisfies a certain bounding condition.

Moreover, we assume that the uncertainty satisfies a so-called *matching condition*:

$$f^*(x) = G_0(x)\Delta_f^*(x) \tag{5.90}$$
$$G^*(x) = G_0(x)\Delta_G^*(x). \tag{5.91}$$

The matching condition implies that the uncertainty terms appear in the same equations as the control inputs u, and as a result they can be handled by the controller.

By substituting (5.88)–(5.89) and (5.90)–(5.91) in (5.87) we obtain

$$\dot{x} = f_0(x) + G_0(x)\left(u + \eta(x,u)\right), \tag{5.92}$$

where η comprises all the uncertainty terms, and is given by

$$\eta(x,u) = \Delta_f^* + \Delta_G^* u.$$

The Lyapunov redesign method addresses the following problem: suppose that the equilibrium of the nominal model $\dot{x} = f_0(x) + G_0(x)u$ can be made uniformly asymptotically stable by using a feedback control law $u = p_0(x)$. The objective is to design a corrective control function $p^*(x)$ such that the augmented control law $u = p_0(x) + p^*(x)$ is able to stabilize the system (5.92) subject to the uncertainty $\eta(x,u)$ being bounded by a known function.

Next, we consider the details of the Lyapunov redesign method, which is thoroughly presented for a more general case in [134]. We assume that there exists a control law $u = p_0(x)$ such that $x = 0$ is a uniformly asymptotically stable equilibrium point of the closed-loop nominal system

$$\dot{x} = f_0(x) + G_0(x)p_0(x)$$

We also assume that we know a Lyapunov function $V_0(x)$ that satisfies

$$\alpha_1(\|x\|) \le V_0(x) \le \alpha_2(\|x\|) \tag{5.93}$$
$$\frac{\partial V_0}{\partial x}\left[f_0(x) + G_0(x)p_0(x)\right] \le -\alpha_3(\|x\|) \tag{5.94}$$

where $\alpha_1, \alpha_2, \alpha_3 : \Re^+ \mapsto \Re^1$ are strictly increasing functions that satisfy $\alpha_i(0) = 0$ and $\alpha_i(r) \to \infty$ as $r \to \infty$. These type of functions are sometimes called class $\mathcal{K}_\infty$ functions [134].

The uncertainty term is assumed to satisfy the bound

$$\|\eta(x,u)\|_\infty \le \bar{\eta}(t,x) \tag{5.95}$$

where the bounding function $\bar{\eta}$ is assumed to be known *apriori* or available for measurement. Now, we will proceed to the design of the "corrective control" component $p^*(x)$ such that $u = p_0 + p^*$ stabilizes the class of systems described by (5.92) and satisfying (5.95). The corrective control term is designed based on a technique following the nominal Lyapunov function V_0, which justifies the name *Lyapunov redesign method.*

Consider the same Lyapunov function V_0 that guarantees the asymptotic stability of the nominal closed-loop system, but now consider the time derivative of V_0 along the solutions of the full system (5.92). We have

$$\begin{aligned}
\dot{V}_0 &= \frac{\partial V_0}{\partial x}\left[f_0(x) + G_0(x)\left(u + \eta(x,u)\right)\right] \\
&= \frac{\partial V_0}{\partial x}\left[f_0(x) + G_0(x)p_0(x)\right] + \frac{\partial V_0}{\partial x}G_0(x)\left(p^*(x) + \eta(x,u)\right) \\
&\le -\alpha_3(\|x\|) + \omega(x)^\top p^*(x) + \omega(x)^\top \eta(x,u),
\end{aligned}$$

where

$$\omega(x) = \left[\frac{\partial V_0}{\partial x} G_0(x)\right]^{\mathrm{T}} \in \Re^m \tag{5.96}$$

which is a known function. By taking bounds we obtain

$$\begin{aligned} \dot{V}_0 &\leq -\alpha_3(\|x\|) + \sum_{i=1}^{m} \omega_i(x) p_i^*(x) + \|\omega(x)\|_1 \|\eta(x,u)\|_\infty \\ &= -\alpha_3(\|x\|) + \sum_{i=1}^{m} \left(\omega_i(x) p_i^*(x) + \bar{\eta}(x,t)|\omega_i(x)|\right). \end{aligned} \tag{5.97}$$

The second term of the right-hand side of (5.97) can be made zero if $p_i^*(x)$ is selected as

$$p_i^*(x) = -\bar{\eta}(x,t)\,\mathrm{sgn}(\omega_i(x)). \tag{5.98}$$

Each component of the corrective control vector $p^*(x)$ is selected to be of the form $p^*(x) = \pm\bar{\eta}(x,t)$, where the sign of $p_i^*(x)$ depends on the sign of $\omega_i(x)$ and, in fact, changes as $\omega_i(x)$ changes sign.

By substituting (5.98) in (5.97) we obtain the desired "stability" property

$$\dot{V}_0 \leq -\alpha_3(\|x\|),$$

which implies that the closed-loop system is asymptotically stable.

The augmented control law $u = p_0(x) + p^*(x)$ is discontinuous since each element $p_i^*(x)$ is discontinuous at $\omega_i(x) = 0$. Moreover, the discontinuity jump $\bar{\eta}(x,t) \mapsto -\bar{\eta}(x,t)$ can be of large magnitude if the uncertainty bound $\bar{\eta}$ is large. As discussed earlier, discontinuities in the control law can cause chattering, therefore it is desirable to smooth the discontinuity and at the same time retain to some degree the nice stability properties of the original discontinuous control law.

This can be achieved by replacing (5.98) with

$$p_i^*(x) = -\bar{\eta}(x,t)\,\tanh\left(\frac{\omega_i(x)}{\varepsilon}\right), \tag{5.99}$$

where $\varepsilon > 0$ is a small design constant. Note that as ε approaches zero, the $\tanh\left(\frac{\omega_i}{\varepsilon}\right)$ function converges to the discontinuous $\mathrm{sgn}(\omega_i)$ function.

By substituting (5.99) in (5.97) we obtain

$$\dot{V}_0 \leq -\alpha_3(\|x\|) + \bar{\eta}(x,t) \sum_{i=1}^{m} \left(|\omega_i(x)| - \omega_i(x)\tanh\left(\frac{\omega_i(x)}{\varepsilon}\right)\right).$$

Using Lemma A.5.1 (see p. 397),

$$\dot{V}_0 \leq -\alpha_3(\|x\|) + \varepsilon m \kappa\, \bar{\eta}(x,t), \tag{5.100}$$

where $\kappa = 0.2785$. Since α_3 is a class $\mathcal{K}_\infty$ function (strictly increasing), for any uniformly bounded function $\bar{\eta}$ and for any $r > 0$, there exists an ε (sufficiently small), such that $\dot{V} \leq 0$ for x outside a region $\mathcal{D}_\varepsilon = \{x \mid V(x) \leq r\}$. Therefore, the trajectory is convergent to the invariant set $\mathcal{D}_\varepsilon$.

The following example illustrates the use of the Lyapunov redesign method.

EXAMPLE 5.9

Consider the nonlinear system

$$\begin{aligned}\dot{x}_1 &= -x_2 - \frac{3}{2}x_1^2 - \frac{1}{2}x_1^3 \\ \dot{x}_2 &= -u - \eta(x),\end{aligned}$$

where η is unknown but is known to satisfy the inequality

$$\|\eta(x)\|_\infty \leq \bar{\eta}(t, x)$$

for some known bound $\bar{\eta}$. This second-order model represents a jet engine compression system with no-stall [139], which is based on the Galerkin approximation of the nonlinear PDE model [176]. The state x_1 corresponds to the mass flow and x_2 is the pressure rise.

The first step is to design the nominal control law $u = p_0(x)$ for the case of $\eta = 0$. This can be accomplished by feedback linearization (note that it can also be accomplished by the backstepping method). Consider the change of coordinates $z = T(x)$ where

$$\begin{aligned}z_1 &= x_1 \\ z_2 &= -x_2 - \frac{3}{2}x_1^2 - \frac{1}{2}x_1^3.\end{aligned}$$

The dynamics in the z-coordinates are described by

$$\begin{aligned}\dot{z}_1 &= z_2 \\ \dot{z}_2 &= u - 3z_1 z_2 - \frac{3}{2}z_1^2 z_2 + \eta_z(z),\end{aligned}$$

where $\eta_z(z) = \eta(x)|_{x=T^{-1}(z)}$. A stabilizing nominal controller is given by

$$u = p_0(z) = -z_1 - 2z_2 + 3z_1 z_2 + \frac{3}{2}z_1^2 z_2.$$

A nominal Lyapunov function associated with the above nominal controller is given by

$$V_0(z) = 2z_1^2 + (z_1 + z_2)^2,$$

whose time derivative is given by

$$\dot{V}_0 = -2(z_1^2 + z_2^2).$$

Since by eqn. (5.96) $\omega(z) = 2(z_1 + z_2)$, the corrective feedback control law obtained using the Lyapunov redesign method is given by

$$p^*(z) = -\bar{\eta}_z(z)\ \mathrm{sgn}(z_1 + z_2),$$

where $\bar{\eta}_z$ is the assumed bound on η. The above control law can be made continuous using the following approximation

$$p^*(z) = -\bar{\eta}_z(z)\ \tanh\left(\frac{z_1 + z_2}{\varepsilon}\right),$$

where $\varepsilon > 0$ is a small design constant. △

5.4.4 Nonlinear Damping

The Lyapunov redesign approach developed in Section 5.4.3 is based on the principle of first designing a nominal controller $u = p_0(x)$, with a Lyapunov function such that the nominal system satisfies some desirable stability properties, and then augmenting the control law using $u = p_0(x) + p^*(x)$ such that the corrective term $p^*(x)$ is designed (using the same nominal Lyapunov function) to address a matched uncertainty term $\eta(x, u)$. One of the key assumptions made in the design methodology described in Section 5.4.3 is that the uncertainty term $\eta(x, u)$ is bounded by a *known* bounding term $\bar{\eta}(t, x)$. The nonlinear damping method developed in this section relaxes somewhat this assumption by not requiring that the bounding term $\bar{\eta}$ is known.

Consider again the system described by (5.92); i.e.,

$$\dot{x} = f_0(x) + G_0(x)\left(u + \eta(x, u)\right). \tag{5.101}$$

The uncertainty function $\eta(x, u)$ is assumed to be of the form

$$\eta(x, u) = \Phi(t, x)\eta_0(x, u), \tag{5.102}$$

where the $m \times m$ matrix Φ is known, and η_0 is unknown but uniformly bounded (i.e., $\|\eta_0(x, u)\|_\infty < M$ for all (x, u)). In this case the bound M does not need to be known. Again, the objective is to design a "corrective" control law $p^*(x)$ that stabilizes the closed-loop system.

Following the same procedure as in Section 5.4.3, we consider a nominal Lyapunov function $V_0(x)$ that satisfies (5.93), (5.94) for some class $\mathcal{K}_\infty$ functions α_1, α_2, α_3. The time derivative of V_0 along the solutions of (5.101), (5.102) is given by

$$\begin{aligned} \dot{V}_0 &= \frac{\partial V_0}{\partial x}\left[f_0(x) + G_0(x)\left(u + \Phi(t, x)\eta_0(x, u)\right)\right] \\ &\leq -\alpha_3(\|x\|) + \omega(x)^\top p^*(x) + \omega(x)^\top \Phi(t, x)\eta_0(x, u), \end{aligned} \tag{5.103}$$

where $\omega(x)$ is the same as defined in (5.96). Now, let us select $p^*(x)$ as

$$p^*(x) = -k\omega(x)\|\Phi(t, x)\|_2^2, \tag{5.104}$$

where $k > 0$ is a scalar. By substituting (5.104) in (5.103) we obtain

$$\dot{V}_0 \leq -\alpha_3(\|x\|) - k\|\omega(x)\|_2^2\,\|\Phi(t, x)\|_2^2 + \omega(x)^\top \Phi(t, x)\eta_0(x, u).$$

Since $\eta_0(x, u)$ is uniformly bounded in (x, u),

$$\omega(x)^\top \Phi(t, x)\eta_0(x, u) \leq \|\omega(x)\|_2\,\|\Phi(t, x)\|_2\,M.$$

The term

$$Q = -k\|\omega(x)\|_2^2\,\|\Phi(t, x)\|_2^2 + \|\omega(x)\|_2\,\|\Phi(t, x)\|_2\,M$$

is of the form $Q(\alpha) = -k\alpha^2 + \alpha M$, where $\alpha = \|\omega(x)\|_2 \cdot \|\Phi(t, x)\|_2$; therefore, Q attains the maximum value of $M/4k$ at $\alpha = M/2k$. Therefore,

$$\dot{V}_0 \leq -\alpha_3(\|x\|) + \frac{M}{4k}.$$

Since $\alpha_3(\|x\|)$ is strictly increasing and approaches ∞ as $\|x\| \to \infty$, there exists a ball $\mathcal{B}_\rho$ of radius ρ such that $\dot{V}_0 \leq 0$ for x outside $\mathcal{B}_\rho$. Therefore, the closed-loop system is uniformly bounded and the trajectory $x(t)$ converges to the invariant set

$$\Omega = \left\{ x \,\middle|\, V_0(x) \leq \max_{\|x\|=\rho} (V_0(x)) \right\},$$

where ρ can be made smaller by increasing the feedback gain k or by decreasing the infinity norm of the model error.

EXAMPLE 5.10

Consider the nonlinear model of Example 5.9. In this case, instead of assuming that $\|\eta_z(z)\| \leq \bar{\eta}_z(z)$ where $\bar{\eta}_z(z)$ is known, we assume that $\eta_z(z) = \Phi(z)\eta_0(z)$ where Φ is known, while η_0 is unknown but uniformly bounded.

The corrective control term obtained using the nonlinear damping method is given by

$$p^*(x) = -2k(z_1 + z_2)\|\Phi(t, x)\|_2^2.$$

It is noted that this control law is not switching as it was in the case of the Lyapunov redesign method. △

5.4.5 Adaptive Bounding Control

Of the four techniques presented in this section, namely bounding control, sliding mode control, Lyapunov redesign, and nonlinear damping, the first three are based on the key assumption of a known bound on the uncertainty. The nonlinear damping technique does not make this bounding assumption; however, the resulting stability property does not guarantee the convergence of the tracking error to zero, but to an invariant set whose radius is proportional to the ∞-norm of the uncertainty. Even though the residual error in the nonlinear damping design can be reduced by increasing the feedback gain parameter k, this is not without drawbacks, since increasing the feedback gain may result in high-gain feedback , with all the undesirable consequences.

In this subsection, we introduce another technique which also relaxes the assumption of a known bound. Specifically, it is assumed that $\eta(x, u)$ is bounded by

$$\|\eta(x, u)\|_\infty \leq \theta^\top \varphi(x, t),$$

where θ is an unknown parameter vector of dimension q and φ is a known vector function. Since $\theta^\top \varphi$ represents a bound on the uncertainty, each element of θ and φ is assumed to be non-negative. Typically, the dimension q is simply equal to one. However, the general case where both θ and φ are vectors, allows the control designer to take advantage of any knowledge where the bound changes for different regions of the state-space x. If a known function $\varphi(x)$ is not available, it can be simply assumed that $\|\eta(x, u)\|_\infty \leq \theta$, where θ is a scalar unknown bounding constant. The adaptive bounding control method was introduced in [215] and was later used in neural control [209].

It is worth noting that the bounding assumption of the adaptive bounding control method is significantly less restrictive than that of the Lyapunov redesign method where the bound is assumed to be known. Even though one may consider simply increasing the bound of the Lyapunov redesign method until the assumed bound holds, this is not always possible, and quite often it is not an astute way to handle the problem since it will increase the feedback gain of the system.

The adaptive bounding control technique is based on the idea of estimating online the unknown parameter vector θ. The feedback controller utilizes the parameter estimate $\hat{\theta}(t)$ instead of the true bounding vector θ. One of the key questions has to do with the design of the adaptive law for generating $\hat{\theta}(t)$. As we will see, this is achieved again by Lyapunov analysis.

Let $\tilde{\theta}(t) = \hat{\theta}(t) - \theta$ denote the parameter estimation error. Consider the augmented Lyapunov function

$$V(x, \tilde{\theta}) = V_0(x) + \frac{1}{2}\tilde{\theta}^\top \Gamma^{-1}\tilde{\theta},$$

where Γ is a positive define matrix of dimension $q \times q$, which represents the adaptive gain. By taking the time derivative of V along the solutions of (5.92), we obtain

$$\begin{aligned}
\dot{V} &= \frac{\partial V_0}{\partial x}\left[f_0(x) + G_0(x)\left(u + \eta(x,u)\right)\right] + \tilde{\theta}^\top \Gamma^{-1}\dot{\tilde{\theta}} \\
&\leq -\alpha_3(\|x\|) + \omega(x)^\top p^*(x) + \omega(x)^\top \eta(x,u) + \tilde{\theta}^\top \Gamma^{-1}\dot{\hat{\theta}} \\
&\leq -\alpha_3(\|x\|) + \sum_{i=1}^{m}\left(\omega_i(x) p_i^*(x) + \hat{\theta}^\top \varphi(x,t)|\omega_i(x)|\right) \\
&\quad - \tilde{\theta}^\top \varphi(x,t)\|\omega(x)\|_1 + \tilde{\theta}^\top \Gamma^{-1}\dot{\hat{\theta}}.
\end{aligned}$$

We choose the corrective control term $p_i(x)$ and the update law for $\hat{\theta}$ as follows:

$$p_i^*(x) = -\hat{\theta}^\top \varphi(x,t)\,\mathrm{sgn}(\omega_i(x)) \tag{5.105}$$

$$\dot{\hat{\theta}} = \Gamma\varphi(x,t)\|\omega(x)\|_1, \tag{5.106}$$

which implies that $\dot{V} \leq -\alpha_3(\|x\|)$. Therefore, both $x(t)$ and $\tilde{\theta}(t)$ remain bounded and $x(t)$ converges to zero (using Barbălat's Lemma).

The feedback control law (5.105) is discontinuous at $\omega_i(x) = 0$. As discussed in Section 5.4.3, the discontinuous sign functions can be smoothed by using the $\tanh(\cdot)$ function:

$$p_i^*(x) = -\hat{\theta}^\top \varphi(x,t)\tanh\left(\frac{\omega_i(x)}{\varepsilon}\right),$$

where $\varepsilon > 0$ is a small design constant. Another issue that arises with adaptive bounding control is the possible parameter drift of the bounding estimate $\hat{\theta}(t)$. This may occur as a consequence of using the smooth approximation $\tanh(\omega_i(x)/\varepsilon)$, which may result in a small residual error. Moreover, in the presence of measurement noise or disturbances, again the bounding parameter estimate $\hat{\theta}$ may not converge. Since the right-hand side of (5.106) is nondecreasing, the presence of such residual errors (even if small) may cause the parameter drift of the estimate, which in turn will cause the feedback control signal to become large. This can be prevented by using a robust adaptive law, as described in Chapter 4. One of the available techniques is the dead-zone, which requires knowledge of the size of the residual error. Another method is the projection modification, which prevents the parameter estimate from becoming larger than a preselected level. Yet another approach is the σ modification.

The adaptive bounding control method is also used in adaptive approximation based control in order to address the issue of having the trajectory leave the approximation region. This is illustrated in Chapters 6 and 7.

5.5 ADAPTIVE NONLINEAR CONTROL

Adaptive control deals with systems where some of the parameters are unknown or slowly time-varying. The basic idea behind adaptive control is to estimate the unknown parameters online using parameter estimation methods (such as those presented in Chapter 4), and then to use the estimated parameters, in place of the unknown ones, in the feedback control law. Most of the research in adaptive control has been developed for linear models, even though in the last decade or so there has been a lot of activity on adaptive nonlinear control as well. Even in the case of adaptive control applied to linear systems, the resulting control law is nonlinear. This is due to the parameter update laws, which render the feedback controller nonlinear.

There are two strategies for combining the control law and the parameter estimation algorithm. In the first strategy, referred to as *indirect adaptive control*, the parameter estimation algorithm is used to estimate the unknown parameters of the plant. Based on these parameter estimates, the control law is computed by treating the estimates as if they were the true parameters, based on the *certainty equivalence principle* [11]. In the second strategy, referred to as *direct adaptive control*, the parameter estimator is used to estimate directly the unknown controller parameters.

It is interesting to note the similarities and difference between so called robust control laws and adaptive control laws. The robust approaches, which were discussed in Section 5.4, treat the uncertainty as an unknown box where the only information available are some bounds. The robust control law is obtained based on these bounds, and in fact is designed to stabilizes the system for any uncertainty within the assumed bounds. As a result, the robust control law tends to be conservative and it may lead to large control input signals or control saturation. On the other hand, adaptive control assumes a special structure for the uncertainty where the nonlinearities are known but the parameters are unknown. In contrast to robust control, in adaptive control the objective is to try to estimate the uncertain (or time-varying) parameters to reduce the level of uncertainty.

In the next chapter, we will start investigating the adaptive approximation control approach where the uncertainty also includes nonlinearities that are estimated online. Hence, adaptive approximation based control can be viewed as an expansion of the adaptive control methodology where instead of having simply unknown parameters we have unknown nonlinearities.

Adaptive control is a well-established methodology in the design of feedback control systems. The first practical attempts to design adaptive feedback control systems go back as far as the 1950s, in connection with the design of autopilots [295]. Stability analysis of adaptive control for linear systems started in the mid-1960s [196] and culminated in 1980 with the complete stability proof for linear systems [69, 177, 180]. The first stability results assumed that the only uncertainty in the system was due to unknown parameters; i.e., no disturbances, measurement noise, nor any other form of uncertainty. In the 1980s, adaptive control research focused on robust adaptive control for linear systems, which dealt with modifications to the adaptive algorithms and the control law in order to address some types of uncertainties [119]. In the 1990s, most of the effort in adaptive control focused on adaptive control of nonlinear systems with some elegant results [139].

To illustrate the use of the adaptive control methodology we consider below two examples of adaptive nonlinear control.

EXAMPLE 5.11

In this example we consider the feedback linearization problem of Section 5.2 with unknown parameters. Consider the n-th order model

$$\begin{aligned}\dot{x}_1 &= x_2\\ \dot{x}_2 &= x_3\\ &\vdots\\ \dot{x}_n &= \theta_1 f_1(x) + \theta_2 f_2(x) + \theta_3 u\end{aligned}$$

where $\theta_1, \theta_2, \theta_3$ are unknown, constant parameters and f_1 and f_2 are known functions. The objective is to design an adaptive controller such that $y(t) = x_1(t)$ tracks a desired signal $y_d(t)$. Let $e = y - y_d$ be the tracking error.

If $\theta_1, \theta_2, \theta_3$ were known and $\theta_3 \neq 0$ then the control law

$$u = \frac{1}{\theta_3}\left[-\theta_1 f_1(x) - \theta_2 f_2(x) + y_d^{(n)} - \lambda_{n-1}e^{(n-1)} - \cdots - \lambda_2 e^{(2)} - \lambda_1 \dot{e} - \lambda_0 e\right]$$

would result in the following tracking error dynamics:

$$e^{(n)} + \lambda_{n-1}e^{(n-1)} + \cdots + \lambda_2 e^{(2)} + \lambda_1 \dot{e} + \lambda_0 e = 0.$$

The coefficients $\{\lambda_1,\ \lambda_2,\ \ldots\ \lambda_{n-1}\}$ would be selected such that the characteristic polynomial

$$s^n + \lambda_{n-1}s^{n-1} + \cdots + \lambda_2 s^2 + \lambda_1 s + \lambda_0 = 0$$

has all its roots in the left-half complex plane.

Since $\theta_1, \theta_2, \theta_3$ are unknown, we replace them in the control law by their corresponding estimates $\hat{\theta}_1(t)$, $\hat{\theta}_2(t)$, $\hat{\theta}_3(t)$, where it is assumed for the time being that $\hat{\theta}_3(t) \neq 0$ for all $t \geq 0$. The adaptive control law is given by

$$u = \frac{1}{\hat{\theta}_3}\left[-\hat{\theta}_1 f_1(x) - \hat{\theta}_2 f_2(x) + y_d^{(n)} - \lambda_{n-1}e^{(n-1)} - \cdots - \lambda_2 e^{(2)} - \lambda_1 \dot{e} - \lambda_0 e\right]$$

which yields the following tracking error dynamics

$$e^{(n)} + \lambda_{n-1}e^{(n-1)} + \cdots + \lambda_2 e^{(2)} + \lambda_1 \dot{e} + \lambda_0 e = -\tilde{\theta}_1 f_1(x) - \tilde{\theta}_2 f_2(x) - \tilde{\theta}_3 u$$

where $\tilde{\theta}_i = \hat{\theta}_i - \theta_i$ for $i = 1, 2, 3.$. If we let $\chi = [e \ \ \dot{e} \ \ e^{(2)} \ \cdots e^{(n-1)}]^\top$ then the tracking error dynamics can be written as

$$\dot{\chi} = A_0\chi - B_0\left(\tilde{\theta}_1 f_1(x) + \tilde{\theta}_2 f_2(x) + \tilde{\theta}_3 u\right)$$

where A_0, B_0 is the canonical form pair given by

$$A_0 = \begin{bmatrix} 0 & 1 & 0 & \cdots & 0\\ 0 & 0 & 1 & & 0\\ \vdots & & & \ddots & \vdots\\ 0 & & & & 1\\ -\lambda_0 & -\lambda_1 & & \cdots & -\lambda_{n-1}\end{bmatrix}, \qquad B_0 = \begin{bmatrix}0\\0\\ \vdots\\ 0\\ 1\end{bmatrix}.$$

Since A_0 is a stability matrix, there exists a positive definite matrix P such that

$$A_0^\top P + PA_0 = -I.$$

We choose the Lyapunov function

$$V = \chi^\top P \chi + \frac{1}{\gamma_1}\tilde{\theta}_1^2 + \frac{1}{\gamma_2}\tilde{\theta}_2^2 + \frac{1}{\gamma_3}\tilde{\theta}_3^2$$

whose time derivative along the solution of the tracking error dynamics is given by

$$\begin{aligned}\dot{V} &= -\chi^\top \chi + \frac{2}{\gamma_1}\tilde{\theta}_1\left(\dot{\hat{\theta}}_1 - \gamma_1 \chi^\top P B_0 f_1(x)\right) + \frac{2}{\gamma_2}\tilde{\theta}_2\left(\dot{\hat{\theta}}_2 - \gamma_2 \chi^\top P B_0 f_2(x)\right)\\ &\quad + \frac{2}{\gamma_1}\tilde{\theta}_3\left(\dot{\hat{\theta}}_3 - \gamma_3 \chi^\top P B_0 u\right).\end{aligned}$$

Therefore, we select the adaptive laws as follows:

$$\begin{aligned}\dot{\hat{\theta}}_1 &= \gamma_1 \chi^\top P B_0 f_1(x)\\ \dot{\hat{\theta}}_2 &= \gamma_2 \chi^\top P B_0 f_2(x)\\ \dot{\hat{\theta}}_3 &= \gamma_3 \chi^\top P B_0 u.\end{aligned}$$

Clearly, this results in

$$\dot{V} = -\chi^\top \chi$$

which implies that the tracking error, its derivatives and the parameter estimates are uniformly bounded and the tracking error converges to zero (by Barbălat's Lemma).

Although it has not been included in the above analysis, projection would be required to maintain $\hat{\theta}_3 > 0$. △

EXAMPLE 5.12

In this example we consider the backstepping control procedure of Section 5.3 for the case where there is an unknown parameter. Consider the second-order system

$$\begin{aligned}\dot{x}_1 &= x_2 + \theta f(x_1)\\ \dot{x}_2 &= u\end{aligned}$$

where θ is an unknown parameter and f is a known function. The parameter estimate for θ is defined as $\hat{\theta}$, while $\tilde{\theta}(t) = \hat{\theta}(t) - \theta$ is the parameter estimation error. The objective is to design an adaptive nonlinear tracking controller such that $y = x_1$ tracks a desired signal $y_d(t)$.

We define the change of coordinates

$$\begin{aligned}z_1 &= x_1 - y_d\\ z_2 &= x_2 - \alpha,\end{aligned}$$

where α is defined as

$$\alpha = -k_1 z_1 - \hat{\theta} f(x_1) + \dot{y}_d, \qquad k_1 > 0.$$

The dynamics of the new coordinates z_1, z_2 are given by

$$\begin{aligned}\dot{z}_1 &= -k_1 z_1 - \tilde{\theta} f(x_1) + z_2 \\ \dot{z}_2 &= u - \dot{\alpha},\end{aligned}$$

where $\dot{\alpha}$ denotes the time derivative of α, which can be computed as follows:

$$\begin{aligned}\dot{\alpha} &= -k_1 \left(x_2 + \hat{\theta} f(x_1) - \dot{y}_d\right) - \dot{\hat{\theta}} f(x_1) - \hat{\theta}\frac{\partial f(x_1)}{\partial x_1}\left(x_2 + \hat{\theta} f(x_1)\right) + \ddot{y}_d \\ &\quad + k_1 \tilde{\theta} f(x_1) + \hat{\theta}\frac{\partial f(x_1)}{\partial x_1}\tilde{\theta} f(x_1),\end{aligned}$$

where the term in the second line cannot be computed. Therefore, the feedback control law is selected as follows:

$$\begin{aligned}u &= -z_1 - k_2 z_2 - k_1 \left(x_2 + \hat{\theta} f(x_1) - \dot{y}_d\right) - \dot{\hat{\theta}} f(x_1) \\ &\quad - \hat{\theta}\frac{\partial f(x_1)}{\partial x_1}\left(x_2 + \hat{\theta} f(x_1)\right) + \ddot{y}_d\end{aligned}$$

where $k_2 > 0$ is a design constant. The resulting closed-loop z dynamics are given by

$$\begin{aligned}\dot{z}_1 &= -k_1 z_1 - \tilde{\theta} f(x_1) + z_2 \\ \dot{z}_2 &= -k_2 z_2 - z_1 - \tilde{\theta} f(x_1)\left(k_1 + \hat{\theta}\frac{\partial f(x_1)}{\partial x_1}\right).\end{aligned}$$

Now, consider the time derivative of the Lyapunov function candidate

$$V = \frac{1}{2}z_1^2 + \frac{1}{2}z_2^2 + \frac{1}{2\gamma}\tilde{\theta}^2,$$

where $\gamma > 0$ is the adaptive gain. We have

$$\begin{aligned}\dot{V} &= -k_1 z_1^2 - \tilde{\theta} f(x_1) z_1 + z_1 z_2 \\ &\quad - k_2 z_2^2 - z_1 z_2 - \tilde{\theta} f(x_1) z_2 \left(k_1 + \hat{\theta}\frac{\partial f(x_1)}{\partial x_1}\right) + \frac{1}{\gamma}\tilde{\theta}\dot{\hat{\theta}} \\ &= -k_1 z_1^2 - k_2 z_2^2 + \frac{1}{\gamma}\tilde{\theta}\left(\dot{\hat{\theta}} - \gamma f(x_1)\left(z_1 + k_1 z_2 + \hat{\theta}\frac{\partial f(x_1)}{\partial x_1} z_2\right)\right).\end{aligned}$$

Based on the above derivative of the Lyapunov function, we select the update law for $\hat{\theta}$ as follows:

$$\dot{\hat{\theta}} = \gamma f(x_1)\left(z_1 + k_1 z_2 + \hat{\theta}\frac{\partial f(x_1)}{\partial x_1} z_2\right).$$

Hence,

$$\dot{V} = -k_1 z_1^2 - k_2 z_2^2,$$

which implies that z_1, z_2 and $\hat{\theta}$ are uniformly bounded and z_1, z_2 both converge to zero. △

5.6 CONCLUDING SUMMARY

In addition to introducing a few of the dominant nonlinear control system design methodologies, this chapter has reviewed methods used to achieve robustness to nonlinear model error and discussed situations in which online approximation might be useful for improving such robustness and tracking performance. As discussed in Chapter 2, online approximation can be achieved only over a compact set denoted by $\mathcal{D}$. Within $\mathcal{D}$, due to the use of the adaptive approximator, the nonlinear model errors should be small. Outside of $\mathcal{D}$, the nonlinear model errors may still be large. Therefore, $\mathcal{D}$ should be defined to contain the set of desired system trajectories. For this reason, the set $\mathcal{D}$ is often referred to as the operating envelope. An important issue in the design of an adaptive approximation based control system, as we will see in Chapter 6, is the design of mechanisms to ensure that, for any initial conditions, the system state converges to and stays within the operating envelope $\mathcal{D}$. In order to prevent the state trajectories from leaving the region $\mathcal{D}$, some bound (possibly state-dependent) on the unknown function will be required. In this chapter, we saw that such bounds were also required for the use of sliding mode control, Lyapunov redesign method and adaptive bounding control.

5.7 EXERCISES AND DESIGN PROBLEMS

Exercise 5.1 Consider the nonlinear system

$$\begin{aligned}\dot{x}_1 &= \frac{-6x_1}{(1+x_1^2)^2} + 2x_2 \\ \dot{x}_2 &= \frac{-2(x_1+x_2)}{(1+x_1^2)^2} + u\end{aligned}$$

1. Linearize the system around $x_1 = 0$, $x_2 = 0$ and $u = 0$.
2. Is the linear model stable in an open-loop mode?
3. Verify that the resulting $(A,\ B)$ of the linear model is stabilizable.
4. Design a feedback controller $u = k_1x_1 + k_2x_2$ such that both poles of the closed-loop system for the linear model are located at $s = -2$.

Exercise 5.2 Consider the nonlinear system

$$\begin{aligned}\dot{x}_1 &= 4x_1^2x_2 + 4(x_1^2 + 2x_2^2 - 4) \\ \dot{x}_2 &= -2x_1^3 - 2(x_1^2 + 2x_2^2 - 4) + u\end{aligned}$$

1. Verify that $x^* = [1\ \ 1]^\top$, $u^* = 0$ is an equilibrium point of the nonlinear system.
2. Perform a change of coordinates $z = x - x^*$ and rewrite the nonlinear system in the z-coordinates.
3. Verify that $z^* = [0\ \ 0]^\top$, $u^* = 0$ is an equilibrium point of the nonlinear system in the z-coordinates.
4. Linearize the system around the equilibrium point $z^* = [0\ \ 0]^\top$, $u^* = 0$.

5. Design a feedback controller $u = k_1 z_1 + k_2 z_2$ such that the poles of the closed-loop system for the linear model are located at $s = -1 \pm j$.

Exercise 5.3 Use a simulation study to investigate the performance of the linear feedback control law

$$u = k_1(x_1 - 1) + k_2(x_2 - 1)$$

developed in Exercise 5.2 when applied to the original nonlinear system. Consider several initial conditions close to the equilibrium point $x = x^*$ to get a rough idea of how large is the region of attraction around the equilibrium point.

Exercise 5.4 Use a simulation study for the satellite example of Example 5.2. Assume that: $\beta = 10$; $x_1(0) = r_0 = 10$; $x_2(0) = \dot{r}(0) = 0$; $x_3(0) = \theta_0 = 0$. Consider the following cases:

(a) $x_4(0) = 0.1, \quad u_1(t) = 0, \quad u_2(t) = 0;$

(b) $x_4(0) = 0.095, \quad u_1(t) = 0, \quad u_2(t) = 0;$

(c) $x_4(0) = 0.105, \quad u_1(t) = 0, \quad u_2(t) = 0;$

(d) $x_4(0) = 0.1, \quad u_1(t) = 0.02, \quad u_2(t) = 0;$

(e) $x_4(0) = 0.1, \quad u_1(t) = 0.1\sin(t), \quad u_2(t) = 0.1\cos(t);$

(f) $x_4(0) = 0.09, \quad u_1(t) = 0, \quad u_2(t) = 0.1\cos(t).$

Simulate the differential equation for about 100 s. Provide plots of the satellite motion in cartesian coordinates instead of polar coordinates. Interpret your results. Compare the solution of the nonlinear differential equation with that of the linearized model (assume that $r_0 = 10$; $\theta_0 = 0$; $\omega = 0.1$). Discuss the accuracy of the linearized model as an approximation of the nonlinear system. Plot the trajectories of the satellite motion of both the linear and nonlinear model on the same diagram for comparison purposes.

Exercise 5.5 Consider the nonlinear state equation

$$\begin{aligned}
\begin{bmatrix} \dot{x}_1(t) \\ \dot{x}_2(t) \\ \dot{x}_3(t) \end{bmatrix} &= \begin{bmatrix} u(t) \\ x_1(t)u(t) - x_3(t) \\ x_2(t) - 2x_3(t) \end{bmatrix} \\
y(t) &= x_2(t) - 2x_3(t)
\end{aligned}$$

with the nominal initial state $x_1^*(0) = 0$, $x_2^*(0) = -3$, $x_3^*(0) = -2$, and the nominal input $u^*(t) = 1$. Show that the nominal output is $y^*(t) = 1$. Linearize the state equation about the nominal solution.

Exercise 5.6 Consider the following second-order model which represents a field controlled DC motor [248]

$$\begin{aligned}
\dot{x}_1 &= -50x_1 - 0.4x_2 u + 40 \\
\dot{x}_2 &= -5x_2 + 50000x_1 u \\
y &= x_2
\end{aligned}$$

where x_1 is the armature current, x_2 is the speed, and u is the field current. It is required to design a speed control system so that $y(t)$ asymptotically tracks a constant reference speed

$y_d = 100$. It is assumed that the domain of operation for the armature current is restricted to $x_1 > 0.2$.

1. Find the steady-state field current u_{ss} and steady-state armature current x_{1ss} (within the domain of operation) such that the output y follows exactly the desired constant speed $y_d = 100$.

2. Verify that the control $u = u_{ss}$ results in an asymptotically stable equilibrium point.

3. Using small-signal linearization techniques, design a state feedback control law to achieve the desired speed control.

4. Using computer simulations, study the performance of the linear controller of part (c) when applied to the nonlinear system. Assume that $y_d = 100$ and at a certain time it increases (step change) to $y_d = 105$. Repeat the simulation experiment while gradually increasing the step change to $y_d = 110, \quad 115, \quad 120, \ldots$.

Exercise 5.7 Consider the same field controlled DC motor of Exercise 5.6. Suppose that the speed x_2 is measurable but the armature current x_1 is not measured for feedback control purposes.

1. Repeat part (d) of Exercise 5.6 using an observer to estimate the current; i.e., instead of using x_1 in the feedback control, use $\hat{x}_1$ where $\hat{x}_1(t)$ is generated by an observer.

2. Design a gain scheduling, observer based controller, where the scheduling variable is the measured speed x_2.

3. Study the performance of the gain scheduling controller using computer simulation. Compare to the performance of the linear controller of part (a) obtained via small-signal linearization and discuss.

Exercise 5.8 Consider the Example 5.4 on page 195, which describes the model of a single-link manipulator with flexible joints.

1. Show that the transformation $z = T(x)$ given by (5.30) is indeed a diffeomorphism, by obtaining the inverse $x = T^{-1}(z)$. What is the region in which this diffeomorphism is valid.

2. Verify the differential equations (5.31).

Exercise 5.9 Consider the system

$$\begin{aligned}
\dot{x}_1 &= x_2 + \frac{1}{2}x_3^2 \\
\dot{x}_2 &= x_3 - 2x_3x_4 \\
\dot{x}_3 &= x_4 \\
\dot{x}_4 &= u \\
y &= x_1
\end{aligned}$$

Convert the system to normal form. Design a feedback linearizing tracking controller so that $y(t)$ tracks the target signal $y_d(t) = \sin(t)$.

Exercise 5.10 For the system given in Exercise 5.9, after converting the system to normal form, use standard backstepping to design a tracking controller so that $y(t)$ tracks the target signal $y_d(t)$.

Exercise 5.11 For the system given in Exercise 5.9, use command filtered backstepping to design a tracking controller so that $y(t)$ tracks the target signal $y_d(t)$.

Exercise 5.12 Consider the system

$$\begin{aligned}\dot{x}_1 &= x_2 + f(x_1, x_2)\\ \dot{x}_2 &= u\\ y &= x_1\end{aligned}$$

1. Is the system input-output linearizable? Under what conditions? Assuming that these conditions are valid, design a tracking controller.
2. Assume that $f = (1+\epsilon)\tilde{f}(x_1, x_2)$, where $\tilde{f}$ is known while ϵ is assumed by the designer to be zero, while in reality it is equal to 0.05. Investigate to what degree this modeling error affects the linearization and the design of the tracking controller.

Exercise 5.13 Design a tracking control algorithm for the system

$$\begin{aligned}\dot{x}_1 &= x_2 - 2x_1^2\\ \dot{x}_2 &= u\\ \dot{x}_3 &= x_1 - x_2^2 - x_3\\ y &= x_1\end{aligned}$$

where the desired output signal is $y_d(t) = \sin(3t)$.

Exercise 5.14 Consider Example 5.7 on page 206. Perform a computer simulation study to illustrate the performance of the control system. Similarly, perform a computer simulation for Example 5.8 and compare the differences.

Exercise 5.15 Consider Example 5.9 on page 218. Assume that the actual uncertainty term η is given by

$$\eta(x) = 1.2\cos(x_1)$$

while the bound is given by $\bar{\eta} = 1.8$. Perform a computer simulation study to illustrate the performance of the control system using both the discontinuous algorithm and the continuous approximation obtained using the tanh function, with $\varepsilon = 0.1$.

Exercise 5.16 Consider Example 5.10 on page 220. As in Example 5.15, assume that the actual uncertainty term η is given by $\eta(x) = 1.2\cos(x_1)$. Let $\phi = 1$ and $\eta_0 = 1.2\cos(x_1)$. Perform a computer simulation study to illustrate the performance of the control system obtained using the nonlinear damping method. Repeat the simulation for various values of k. Compare the control performance and control effort with the Lyapunov redesign method of Example 5.15.

Exercise 5.17 Consider Example 5.12 on page 224. Let $f(x_1) = x_1^2$ and $\theta = 1$. Simulate this example for $k_1 = k_2 = \gamma = 2$. Plot the tracking error, the control effort and the parameter estimation error. Discuss your results.

Exercise 5.18 For the bounding control of Section 5.4.1 that uses the smoothing approximation, show that $e(t)$ ultimately converges to the set $|e| < \delta$. Also show that $|e(t)| \leq \delta$ for $t > \frac{V(0)}{a_m \delta^2}$.

CHAPTER 6

ADAPTIVE APPROXIMATION: MOTIVATION AND ISSUES

Chapters 2 and 3 have presented approximator properties and structures. Chapter 4 discussed and analyzed methods for parameter estimation and issues related to adaptive approximation. Chapter 5 reviewed various nonlinear control design methods. The objective of this chapter is to bring these different topics together in the synthesis and analysis of adaptive approximation based control systems. An additional objective of this chapter is to clearly state and intuitively explain certain issues that must be addressed in adaptive approximation based control problems. To allow the reader to focus on these issues without the distraction of mathematical complexities, in the majority of this chapter we will restrict our discussion to scalar systems. Adaptive approximation based control for higher order dynamical systems will be considered in Chapter 7.

In addition to presenting nonlinear control design methods, Chapter 5 also discussed the effect of nonlinear model errors on the controller performance. Nonlinear damping, Lyapunov redesign, high-gain, and adaptive approximation were discussed as possible methods to address modeling error. The first three approaches rely on bounds on the model error to develop additional terms in the control law that dominate the model error. Typically, these terms are large in magnitude and may involve high frequency switching. Neither of these characteristics is desirable in a feedback control system.

The role of adaptive approximation based control will be to estimate unknown nonlinear functions and cancel their effect using the feedback control signal. Cancelling the estimated nonlinear function allows accurate tracking to be achieved with a smoother control signal. The tradeoff is that the adaptive approximation based controller will typically have much higher state dimension (with the approximator adaptive parameters considered as states).

Adaptive Approximation Based Control: Unifying Neural, Fuzzy and Traditional Adaptive Approximation Approaches. By Jay A. Farrell and Marios M. Polycarpou

This tradeoff has become significantly more feasible over the past few decades, since controllers are frequently implemented via digital computers which have increased remarkably in memory and computational capabilities over this recent time span.

The chapter starts with a general perspective for motivating the use of adaptive approximation based control. Then we develop a set of intuitive design and analysis tools by considering the stabilization of a simple scalar example with an unknown nonlinearity. Some more advanced tools are then motivated and developed based on the tracking problem for a scalar system with two unknown nonlinearities.

6.1 PERSPECTIVE FOR ADAPTIVE APPROXIMATION BASED CONTROL

The techniques developed in this book are suitable for systems with uncertain nonlinearities. As motivated in Chapter 1, adaptive approximation methods rely on the approximation of the uncertain nonlinearities. In this subsection, we present a general perspective for adaptive approximation based control to help the reader obtain a firmer understanding and better intuition behind the use of this control methodology.

The need to address uncertain nonlinearities in feedback control problems is well known. As illustrated in many simulation and experimental studies, uncertain nonlinearities that have not been accounted for in the feedback control design can cause instability or severe transient and steady state performance degradation. Such uncertain nonlinearities may arise due to several reasons:

- **Modeling errors.** The design of the feedback control system typically depends on a mathematical model which should represent the real system/process. Naturally, there are discrepancies between the dynamic behavior of the real system and the assumed mathematical models. These discrepancies arise due to several factors, but mainly due to difficulties in capturing in a mathematical model the behavior of a real system under different conditions. Thus, modeling errors are an important component of mathematical representations, and accordingly should also be considered in the feedback control design.

- **Modeling simplifications.** In some applications, the derived mathematical model may be too complex to allow for a feedback control design. In other words, the full mathematical model may be quite accurate, but its complexity is to the point where the designer cannot use the full model to derive a suitable feedback control law. Therefore, there is a need to derive a simplified mathematical model that captures the "crucial" dynamics of the real system and, at the same time, it allows the design of a feedback control law. Modeling simplification is usually achieved by reducing the dynamic order of the model, by ignoring certain nonlinear functions, by assuming that certain slowly time-varying parameters are constant, or by ignoring the effect of certain external factors.

As illustrated in Figure 6.1, the modeling procedure typically consists of first creating a possibly complex mathematical model, which attempts to capture all the details of the dynamic system under various operating conditions; this model is later simplified for the purpose of control design. Usually, the advanced (complex) model is used for simulation purposes, for predicting future behavior of the process, as well as for fault monitoring and diagnosis purposes. The simplified model is typically used for the control design and for analytical studies.

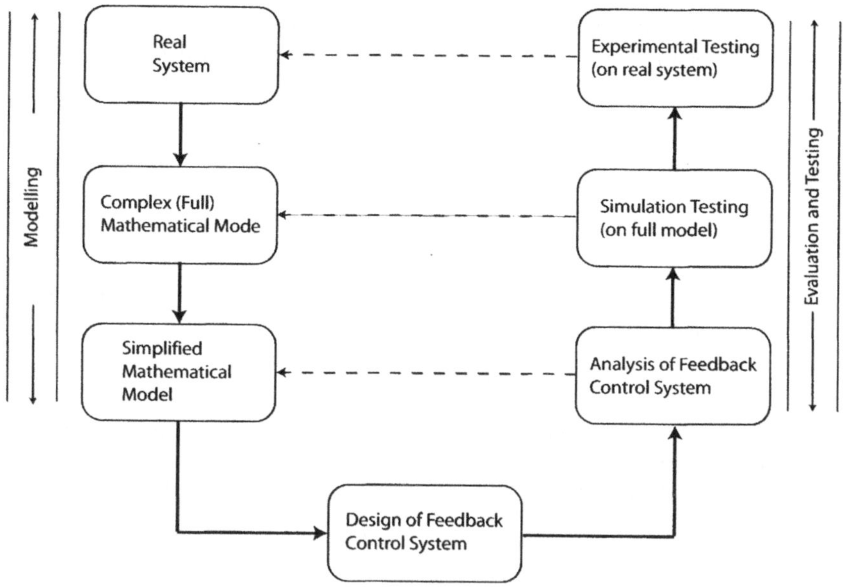

Figure 6.1: Flow chart of the modeling, feedback control design, and evaluation and testing procedure.

In general, the feedback control evaluation procedure consists of the following three steps: (i) stability and convergence analysis; (ii) simulation studies; and (iii) experimental testing and evaluation. As shown in Figure 6.1, typically the stability and convergence analysis is performed on the simplified model. The simulation studies are based on the advanced (complex) model, while the experimental testing studies are based on the real system (or a simplified and possibly less expensive version of the real system).

It is important to note a key special case in the above general methodology for modeling dynamic systems and for designing and testing feedback control algorithms. In many applications, the simplified model used for the control design is a linear model, which is accurate at (and, possibly, near) a nominal operating point in the state space, but possibly inaccurate at operating conditions away from the nominal point. As discussed in other sections of this book, linear models are convenient for feedback control design and analysis due to the plethora of analytic tools that are available for linear systems.

As discussed in Chapter 1 and illustrated again in this chapter, one of the key motivations for using adaptive approximation methods is to estimate the unknown nonlinearities during operation. In view of the above framework for modeling and controlling dynamical systems, the key concept behind adaptive approximation based control is to start with a feedback control design that is based on the simplified model and end up (after adjustment of the adaptable parameters during operation) with a feedback controller suitable for the advanced (complex) model. Another way to view the adaptive approximation based control approach is that of a general parameterized controller, which, depending on the value of some adjustable parameters, is suitable for the nominal simplified model as well as a family of other nonlinear models, including (hopefully) an accurate model of the real system. By adjusting the adaptable parameters, the objective is to fine-tune the feedback controller such that the closed-loop dynamics for the real system follow a desired trajectory.

EXAMPLE 6.1

Consider a system described by

$$\dot{x} = f(x) + G(x)u, \tag{6.1}$$

where $x \in \Re^n$ is the state and $u \in \Re^m$ is the controlled input. The vector field $f : \Re^n \mapsto \Re^n$ is of dimension $n \times 1$ and the matrix $G : \Re^n \mapsto \Re^{n\times m}$ is of dimension $n \times m$. We assume that the nominal model is given by

$$\dot{x}_n = f_o(x_n) + G_o(x_n)u \tag{6.2}$$

where we are using the symbol $x_n \in \Re^n$ to denote the state vector for the nominal model. If the control objective is to achieve stabilization of x to zero, then based on the nominal model we design a nominal feedback control law of the form

$$u = u_o = k_0(x_n) + B_o(x_n)v \tag{6.3}$$

where $k_0(x)$ is of dimension $m \times 1$, $B_o(x)$ is of dimension $m \times m$, and v is an m-dimensional intermediate control variable that can be chosen to achieve the control objective. In the framework of Figure 6.1, the full mathematical model is described by eqn. (6.1), while eqn. (6.2) represents the simplified mathematical model.

Next, let us consider the evaluation and testing of the closed-loop system, which will lead to the motivation for using adaptive approximation based approaches.

Typically, the standard stability analysis is performed on the nominal (simplified) model. If we apply the nominal control law of eqn. (6.3) to the nominal model of eqn. (6.2), we obtain the following closed-loop dynamics

$$\dot{x}_n = f_o(x_n) + G_o(x_n)k_0(x_n) + G_o(x_n)B_o(x_n)v.$$

As described in Chapter 5, feedback linearization approaches rely on the use of a local diffeomorphism $z = T(x_n)$, with $T(0) = 0$, such that in the z-coordinates the closed-loop dynamics are given by

$$\dot{z} = Az + Bv$$

where (A, B) is a controllable pair. Therefore, by selecting $v = K_z z = K_z T(x_n)$, where K_z is a $m \times n$ constant matrix, we obtain the following closed-loop dynamics:

$$\dot{z} = (A + BK_z)z.$$

Since (A, B) is controllable, there does exist K_z such that the closed-loop system is stable with designer-specified pole locations. This results in the following closed-loop system in the x-domain:

$$\dot{x}_n = f_o(x_n) + G_o(x_n)k_0(x_n) + G_o(x_n)B_o(x_n)K_zT(x_n).$$

Note that the above control law was designed to ensure the stability of the nominal model, not the full model of eqn. (6.1).

If the derived nominal control law

$$u = k_0(x) + B_o(x)K_zT(x)$$

is applied to the full model of eqn. (6.1), then the closed-loop dynamics will be different. Specifically, if we let $f^*(x) = f(x) - f_o(x)$ and $G^*(x) = G(x) - G_o(x)$ then we obtain

$$\dot{x} = f_o(x) + G_o(x)k_0(x) + G_o(x)B_o(x)K_zT(x) + \Delta^*(x)$$

where

$$\Delta^*(x) = f^*(x) + G^*(x)k_0(x) + G^*(x)B_o(x)K_zT(x).$$

In the z-domain, the closed-loop system is given by

$$\dot{z} = (A + BK_z)z + \left[\frac{\partial T}{\partial x}\Delta^*(x)\right]\Bigg|_{x=T^{-1}(z)}.$$

The motivation for using adaptive approximation can be viewed as a way to estimate during operation the unknown functions $f^*(x)$ and $G^*(x)$ by $\hat{f}(x;\hat{\theta}_f,\hat{\sigma}_f)$ and $\hat{G}(x;\hat{\theta}_G,\hat{\sigma}_G)$, respectively, and use these approximations to improve the performance of the controlled system. If the initial weights of the adaptive approximators are chosen such that $\hat{f}(x;\hat{\theta}_f(0),\hat{\sigma}_f(0)) = 0$ and $\hat{G}(x;\hat{\theta}_G(0),\hat{\sigma}_G(0)) = 0$ for all x, then at $t = 0$ the control law is the same as the nominal control law u_o. During operation, the objective is for the adaptive approximators $\hat{f}(x;\hat{\theta}_f(t),\hat{\sigma}_f(t))$ and $\hat{G}(x;\hat{\theta}_G(t),\hat{\sigma}_G(t))$ to learn the unknown functions such that they can be used in the feedback control law. The sought enhancement in performance can be in the form of a larger region of attraction (i.e., loosely speaking, a larger region of attraction implies that initial conditions further away from the equilibrium still converge to the equilibrium), faster convergence, or more robustness in the presence of modeling errors and disturbances. △

To illustrate some key concepts in adaptive approximation based control, it is useful to consider the stability properties of the equilibrium of the closed-loop system in terms of the size of the region of attraction. Let us define the following type of stability results [134]. For better understanding, the definitions are provided for a scalar system with a single state $y(t)$. We let the initial condition y(0) be denoted by y_0.

- *Local Stability.* The results hold only for some initial conditions $y_0 \in [-a,\ b]$, where a, b are positive constants, whose magnitude may be arbitrarily small. In addition, the values of a and b are determined by f^* which is unknown; hence a and b are unknown.

- *Regional Stability.* The results hold only for some initial conditions that belong to a known and predetermined range $y_0 \in [-a,\ b]$. Typically, the magnitude of a and b is not "too small."

- *Semi-global Stability.* In this case, the stability results are valid for any initial conditions $y_0 \in [-a,\ a]$, where a is a finite constant that can be arbitrarily large. The value of a is determined by the designer.

- *Global Stability.* The stability results hold for any initial condition $y_0 \in \Re$.

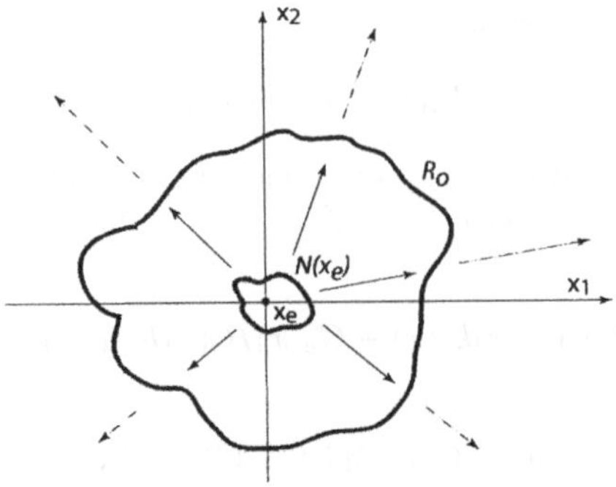

Figure 6.2: Diagram to illustrate local stability, regional stability, and expansion to global stability.

Although the above definitions of stability may be a bit subjective, as we will see, each case corresponds to certain design techniques and assumptions. In general, linear control techniques applied to nonlinear systems result in local stability. Adaptive approximation methods are based on approximating the unknown functions within a compact region of the state space; therefore, typically it results in regional stability. Later in this chapter we will develop an adaptive bounding technique, which if augmented to adaptive approximation based control may yield global stability results.

While the definitions of local stability, semi-global stability, and global stability are well established in the nonlinear systems literature [134], the definition of regional stability is added here to emphasize the ability of adaptive approximation based control to establish closed-loop stability over a larger region as compared to local stability, which is typically associated with linear systems. Moreover, the region of attraction can be expanded by the use of a larger number of basis functions (resulting in more weights), and can be made global by using bounding or adaptive bounding techniques.

Let $x_e \in \Re^2$ be an equilibrium point in a 2-dimensional space. Figure 6.2 shows an example of a local stability region $N(x_e)$ and a regional stability region R_o, in a 2-dimensional space. One perspective for the utilization of adaptive approximation based control is to increase the region of attraction from $N(x_e)$ to R_0. The region of attraction can be further expanded by the use of adaptive bounding techniques.

6.2 STABILIZATION OF A SCALAR SYSTEM

In this section, we consider the problem of controlling simple dynamical systems with unknown nonlinearities. Specifically, we consider scalar systems, described by first-order differential equations. These examples help to illustrate some of the key issues that arise in adaptive approximation based control, without some of the complex mathematics that is required for higher order systems. To facilitate the presentation of certain illustrative figures, this section will focus on regulation as opposed to trajectory tracking.

This section considers in detail the benefits, drawbacks, and provable performance that applies to alternative mechanisms available for addressing unknown nonlinearities. It is intended to provide the reader with an intuitive understanding of the key issues, which have also been discussed in the previous section. The ideas and techniques developed in this section will be expanded to the tracking problem in Section 6.3 and then extended to more realistic higher order systems in Chapter 7.

Consider the scalar system described by

$$\dot{y} = f(y) + u, \qquad y(0) = y_0 \tag{6.4}$$

where $u \in \Re^1$ is the controlled input, $y \in \Re^1$ is the measured output, and $f(y)$ is an unknown nonlinearity. Without loss of generality we assume that $f(0) = 0$. To allow the possibility of incorporating any available information into the control design, we assume that f is made of two components:

$$f(y) = f_o(y) + f^*(y),$$

where $f_o(y)$ is a known function representing the nominal dynamics of the system, and $f^*(y)$ is an unknown function representing the nonlinear uncertainty. The control objective is to design a control law (possibly dynamic) such that $u(t)$ and $y(t)$ remain bounded and $y(t)$ converges to zero (or to a small neighborhood of zero) as $t \to \infty$.

The following subsections will lead the reader through a series of different assumptions and design techniques that will yield different stability results and will provide some intuition about the achievable levels of performance and the trade-offs between different techniques.

6.2.1 Feedback Linearization

First consider the case where the nonlinear function f is known (i.e., assume that $f^*(y) = 0$ for all $y \in \Re$). In this simple case, we saw in Chapter 5 that the control law

$$u = -a_m y - f(y) = -a_m y - f_o(y) \quad \text{with } a_m > 0 \tag{6.5}$$

achieves the desired control objective since the closed-loop dynamics $\dot{y} = -a_m y$ make the equilibrium point $y = 0$ asymptotically stable. In fact, $y(t)$ converges to zero exponentially fast.

Obviously, if the function $f(y)$ is known exactly for all $y \in \Re^1$, the stability results are global. On the other hand, if $f(y)$ is known only for $y \in [-a,\ b]$ then the stability results are regional, assuming that we use the same control law as in eqn. (6.5). Specifically, if the initial condition y_0 belongs in the range $[-a,\ b]$, then $y(t)$ converges to zero exponentially fast; otherwise it may not converge.

The reader will recall from Chapter 5 that this control strategy is known as feedback linearization. It is based on the simple idea that if all the nonlinearities of the system are known and of a certain structure such that they can be "cancelled" by the controlled variable u, then the feedback control law is used to make the closed-loop dynamics linear. Once this is achieved, then standard linear control techniques can be used to achieve the desired transient and steady-state objectives.

As discussed in Chapter 5, the practical use of feedback linearization techniques faces some difficulties. First, not all nonlinear systems are feedback linearizable. Second, in many practical nonlinear systems some of the nonlinearities are "useful" (in the sense that they have a stabilizing effect) and therefore it is not advisable to employ significant control effort for cancelling stabilizing components of the system. Thirdly, and perhaps most importantly, in most practical systems the nonlinearities are not known exactly, or they may change unpredictably during operation. Therefore, in general, it is not possible to achieve perfect cancellation of the nonlinearities, which motivates the use of more advanced control approaches to handle uncertainty. The effect on the closed-loop performance of inaccurate cancellation of the nonlinearities was illustrated in the simple example discussed in Chapter 1.

6.2.2 Small-Signal Linearization

Next, suppose that we linearize the nonlinear system around the equilibrium point $y = 0$, and then employ a linear control law. In this case, the linearized system is given

$$\dot{y}_l = a^* y_l + u \quad \text{where } a^* = \frac{df(0)}{dy} = \left.\frac{df}{dy}\right|_{y=0}.$$

Therefore, the linear control law $u = -(a^* + a_m)y$ results in the closed-loop dynamics

$$\dot{y} = -a_m y + \left(f(y) - \frac{df(0)}{dy} y \right). \tag{6.6}$$

Consider the Lyapunov function $V = \frac{1}{2}y^2$. The time derivative of V along the solutions of eqn. (6.6) is

$$\frac{dV}{dt} = -a_m y^2 + \left(f(y)y - \frac{df(0)}{dy} y^2 \right).$$

Thus, the region of convergence is

$$\mathcal{A} = \left\{ y \;\middle|\; a_m y^2 > \left(f(y)y - \frac{df(0)}{dy} y^2 \right) \right\}.$$

Therefore, the linear control law, applied to the nonlinear system, results in local stability results. Specifically, if the initial condition y_0 is in $\mathcal{A}$, then we have asymptotic convergence to zero. According to the fundamental theorem of stability (see Chapter 5), the size of the region $\mathcal{A}$ can be arbitrarily small, depending on the nature of $f(y)$ relative to that of its linearization. In this case, we cannot quantify a specific range $[-a,\ b]$ without additional assumptions about the nature of f. The designer can increase the size of the set $\mathcal{A}$ by increasing a_m (i.e., high-gain control); however, this is an undesirable approach to increasing the domain of attraction, as the parameter a_m determines the bandwidth of the control system. Increasing a_m to enlarge the theoretical domain of attraction would necessitate faster (more expensive actuators) and might result in excitation of previously unmodeled higher frequency dynamics. The use of high-gain feedback is particularly problematic in the presence of measurement noise. For special classes of nonlinear systems, it may also result in large transient errors, which is known as the *peaking phenomenon* [134, 263].

■ **EXAMPLE 6.2**

Consider the scalar example

$$\dot{y} = ky^2 + u$$

The linearized system is given by $\dot{y}_l = u$, which can be easily controlled by a linear control law of the form $u = -a_m y_l$, $a_m > 0$. If we apply the same linear control law to the original (nonlinear) system, the resulting closed-loop system is given by

$$\dot{y} = -a_m y + ky^2$$

The resulting system is locally asymptotically stable, with the region of attraction $\mathcal{A}$ given by

$$\mathcal{A} = \begin{cases} y \in [-\infty, \frac{a_m}{k}) & \text{if } k > 0, \\ y \in (\frac{a_m}{k}, \infty] & \text{if } k < 0. \end{cases}$$

Independent of the sign of k, the set $\{|y| < |\frac{a_m}{k}|\}$ is in the domain of attraction. However, we notice that, depending on the value of k, the region of attraction can come arbitrarily close to the equilibrium point. This illustrates the local nature of stability for controllers designed using small-signal linearization. △

6.2.3 Unknown Nonlinearity with Known Bounds

Now consider the situation where the function f is unknown due to f^* being unknown. However, we assume that the unknown function f^* belongs to a certain known range as follows:

$$f_L(y) \leq f^*(y) \leq f_U(y)$$

where f_L and f_U are known lower and upper bounds, respectively, of the uncertainty f^*. Since f_o is the assumed nominal function representing f and f^* characterizes the uncertainty around f_o, typically the lower bound f_L will be negative and the upper bound f_U will be positive for all $y \in \Re^1$. However, the design and analysis procedure is valid even if this is not true.

In this case we use the control law

$$u = -a_m y - f_o(y) - v(y) \tag{6.7}$$

$$v(y) = \begin{cases} f_U(y) & \text{if } y > 0 \\ f_L(y) & \text{if } y < 0, \end{cases} \tag{6.8}$$

which yields the closed-loop dynamics

$$\dot{y} = -a_m y + f^* - v(y).$$

In general, the above control law of eqn. (6.7) is discontinuous at $y = 0$ (unless $f_L(0) = f_U(0)$; i.e., there is no uncertainty at $y = 0$). When the control law is discontinuous at $y = 0$, the discontinuity may cause the trajectory $y(t)$ to keep changing signs, causing the control law to switching back and forth, thus creating chattering problems. The chattering can be remedied by using a smooth approximation of the form

$$v(y) = \begin{cases} f_U(y) & \text{if } y > \varepsilon \\ \frac{1}{2\varepsilon}\left[(\varepsilon - y) f_L(-\varepsilon) + (\varepsilon + y) f_U(\varepsilon)\right] & \text{if } |y| \leq \varepsilon \\ f_L(y) & \text{if } y < -\varepsilon. \end{cases} \tag{6.9}$$

This smooth approximation of $v(y)$ is illustrated by an example in Figure 6.3, where both the upper bound $f_U(y)$ and lower bound $f_L(y)$ are also shown.

By using the Lyapunov function $V = \frac{1}{2}y^2$ we see that for y in the region $\mathcal{A}_1 = \{y \mid |y| \geq \varepsilon\}$ the time derivative of V satisfies $\dot{V} \leq -a_m y^2$, which implies that $|y(t)|$ decreases monotonically. For y in the region $\mathcal{A}_2 = \{y \mid |y| < \varepsilon\}$ the time derivative of V

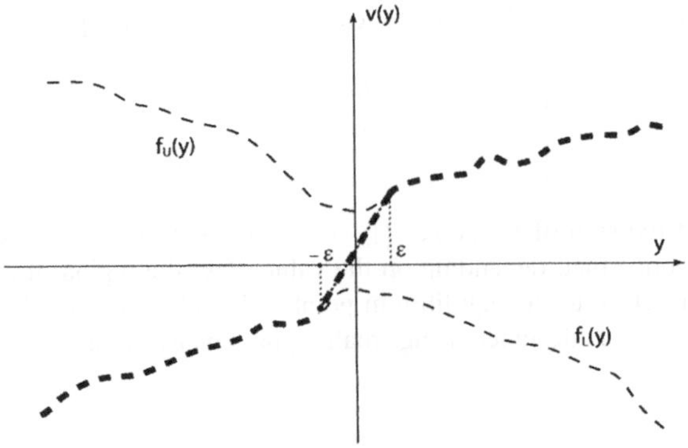

Figure 6.3: Plot illustrating the smooth approximation of $v(y)$. The upper bound $f_U(y)$ is plotted above the y-axis, while the lower bound $f_L(y)$ is plotted below the y-axis. The function $v(y)$ of eqn. (6.9) is plotted as the bold portion of f_L and f_U along with the bold dashed line for $y \in [-\varepsilon, \varepsilon]$.

satisfies

$$\begin{aligned}
\dot{V} &= -a_m y^2 + y f^*(y) - y v(y) \\
&\leq -a_m y^2 + |y|\kappa, \quad \text{where } \kappa = \max_{|y|\leq\varepsilon} |f_U(y) - f_L(y)| \\
&\leq -2a_m V + \varepsilon\kappa.
\end{aligned}$$

Therefore, using Lemma A.3.2, given any $\mu > \frac{\varepsilon\kappa}{2a_m}$ there exists a time T_μ such that for all $t \geq T_\mu$ we have $V(t) \leq \mu$. This implies that asymptotically (as $t \to \infty$), the output $y(t)$ satisfies $|y(t)| \leq \sqrt{\varepsilon\kappa/a_m}$.

Therefore, by combining the stability analysis for both regions $\mathcal{A}_1$ and $\mathcal{A}_2$ we obtain that asymptotically the output $y(t)$ goes within a bound which is the minimum of ε and $\sqrt{\varepsilon\kappa/a_m}$. We notice that as ε becomes smaller, the residual regulation error $y(t)$ also becomes smaller; however, the control switching frequency increases. In the limit, as ε approaches zero, the control law becomes discontinuous and the output $y(t)$ converges to zero asymptotically.

The feedback control law used in this subsection employs the known bounding functions $f_L(y)$, $f_U(y)$ to guarantee that the feedback system is able to handle the worst-case scenario of the unknown nonlinearity. However, this may result in unnecessarily large control efforts (high-gain feedback), and also possibly degraded transient behavior. Although the closed-loop stability, as we saw earlier, can be guaranteed, from a practical perspective there are some other issues that the designer needs to be aware of:

- Large control efforts may be undesirable due to the additional cost.
- In practice, the control input generated by the controller can be implemented only if it is within a certain range. High-gain feedback may cause saturation of the controller, which can degrade the performance or even cause instability.
- In the presence of noise or disturbances, high-gain feedback may perform poorly and can result in instability. The robustness issue is quite critical in practice because measurement noise is inherently present in most feedback systems. Intuitively, we

can see that measurement noise will appear to the controller as small tracking errors, which with a high-gain control scheme can cause large actuation signals that may result in significant tracking errors.

- Typically, the mathematical model on which the control design is based is the result of a reduced-order simplification of the actual plant. For example, the real plant may be of order 20, while the model used for control design may be 3rd order. Such model reduction is achieved by ignoring the so-called fast dynamics. Unfortunately, high-gain feedback may excite these fast dynamics, possibly degrading the performance of the closed-loop system.

The extend to which these are critical problems depends on the specific application and the amount of uncertainty. In some applications the plant is quite susceptible to the problems of high-gain feedback and switching, while in others there is significant margin of tolerance. The magnitude of uncertainty also plays a key role. The level of uncertainty is represented by the difference between $f_L(y)$ and $f_U(y)$. If the difference is "large" then that is an indication that the range in which the uncertainty f^* may vary is large, and thus the control design team would need to be conservative, which results in larger control effort than necessary. On the other hand, if the bounding functions selected do not hold in practice, then stability of the closed-loop system cannot be guaranteed. Two methods to decrease the conservatism are to approximate the nonlinearity f^* and to estimate the bounding functions $f_L(y)$ and $f_U(y)$. These methods are considered in the sequel.

6.2.4 Adaptive Bounding Methods

One approach to try to reduce the amount of uncertainty, and thus have a less conservative control algorithm, is to use adaptive bounding methods [215]. According to this approach, the unknown function f^* is assumed to belong to a *partially* known range as follows:

$$\alpha_l f_l(y) \;\leq\; f^*(y) \;\leq\; \alpha_u f_u(y)$$

where f_l and f_u are known positive lower and upper bounding functions, respectively, while α_l and α_u are unknown constant parameters multiplying the bounding functions. The unknown parameters α_l, α_u can be positive or negative depending on the nature of the bounding functions $f_l(y)$ and $f_u(y)$. The procedure that we will follow is based on estimating online the unknown parameters α_l, α_u and using the estimated parameters in the feedback control law. It is noted that the above condition is similar to the sector nonlinearity condition which has been considered in terms of absolute stability of nonlinear systems [134].

The advantage of this formulation over a fixed bounding method is that it allows the design of control algorithms for the case where the bounds are not known. The function $f_l(y)$ (correspondingly $f_u(y)$) represents the general structure of the uncertainty, however the level of uncertainty is characterized by the unknown parameter α_l (correspondingly α_u). In the absence of any information about the uncertainty, the bounding functions can both be taken to be $f_l(y) = f_u(y) = 1$.

Now, the control law is given by

$$u = -a_m y - f_o(y) - v(y) \tag{6.10}$$

$$v(y) = \begin{cases} \hat{\alpha}_u f_u(y) & \text{if } y > 0 \\ \hat{\alpha}_l f_l(y) & \text{if } y < 0. \end{cases} \tag{6.11}$$

The parameter bounding estimates $\hat{\alpha}_l(t)$ and $\hat{\alpha}_u(t)$ are generated according to the following adaptive laws:

$$\dot{\hat{\alpha}}_u = \begin{cases} \gamma_u y f_u(y) & \text{if } y > 0 \\ 0 & \text{if } y < 0 \end{cases} \tag{6.12}$$

$$\dot{\hat{\alpha}}_l = \begin{cases} 0 & \text{if } y > 0 \\ \gamma_l y f_l(y) & \text{if } y < 0 \end{cases} \tag{6.13}$$

where γ_u, γ_l are positive constants representing the adaptive gain of the update laws for $\hat{\alpha}_u$ and $\hat{\alpha}_l$, respectively.

The stability analysis of this scheme can be derived by considering the Lyapunov function candidate

$$V = \frac{1}{2}y^2 + \frac{1}{2\gamma_l}(\hat{\alpha}_l - \alpha_l)^2 + \frac{1}{2\gamma_u}(\hat{\alpha}_u - \alpha_u)^2.$$

First let us consider the case of $y > 0$. The time derivative of V along the solutions of the differential equations for y, $\hat{\alpha}_u$ and $\hat{\alpha}_l$ is given by

$$\begin{aligned} \dot{V} &= -a_m y^2 + y f^*(y) - y\hat{\alpha}_u f_u(y) + (\hat{\alpha}_u - \alpha_u) y f_u(y) \\ &\leq -a_m y^2 + y\alpha_u f_u(y) - y\hat{\alpha}_u f_u(y) + (\hat{\alpha}_u - \alpha_u) y f_u(y) \\ &= -a_m y^2. \end{aligned}$$

If $y < 0$ then we get similar results:

$$\begin{aligned} \dot{V} &= -a_m y^2 + y f^*(y) - y\hat{\alpha}_l f_l(y) + (\hat{\alpha}_l - \alpha_l) y f_l(y) \\ &\leq -a_m y^2 + y\alpha_l f_l(y) - y\hat{\alpha}_l f_l(y) + (\hat{\alpha}_l - \alpha_l) y f_l(y) \\ &= -a_m y^2. \end{aligned}$$

Since $\dot{V}$ is negative semidefinite, we conclude that $y(t)$, $\hat{\alpha}_u(t)$ and $\hat{\alpha}_l(t)$ are uniformly bounded (e.g., $y(t) \leq V(0)$ for all $t \geq 0$). Furthermore, using the standard Lyapunov analysis procedure, based on Barbălat's Lemma, it can be readily shown that $\lim_{t\to\infty} y(t) = 0$.

Again, as discussed before, the above control law is, in general, discontinuous at $y = 0$, which may cause chattering problems. This problem can again be remedied by using smooth approximations to the discontinuous sign function.

In the special case that $f_u(0) = f_l(0) = 0$, the feedback control becomes continuous. It turns out that this special case is an important one in stabilization tasks: in regulation problems it is often the case that the model is obtained after small-signal linearization around the desired setpoint. Therefore, in such situations, it is reasonable to assume that the uncertainty at the setpoint $y = 0$ is zero, causing $f_u(0) = f_l(0) = 0$. As we will see in the next section, in tracking control applications the switching will be a function of the tracking error.

Note that the parameter estimation differential equations (6.12)–(6.13) are monotonically positive and negative, respectively (with respect to y). Therefore, they may not be robust with respect to measurement noise and disturbances, in the sense that a small disturbance or noise term will cause the parameter estimate to keep on increasing in magnitude. In practical applications, $\hat{\alpha}_u$ and $\hat{\alpha}_u$ may wander towards ∞ and $-\infty$, respectively, which is known as parameter drift. To remedy this situation, the bounding parameter update laws would have to be modified as discussed in Section 4.6.

6.2.5 Approximating the Unknown Nonlinearity

So far the control design has been based on using some known (or partially known) lower and upper bounds on the modeling uncertainty. In the context of "learning" the uncertainty we now use adaptive approximation methods. The idea is to use an adaptive approximation model of the general form $\hat{f}(y;\theta,\sigma)$ to learn the uncertain component $f^*(y)$.

We represent $f^*(y)$ as $f^*(y) = \hat{f}(y;\theta^*,\sigma^*) + \bar{e}_f(y)$ where (θ^*,σ^*) are the optimal weights of the adaptive approximation model and the quantity

$$\bar{e}_f(y) = f^*(y) - \hat{f}(y;\theta^*,\sigma^*)$$

is the minimum functional approximation error (MFAE), which is a function of y. Similar to the way it was defined in Section 3.1.3, the MFAE represents the minimum possible deviation between the unknown function f^* and the adaptive approximator $\hat{f}$ that can be achieved by selection of θ, σ, where the minimum is interpreted with respect the the infinity norm over a compact set $\mathcal{D}$. Specifically, the optimal weights (θ^*,σ^*) are defined as:

$$(\theta^*,\sigma^*) = \arg \min_{(\theta,\sigma)\in\Omega} \left\{ \sup_{y\in\mathcal{D}} |f^*(y) - \hat{f}(y;\theta,\sigma)| \right\},$$

where Ω is a convex set representing the allowable parameter space.

The extent to which the MFAE can be made small over the region $\mathcal{D}$ depends on many factors, including the type of approximation model used, the number of adjustable parameters, as well as the size of the parameter space Ω. For example, the constraint that the optimal weights (θ^*,σ^*) belong to the set Ω may increase the size of the MFAE. However, if the size of the set Ω is large then any increase in MFAE due to the parameter space constrained in Ω will be small. Typically, the function approximation cannot be expected to be global with respect to y. For analysis, it is useful to define

$$e_f(t) = \bar{e}_f(y(t)) = f^*(y(t)) - \hat{f}(y(t),\theta^*,\sigma^*).$$

If we follow the same feedback control structure as in eqn. (6.5) we obtain:

$$u = -a_m y - f_o(y) - \hat{f}(y;\hat{\theta},\hat{\sigma}), \tag{6.14}$$

where $\hat{\theta}$ and $\hat{\sigma}$ represent the adjustable parameters (weights) of the adaptive approximation network. Therefore, now we are seeking to derive adaptive laws for updating the weights of the adaptive approximation network. By substituting the feedback control law of eqn. (6.14) in the plant eqn. (6.4) we obtain

$$\begin{aligned} \dot{y} &= -a_m y + f^*(y) - \hat{f}(y;\hat{\theta},\hat{\sigma}) \\ &= -a_m y + \hat{f}(y;\theta^*,\sigma^*) - \hat{f}(y;\hat{\theta},\hat{\sigma}) + \bar{e}_f(y). \end{aligned} \tag{6.15}$$

First, we consider the case where the adaptive approximation network is linearly parameterized (i.e., $\hat{f}(y;\theta,\sigma) = \phi(y)^\top\theta = \theta^\top\phi(y)$, where ϕ are the basis functions, θ are the adjustable parameters and σ are fixed *a priori*).

To derive an update algorithm for $\hat{\theta}(t)$ and to investigate analytically the stability properties of the feedback system, we consider the following Lyapunov function candidate:

$$V = \frac{1}{2}y^2 + \frac{1}{2}\left(\hat{\theta} - \theta^*\right)^\top \Gamma^{-1}\left(\hat{\theta} - \theta^*\right).$$

By this time the reader can recognize this as a rather standard adaptive scheme with the time derivative of V satisfying

$$\dot{V} = -a_m y^2 + y\bar{e}_f(y) + \left(\hat{\theta} - \theta^*\right)^\top \Gamma^{-1}\left(\dot{\hat{\theta}} - \Gamma\phi(y)y\right). \tag{6.16}$$

Therefore, if we select the adaptive update algorithm for $\hat{\theta}(t)$ as

$$\dot{\hat{\theta}} = \Gamma\phi(y)y, \tag{6.17}$$

then the Lyapunov function derivative satisfies

$$\dot{V} = -a_m y^2 + y\bar{e}_f(y). \tag{6.18}$$

Let us try to understand better the effect of $\bar{e}_f(y)$ on stability. First, assume that the MFAE satisfies $\bar{e}_f(y) = 0$ for all $y \in \mathcal{D} = \{|y| \leq \alpha\}$. In this case, $\dot{V} = -a_m y^2$ for y in the region $|y| \leq \alpha$. If $|y(t)| > \alpha$ for some $t > 0$, then nothing can be said about the stability of the present feedback system, as the present controller does not address $\bar{e}_f(y)$ outside of $\mathcal{D}$. If the initial condition $y(0) = y_0$ satisfies $|y_0| \leq \alpha$ there is no guarantee that $|y(t)| \leq \alpha$ for all $t \geq 0$, since the Lyapunov function V depends on both y and $\hat{\theta}$. Whether or not $y(t)$ remains within $[-\alpha,\ \alpha]$ depends on the initial parameter estimation error $\hat{\theta}(0) - \theta^*$, in addition to the initial condition y_0. Moreover, the closed-loop stability properties depend critically on design variables such as the learning rate matrix Γ and the selected feedback pole location a_m.

This type of situation is commonly found in the use of approximators in feedback systems. In general, adaptive approximation methods provide reasonably accurate approximation of the uncertainty over a certain region of the state space denoted by $\mathcal{D}$, while not providing accurate approximation in the rest of the state space (outside the approximation region). Therefore, it is worthwhile taking a closer look at the parameters that influence stability and performance. We start with a simple example of a scalar parameter estimate and then extend the results to vector parameter estimates.

■ **EXAMPLE 6.3**

Consider a simple scalar example where the modeling uncertainty is approximated by a single basis function $\phi(y)$. In this case, the dynamics of the closed-loop feedback system are described by the second-order system

$$\dot{y} = -a_m y - (\hat{\theta} - \theta^*)\phi(y) \qquad y(0) = y_0 \tag{6.19}$$

$$\dot{\hat{\theta}} = \gamma\phi(y)y \qquad \hat{\theta}(0) = \hat{\theta}_0. \tag{6.20}$$

We are looking for conditions on the initial parameters y_0, $\hat{\theta}_0$ and design parameters γ, a_m, under which $y(t)$ remains within the region $[-\alpha,\ \alpha]$ for all $t \geq 0$, where α is some prespecified bound within which the approximation of the uncertainty is valid. Using standard stability methods it can be readily shown that if $y(t)$ remains within the region $[-\alpha,\ \alpha]$ then it will converge to zero asymptotically.

By using the Lyapunov function $V = \frac{1}{2}y^2 + \frac{1}{2\gamma}(\hat{\theta} - \theta^*)^2$ we see that in order to guarantee that $|y(t)| \leq \alpha$ we need the initial conditions y_0 and $\hat{\theta}_0$ such that

$$\frac{1}{2}y_0^2 + \frac{1}{2\gamma}(\hat{\theta}_0 - \theta^*)^2 \leq \frac{1}{2}\alpha^2. \tag{6.21}$$

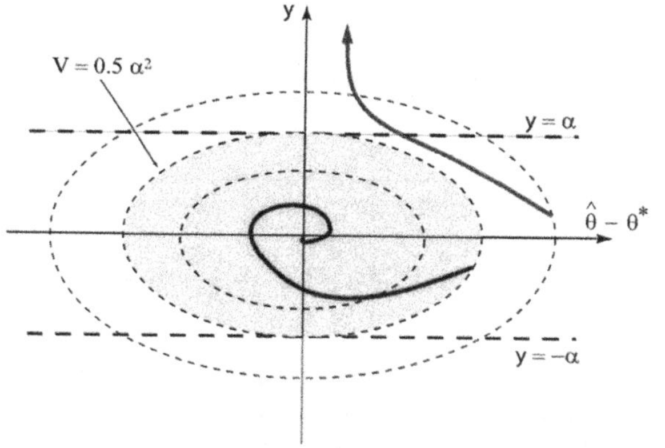

Figure 6.4: Plot of y versus $\hat{\theta} - \theta^*$ to illustrate the derivation of initial conditions (shaded region) which guarantee that the trajectory $y(t)$ remains within $|y(t)| \leq \alpha$ for all $t \geq 0$.

This corresponds to the trajectory being inside the iso-distance curve $V = \frac{1}{2}\alpha^2$. If this condition is not satisfied then it is possible for the trajectory to leave the region $|y(t)| \leq \alpha$. This is illustrated in Figure 6.4, which shows three oval curves $V = k$ for different values of k. The important curve is the largest oval curve that does not cross the line $|y(t)| \leq \alpha$ (in the diagram this is shown as the shaded oval). If the trajectory is within this region then we know from the Lyapunov analysis that $\dot{V} \leq 0$; this, together with Barbălat's Lemma, implies that the trajectories are attracted to the origin. If the initial conditions do not satisfy eqn. (6.21) then, even if $|y_0| \leq \alpha$, it is possible for $y(t)$ to cross the line $|y(t)| = \alpha$, as shown in the diagram. Once $|y(t)| > \alpha$, the $\bar{e}_f(y)$ term could cause divergence.

From eqn. (6.21) we obtain that to guarantee stability the initial parameter estimation error $\hat{\theta}_0 - \theta^*$ needs to satisfy:

$$|\hat{\theta}_0 - \theta^*| \leq \sqrt{\gamma}(\alpha^2 - y_0^2)^{1/2}. \tag{6.22}$$

Therefore, for a given α and initial condition y_0, increasing the value of γ increases the maximum allowable parameter estimation error $\hat{\theta}_0 - \theta^*$. Diagrammatically, from the definition of the Lyapunov function it is also easy to see that as the learning rate γ is made larger then the oval region which guarantees that the trajectory remains within $|y(t)| \leq \alpha$ becomes more wide, thereby allowing larger initial parametric estimation errors $\hat{\theta}_0 - \theta^*$. This is illustrated in Figure 6.5, which shows the attractive region for different values of γ. Intuitively, this can be explained by the fact that larger γ implies faster adaptation, which allows $|\hat{\theta}_0 - \theta^*|$ to be larger and still manage to keep the trajectory within $|y(t)| \leq \alpha$. In the limit, as γ becomes very large then the region of attraction approaches the whole region $\{y \mid |y(t)| \leq \alpha\}$. However, there is a crucial trade-off that the designer needs to keep in mind: in the presence of measurement noise (or some other type of uncertainty) a larger adaptation gain causes greater reaction to small errors which may result in deteriorated performance, or even instability.

As we see from the Lyapunov argument, the design parameter a_m does not effect how large is the region of attraction. However, the selection of a_m does influence the behavior of the trajectories, especially the way $y(t)$ converges to zero.

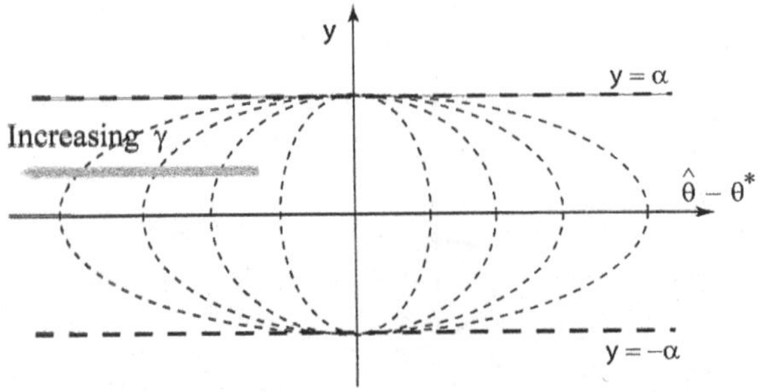

Figure 6.5: Plot of y versus $\hat{\theta} - \theta^*$ to illustrate the effect of the adaptation rate parameter γ on the set of initial conditions which guarantee that the trajectory $y(t)$ remain within $|y(t)| \leq \alpha$ for all $t \geq 0$.

Finally, it should be noted that the above arguments are based on deriving *sufficient* conditions under which it is guaranteed that the trajectory does not leave the approximation region $\{y \mid |y(t)| \leq \alpha\}$. However, the derived conditions are by no means *necessary* conditions. Indeed, it can be readily verified that it is possible for the unknown nonlinearity (modeling uncertainty) to steer the system towards the stability region even if the inequality (6.22) is not satisfied. △

The conditions derived above for the case of a scalar parametric approximator can readily be extended to the more realistic case of a vector parametric estimate, which yields the following inequality for the initial conditions

$$\frac{1}{2}y_0^2 + \frac{1}{2}(\hat{\theta}_0 - \theta^*)^\top \Gamma^{-1}(\hat{\theta}_0 - \theta^*) \leq \frac{1}{2}\alpha^2. \tag{6.23}$$

Using the inequality [99]

$$x^\top \Gamma^{-1} x \leq \lambda_{\max}\left(\Gamma^{-1}\right) \|x\|_2^2 = \frac{1}{\lambda_{\min}\left(\Gamma\right)} \|x\|_2^2$$

we obtain that the initial parameter estimation error needs to satisfy the following inequality to guarantee that $|y(t)| \leq \alpha$ for all $t \geq 0$:

$$\|\hat{\theta}_0 - \theta^*\|_2 \leq \sqrt{\lambda_{\min}\left(\Gamma\right)}\left(\alpha^2 - y_0^2\right)^{1/2}. \tag{6.24}$$

Similar conclusions apply to the learning rate matrix Γ as applied to the scalar parameter γ.

Trajectories Outside the Approximation Region. So far we have considered what happens when the plant output $y(t)$ remains within the approximation region $\mathcal{D} = \{y \mid |y| \leq \alpha\}$ and under what conditions it is guaranteed that $y(t) \in \mathcal{D}$. Next, we investigate what happens if $y(t)$ leaves the region $\mathcal{D}$.

From eqn. (6.18) it is easy to see that if the MFAE, denoted by $\bar{e}_f(y)$, outside the approximation region $\mathcal{D}$ grows faster than a certain rate then the trajectory may become unbounded. For example, if $\bar{e}_f(y) = k_e y$ outside of $\mathcal{D}$, where $k_e > a_m$, then the derivative $\dot{V}$ of the Lyapunov function becomes positive. This implies that at least one (and possibly both) of the two variables $|y(t)|^2$ and $|\tilde{\theta}(t)|^2 = |\hat{\theta}(t) - \theta^*|^2$ grows with time. It is important to note that if $y(t)$ moves further away from the approximation region, naturally the approximation capability of the network may become even worse, possibly leading to further instability problems. The reader may recall that in the case of localized basis function (see Chapter 2) the approximation holds only within a certain region, and beyond this region the approximator gives a zero, or some other constant, value.

To derive some intuition, let us consider the case where $|\bar{e}_f(y)| \leq k_e$ for $|y| > \alpha$ and, as before, it assumed that $|\bar{e}_f(y)| = 0$ for $|y| \leq \alpha$. Therefore, when $|y| > \alpha$, $\dot{V}$ satisfies

$$\dot{V} \leq -a_m y^2 + k_e|y|,$$

which implies that for $\alpha < |y| \geq k_e/a_m$ (for the time being we assume that $\alpha < k_e/a_m$), the Lyapunov derivative satisfies $\dot{V} \leq 0$, while for $|y| < k_e/a_m$, the Lyapunov derivative is indefinite (can be positive or negative). This observation, combined with the assumption that for $|y| \leq \alpha$, the approximation error (MFAE) is zero and thus $\dot{V} \leq 0$, yields the following general description for the behavior of the Lyapunov function derivative:

$$\dot{V} = \begin{cases} \leq 0 & \text{if } |y| \leq \alpha \\ \text{indefinite} & \text{if } \alpha \leq |y| \leq k_e/a_m \\ \leq 0 & \text{if } |y| \geq k_e/a_m. \end{cases} \tag{6.25}$$

Another key observation regarding the stability properties is that during time periods when $\dot{V}$ is indefinite there is nothing to prevent the parameter estimation error $\tilde{\theta}(t)$ from growing indefinitely. Specifically, if $|y| \leq k_e/a_m$ and $\tilde{\theta}(t)$ does not satisfy the inequality (6.24) then it is possible for $|\hat{\theta}| \to \infty$. This type of scenario was encountered earlier in Chapter 4, where it was referred to as *parameter drift*. As discussed in that chapter, parameter drift can be prevented by using so-called robust adaptive laws as described in Section 4.6. If we use the projection modification in this case, we can guarantee that $|\hat{\theta}(t)| \leq \hat{\theta}_m$, where $\hat{\theta}_m$ is the maximum allowed magnitude for the parameter estimate. As we saw earlier in Chapter 4, with the projection modification the closed-loop stability properties are retained if $\hat{\theta}_m$ is large enough such that $|\theta^*| \leq \hat{\theta}_m$.

The stability properties are illustrated in Figure 6.6 for the case where both y and $\hat{\theta}$ are scalar. The dark shaded region $\mathcal{R}_1$ corresponds to the asymptotic stability region that guarantees that $y(t) \to 0$. In other words, if the initial condition $(y_0, \hat{\theta}_0) \in \mathcal{R}_1$ (or if at some time $t = t^*$, we have $(y(t^*), \hat{\theta}(t^*)) \in \mathcal{R}_1$, then it is guaranteed that the trajectory will remain in $\mathcal{R}_1$ and $\lim_{t\to\infty} y(t) = 0$. The property that $(y_0, \hat{\theta}_0) \in \mathcal{R}_1$ implies $(y(t), \hat{\theta}(t)) \in \mathcal{R}_1$ for all $t \geq 0$ makes $\mathcal{R}_1$ a *positively invariant set* [134]. If $(y_0, \hat{\theta}_0) \in \mathcal{R}_2$ or $(y_0, \hat{\theta}_0) \in \mathcal{R}_3$ (medium shaded region), then the Lyapunov function derivative $\dot{V}$ is still negative semidefinite; however, in this case the trajectory may go into $\mathcal{R}_4$ (lightly shaded region), where $\dot{V}$ is indefinite (can be either positive or negative). For example, a trajectory that starts in $\mathcal{R}_2$ may go to $\mathcal{R}_1$, or it may go to $\mathcal{R}_4$. From $\mathcal{R}_4$ (indefinite region) it may go to $\mathcal{R}_3$, it may go back to $\mathcal{R}_2$, or it may even go to $\mathcal{R}_1$. In summary, a trajectory in $\mathcal{R}_1$ will remain there and cause $y(t)$ to converge to zero, while a trajectory in $\mathcal{R}_2 \bigcup \mathcal{R}_3 \bigcup \mathcal{R}_4$ will remain bounded but it may not go to the convergent set $\mathcal{R}_1$.

From the diagram, assuming that $|y(0)| < \alpha$ and $\hat{\theta} < \hat{\theta}_m$, we see that the maximum value that $y(t)$ can take (let us refer to it as y_m; i.e., $|y(t)| \leq y_m$ for all $t \geq 0$), can be obtained by looking at the Lyapunov curve passing through the point $(y, \tilde{\theta}) = (k_e/a_m, \bar{\theta})$, where $\bar{\theta}$ is given by

$$\bar{\theta} = \max\{\hat{\theta}_m - \theta^*, \ \ -\hat{\theta}_m - \theta^*\}.$$

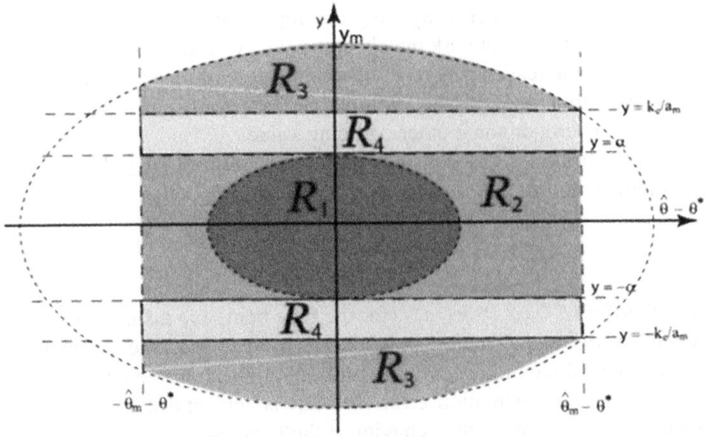

Figure 6.6: Plot of y versus $\hat{\theta} - \theta^*$ to illustrate the stability regions for the case where $\alpha < \frac{k_e}{a_m}$ and the approximation error is zero for $|y| \leq \alpha$ and bounded by k_e for $|y| \geq \alpha$.

This curve is therefore given by

$$V_0 = \frac{1}{2}\left(\frac{k_e}{a_m}\right)^2 + \frac{1}{2\gamma}\bar{\theta}^2.$$

To compute y_m, we find the maximum point that y can take on this curve. Therefore, $V_0 = \frac{1}{2}y_m^2$, which implies that

$$y_m = \left(\left(\frac{k_e}{a_m}\right)^2 + \frac{1}{\gamma}\bar{\theta}^2\right)^{1/2}.$$

In the case of a parameter vector (instead of a scalar) we obtain

$$y_m = \left(\left(\frac{k_e}{a_m}\right)^2 + \bar{\theta}^\top \Gamma^{-1} \bar{\theta}\right)^{1/2}. \tag{6.26}$$

The maximum value that $y(t)$ can take can be thought of as a stability region in the sense that, if $|y_0| \leq y_m$, then it is guaranteed that $|y(t)| \leq y_m$ for all $t \geq 0$. However, other than uniform boundedness, nothing can be concluded about the trajectory, unless it is assumed that $|y_0| \leq \alpha$ and condition (6.24) is satisfied, in which case we can conclude that the trajectory is uniformly stable (in the sense of Lyapunov) and $y(t)$ converges to zero asymptotically.

From eqn. (6.26) we can make some key observations:

- As k_e increases, y_m also increases. Intuitively this should make sense since as the maximum approximation error k_e increases it is expected that the maximum value that $y(t)$ can take also increases.

- As Γ increases, y_m decreases. This implies that increasing the learning rate can decrease the maximum value that $y(t)$ can take. In the limit, as Γ becomes very

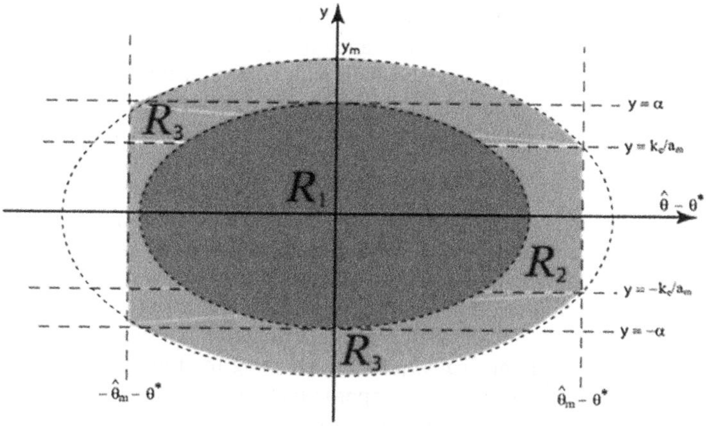

Figure 6.7: Plot of y versus $\hat{\theta} - \theta^*$ to illustrate the stability regions for the case where the approximation error is zero for $|y| \leq \alpha$ and bounded by k_e for $|y| \geq \alpha$, and $\alpha \geq k_e/a_m$.

large, $y_m \rightarrow k_e/a_m$. However, as discussed earlier, increasing the learning rate may create some serious problems in the presence of measurement noise.

- As a_m increases, y_m decreases. This is another method for decreasing y_m. As with the increase of the learning rate, there is a trade-off here because increasing a_m causes a greater control effort, which requires more "energy" and may lead to some of the problems associated with high-gain feedback.

In the above analysis and in the diagram of Figure 6.6 we have assumed for convenience that $\alpha < k_e/a_m$. In the case that $\alpha > k_e/a_m$ the diagram changes to Figure 6.7. In comparing Figures 6.6 and 6.7, we see that the indefinite region $\mathcal{R}_4$ is not present anymore. Therefore, a trajectory in either $\mathcal{R}_2$ or $\mathcal{R}_3$ will end up in $\mathcal{R}_1$, causing $y(t)$ to converge to zero. Clearly, in this case there is a larger region of convergence since initial conditions from the union of the regions $\mathcal{R}_1$, $\mathcal{R}_2$, and $\mathcal{R}_3$ result in trajectories convergent to the origin. It is also worth noting that as the approximation region $\mathcal{D}$ becomes sufficiently large such that

$$\alpha \geq \left(\left(\frac{k_e}{a_m} \right)^2 + \bar{\theta}^\top \Gamma^{-1} \bar{\theta} \right)^{1/2},$$

then it is guaranteed that for any initial feasible initial condition satisfying $\{|y_0| \leq \alpha;\ |\hat{\theta}_0| \leq \hat{\theta}_m\}$ the trajectory remains in the region $\mathcal{R}_1$ and $y(t)$ converges to zero. This case, of course, corresponds to the inequality (6.24) being valid by assumption. This situation may arise if $\mathcal{D}$ is very large (e.g., a large number of basis functions are used) or if there is sufficient prior information on the uncertainty such that the maximum value for $\tilde{\theta}$ is small.

Appraising Remark. At this point it is useful to pause and summarize what this detailed example has discussed so far. Section 6.2.1 showed that feedback linearization achieved exponential stability within the region for which the model error was zero. Outside that region, nothing general could be said. Section 6.2.3 considered the case where bounds were known for the unknown dynamics. In that case we were able to derive a control law of the form of eqns. (6.7)–(6.8), which utilized the known upper and lower bounds on the uncertainty. Asymptotic convergence to a region of uniform boundedness was shown, but required a control signal that may be high gain with high-frequency switching. To decrease the conservatism due to the use of prior bounds, Section 6.2.4 consider the case where the known bounds on the uncertainty were multiplied by unknown coefficients. These unknown coefficients were estimated online to derive the adaptive bounding control scheme described by eqns. (6.10)–(6.13). Section 6.2.5 considered an alternative approach that attempts to approximate the unknown nonlinearities and cancel their effects in the sense of feedback linearization, thus avoiding the high-gain, high-frequency switching required for (adaptive) bounding methods. However, as we saw adaptive approximation methods are, in general, valid only in a finite region, which depends on both the state and parameter error. If the trajectory leaves the so-called "approximation region" $\mathcal{D}$, then the approximation accuracy may deteriorate dramatically, possibly allowing the trajectory into an unstable region. Even in the mild case where the approximation error is bounded by a constant (outside the approximation region), we saw that once the vector of state and parameter errors leaves the region $\mathcal{R}_1$, the trajectory may never return back. Therefore, we need methods to cause the trajectory to return back to the approximation region $\mathcal{R}_1$. This can be achieved by combining the adaptive approximation techniques with the bounding methods.

6.2.6 Combining Approximation with Bounding Methods

We consider the adaptive approximation based control law of eqn. (6.14) augmented by an additional term $v_0(y)$, which will be used to address the presence of the approximation error (formally defined as minimum functional approximation error (MFAE)). First, we assume that the MFAE $\bar{e}_f(y)$ satisfies

$$\bar{e}_L(y) \leq \bar{e}_f(y) \leq \bar{e}_U(y), \quad \forall y \in \Re^1 \tag{6.27}$$

where $\bar{e}_L(y)$ and $\bar{e}_U(y)$ are known lower and upper bounds, respectively, on the MFAE. Due to the use of adaptive approximation, it is reasonable to define $\bar{e}_L(y)$ and $\bar{e}_U(y)$ very small (even zero) for $y \in \mathcal{D}$, and larger for y outside of $\mathcal{D}$.

The overall feedback control law is given by

$$\begin{aligned} u &= -a_m y - f_o(y) - \phi(y)^\top \hat{\theta} - v_0(y) \\ v_0(y) &= \begin{cases} \bar{e}_U(y) & \text{if} \quad y > 0 \\ \bar{e}_L(y) & \text{if} \quad y < 0 \end{cases} \\ \dot{\hat{\theta}} &= \Gamma \phi(y) y. \end{aligned} \tag{6.28}$$

The feedback control law described by eqn. (6.28) is of the same form as the bounding control law of eqn. (6.7) with the key difference that the adaptive approximation scheme is used to handle the major part of the uncertainty $f^*(y)$ for $y \in \mathcal{D}$. The bounding term $v_0(y)$ ensures that all trajectories return to and stay within $\mathcal{D}$ (i.e., $\mathcal{D}$ is positively invariant).

Within $\mathcal{D}$, the bounding term $v_0(y)$ is used only for handling the residual approximation error $\bar{e}_f(y)$ which is small (or zero). Previously we had assumed that the approximation error $\bar{e}_f(y)$ was zero for $y \in \mathcal{D}$. This assumption can easily be incorporated into the control law of eqn. (6.28) by having both $\bar{e}_U(y)$ and $\bar{e}_L(y)$ be zero for $y \in \mathcal{D}$. This will cause the control component $v_0(y)$ to be activated only if $y(t)$ leaves the region $y \in \mathcal{D}$. However, the above scheme is more general in allowing the MFAE to be nonzero even within the approximation region $y \in \mathcal{D}$ (as long as we have upper/lower bounds for it).

Global Stability Proof. With the combined adaptive approximation and bounding control scheme we can now obtain global stability results.

Lemma 6.2.1 *The closed-loop system described by the scalar plant (6.4) and the control law (6.28) guarantees that for any initial condition* $(y_0, \hat{\theta}_0)$, *the trajectories* $y(t)$ *and* $\hat{\theta}(t)$ *are uniformly bounded and* $\lim_{t\to\infty} y(t) = 0$.

Proof: Consider the Lyapunov function candidate

$$V = \frac{1}{2}y^2 + \frac{1}{2}\left(\hat{\theta} - \theta^*\right)^\top \Gamma^{-1}\left(\hat{\theta} - \theta^*\right).$$

By using (6.28), the time derivative of V satisfies

$$\begin{aligned}\dot{V} &= -a_m y^2 + y\bar{e}_f(y) - yv_0(y) + \left(\hat{\theta} - \theta^*\right)^\top \Gamma^{-1}\left(\dot{\hat{\theta}} - \Gamma\phi(y)y\right) \\ &= -a_m y^2 + y\bar{e}_f(y) - yv_0(y).\end{aligned}$$

By using the inequality (6.27) it can be easily shown that $y\bar{e}_f(y) - yv_0(y) \le 0$. Therefore, we conclude that

$$\dot{V} \le -a_m y^2,$$

which implies that $y(t), \hat{\theta}(t) \in \mathcal{L}_\infty$, and the equilibrium $(y, \hat{\theta}) = (0, \theta^*)$ is uniformly stable in the sense of Lyapunov. Furthermore, using Barbălat's Lemma it can be shown that $\lim_{t\to\infty} y(t) = 0$ (see Example A.7 on p. 389 in Appendix A). ■

The control law component $v_0(y)$ in eqn. (6.28) is possibly discontinuous with respect to y. As discussed earlier discontinuous control laws may cause chattering problems, which are characterized by the $y(t)$ trajectory going back and forth across the line $y = 0$ at a fast rate.

In the special case that $\bar{e}_U(0) = \bar{e}_L(0)$ then the control component $v_0(y)$ is continuous at $y = 0$ and therefore the issue does not arise. From a practical perspective, with adaptive approximation, the assumption that $\bar{e}_U(0) = \bar{e}_L(0) = 0$ is quite reasonable for the following reason: even if $f^*(y)$ is unknown at $y = 0$, for $\bar{e}_U(0) = \bar{e}_L(0) = 0$ to be valid all that is required is that there exists a (not necessarily known) parameter vector $\bar{\theta}^*$ such that $f^*(0) = \phi(0)^\top \bar{\theta}^*$. In general, this is an easy condition to satisfy.

If the condition $\bar{e}_U(0) = \bar{e}_L(0)$ is not satisfied the designer has the option of modifying the control component $v_0(y)$ to be continuous at $y = 0$ using the same dead-zone smoothing techniques as described earlier. One way to make $v_0(y)$ continuous is a modification of the form

$$v_0(y) = \begin{cases} \bar{e}_U(y) & \text{if } \; y > \varepsilon \\ \frac{1}{2\varepsilon}\left[(\varepsilon - y)\bar{e}_L(-\varepsilon) + (\varepsilon + y)\bar{e}_U(\varepsilon)\right] & \text{if } \; |y| \le \varepsilon \\ \bar{e}_L(y) & \text{if } \; y < -\varepsilon \end{cases}$$

where $\varepsilon > 0$ is a small design constant. This modification will introduce a positive constant term of the form $\kappa\varepsilon$ in the derivative of the Lyapunov function $\dot{V}$. Even though this term is small in magnitude (since κ is proportional to the approximation error), unless addressed, it may cause problems in the stability analysis because in this case we can no longer guarantee that the parameter estimate vector $\hat{\theta}(t)$ remains bounded while $|y| < \varepsilon$. This can again be remedied by using the robust parameter estimation methods of Section 4.6. In particular, a dead-zone that stops parameter adaptation for $|y| < \varepsilon$ would eliminate the issue.

6.2.7 Combining Approximation with Adaptive Bounding Methods

Finally, in the case where the bounds of eqn. (6.27) are not known, we can combine the adaptive bounding techniques of Section 6.2.4 with the adaptive approximation techniques of Section 6.2.5. Assume that the MFAE function $\bar{e}_f(y)$ (associated with the unknown function f^*) belongs to a partially known range described by

$$\alpha_l \bar{e}_l(y) \leq \bar{e}_f(y) \leq \alpha_u \bar{e}_u(y),$$

where $\bar{e}_l(y)$ and $\bar{e}_u(y)$ are known positive lower and upper bounding functions, respectively, while α_l and α_u are unknown parameters multiplying the bounding functions. The unknown bounding parameters α_l, α_u will be estimated online by a standard parameter estimation method, which will generate the parameter estimates $\hat{\alpha}_l$ and $\hat{\alpha}_u$, respectively.

The overall feedback control scheme in this case is given by

$$u = -a_m y - f_o(y) - \phi(y)^\top \hat{\theta} - v_0(y) \tag{6.29}$$

$$v_0(y) = \begin{cases} \hat{\alpha}_u \bar{e}_u(y) & \text{if } y > 0 \\ \hat{\alpha}_l \bar{e}_l(y) & \text{if } y < 0 \end{cases} \tag{6.30}$$

$$\dot{\hat{\theta}} = \Gamma \phi(y) y \tag{6.31}$$

$$\dot{\hat{\alpha}}_u = \begin{cases} \gamma_u y \bar{e}_u(y) & \text{if } y > 0 \\ 0 & \text{if } y < 0 \end{cases} \tag{6.32}$$

$$\dot{\hat{\alpha}}_l = \begin{cases} 0 & \text{if } y > 0 \\ \gamma_l y \bar{e}_l(y) & \text{if } y < 0. \end{cases} \tag{6.33}$$

The stability properties of this control scheme are similar to those of Lemma 6.2.1, but with robustness (and less conservatism) to the size of the model error outside of $\mathcal{D}$. The details of the stability proof are left as an exercise for the reader (see Exercise 6.3).

Note that in the case where perfect approximation is possible for $y \in \mathcal{D}$, then $\bar{e}_l(y)$ and $\bar{e}_u(y)$ are zero for $y \in \mathcal{D}$. In this case, chattering near $y = 0$ does not occur.

6.2.8 Summary

At this point, at least for the simple example, we should be quite content. For the stabilization problem, we have developed a control law that has global stability properties and high fidelity control within the region $\mathcal{D}$.

In terms of the original simple problem of example 6.3, the region diagram would look similar to Figure 6.7, but without the specific assumptions about the form of the unmodeled nonlinearity. If a trajectory started outside of $\mathcal{D}$, the $v_0(y)$ term would force the trajectory to the boundary of $\mathcal{D}$. Trajectories starting in $\mathcal{D}$ with sufficiently small parameter error, call this region $\mathcal{R}_1$, would stay within $\mathcal{D}$. Trajectories starting within $\mathcal{D}$, but with too large parameter error, call this region $\mathcal{R}_2$, would either converge directly to $\mathcal{R}_1$ or reach the boundary of $\mathcal{D}$. Trajectories at the boundary of $\mathcal{D}$ are not allowed to leave $\mathcal{D}$ due to the $v_0(y)$ term and eventually enter $\mathcal{R}_1$ due to the function approximation on $\mathcal{D}$ and the negative definiteness of the Lyapunov function.

To simplify the presentation and to allow a very clear statement of issues with minimal complicating factors, this section used two major simplifying assumptions. First, we assumed that the control multiplier $g(y) = 1$. Second, we considered stabilization (regulation) instead of tracking problems. The following section considers tracking control for the more general scalar system $\dot{y} = f(y) + g(y)u$.

6.3 ADAPTIVE APPROXIMATION BASED TRACKING

In this section, we consider the more general scalar system

$$\dot{y} = f(y) + g(y)u \tag{6.34}$$

where $f(y)$ and $g(y)$ are unknown nonlinear functions. The tracking control objective is to design a control law that generates u such that $u(t)$ and $y(t)$ remain bounded and $y(t)$ tracks a desired function $y_d(t)$. The control design approach assumes knowledge and boundedness of y_d and all necessary derivatives. This assumption can always be achieved through prefiltering, as discussed in Section A.4. In addition to solving the tracking control problem, the objective of this section is to highlight the issues that differ between adaptive approximation based stabilization and tracking. In the next subsections, we consider different approaches for the design of feedback control algorithms for tracking, depending on our partial knowledge (if any) of the nonlinear functions f and g.

6.3.1 Feedback Linearization

We start by first considering the case where both f and g are completely known. In this case, it is straightforward to see that the control law

$$u = \frac{1}{g(y)}\left[-a_m(y - y_d) + \dot{y}_d - f(y))\right] \tag{6.35}$$

where $a_m > 0$ is a design constant, achieves the control objective for $g(y) \neq 0$. Specifically, with the above feedback control algorithm, the tracking error $e(t) = y(t) - y_d(t)$ satisfies $\dot{e} = -a_m e$. Hence, the tracking error converges to zero exponentially fast from any initial condition (global stability results). The reader will recall from the recent comments in Section 6.2.1 that the standard feedback linearizing control procedure relies on exact cancellation of all the nonlinearities. In the presence of uncertainties, exact cancellation is not possible.

In the system considered in this section, due to $g(y)$, implementation of the feedback control algorithm (6.35) is feasible only if the function $g(y) \neq 0$ for all $y \in \Re$. In practice, $g(y)$ should not only be away from the point $y = 0$, but also should be away from a neighborhood of $y = 0$; otherwise, if $g(y)$ approaches zero then the control effort becomes large, causing saturation of the control input and possibly leading to instability. As discussed in Chapter 5, this is known as the *stabilizability* or *controllability* problem. While in the case of no uncertainty it is reasonable to assume that exact cancellation of $g(y)$ is feasible, the issue becomes more difficult in the presence of uncertainty. As we will see later in this section, it will be required that the adaptive approximator of $g(y)$, denoted as $\hat{g}(y(t); \theta_g(t), \sigma_g(t))$, remains away from zero for all $t \geq 0$. In other words, it is required that the adaptive approximator that is used as an estimator of g remains away from zero while it adapts its weights.

6.3.2 Tracking via Small-Signal Linearization

Standard techniques in linear control systems are based on linearizing the nonlinear system (6.34) around some equilibrium point or around a reference trajectory. If the nonlinear system is linearized around $y = 0$ then the linearized system is described by

$$\dot{y}_l = a^* y_l + b^* u_l \tag{6.36}$$

where y_l is the state of the linearized model, the control signal for the linearized model is

$$u_l = u + \frac{f(0)}{g(0)}$$

and the parameters a^* and b^* are given by

$$a^* = \frac{df}{dy}(y)\Big|_{y=0} = \frac{df(0)}{dy}\,; \quad \text{and,} \qquad b^* = g(0).$$

If we select the linear control law

$$u_l = -\left(\frac{a^* + a_m}{b^*}\right)(y_l - y_d) + \frac{1}{b^*}\dot{y}_d - \frac{a^*}{b^*}y_d$$

and apply it to the linear model of eqn. (6.36), it can be readily shown that it results in

$$\frac{d}{dt}\left[y_l(t) - y_d(t)\right] = -a_m\left(y_l(t) - y_d(t)\right).$$

Hence the linearizing control law is designed to make the tracking error for the linear model converge to zero exponentially fast.

Now, by considering the control offset $u = u_l - \frac{f(0)}{g(0)}$ and applying the linearizing control law to the original nonlinear system, it becomes

$$u = -\frac{a^* + a_m}{b^*}(y - y_d) + \frac{1}{b^*}\dot{y}_d - \frac{a^*}{b^*}y_d - \frac{f(0)}{g(0)}.$$

With the above feedback control law, the closed-loop dynamics for the tracking error $e = y - y_d$ is given by

$$\dot{e} = -a_m e + \left(f(y) - f(0) - \frac{df(0)}{dy}y\right) + \left(g(y) - g(0)\right)u.$$

This closed-loop system (for $y_d = 0$) can be shown to be locally asymptotically stable (in fact, it is locally exponentially stable). However, this theoretical result is not very satisfying, as it holds in an neighborhood of the $y = 0$ that may be arbitrarily small. This "small neighborhood" limitation is at odds with the tracking objective, which requires that $y(t)$ follows $y_d(t)$.

The tracking objective may be more suitably addressed by a control that incorporates linearization about the desired trajectory $y_d(t)$. In this case the linearizing feedback control is given by

$$u = -\frac{1}{g(y_d)}\left(\frac{df}{dy}(y)\Big|_{y=y_d} + a_m\right)(y - y_d) + \frac{1}{g(y_d)}\left(\dot{y}_d - f(y_d)\right). \tag{6.37}$$

Although the above feedback control law may appear rather complex, once $y_d(t)$ is replaced by its corresponding function of time, it becomes a linear time-varying control law of the form

$$u = -k_1(t)e + k_2(t).$$

Similar to the earlier derivation for linearization around the fixed point $y = 0$, for the time-varying tracking function y_d, the resulting closed-loop tracking error dynamics are given by

$$\dot{e} = -a_m e + \left(f(y) - f(y_d) - \frac{df}{dy}(y)\bigg|_{y=y_d} (y - y_d) \right) + (g(y) - g(y_d))\, u. \tag{6.38}$$

Again, the stability analysis is only local, but now it is local in a neighborhood of $e = 0$. The following example illustrates some of the concepts developed in this subsection.

EXAMPLE 6.4

Consider the scalar system

$$\dot{y} = 2y - \frac{1}{4}y^4 + (2 + y)u$$

which is controllable for $y \neq -2$. The objective is to linearize the system and design a linear control law for forcing the system to track the desired trajectory $y_d = \frac{1}{2}\sin t$. Let us first linearize around the fixed point $y = 0$. In this case, the linearized system is given by

$$\dot{y}_l = 2y_l + 2u_l.$$

Let a_m be chosen as $a_m = 1$. The linear control law obtained based on the derived linear system is

$$u = -\frac{3}{2}(y - y_d) + \frac{1}{2}\dot{y}_d - y_d$$

resulting in the closed-loop error dynamics

$$\dot{e} = -e - \frac{1}{4}y^4 + yu. \tag{6.39}$$

Next, let us consider the linearization around the desired trajectory $y_d = \frac{1}{2}\sin t$. The linearized model is given by

$$\begin{aligned} \dot{e}_l &= (2 - y_d^3)e_l + (2 + y_d)u_l \\ &= a^*(t)e_l + b^*(t)u_l. \end{aligned}$$

Following the linearizing feedback control described by of eqn. (6.37), the resulting control law is given by

$$u = -\frac{3 - y_d^3}{2 + y_d}(y - y_d) + \frac{1}{2 + y_d}\left(\dot{y}_d - 2y_d + \frac{1}{4}y_d^4\right).$$

In this case the closed-loop dynamics are given by

$$\dot{e} = -e - \frac{1}{4}\left(y^4 - y_d^4\right) + y_d^3(y - y_d) + (y - y_d)u. \tag{6.40}$$

It is noted that if $y_d = 0$ then the tracking error dynamics of eqn. (6.40) becomes of eqn. (6.39). We also note that $e = 0$ is an equilibrium of eqn. (6.40); this implies that if $y(0) = y_d(0)$ then $y(t) = y_d(t)$ for all $t \geq 0$. △

6.3.3 Unknown Nonlinearities with Known Bounds

Here, we assume that

$$\begin{aligned} f(y) &= f_o(y) + f^*(y) \\ g(y) &= g_o(y) + g^*(y) \end{aligned}$$

where f_o and g_o are known functions, representing the nominal dynamics of the system, while f^* and g^* are unknown functions representing the nonlinear uncertainty. It is assumed that the unknown functions f^* and g^* are within certain known bounds as follows:

$$\begin{aligned} f_L(y) \le \; & f^*(y) \; \le f_U(y) \\ g_L(y) \le \; & g^*(y) \; \le g_U(y) \end{aligned}$$

where f_L, g_L are lower bounds and f_U, g_U are upper bounds on the corresponding uncertain functions. To avoid any stabilizability problems, we assume that $g(y) > 0$ for all y, which implies that the lower bound should satisfy $g_L(y) > -g_o(y)$. A similar framework can be developed if $g(y) < 0$ for all y.

The control law is chosen as follows:

$$u = \frac{1}{g_o(y) + v_g(y, e, u)} \left[-a_m e + \dot{y}_d - f_o(y) - v_f(y, e)\right] \quad (6.41)$$

$$v_f(y, e) = \begin{cases} f_U(y) & \text{if } e \ge 0 \\ f_L(y) & \text{if } e < 0 \end{cases} \quad (6.42)$$

$$v_g(y, e, u) = \begin{cases} g_U(y) & \text{if } eu \ge 0 \\ g_L(y) & \text{if } eu < 0. \end{cases} \quad (6.43)$$

It may not be obvious at first sight, however, in the above feedback control definition for u there exists the possibility of an algebraic loop singularity. This is due to the fact that the right-hand side of eqn. (6.41) depends on u as a result of the switching present in eqn. (6.43) that depends on the sign of u. This algebraic loop singularity will be eliminated later by slightly modifying the definition of $v_g(y, e, u)$.

Next, we proceed to derive the stability properties of the above feedback control scheme. By substituting the control law of eqn. (6.41) into the original system of eqn. (6.34), the tracking error dynamics satisfy

$$\begin{aligned} \dot{e} &= \dot{y} - \dot{y}_d \\ &= f_o(y) + f^*(y) + (g^*(y) - v_g(y, e, u))\, u - \dot{y}_d \\ &\quad - a_m e + \dot{y}_d - f_o(y) - v_f(y, e) \\ &= -a_m e + (f^*(y) - v_f(y, e)) + (g^*(y) - v_g(y, e, u))\, u. \end{aligned} \quad (6.44)$$

Now, let us analyze the closed-loop stability properties by using the quadratic Lyapunov function $V = \frac{1}{2}e^2$. The derivative of V along the solution of eqn. (6.44) is given by

$$\dot{V} = -a_m e^2 + e\,(f^*(y) - v_f(y, e)) + eu\,(g^*(y) - v_g(y, e, u)).$$

Based on the definition of $v_f(y, e)$ and $v_g(y, e, u)$, as given in eqns. (6.42) and (6.43), respectively, it can be readily shown that

$$\begin{aligned} e\,(f^*(y) - v_f(y, e)) &\le 0 \\ eu\,(g^*(y) - v_g(y, e, u)) &\le 0, \end{aligned}$$

which implies that $\dot{V} \leq a_m e^2 = 2a_m V$. Therefore, the tracking error $e(t) = y(t) - y_d(t)$ converges to zero exponentially fast. If the assumed bounds on the uncertainty are global, then the stability results will also be global.

The algebraic singularity introduced by the definition of $v_g(y, e, u)$ can be eliminated as follows. The control law of eqn. (6.41) can be rewritten as

$$u = \frac{u_a}{g_o(y) + v_g(y, e, u_a)},$$

where the intermediate control variable u_a is given by

$$u_a = -a_m e + \dot{y}_d - f_o(y) - v_f(y, e).$$

Since $g_o(y) + v_g$ is assumed to be positive for all y (for stabilizability purposes), the sign of u is the same as the sign of u_a. Therefore the definition of $v_g(y, e, u)$ can be modified as follows without losing any of the stability properties, and at the same time eliminating the algebraic singularity:

$$v_g(y, e, u_a) = \begin{cases} g_U(y) & \text{if } \; e u_a \geq 0 \\ g_L(y) & \text{if } \; e u_a < 0. \end{cases} \tag{6.45}$$

The above feedback control law is, in general, discontinuous at $e = 0$ and at $u_a = 0$. This may cause chattering at the switching surfaces. As discussed earlier, this problem can be remedied by using a smooth approximation of the form described by eqn. (6.9), as shown diagrammatically in Figure 6.3. As before, the main idea is to create a smooth transition of v_f and v_g at the switching surfaces $e = 0$ and $u_a = 0$. In this case, the design and stability derivation is a bit more tricky because the bounds f_L, f_U, g_L, and g_U are functions of y while the switching is a function of the tracking error $e(t)$ and the signal u_a.

■ EXAMPLE 6.5

Consider the scalar system model

$$\dot{y} \;=\; f^*(y) + (g_o + g^*(y))u$$

where the only available *a priori* information is $g_o = 2$ and

$$-y^2 \;\leq f^*(y) \leq\; y^2$$
$$-1 \;\leq g^*(y) \leq\; \left(|y| + \frac{1}{2}\right)^2$$

for all $y \in \Re^1$. The feedback control specification is to track reference inputs $y_c(t)$ with bandwidth up to 2 $\frac{rad}{s}$ and reject initial condition errors with a time constant of approximately 0.1 s.

For the tracking control design process we require a reference trajectory $y_d(t)$ and its derivative $\dot{y}_d(t)$. If the derivative of y_c is not available, then as discussed in Section A.4, we can design a prefilter with $y_c(t)$ as its input and $[y_d(t), \dot{y}_d(t)]$ as outputs. To ensure that the error between $y_d(t)$ and $y_c(t)$ is small, the prefilter should be stable, with unity gain at low frequencies, and with bandwidth of at least 2 $\frac{rad}{s}$. Such a prefilter, as discussed in Example A.9 of Section A.4, is given by

$$\begin{bmatrix} \dot{x}_1 \\ \dot{x}_2 \end{bmatrix} = \begin{bmatrix} 0 & 1 \\ -4 & -2.8 \end{bmatrix} \begin{bmatrix} x_1 \\ x_2 \end{bmatrix} + \begin{bmatrix} 0 \\ 4 \end{bmatrix} y_c \tag{6.46}$$

$$\begin{bmatrix} y_d \\ \dot{y}_d \end{bmatrix} = \begin{bmatrix} 1 & 0 \\ 0 & 1 \end{bmatrix} \begin{bmatrix} x_1 \\ x_2 \end{bmatrix} \tag{6.47}$$

which provides continuous and bounded signals $(y_d, \dot{y}_d)$ for any bounded input signal y_c.

Following the design procedure of this subsection, we define

$$\begin{aligned}
e &= y - y_d \\
v_f(y,e) &= \begin{cases} y^2 & \text{if } e \geq 0 \\ -y^2 & \text{if } e < 0 \end{cases} \\
u_a &= -10e + \dot{y}_d - v_f(y,e) \\
v_g(y,e,u_a) &= \begin{cases} \left(|y| + \frac{1}{2}\right)^2 & \text{if } eu_a \geq 0 \\ -1 & \text{if } eu_a < 0 \end{cases} \\
u &= \frac{u_a}{2 + v_g(y,e,u_a)}.
\end{aligned}$$

By the analysis of this section, this controller achieves global asymptotic tracking of y_d by y. In an ideal continuous-time implementation, the trajectory would reach the discontinuity surfaces $e = 0$ and $eu_a = 0$ and remain at some equilibrium state in order to retain the tracking error at zero. One approach for achieving this equilibrium state at the discontinuity surface is Filippov's method [81, 213]. In the presence of noise or for a discrete-time implementation with a finite (non-zero) sampling period, the control signal u would be discontinuous. The magnitude of the switching would be especially large when y_c is not near the origin.

The discontinuity of the switching due to noise could be addressed by modifying the signals v_f and v_g as follows:

$$\begin{aligned}
v_f(y,e) &= \begin{cases} y^2 & \text{if } e > \varepsilon \\ \frac{1}{2\varepsilon}\left(y^2(e+\varepsilon) - y^2(\varepsilon - e)\right) & \text{if } |e| \leq \varepsilon \\ -y^2 & \text{if } e < -\varepsilon \end{cases} \\
v_g(y,e,u_a) &= \begin{cases} \left(|y| + \frac{1}{2}\right)^2 & \text{if } eu_a > \varepsilon \\ \frac{1}{2\varepsilon}\left(\left(|y| + \frac{1}{2}\right)^2 (eu_a + \varepsilon) - 1.0(\varepsilon - eu_a)\right) & \text{if } |eu_a| \leq \varepsilon \\ -1.0 & \text{if } eu_a < -\varepsilon. \end{cases}
\end{aligned}$$

The smoothing parameter $\varepsilon > 0$ should be selected at least as large as the magnitude of the measurement noise so that noise cannot cause switching in v_f. The drawback to increasing the size of ε is that convergence of e is only guaranteed to a radius of the neighborhood of $e = 0$ which is proportional to ε. This modification does not alter the fact that switching outside this neighborhood, in part due to non-zero sampling time in discrete-time implementations, may occur and may be of large magnitude. A simulation example of this controller is included in Example 6.9 on page 272. △

6.3.4 Adaptive Bounding Design

In Section 6.3.3, the control design was based on the assumption that the unknown nonlinearities of the system lie within certain known bounds. In this section, we consider the case where the unknown nonlinearities lie within bounds that are only partially known. Specifically, each bound is composed of an unknown parameter multiplied by a known nonlinear function. The adaptive bounding method developed here allows for a less conservative control design, which is achieved through adaptive estimation of the bounding functions.

We develop the adaptive bounding design based on the assumption that the uncertainty bounds are only partially known as follows:

$$\begin{aligned} \alpha_l f_l(y) &\leq f^*(y) \leq \alpha_u f_u(y) \\ \beta_l g_l(y) &\leq g^*(y) \leq \beta_u g_u(y) \end{aligned}$$

where f_l, g_l are known lower functional bounds and f_u, g_u are known upper bounds, while α_l, α_u, β_l and β_u are unknown parameters multiplying the bounding functions. Since f^* and g^* represent the uncertain part of the plant, the lower bound is assumed to be negative and the upper bound is assumed to be positive. Without loss of generality, the functional bounds $f_u(y)$ and $g_u(y)$ are positive for all y and the lower bounds $f_l(y)$, $g_l(y)$ are negative; this implies that the unknown bounding parameters α_l, α_u, β_l and β_u are all positive.

The control law is chosen as follows:

$$u = \frac{u_a}{g_o(y) + v_g(y, e, u_a)} \tag{6.48}$$

$$u_a = -a_m e + \dot{y}_d - f_o(y) - v_f(y, e) \tag{6.49}$$

$$v_f(y, e) = \begin{cases} \hat{\alpha}_u f_u(y) & \text{if } e \geq 0 \\ \hat{\alpha}_l f_l(y) & \text{if } e < 0 \end{cases} \tag{6.50}$$

$$v_g(y, e, u_a) = \begin{cases} \hat{\beta}_u g_u(y) & \text{if } e u_a \geq 0 \\ \hat{\beta}_l g_l(y) & \text{if } e u_a < 0. \end{cases} \tag{6.51}$$

The derivation of the update laws for the bounding parameter estimates is obtained by the use of a Lyapunov function. The derivative of the Lyapunov function is used to design the update laws such that the derivative along the solutions is negative semidefinite. We consider the Lyapunov function candidate

$$V = \frac{1}{2} e^2 + \frac{1}{2\gamma} \left((\hat{\alpha}_l - \alpha_l)^2 + (\hat{\alpha}_u - \alpha_u)^2 + (\hat{\beta}_l - \beta_l)^2 + (\hat{\beta}_u - \beta_u)^2 \right).$$

Based on the above Lyapunov function we derive the following adaptive laws for the parameter bounding estimates $\hat{\alpha}_l(t)$, $\hat{\alpha}_u(t)$, $\hat{\beta}_l(t)$, and $\hat{\beta}_u(t)$:

$$\dot{\hat{\alpha}}_u = \begin{cases} \gamma e f_u(y) & \text{if } e \geq 0 \\ 0 & \text{if } e < 0 \end{cases} \tag{6.52}$$

$$\dot{\hat{\alpha}}_l = \begin{cases} 0 & \text{if } e \geq 0 \\ \gamma e f_l(y) & \text{if } e < 0 \end{cases} \tag{6.53}$$

$$\dot{\hat{\beta}}_u = \begin{cases} \gamma e g_u(y) u & \text{if } e u_a \geq 0 \\ 0 & \text{if } e u_a < 0 \end{cases} \tag{6.54}$$

$$\dot{\hat{\beta}}_l = \begin{cases} 0 & \text{if } e u_a \geq 0 \\ \gamma e g_l(y) u & \text{if } e u_a < 0. \end{cases} \tag{6.55}$$

The stability analysis can be obtained by considering the following four cases corresponding to the switching of the feedback control and adaptive laws:

- $e \geq 0$ and $u_a \geq 0$;
- $e \geq 0$ and $u_a < 0$;

- $e < 0$ and $u_a \geq 0$;
- $e < 0$ and $u_a < 0$.

We illustrate the stability analysis for one of the above four cases and leave the remaining three cases as an exercise for the reader. See Exercise 6.12.

Let us consider the fourth case, where $e < 0$ and $u_a < 0$. In this case, the update laws for $\hat{\alpha}_u$ and $\hat{\beta}_l$ are zero, i.e., $\dot{\hat{\alpha}}_u = 0$ and $\dot{\hat{\beta}}_l = 0$. Therefore, after some algebraic manipulation, the time derivative of the Lyapunov function satisfies:

$$\begin{aligned}\dot{V} &= -a_m e^2 + (f^*(y) - v_f(y,e))\,e + (g^*(y) - v_g(y,e,u_a))\,eu \\ &\quad + (\hat{\alpha}_l - \alpha_l)\,f_l e + \left(\hat{\beta}_u - \beta_u\right) g_u eu \\ &= -a_m e^2 + (f^*(y) - \alpha_l f_l)\,e + (g^*(y) - \beta_u g_u)\,eu \\ &\leq -a_m e^2.\end{aligned}$$

Using a similar analysis procedure, it can be shown that in each of the four cases the time derivative of the Lyapunov function satisfies $\dot{V} \leq -a_m e^2$. This implies that the tracking error $e(t)$ and the parameter bounding estimates $\hat{\alpha}_l(t)$, $\hat{\alpha}_u(t)$, $\hat{\beta}_l(t)$, $\hat{\beta}_u(t)$ are uniformly bounded. Moreover, using Barbălat's lemma it can be shown that the tracking error converges to zero asymptotically. The convergence is not exponential anymore, as it was in the case of completely known bounds; however, it can be shown that $e(t) \in \mathcal{L}_2$.

The adaptive bounding control design described by eqns. (6.48)–(6.51) and (6.52)–(6.55) is to be treated as the nominal control scheme. In practice, three issues must be addressed to ensure that the closed-loop system operates smoothly. The first is the smoothing of the discontinuity at the switching surfaces of $e = 0$ or $u_a = 0$, for both the functions v_f and v_g, as well as the parameter estimation equations. The second issue is to ensure the stabilizability property during the adaptation of the bounding parameter $\hat{\beta}_l(t)$. Finally, the third issue arises due to update equations for the bounding parameters $(\hat{\alpha}_u, \hat{\alpha}_l, \hat{\beta}_l, \hat{\beta}_u)$ each changing monotonically in one direction. This may lead to parameter drift problems in the presence of noise or disturbances.

Next, we discuss ways to address the above three issues:

- **Smoothing of the discontinuity.** There are two discontinuity issues to be considered here. The first one is the discontinuity of v_f and v_g and the second one is the discontinuity with regards to the update laws of $\hat{\alpha}_l(t)$, $\hat{\alpha}_u(t)$, $\hat{\beta}_l(t)$, $\hat{\beta}_u(t)$. Smoothing the discontinuity of v_f and v_g can be done in the same way as in the previous section, by creating an ε-wide smooth transition between the upper and lower bounds:

$$v_f(y,e) = \begin{cases} \hat{\alpha}_u f_u(y) & \text{if } e > \varepsilon \\ \dfrac{1}{2\varepsilon}\left(\hat{\alpha}_l f_l(y)\,(\varepsilon - e) + \hat{\alpha}_u f_u(y)\,(e + \varepsilon)\right) & \text{if } |e| \leq \varepsilon \\ \hat{\alpha}_l f_l(y) & \text{if } e < -\varepsilon \end{cases}$$

$$v_g(y,e,u_a) = \begin{cases} \hat{\beta}_u g_u(y) & \text{if } eu_a > \varepsilon \\ \dfrac{1}{2\varepsilon}\left(\hat{\beta}_l g_l(y)\,(\varepsilon - eu_a) + \hat{\beta}_u g_u(y)\,(eu_a + \varepsilon)\right) & \text{if } |eu_a| \leq \varepsilon \\ \hat{\beta}_l g_l(y) & \text{if } eu_a < -\varepsilon. \end{cases}$$

In the case of the update laws, the discontinuity at $e = 0$ and at $eu_a = 0$ causes switching of the update laws between the upper bound estimated parameters $\hat{\alpha}_u$, $\hat{\beta}_u$

and the lower bound estimated parameters $\hat{\alpha}_l$, $\hat{\beta}_l$, respectively. One approach to avoid these switchings between the update parameters is to create a small dead-zone in which none of the parameters gets updated. Therefore, the update laws of eqns. (6.52)–(6.55) can be modified as follows:

$$\dot{\hat{\alpha}}_u = \begin{cases} \gamma e f_u(y) & \text{if } e \geq \varepsilon \\ 0 & \text{if } e < \varepsilon \end{cases} \tag{6.56}$$

$$\dot{\hat{\alpha}}_l = \begin{cases} 0 & \text{if } e \geq -\varepsilon \\ \gamma e f_l(y) & \text{if } e < -\varepsilon \end{cases} \tag{6.57}$$

$$\dot{\hat{\beta}}_u = \begin{cases} \gamma e g_u(y) u & \text{if } e u_a \geq \varepsilon \\ 0 & \text{if } e u_a < \varepsilon \end{cases} \tag{6.58}$$

$$\dot{\hat{\beta}}_l = \begin{cases} 0 & \text{if } e u_a \geq -\varepsilon \\ \gamma e g_l(y) u & \text{if } e u_a < -\varepsilon, \end{cases} \tag{6.59}$$

where $\varepsilon > 0$ is a small design constant. By introducing an ε-wide smooth transition between the upper and lower bounds for v_f and v_g and by introducing a dead-zone in the update laws for $\hat{\alpha}_u$, $\hat{\beta}_u$, $\hat{\alpha}_l$, $\hat{\beta}_l$, we have created some additional terms (proportional to ε) in the derivative of the Lyapunov function. Specifically, smoothing the discontinuities introduces an additional term resulting in the inequality $\dot{V} \leq a_m e^2 + k\varepsilon$, where $k > 0$ is a constant. Even this small term, $k\varepsilon$, can cause parameter drift of the adaptive bounding parameters. As we will see below, parameter drift can be prevented by one of the available robust parameter estimation techniques, such as σ-modification, projection modification, etc.

- **Stabilizability during adaptation.** For stabilizability purposes it is important that the denominator of the control signal u does not cross the zero point. If we assume that $g_o(y) + g^*(y) > 0$ for all y, it is important that $g_o(y) + v_g(y, e, u_a) > 0$. Since v_g depends on the update parameters $\hat{\beta}_u$ and $\hat{\beta}_l$, a projection modification is required to ensure that $g_o(y) + v_g(y, e, u_a)$ remains away from zero. A closer look reveals that $\hat{\beta}_u(t) g_u(y(t)) \geq 0$, therefore in this case the denominator is not at risk of approaching zero. On the other hand, since $\hat{\beta}_l(t) \geq 0$ and $g_l(y) \leq 0$, it is possible for large values of $\hat{\beta}_l(t)$ for the denominator $g_o(y(t)) + \hat{\beta}_l(t) g_l(y(t))$ to become zero. This can be prevented if an upper bound $\bar{\beta}_l$ is imposed on the value of $\hat{\beta}_l(t)$ as follows:

$$\dot{\hat{\beta}}_l = \begin{cases} 0 & \text{if } e u_a \geq 0 \\ \gamma e g_l(y) u & \text{if } e u_a < 0 \quad \text{and } \hat{\beta}_l(t) < \bar{\beta}_l \\ & \qquad \text{or } \{\hat{\beta}_l(t) = \bar{\beta}_l \text{ and } \dot{\hat{\beta}}_l \leq 0\} \\ 0 & \text{if } e u_a < 0 \quad \text{and } \{\hat{\beta}_l(t) = \bar{\beta}_l \text{ and } \dot{\hat{\beta}}_l > 0\}. \end{cases}$$

This modification to the adaptive law of eqn. (6.55) is known as the projection modification and was presented more extensively in Section 4.6. In this case it is used to ensure that $\hat{\beta}_l(t)$ remains within a certain region to guarantee that the denominator of the control law does not approach zero.

- **Parameter drift of the bounding parameters.** In the presence of noise or even small disturbances, the adaptive laws for the updated bounding parameters ($\hat{\alpha}_u$, $\hat{\alpha}_l$, $\hat{\beta}_l$, $\hat{\beta}_u$) may cause the parameter estimate to drift to infinity. For example, consider the case of the parameter estimate $\hat{\alpha}_u$. With a positive bounding function $f_u(y)$, the right-hand side $\gamma e f_u(y)$ is strictly positive for $e > 0$, which may cause $\hat{\alpha}_u(t) \to \infty$ unless the tracking error $e(t)$ converges to zero. Now, in the presence of even small disturbances or measurement noise, the tracking error will not converge to zero, therefore the parameter estimate will continue to increase with time. This problem, which is well understood in the adaptive control literature, is known as *parameter drift*, and has been discussed in Section 4.6. Parameter drift can be prevented by using one of the available robust parameter estimation techniques that have been discussed in Chapter 4, such as projection modification, σ-modification, dead-zone, etc. For example, if we use the σ-modification, the update law for $\hat{\alpha}_u$ will become:

$$\dot{\hat{\alpha}}_u = \begin{cases} \gamma e f_u(y) - \gamma\sigma(t)(\hat{\alpha}_u - \hat{\alpha}_u^0) & \text{if } e \geq 0 \\ 0 & \text{if } e < 0 \end{cases}$$

where $\hat{\alpha}_u^0$ is a design constant and $\sigma(t)$ is a parameter that adjusts the magnitude of the leakage term (the second term of the right-hand side of the adaptive law). For simplicity, $\sigma(t)$ is often chosen to be a constant $\sigma(t) = \sigma$. However, it is also possible to select a more advanced leakage term, where $\sigma(t) = 0$ for $\hat{\alpha}_u \leq M$, where M is a design parameter, and $\sigma(t) = \sigma$ for $\hat{\alpha}_u > M$. If instead of the σ-modification we use a dead-zone, then the resulting adaptive laws will look similar to those described by eqns. (6.56)–(6.59). Therefore, the adaptive laws of eqns. (6.56)–(6.59) address both the issue of parameter drift and smoothing the discontinuity in the update law. However, the designer needs to be careful in selecting the size of the dead-zone, which is denoted by ε.

The feedback control design of this section illustrates an important component of adaptive control as well as adaptive approximation based approaches: first, the designer proceeds to derive an adaptive scheme (including both the feedback control law and the parameter updates laws), which is stable under certain assumptions (typically, under ideal operating conditions). Then, in order to address the nonideal case, a set of modifications are proposed. These modifications may include smoothing the feedback control law, making the adaptive law robust with respect to disturbances and measurement noise, or using the projection algorithm in order to prevent certain parameters from entering an undesired region (for example, a region that makes the denominator of the feedback control function approach zero). In the literature, these modifications are sometimes developed in an ad hoc fashion but often they are rigorously designed and analyzed.

6.3.5 Adaptive Approximation of the Unknown Nonlinearities

Now, we proceed to approximating the unknown nonlinearities $f^*(y)$ and $g^*(y)$ using adaptive approximation models and employing learning methods. In this section, we consider a slightly more general tracking objective where the feedback control law is designed to track the filtered tracking error

$$e_F(t) = e(t) + c\int_o^t e(\tau)d\tau,$$

where $c \geq 0$ is a design constant. As discussed in Chapter 5, the filtered error can be thought of providing a proportional integral (PI) control objective. In the special case that $c = 0$, then the filtered error is equal to the standard tracking error $e = y - y_d$.

To illustrate some of the stability issues that may arise, we first consider the simpler case where both $f^*(y)$ and $g^*(y)$ can be approximated exactly by linearly parameterized approximators. Therefore, the system under consideration is described by

$$\dot{y} = f_o(y) + f^*(y) + (g_o(y) + g^*(y))\,u \tag{6.60}$$

where the unknown functions $f^*(y)$, $g^*(y)$ can be represented by

$$\begin{aligned} f^*(y) &= \phi_f(y)^\top \theta_f^* \\ g^*(y) &= \phi_g(y)^\top \theta_g^* \end{aligned}$$

for some unknown parameters θ_f^*, θ_g^*. In the feedback control law we replace the unknown functions $f^*(y)$ and $g^*(y)$ by the adaptive approximations $\hat{f}(y, \hat{\theta}_f) = \phi_f(y)^\top \hat{\theta}_f$ and $\hat{g}(y, \hat{\theta}_g) = \phi_g(y)^\top \hat{\theta}_g$, respectively. This yields the feedback controller

$$u = \frac{1}{g_o(y) + \phi_g(y)^\top \hat{\theta}_g}\left[-a_m e_F - c(y - y_d) + \dot{y}_d - f_o(y) - \phi_f(y)^\top \hat{\theta}_f\right]. \tag{6.61}$$

For the time being we assume that parameter estimate $\hat{\theta}_g$ is such that the denominator $g_o(y) + \phi_g(y)^\top \hat{\theta}_g$ is bounded away from zero. Later, we will include conditions to ensure that this is true.

If we substitute the feedback control of eqn. (6.61) into eqn. (6.60), then the filtered tracking error dynamics satisfy

$$\dot{e}_F = -a_m e_F - \phi_f(y)^\top \left(\hat{\theta}_f - \theta_f^*\right) - \phi_g(y)^\top \left(\hat{\theta}_g - \theta_g^*\right) u.$$

These tracking error dynamics are rather standard in the adaptive control literature. The adaptive laws can be derived by considering the Lyapunov function

$$V = \frac{1}{2}e_F^2 + \frac{1}{2}\left(\hat{\theta}_f - \theta_f^*\right)^\top \Gamma_f^{-1}\left(\hat{\theta}_f - \theta_f^*\right) + \frac{1}{2}\left(\hat{\theta}_g - \theta_g^*\right)^\top \Gamma_g^{-1}\left(\hat{\theta}_g - \theta_g^*\right).$$

The time derivative of the Lyapunov function satisfies

$$\begin{aligned} \dot{V} &= -a_m e_F^2 + \left(\hat{\theta}_f - \theta_f^*\right)^\top \Gamma_f^{-1}\left(\dot{\hat{\theta}}_\phi - \Gamma_f \phi_f(y) e_F\right) \\ &\quad + \left(\hat{\theta}_g - \theta_g^*\right)^\top \Gamma_g^{-1}\left(\dot{\hat{\theta}}_\kappa - \Gamma_g \phi_g(y) e_F u\right). \end{aligned}$$

Therefore, the adaptive update algorithms for generating the parameter estimates $\hat{\theta}_f(t)$, $\hat{\theta}_g(t)$ are given by

$$\dot{\hat{\theta}}_f = \Gamma_f \phi_f(y) e_F \tag{6.62}$$

$$\dot{\hat{\theta}}_g = \Gamma_g \phi_g(y) e_F u. \tag{6.63}$$

Based on the feedback control law and adaptive laws, the derivative of the Lyapunov function satisfies $\dot{V} = -a_m e_F^2$, which implies that the closed-loop system is stable and the filtered tracking error converges to zero with $e_F(t) \in \mathcal{L}_2$.

The above design and analysis of an adaptive approximation based control scheme for tracking was based on some key assumptions. For example, it was assumed that there are no modeling errors within the approximation region $\mathcal{D}$, nor any disturbances or noise components. Another key assumption was that the adaptation of $\hat{\theta}_g(t)$ is such that the denominator of the control law in eqn. (6.61) never approaches zero. Finally, it was assumed that $x(t) \in \mathcal{D}$ for any $t \geq 0$. In the next subsection we will examine in more detail some of the potential instability problems in adaptive approximation based control, and we will develop modifications to the standard control scheme to prevent such instability mechanisms.

6.3.6 Robust Adaptive Approximation

Historically, the development of adaptive approximation based control algorithms in the context of neural networks started around 1990 with the design of neural control schemes under certain assumptions, as presented in the previous subsection. The stability analysis under these assumptions could be carried out following standard techniques of adaptive linear control [11, 119, 179] or techniques of adaptive nonlinear control [134, 139, 159]. However, the adaptive linear control methodology is based on an assumed linear model, represented by a transfer function with some unknown coefficients, which are estimated using parameter estimation techniques. Therefore, adaptive linear control does not deal directly with an approximation subregion of the state-space and what happens if the trajectory reaches the boundary of that region. There are also no explicit concerns of an approximation error within the coverage region $\mathcal{D}$.

Adaptive approximation based control has some special stability and robustness issues that require special attention. Examination of the instability mechanisms for adaptive approximation based control, in the context of neural networks, was first presented in [211, 212]. Next, we discuss these potential instability mechanisms and ways to address these issues.

Stabilizability. The stability results of Section 6.3.5 were obtained under the crucial assumption that the feedback control law is well defined and remains bounded for all time $t \geq 0$. In general, the adaptive law for $\hat{\theta}_g(t)$ does not guarantee that the denominator in the feedback control law will remain away from zero. Specifically, it is required that $\phi_g(y(t))^\top \hat{\theta}_g(t) > -g_o(y(t))$ for all $t \geq 0$. In practice, the denominator in the feedback control law cannot be allowed to come arbitrarily close to zero since in that case the control effort becomes infinitely large. Let ε_g be a small positive number such that $\phi_g^\top \hat{\theta}_g(t) + g_o(y(t)) > \varepsilon_g$, denotes a safe distance for the denominator from the point of singularity. Therefore, it is required that

$$\phi_g(y(t))^\top \hat{\theta}_g(t) \geq -g_o(y(t)) + \varepsilon_g. \tag{6.64}$$

For general approximators, this condition can be difficult to ensure; however, as shown in the following example, if the approximator is linear in the parameters (LIP) with positive basis functions forming a partition of unity (see Section 2.4.8.1), then the condition is straightforward to ensure using projection.

EXAMPLE 6.6

Consider the adaptive law described by eqn. (6.63) where Γ_g is positive definite and diagonal, with elements denoted by γ_i. Let the approximator for $g^*(y)$ be $\phi_g(y)^\top \hat{\theta}_g$ where $\phi_g(y)^\top$ form a partition of unity. Then to satisfy the condition that $\phi_g(y)^\top \hat{\theta}_g(t) \geq -g_o(y) + \varepsilon_g$, it is sufficient that for each i

$$\hat{\theta}_{g_i} > \varepsilon_g - \min_{y \in Supp(\phi_i)} \{g_o(y)\}$$

where $Supp(\phi_i) = \{y \,|\, \phi_i(y) > 0\}$ is the support of ϕ_i. This condition is sufficient since, by the partition of unity,

$$\phi_g(y)^\top \hat{\theta}_g = \sum_i \phi_{g_i}(y) \hat{\theta}_{g_i}$$

$$\begin{aligned}
&\geq \sum_i \left(\phi_{g_i}(y)\left(\varepsilon_g - \min_{y\in Supp(\phi_i)}\{g_o(y)\}\right)\right)\\
&= \varepsilon_g \sum_i \phi_{g_i}(y) - \sum_i \left(\phi_{g_i}(y) \min_{y\in Supp(\phi_i)}\{g_o(y)\}\right) \geq (\varepsilon_g - g_o(y)).
\end{aligned}$$

The set defined by $\left\{\hat{\theta}_g \,\middle|\, \kappa(\hat{\theta}_{g_i}) < 0 \text{ for } i = 1\ldots, N\right\}$ is convex where $\kappa(\hat{\theta}_{g_i}) = \varepsilon_g - \min_{y\in Supp(\phi_i)}\{g_o(y)\} - \hat{\theta}_{g_i}$. Therefore, the projection algorithm of Section 4.6, yields for each i

$$\dot{\hat{\theta}}_{g_i} = \begin{cases} \gamma_i \phi_{g_i}(y) e_F u & \text{if } \hat{\theta}_{g_i} > \varepsilon_g - \min_{y\in Supp(\phi_i)}\{g_o(y)\} \\ & \text{or } \dot{\hat{\theta}}_{g_i} > 0 \\ 0 & \text{otherwise.} \end{cases}$$

which will ensure the stabilizability condition. Note that when each ϕ_i is locally supported, it is particularly easy to evaluate $\min_{y\in Supp(\phi_i)}\{g_o(y)\}$. △

Robust Parameter Adaptation. The parameter adaptive laws of eqns. (6.62)–(6.63) were developed under the assumption of no disturbances, modeling error or measurement noise. In the presence of such perturbations, it is possible for the trajectory $y(t)$ to leave the approximation region $\mathcal{D}$, in which case adaptive approximation is not possible and hence it may lead to instability. This issue was illustrated with a simple regulation example in Section 6.2.5. It will be further discussed and addressed in the next section, where an adaptive bounding technique will be developed to ensure that the trajectory remains with a certain region. However, even if the trajectory remains with $\mathcal{D}$, we may still encounter another problem related to the drifting of the parameter estimates to infinity. To address the problem of parameter drift, the parameter adaptive laws of eqns. (6.62)–(6.63) should be modified as discussed in detail in Section 4.6. To illustrate the need and design of robust parameter adaptation, we consider the presence of a residual approximation error and an additive disturbance term in the system dynamics.

Suppose that the unknown functions f^* and g^* are represented in the region $\mathcal{D}$ by their corresponding adaptive approximators as follows:

$$\begin{aligned}
f^*(y) &= \phi_f(y)^\top \theta_f^* + e_f(y)\\
g^*(y) &= \phi_g(y)^\top \theta_g^* + e_g(y)
\end{aligned}$$

where e_f and e_g are the corresponding residual approximation error functions. Moreover, we assume that there is a disturbance term $d(t)$ in the system under consideration. Therefore, the plant described by (6.60) is now described by

$$\begin{aligned}
\dot{y} &= f_o(y) + f^*(y) + (g_o(y) + g^*(y))\,u + d(t)\\
&= f_o(y) + \phi_f(y)^\top \theta_f^* + e_f(y) + \left(g_o(y) + \phi_g(y)^\top \theta_g^* + e_g(y)\right) u + d(t)\\
&= f_o(y) + \phi_f(y)^\top \theta_f^* + \left(g_o(y) + \phi_g(y)^\top \theta_g^*\right) u + \omega(t)
\end{aligned}$$

where $\omega(t) = e_f(y(t)) + e_g(y(t))u(t) + d(t)$ represents the total modeling error. Using the standard adaptive laws (6.62)-(6.63) in the Lyapunov function, we obtain the following Lyapunov time derivative:

$$\dot{V} = -a_m e_F^2 + e_F \omega. \tag{6.65}$$

Suppose $|\omega(t)| \leq \bar{\omega}$, where $\bar{\omega}$ is a constant. It can be readily seen that the derivative of the Lyapunov function satisfies $\dot{V} \leq 0$ when $|e_F(t)| > \bar{\omega}/a_m$. However, if $|e_F(t)| < \bar{\omega}/a_m$ then $\dot{V}$ may become positive, which implies the parameter estimates may grow unbounded. In other words, for small enough values of $e_F(t)$ the parameter estimates may keep on increasing (or decreasing), leading to the phenomenon of parameter drift.

As discussed in Section 4.6, there are several techniques for modifying the adaptive laws such that they are robust with respect to modeling errors. Such modifications include the dead-zone, the σ-modification, the projection modification.

Robustness to large initial parameter errors. The proof of stability of the previous section implicitly assumes that the state stays within the domain of approximation $\mathcal{D}$. This issue was thoroughly discussed in Section 6.2.5 relative to the regulation problem. Similar issues arise in the tracking problem; the essential summary is that if the initial parameter errors are sufficiently large, then the state could leave the region $\mathcal{D}$, unless the designer anticipates this contingency and adds a term to the control law to ensure against it. The following two sections address this issue.

Note that the issue of the state leaving the region $\mathcal{D}$ has additional importance for the tracking problem, since the desired trajectory y_d may take the state near the boundary of $\mathcal{D}$.

6.3.7 Combining Adaptive Approximation with Adaptive Bounding

In this section, we present several complete adaptive approximation based designs that contain all the required elements. The required elements include the control algorithm, a robust parameter estimation algorithm, and a bounding term to ensure that the state remains within the approximation region.

Assume that for the system

$$\dot{y} = f(y) + g(y)u$$

the objective is to track a signal $y_d(t)$ that is continuous, differentiable, and bounded. The derivative $\dot{y}_d(t)$ is also assumed to be available and bounded. Let the operating region (or approximation region) be denoted by $\mathcal{D} = \{y \mid |y| \leq \alpha\}$ where $\alpha > 0$, which is a compact set. Define $0 < \mu < \alpha$, and assume that y_d, μ, and α are selected such that $|y_d(t)| < \alpha - \mu$, for all $t > 0$. Therefore, the desired signal is at least a distance μ from the boundary of $\mathcal{D}$ at any time. Since $\mathcal{D}$ is the approximation region, μ can be viewed as the radius of a safety region that allows a certain level of tracking error while still having $y \in \mathcal{D}$. Note that in general, the approximation region will be of the form $\mathcal{D} = \{y \mid -\beta \leq y \leq \alpha;\ \alpha, \beta > 0\}$. For notational simplicity, and without any loss of generality, in this section we assume that $\beta = \alpha$.

Let

$$\begin{aligned} f(y) &= f_o(y) + f^*(y) \\ g(y) &= g_o(y) + g^*(y) \end{aligned}$$

where f_o and g_o are known functions, representing the nominal dynamics of the system, while f^* and g^* are unknown functions representing the nonlinear functional uncertainty. It is assumed that the unknown functions are within certain known bounds:

$$\begin{aligned} f_L(y) &\leq f^*(y) \leq f_U(y) \\ -g_o(y) < g_L(y) &\leq g^*(y) \leq g_U(y) \end{aligned}$$

for any $y \in \Re^1$.

Next, we select a set of basis functions $\phi_f(y)$ and $\phi_g(y)$ such that the unknown functions $f^*(y)$, $g^*(y)$ can be represented within $\mathcal{D}$ by

$$\begin{aligned} f^*(y) &= \phi_f(y)^\top \theta_f^* + e_f(y) \\ g^*(y) &= \phi_g(y)^\top \theta_g^* + e_g(y) \end{aligned}$$

for some unknown optimal parameters θ_f^*, θ_g^*. The basis functions $\phi_f(y), \phi_g(y)$ are defined such that they have zero value outside the region $\mathcal{D}$. Hence, $\phi_f(y)^\top \theta_f = 0$ for all $y \in \{\Re^1 - \mathcal{D}\}$ and for any θ_f (similarly for $\phi_g(y)$). Let

$$\begin{aligned} \bar{e}_f &= \max_{y \in \mathcal{D}} |e_f(y)| \\ \bar{e}_g &= \max_{y \in \mathcal{D}} |e_g(y)|. \end{aligned}$$

Since the least upper bounds $\bar{e}_f$ and $\bar{e}_g$ are unknown they will be adaptively estimated. The bounding estimates, denoted by $\hat{e}_f(t)$ and $\hat{e}_g(t)$, respectively, will be used to address via adaptive bounding techniques the presence of the minimum functional approximation errors within $\mathcal{D}$. Therefore, we define

$$F_u(y, \hat{e}_f) = \begin{cases} \hat{e}_f & \text{if} \quad |y| < \alpha - \mu \\ \hat{e}_f \dfrac{\alpha - |y|}{\mu} + f_U(y) \dfrac{|y| - \alpha + \mu}{\mu} & \text{if} \quad \alpha - \mu < |y| < \alpha \\ f_U(y) & \text{if} \quad \alpha < |y| \end{cases}$$

$$F_l(y, \hat{e}_f) = \begin{cases} -\hat{e}_f & \text{if} \quad |y| < \alpha - \mu \\ -\hat{e}_f \dfrac{\alpha - |y|}{\mu} + f_L(y) \dfrac{|y| - \alpha + \mu}{\mu} & \text{if} \quad \alpha - \mu < |y| < \alpha \\ f_L(y) & \text{if} \quad \alpha < |y|. \end{cases}$$

It is easy to verify that

$$F_l(y, \hat{e}_f)|_{\hat{e}_f = \bar{e}_f} \le \left(f^*(y) - \phi_f(y)^\top \theta_f^*\right) \le F_u(y, \hat{e}_f)|_{\hat{e}_f = \bar{e}_f}, \qquad \forall y \in \Re^1.$$

The various quantities discussed in this paragraph are illustrated in Figure 6.8.

Similarly, we define

$$G_u(y, \hat{e}_g) = \begin{cases} \hat{e}_g & \text{if} \quad |y| < \alpha - \mu \\ \hat{e}_g \dfrac{\alpha - |y|}{\mu} + g_U(y) \dfrac{|y| - \alpha + \mu}{\mu} & \text{if} \quad \alpha - \mu < |y| < \alpha \\ g_U(y) & \text{if} \quad \alpha < |y| \end{cases}$$

$$G_l(y, \hat{e}_g) = \begin{cases} -\hat{e}_g & \text{if} \quad |y| < \alpha - \mu \\ -\hat{e}_g \dfrac{\alpha - |y|}{\mu} + g_L(y) \dfrac{|y| - \alpha + \mu}{\mu} & \text{if} \quad \alpha - \mu < |y| < \alpha \\ g_L(y) & \text{if} \quad \alpha < |y| \end{cases}$$

which satisfies

$$G_l(y, \hat{e}_g)|_{\hat{e}_g = \bar{e}_g} \le \left(g^*(y) - \phi_g(y)^\top \theta_g^*\right) \le G_u(y, \hat{e}_g)|_{\hat{e}_g = \bar{e}_g}, \qquad \forall y \in \Re^1.$$

In the following, we assume that $\bar{e}_g \le \varepsilon_g$ where ε_g is a constant that satisfies $\varepsilon_g < g_o(y)$. This condition is necessary to ensure that $g_o(y) + \phi_g(y)^\top \hat{\theta}_g + G_l(y, \bar{e}_g) > 0$.

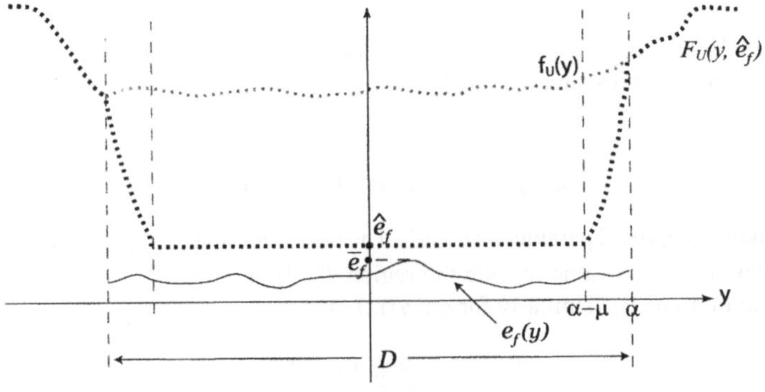

Figure 6.8: Diagram to illustrate the approximation error $e_f(y)$, the upper bound $\bar{e}_f$, its estimate $\hat{e}_f$, the approximation region $\mathcal{D}$ and the derivation of $F_u(y, \hat{e}_f)$.

Define the online estimates of $f^*(y)$ and $g^*(y)$ as $\hat{f}(y;\hat{\theta}_f) = \phi_f(y)^\top \hat{\theta}_f$ and $\hat{g}(y;\hat{\theta}_g) = \phi_g(y)^\top \hat{\theta}_g$, for $y \in \mathcal{D}$, respectively. When $y \in \{\Re^1 - \mathcal{D}\}$, the parameters are not adjusted. When $y \in \mathcal{D}$, the parameters are adapted according to

$$\dot{\hat{\theta}}_f = \Gamma_f \phi_f e_d \qquad \dot{\hat{\theta}}_g = \mathcal{P}_1\left(\Gamma_g \phi_g e_d u\right) \tag{6.66}$$

$$\dot{\hat{e}}_f = \gamma |e_d| \qquad \dot{\hat{e}}_g = \mathcal{P}_2\left(\gamma |e_d u|\right), \tag{6.67}$$

where $\gamma > 0$, Γ_f and Γ_g are positive definite, $\mathcal{P}_1$ is a projection operator that will be used to ensure the stabilizability condition of eqn. (6.64), $\mathcal{P}_2$ is a projection operator that will be used to ensure $\hat{e}_g < \varepsilon_g$ and

$$e_d = d(e) = \begin{cases} 0 & \text{if } |e| < \varepsilon \\ e & \text{if } |e| \geq \varepsilon \end{cases}$$

denotes the tracking error $e = y - y_d$ processed by a dead-zone (see Section 4.6). The dead-zone is included to prevent against parameter drift due to noise, disturbances, or MFAE. Note that the positive design parameter ε is small and independent of the control gain a_m.

Finally, define the feedback controller

$$u_a = -a_m e + \dot{y}_d - f_o(y) - \phi_f(y)^\top \hat{\theta}_f - v_f(y, e, \hat{e}_f) \tag{6.68}$$

$$u = \frac{u_a}{g_o(y) + \phi_g(y)^\top \hat{\theta}_g + v_g(y, e, u_a, \hat{e}_g)}, \tag{6.69}$$

with $a_m > 0$ and

$$v_f(y, e, \hat{e}_f) = \begin{cases} F_u(y, \hat{e}_f) & \text{if } e > \varepsilon \\ F_l(y, \hat{e}_f)\dfrac{\varepsilon - e}{2\varepsilon} + F_u(y, \hat{e}_f)\dfrac{\varepsilon + e}{2\varepsilon} & \text{if } |e| < \varepsilon \\ F_l(y, \hat{e}_f) & \text{if } e < -\varepsilon \end{cases} \tag{6.70}$$

$$v_g(y, e, u_a, \hat{e}_g) = \begin{cases} G_u(y, \hat{e}_g) & \text{if } e u_a > 0 \\ G_l(y, \hat{e}_g) & \text{if } e u_a < 0. \end{cases} \tag{6.71}$$

As we will see, the auxiliary terms v_f and v_g in the control law (6.68), (6.69) are used to enhance the robustness of the closed-loop system. When the control law is substituted into the system dynamics, the resulting tracking error dynamics are

$$\begin{aligned}
\dot{y} &= f_o + \hat{f} + (g_o + \hat{g})u + (f^* - \hat{f}) + (g^* - \hat{g})u \\
&= f_o + \hat{f} + u_a - uv_g + (f^* - \phi_f^\top \hat{\theta}_f) + (g^* - \phi_f^\top \hat{\theta}_f)u \\
\dot{e} &= -a_m e - v_f - uv_g + (f^* - \phi_f^\top \hat{\theta}_f) + (g^* - \phi_f^\top \hat{\theta}_f)u.
\end{aligned}$$

Assume that the state is outside of $\mathcal{D}$ at some time $t_1 \geq 0$ (i.e., $|y(t_1)| > \alpha$). While the state is outside of $\mathcal{D}$, parameter adaptation is off and $\phi_f = 0$ and $\phi_g = 0$; therefore, we can consider the simple Lyapunov function

$$V_1 = \frac{1}{2}e^2.$$

The derivative of V_1 reduces to

$$\frac{dV_1}{dt} = -a_m e^2 + e\,(-v_f + f^*) + eu\,(-v_g + g^*)\,. \tag{6.72}$$

Using the definitions of v_f and v_g for y outside $\mathcal{D}$, we obtain that $e\,(-v_f + f^*) \leq 0$ and $eu\,(-v_g + g^*) \leq 0$. Therefore,

$$\frac{dV_1}{dt} \leq -a_m e^2.$$

Since $|y_d(t)| < \alpha - \mu$ and $|y| > \alpha$, we have $|e| = |y - y_d| > \mu > 0$. Hence, for $t \leq t_1$, as long as $y(t)$ is outside $\mathcal{D}$, $|e(t)|$ is decreasing exponentially. This implies that y returns to $\mathcal{D}$ in finite time. Note that in this scalar example, large magnitude switching of v_f will not occur for y outside of $\mathcal{D}$, because it is not possible for e to switch signs without passing through $\mathcal{D}$. Within $\mathcal{D}$, the v_f term may switch signs, but its magnitude is only $\hat{e}_f$.

Within $\mathcal{D}$, consider the Lyapunov function

$$V = \frac{1}{2}\left(e^2 + \frac{1}{\gamma}\left((\hat{e}_f - \bar{e}_f)^2 + (\hat{e}_g - \bar{e}_g)^2\right) + \tilde{\theta}_f^\top \Gamma_f^{-1} \tilde{\theta}_f + \tilde{\theta}_g^\top \Gamma_g^{-1} \tilde{\theta}_g\right).$$

The time derivative of V along the solution of the system model is given by

$$\begin{aligned}
\frac{dV}{dt} &= -a_m e^2 - ev_f - euv_g + e(f^* - \phi_f^\top \hat{\theta}_f) + e(g^* - \phi_g^\top \hat{\theta}_g)u \\
&\quad + \frac{1}{\gamma}\left((\hat{e}_f - \bar{e}_f)\dot{\hat{e}}_f + (\hat{e}_g - \bar{e}_g)\dot{\hat{e}}_g\right) + \tilde{\theta}_f^\top \Gamma_f^{-1} \dot{\hat{\theta}}_f + \tilde{\theta}_g^\top \Gamma_g^{-1} \dot{\hat{\theta}}_g \\
&= -a_m e^2 + e\left(-v_f + f^* - \phi_f^\top \hat{\theta}_f\right) + eu\left(-v_g + g^* - \phi_g^\top \hat{\theta}_g\right) \\
&\quad + \frac{1}{\gamma}\left((\hat{e}_f - \bar{e}_f)\dot{\hat{e}}_f + (\hat{e}_g - \bar{e}_g)\dot{\hat{e}}_g\right) + \tilde{\theta}_f^\top \Gamma_f^{-1} \dot{\hat{\theta}}_f + \tilde{\theta}_g^\top \Gamma_g^{-1} \dot{\hat{\theta}}_g.
\end{aligned}$$

For $|e| \geq \varepsilon$, and in the absence of projection, the time derivative of V becomes

$$\begin{aligned}
\frac{dV}{dt} &= -a_m e^2 + e\,(-v_f + e_f) + eu\,(-v_g + e_g) + \frac{1}{\gamma}(\hat{e}_f - \bar{e}_f)\dot{\hat{e}}_f + \frac{1}{\gamma}(\hat{e}_g - \bar{e}_g)\dot{\hat{e}}_g \\
&\quad + \tilde{\theta}_f^\top \Gamma_f^{-1}\left(\dot{\hat{\theta}}_f - \Gamma_f \phi_f e\right) + \tilde{\theta}_g^\top \Gamma_g^{-1}\left(\dot{\hat{\theta}}_g - \Gamma_g \phi_g eu\right) \qquad (6.73) \\
&\leq -a_m e^2 - ev_f + \bar{e}_f|e| - v_g eu + \bar{e}_g|eu| + (\hat{e}_f - \bar{e}_f)|e| + (\hat{e}_g - \bar{e}_g)|eu| \\
&= -a_m e^2 - ev_f + \hat{e}_f|e| - v_g eu + \hat{e}_g|eu| \\
&\leq -a_m e^2,
\end{aligned}$$

which is negative semidefinite. When projection occurs, its beneficial effects have been discussed in Section 4.6. Therefore, we have shown that $e(t)$ will converge regardless of initial condition to the set $|e| \leq \varepsilon$ within which all parameter adaptation stops. The designer can independently specify the desired tracking accuracy (i.e., ε) and the rate of decay of errors due to disturbances or initial conditions (i.e., a_m).

■ EXAMPLE 6.7

Consider the scalar system first considered in Example 6.5 on page 257. The assumed a priori information, control specification, and prefilter will be the same as in that example. The only additional required information is that the desired trajectory y_c is designed such that for all $t \geq 0$, $y_c(t) \in \mathcal{D}_c = \{y_c \,|\, -9 \leq y_c \leq 9\}$.

Let $\alpha = 10$, $\mu = 1$ and define $\mathcal{D} = \{y \,|\, -10 \leq y \leq 10\}$, which contains $\mathcal{D}_c$. Following the design procedure of this subsection, we define (for $c = 0$, *i.e.*, $e_F = e$)

$$
\begin{aligned}
v_f(y, e, \hat{e}_f) &= \begin{cases} F_u(y, \hat{e}_f) & \text{if} \quad e \geq \varepsilon \\ F_l(y, \hat{e}_f)\frac{\varepsilon - e}{2\varepsilon} + F_u(y, \hat{e}_f)\frac{\varepsilon + e}{2\varepsilon} & \text{if} \quad |e| \leq \varepsilon \\ F_l(y, \hat{e}_f) & \text{if} \quad e < -\varepsilon \end{cases} \\
u_a &= -10e + \dot{y}_d - \phi_f(y)^\top \hat{\theta}_f - v_f(y, e, \hat{e}_f) \\
v_g(y, e, u_a, \hat{e}_g) &= \begin{cases} G_u(y, \hat{e}_g) & \text{if} \quad e u_a > 0 \\ G_l(y, \hat{e}_g) & \text{if} \quad e u_a < 0. \end{cases} \\
u &= \frac{u_a}{2 + \phi_g(y)^\top \hat{\theta}_g + v_g(y, e, u_a, \hat{e}_g)},
\end{aligned}
$$

where the upper and lower functional bounds are defined as

$$
\begin{aligned}
F_u(y, \hat{e}_f) &= \begin{cases} \hat{e}_f & \text{if} \quad |y| < 9 \\ \hat{e}_f \frac{\alpha - |y|}{\mu} + (y^2)\frac{|y| - \alpha + \mu}{\mu} & \text{if} \quad 9 \leq |y| < 10 \\ y^2 & \text{if} \quad \alpha \leq |y| \end{cases} \\
F_l(y, \hat{e}_f) &= \begin{cases} -\hat{e}_f & \text{if} \quad |y| < 9 \\ -\hat{e}_f \frac{\alpha - |y|}{\mu} - (y^2)\frac{|y| - \alpha + \mu}{\mu} & \text{if} \quad 9 \leq |y| < 10 \\ -y^2 & \text{if} \quad 10 \leq |y| \end{cases} \\
G_u(y, \hat{e}_g) &= \begin{cases} \hat{e}_g & \text{if} \quad |y| < \alpha - \mu \\ \hat{e}_g \frac{\alpha - |y|}{\mu} + \left(|y| + \frac{1}{2}\right)^2 \frac{|y| - \alpha + \mu}{\mu} & \text{if} \quad \alpha - \mu \leq |y| < \alpha \\ \left(|y| + \frac{1}{2}\right)^2 & \text{if} \quad \alpha \leq |y| \end{cases} \\
G_l(y, \hat{e}_g) &= \begin{cases} -\hat{e}_g & \text{if} \quad |y| < \alpha - \mu \\ -\hat{e}_g \frac{\alpha - |y|}{\mu} - 1.0\frac{|y| - \alpha + \mu}{\mu} & \text{if} \quad \alpha - \mu \leq |y| < \alpha \\ -1.0 & \text{if} \quad \alpha \leq |y|. \end{cases}
\end{aligned}
$$

The basis elements in $\phi(y)$ are selected to be positive, forming a partition of unity that covers $\mathcal{D}$. Finally, when $y \in \mathcal{D}$, the approximator parameters $\hat{\theta}_f$ and $\hat{\theta}_g$, and bounding parameters $\hat{e}_f$ and $\hat{e}_g$, are adapted according to eqn. (6.66) and (6.67), with projection $\mathcal{P}_1$ maintaining $\hat{\theta}_{g_i} > -1$ for $i = 1, \ldots, N$ and projection $\mathcal{P}_2$ maintaining $0 < \hat{e}_g < \varepsilon_g = 1$.

By the analysis of this section, regardless of the initial values of y, $\hat{\theta}_f$, $\hat{\theta}_g$, $\hat{e}_f$ and $\hat{e}_g$, the tracking error will asymptotically converge to $|e| \leq \varepsilon$ for $y_c \in \mathcal{D}_c$. For $y \in \mathcal{D}$, if switching does occur due to the v_f term in u_a, it will have magnitude of only $\hat{e}_f$.

Note, however, that there are choices of either the initial conditions or the adaptation parameters that could allow $\hat{e}_f$ to become large. If this occurs, then the asymptotic convergence of e will still be achieved; however, the closed-loop tracking performance will be due to high-gain feedback (resulting in large amplitude switching), not due to adaptive approximation. This issue is further discussed in Section 6.3.8. A simulation example of an extension of this controller is included in Example 6.9 on page 272. $\triangle$

6.3.8 Advanced Adaptive Approximation Issues

The adaptive approximation control scheme developed in the previous section consists of two main components: (i) the adaptive approximation based control, which operates within the coverage region $\mathcal{D}$ with the objective of causing the tracking error to converge to zero, or to a neighborhood of zero; and (ii) the adaptive bounding control, which operates mostly on the boundary of $\mathcal{D}$ with the objective of preventing the trajectory from leaving the approximation region $\mathcal{D}$. A secondary objective of the adaptive bounding control component is to estimate and cancel the effect of any approximation error within the region $\mathcal{D}$.

There are several interesting issues and trade-offs that arise as the two control components combine to form the overall controller. In this section we consider two such issues: (i) ensuring the benefits of adaptive approximation by reducing the effect of adapting bounding inside the approximation region $\mathcal{D}$; and (ii) introducing advanced methods for designing the adaptive bounding functions.

Ensuring the Benefits of Adaptive Approximation. In the approach just presented, if the adaptive gain γ of the bounding parameter estimate is large relative to the adaptive gain Γ of the approximator parameter estimates, then it may be the case that the tracking performance is attained predominantly through the adaptive bounding terms. This would be the case if the adaptive bounds quickly increased prior to the adaptive approximators converging. In this case, the switching term would be large even within $\mathcal{D}$, which would eliminate the benefits of including the adaptive approximators.

If alternatively, eqn. (6.67) is changed to include leakage terms

$$\dot{\hat{e}}_f = \gamma\left(|e_d| - \sigma_f \hat{e}_f\right) \qquad \dot{\hat{e}}_g = \gamma\left(|e_d u| - \sigma_g \hat{e}_g\right) \tag{6.74}$$

for $y \in \mathcal{D}$, with $\gamma,\ \sigma_g,\ \sigma_g > 0$, then bounds within $\mathcal{D}$ would be allowed to decay over time. The Lyapunov analysis of the previous approach remains the same until eqn. (6.73). Therefore, we start the analysis from that point.

Within $\mathcal{D}$, for $|e| \geq \varepsilon$, and in the absence of projection, the time derivative of V satisfies

$$\begin{aligned}
\frac{dV}{dt} &= -a_m e^2 + e\left(-v_f + e_f\right) + eu\left(-v_g + e_g\right) + \frac{1}{\gamma}(\hat{e}_f - \bar{e}_f)\dot{\hat{e}}_f + \frac{1}{\gamma}(\hat{e}_g - \bar{e}_g)\dot{\hat{e}}_g \\
&\quad + \tilde{\theta}_f^\top \Gamma_f^{-1}\left(\dot{\hat{\theta}}_f - \Gamma_f \phi_f e\right) + \tilde{\theta}_g^\top \Gamma_g^{-1}\left(\dot{\hat{\theta}}_g - \Gamma_g \phi_g e u\right) \\
&\leq -a_m e^2 - e v_f + \bar{e}_f|e| - v_g e u + \bar{e}_g|eu| + (\hat{e}_f - \bar{e}_f)(|e| - \sigma_f \hat{e}_f) \\
&\quad + (\hat{e}_g - \bar{e}_g)(|eu| - \sigma_g \hat{e}_g) \\
&= -a_m e^2 - e v_f + \hat{e}_f|e| - v_g e u + \hat{e}_g|eu| - \sigma_f(\hat{e}_f - \bar{e}_f)\hat{e}_f - \sigma_g(\hat{e}_g - \bar{e}_g)\hat{e}_g
\end{aligned}$$

$$
\begin{aligned}
&\leq -a_m e^2 - \sigma_f \left(\left(\hat{e}_f - \frac{\bar{e}_f}{2} \right)^2 - \frac{\bar{e}_f^2}{4} \right) - \sigma_g \left(\left(\hat{e}_g - \frac{\bar{e}_g}{2} \right)^2 - \frac{\bar{e}_g^2}{4} \right) \\
&\leq -a_m e^2 + \sigma_f \frac{\bar{e}_f^2}{4} + \sigma_g \frac{\bar{e}_g^2}{4},
\end{aligned}
$$

which is negative for $a_m e^2 > \rho^2$, where $\rho^2 = \sigma_f \frac{\bar{e}_f^2}{4} + \sigma_g \frac{\bar{e}_g^2}{4}$. Therefore, assuming that $\varepsilon > \frac{\rho}{\sqrt{a_m}}$, we have shown that $e(t)$ will converge to $|e(t)| < \varepsilon$. Note that while trying to ensure the condition $\varepsilon > \frac{\rho}{\sqrt{a_m}}$, the designer should not increase a_m since that increases the system bandwidth. Instead, the designer could decrease σ_f, decrease σ_g, or change the basis functions to decrease $\bar{e}_f$ or $\bar{e}_g$.

■ **EXAMPLE 6.8**

Consider the scalar system first considered in Example 6.5 on page 257 and subsequently reconsidered in Example 6.7. This example will use the same prior information, specification, and prefilter as in Example 6.7. The only change that is required to the controller of Example 6.7 to ensure that (ultimately) the tracking performance is achieved through adaptive approximation is that the parameters $\hat{e}_f$ and $\hat{e}_g$ be estimated using eqn. (6.74) instead of eqn. (6.67). △

■ **EXAMPLE 6.9**

Consider the system

$$
\dot{y} = f^* + (2 + g^*)u
$$

with $f^* = \frac{1}{8} y^2 \sin(0.3\pi y)$ and $g^* = \frac{1}{4}\left(y^2 + |y|\right)\cos(0.05\pi y)$. Note that the functions f^* and g^*, which are not known to the designer, satisfy all conditions stated in Example 6.5. This example compares results of simulations of the closed-loop systems designed in Examples 6.5 and 6.8. As much as is possible, the simulations corresponding to these two examples use the same parameters. We specify all parameters necessary to allow interested readers to replicate the simulation. The commanded input is $y_c(t) = 9\sin(0.2\pi t)$, which is applied as an input to the prefilter of eqns. (6.46)–(6.47) to obtain $y_d(t)$ and its derivative $\dot{y}(t)$. Even though this particular y_c is simple enough to be differentiated analytically, we use the prefilter approach because of its generality (i.e., if y_c were changed, no new control law derivations or programming would be required). For both controller implementations we select control parameter $a_m = 10.0$ and smoothing and dead-zone parameter $\varepsilon = 0.05$.

Variables indicating the performance of the closed-loop system using the bounding controller of Example 6.5 are shown in Figure 6.9. Only the first 20 s are shown as the controller has no state; hence, the performance is not time varying after the initial condition errors decay. The performance is quite good. However, to achieve this level of performance required Simulink to use the ODE45 integration routine with a maximum step size of 1.0e-4 s and a relative tolerance of 1.0e-6. For larger step size or higher tolerance, the control signal contained large-magnitude, high-frequency switching and the tracking error increases significantly. Such stringent settings of the numeric integration parameters indicate that a discrete-time implementation of this controller would require a very high sampling frequency. In fact, simulation analysis

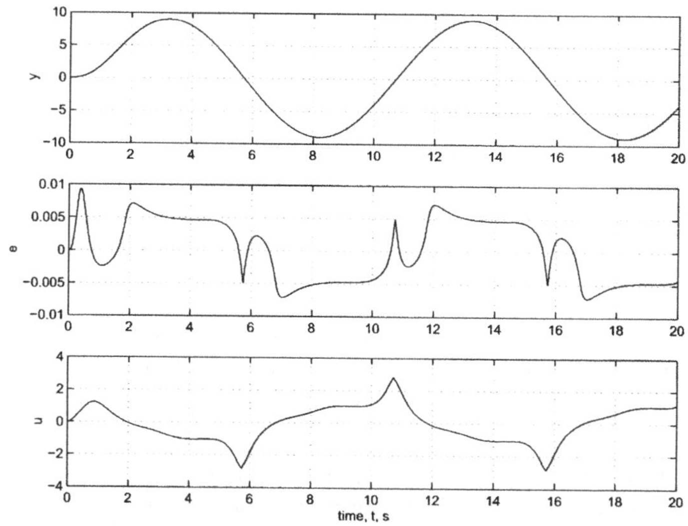

Figure 6.9: Simulation performance of the bounding controller described in Example 6.5. The output is y. The tracking error is $e = y - y_d$. The control signal is u.

of the controller with a sampling time T_s and zero-order-hold control signals between the sampling instants showed very large magnitude switching in the control signal for $T_s > 0.005$ s.

Variables indicating the performance of the approximation based controller of Example 6.7 are show in Figures 6.10-6.13. For implementation of this controller, the approximation functions $\hat{f}$ and $\hat{g}$ are each implemented using normalized radial basis functions using

$$g_i(y) = exp\left(-\frac{1}{2}\left(\frac{y - c_i}{1.2}\right)^2\right)$$

with centers located at $c_i = -10.0 + i$ for $i = 0, \ldots, 20$. Initially, $\theta_f = \theta_g = 0.0$. For function approximation by eqn. (6.66) the learning rates were $\Gamma_f = \Gamma_g = 100I$. For bound estimation by eqn. (6.74), $\gamma = 1.0$ and $\sigma_f = \sigma_g = 1.0$. Figure 6.10 shows the output $y(t)$, tracking error $e(t)$, and control signal $u(t)$ for the first 20 s of the simulation. This figure is included to show that the initial tracking error and control signal transients are reasonable. Note that the tracking error significantly exceeds the dead-zone ($\varepsilon = 0.05$). Therefore, function approximation is occurring. Figure 6.11 shows the output $y(t)$, tracking error $e(t)$, and control signal $u(t)$ for $t \in [80, 100]$. This figure shows that as learning progresses, the increased function approximation accuracy results in improved control performance as exhibited be the decreased tracking error. Note that for the majority of the time shown in Figure 6.11, the tracking error is within the dead-zone $|y| < \varepsilon = 0.05$ for which parameter adaptation no longer occurs. Only for short time intervals at specific ranges of y does the tracking error leave the dead-zone. Therefore, function approximation only is still occurring at those specific ranges of y. If the simulation were continued for a longer duration the tracking error would ultimately enter and remain in the dead-zone. The approximated functions are plotted at 10-s intervals in Figure 6.12. In this figure, the actual functions f^* and g^* are shown with dotted lines. The function approximation errors are shown in Figure 6.13. Several features of these figures are worth noting.

1. The initial $\hat{f}(y) = 0$. The subsequent sequence of approximated functions is ordered from top to bottom at $y = -9$.

2. The initial $\hat{g}(y) = 0$. The subsequent sequence of approximated functions is ordered from bottom to top at $y = -6$.

3. As time progresses and training samples are accumulated, the approximated functions appear to converge toward f^* and g^*. However, this should be interpreted with caution since the analysis guaranteed boundedness, but not convergence of the approximator parameters. Note for example that for the $t = 10$ plot of the functions that while the approximation error for $\hat{g}$ has decreased at $|y| = 2$ it has increased at $y = 0$.

4. As time increase, the approximation error for $|y| \in [9, 10]$ is increasing. This is due to the fact that very few training samples are available in that range. However, these parameters are not diverging. Instead, the parameters are being jointly adjusted to approximate the functions using the available training samples. Throughout the process, the Lyapunov function is decreasing.

These Simulink results were achieved using a maximum step size of 0.005 s and the default relative tolerance of 1e-3. A discrete-time implementation using a zero-order-hold control signal between control samples computed 100Hz yield essentially identical performance to that shown in Figures 6.10–6.13. △

Bounding Function Development. So far, we have estimated parameters $\bar{e}_f$ and $\bar{e}_g$ that bound the function approximation error over the entire region $\mathcal{D}$. In some applications, it is of interest to estimate functions that bound $|e_f(y)|$ and $|e_g(y)|$ over the entire region $\mathcal{D}$. When this is the case, we define $\hat{e}_f(y) = \psi_f^\top \phi(y)$ and $\hat{e}_g(y) = \psi_g^\top \phi(y)$ where each element of each vector ψ_f and ψ_g is positive. Define ψ_f^* and ψ_g^* as the vectors with the smallest elements such that $|e_f(y)| \leq (\psi_f^*)^\top \phi(y)$ and $|e_g(y)| \leq (\psi_g^*)^\top \phi(y)$ for any $y \in \mathcal{D}$. Then, we can change eqn. (6.74) to

$$\dot{\psi}_{f_i} = \gamma \phi_i(y) \left(|e_d| - \sigma_f \psi_{f_i} \right) \qquad \dot{\psi}_{g_i} = \gamma \phi_i(y) \left(|e_d u| - \sigma_g \psi_{g_i} \right) \tag{6.75}$$

for $y \in \mathcal{D}$ and $i = 1, \ldots, N$. With these changes, both the Lyapunov function and its derivative analysis will change, but the underlying ideas are the same.

Consider the Lyapunov function

$$V = \frac{1}{2}\left(e^2 + \frac{1}{\gamma} \sum_i \left((\psi_{f_i} - \psi_{f_i}^*)^2 + (\psi_{g_i} - \psi_{g_i}^*)^2 \right) + \tilde{\theta}_f^\top \Gamma_f^{-1} \tilde{\theta}_f + \tilde{\theta}_g^\top \Gamma_g^{-1} \tilde{\theta}_g \right).$$

The time derivative is

$$\begin{aligned} \frac{dV}{dt} &= -a_m e^2 - e v_f - e u v_g + e(f^* - \phi_f^\top \hat{\theta}_f) + e(g^* - \phi_g^\top \hat{\theta}_g) u \\ &\quad + \frac{1}{\gamma} \sum_i \left((\psi_{f_i} - \psi_{f_i}^*) \dot{\psi}_{f_i} + (\psi_{g_i} - \psi_{g_i}^*) \dot{\psi}_{g_i} \right) + \tilde{\theta}_f^\top \Gamma_f^{-1} \dot{\hat{\theta}}_f + \tilde{\theta}_g^\top \Gamma_g^{-1} \dot{\hat{\theta}}_g \end{aligned}$$

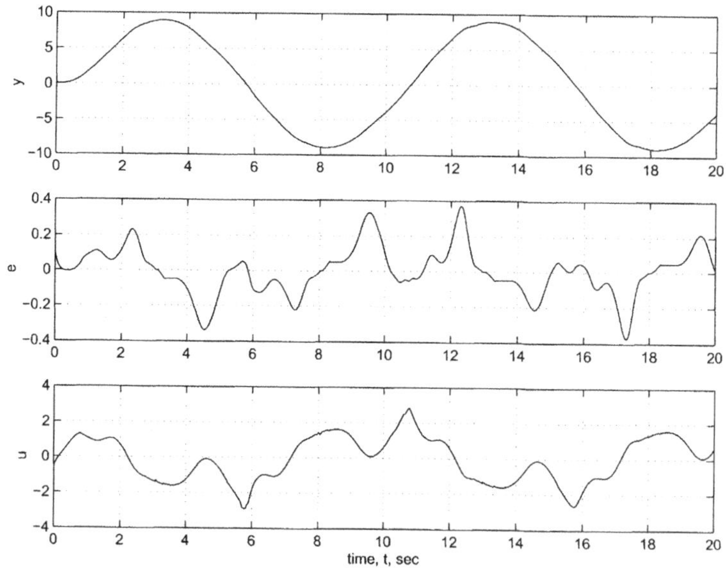

Figure 6.10: Initial simulation performance of the approximation based controller described in Example 6.7. The output is y. The tracking error is $e = y - y_d$. The control signal is u.

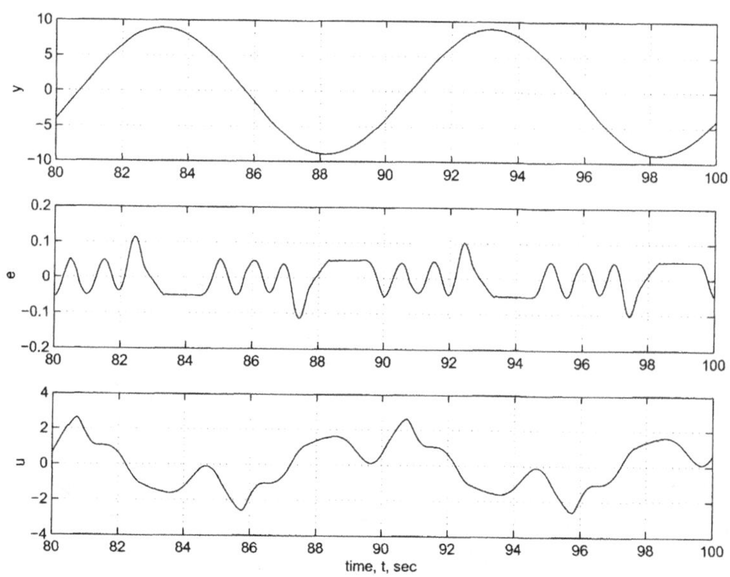

Figure 6.11: Simulation performance of the approximation based controller described in Example 6.7 for $t \in [80, 100]$. The output is y. The tracking error is $e = y - y_d$. The control signal is u.

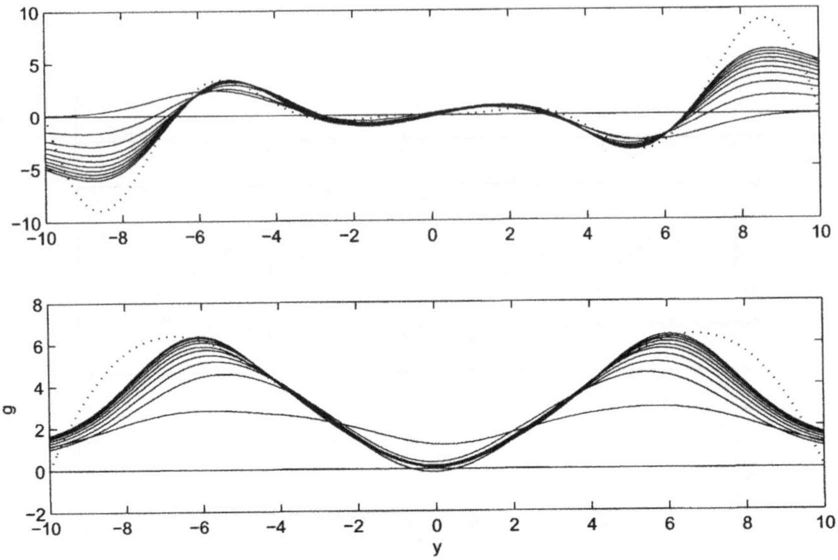

Figure 6.12: The functions f and g (dotted line) and their online approximations (solid lines) at 10-s intervals. At $t = 0$, $\hat{f}(y) = \hat{g}(y) = 0$.

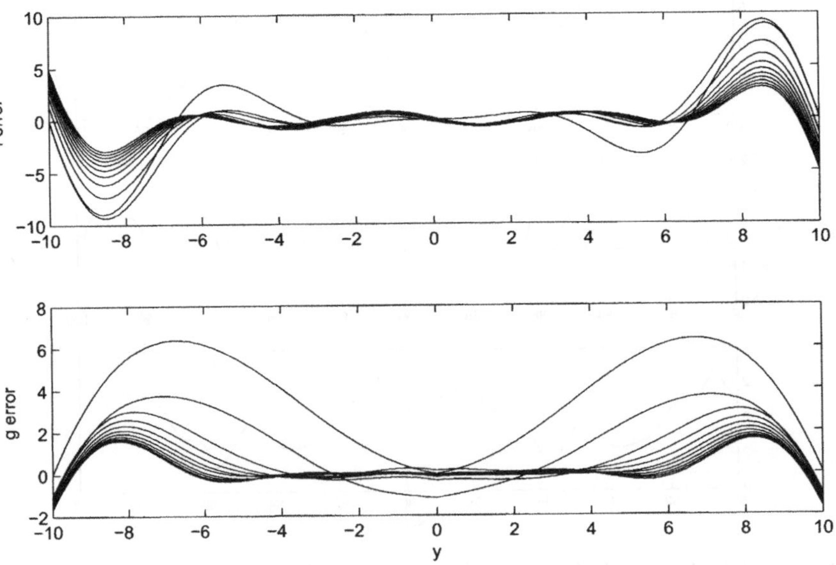

Figure 6.13: The function approximation errors at 10-s intervals.

$$\begin{aligned}
&= -a_m e^2 + e\left(-v_f + f^* - \phi_f^\top \hat{\theta}_f\right) + eu\left(-v_g + g^* - \phi_g^\top \hat{\theta}_g\right) \\
&\quad + \frac{1}{\gamma}\sum_i \left((\psi_{f_i} - \psi_{f_i}^*)\dot{\psi}_{f_i} + (\psi_{g_i} - \psi_{g_i}^*)\dot{\psi}_{g_i}\right) + \tilde{\theta}_f^\top \Gamma_f^{-1}\dot{\hat{\theta}}_f + \tilde{\theta}_g^\top \Gamma_g^{-1}\dot{\hat{\theta}}_g.
\end{aligned}$$

For y outside the approximation region $\mathcal{D}$, using the fact that the parameter adaptation is off, reduces the derivative of V to

$$\frac{dV}{dt} = -a_m e^2 + e\left(-v_f + f^*\right) + eu\left(-v_g + g^*\right) \le -a_m e^2,$$

which is negative semidefinite. In fact, for $y \in \Re^1 - \mathcal{D}$,

$$\frac{dV}{dt} \le -a_m \mu^2,$$

which ensures that y returns to $\mathcal{D}$ in finite time.

Within $\mathcal{D}$, for $|e| \ge \varepsilon$, and in the absence of projection,

$$\begin{aligned}
\frac{dV}{dt} &= -a_m e^2 + e\left(-v_f + e_f\right) + eu\left(-v_g + e_g\right) \\
&\quad + \frac{1}{\gamma}\sum_i \left((\psi_{f_i} - \psi_{f_i}^*)\dot{\psi}_{f_i} + (\psi_{g_i} - \psi_{g_i}^*)\dot{\psi}_{g_i}\right) \\
&\quad + \tilde{\theta}_f^\top \Gamma_f^{-1}\left(\dot{\hat{\theta}}_f - \Gamma_f \phi_f e\right) + \tilde{\theta}_g^\top \Gamma_g^{-1}\left(\dot{\hat{\theta}}_g - \Gamma_g \phi_g eu\right) \\
&\le -a_m e^2 - ev_f + |e_f||e| - v_g eu + |e_g||eu| \\
&\quad + \sum_i \left((\psi_{f_i} - \psi_{f_i}^*)\phi_i(y)\left(|e| - \sigma_f \psi_{f_i}\right) + (\psi_{g_i} - \psi_{g_i}^*)\phi_i(y)\left(|eu| - \sigma_g \psi_{g_i}\right)\right) \\
&\le -a_m e^2 - ev_f + \psi_f^\top \phi(y)|e| + \left(|e_f| - (\psi_f^*)^\top \phi(y)\right)|e| + \psi_g^\top \phi(y)|eu| \\
&\quad - v_g eu + \left(|e_g| - (\psi_g^*)^\top \phi(y)\right)|eu| \\
&\quad - \sum_i \phi_i(y)\left(\sigma_f\left(\psi_{f_i} - \psi_{f_i}^*\right)\psi_{f_i} + +\sigma_g\left(\psi_{g_i} - \psi_{g_i}^*\right)\psi_{g_i}\right) \\
&\le -a_m e^2 - ev_f + \hat{e}_f|e| - v_g eu + \hat{e}_g|eu| + \sum_i \phi_i(y)\rho_i^2 \\
&\le -a_m e^2 + \rho(y)^2
\end{aligned}$$

where

$$\begin{aligned}
\rho(y)^2 &= \sum_i \phi_i(y)\rho_i^2 \le \max_i(\rho_i^2) \\
\rho_i^2 &= \left(\sigma_f\left(\frac{\psi_{f_i}^*}{2}\right)^2 + \sigma_g\left(\frac{\psi_{g_i}^*}{2}\right)^2\right).
\end{aligned}$$

Therefore, the Lyapunov derivative is negative for $|e| > \varepsilon$, if $\varepsilon > \sqrt{\frac{\max_i(\rho_i^2)}{a_m}}$.

When projection occurs, its beneficial effects have been discussed in Section 4.6. Therefore, we have shown that $e(t)$ will converge, for any initial condition, to the set $|e| \le \varepsilon$, within which all parameter adaptation stops. The design parameter $\varepsilon > 0$ is selected by the designer independent of the control gain a_m. The parameter ε is small, but must be larger enough to satisfy the conditions stated in the analysis.

6.4 NONLINEAR PARAMETERIZED ADAPTIVE APPROXIMATION

The differences and trade-offs between linearly and nonlinearly parameterized approximators were discussed in Chapter 2. In the case of linearly parameterized approximators, the parameters σ are kept fixed, therefore $\hat{f}(y;\theta,\sigma) = \phi(y;\sigma)^\top\theta$ can be conveniently written as

$$\hat{f}(y;\theta,\sigma) = \theta^\top\phi(y) = \phi(y)^\top\theta$$

where the dependence of ϕ on the fixed σ vector can be dropped altogether. The synthesis and analysis of adaptive approximation based control systems developed so far in this chapter, were based on linearly parameterized approximators. In this section, we consider the case of nonlinearly parameterized approximation models and derive adaptive laws for updating not only the $\hat{\theta}$ parameters of the adaptive approximators but also the $\hat{\sigma}$ parameters.

Let us consider the adaptive approximation of the unknown function $f^*(y)$ by a nonlinearly parameterized approximator. For notational convenience, define $w := [\theta^\top \ \sigma^\top]^\top$. We have

$$\begin{aligned} f^*(y) &= \hat{f}(y;w^*) + \bar{e}_f(y) \\ &= \hat{f}(y;w) + \left[\hat{f}(y;w^*) - \hat{f}(y;w)\right] + \bar{e}_f(y) \end{aligned} \tag{6.76}$$

where $e_f(y)$ is the MFAE and w^* is the optimal weight vector that minimizes the MFAE within a compact set $\mathcal{D}$, which typically represents the approximation region (see Section 3.1.3).

If we assume that $\hat{f}(y;w)$ is a smooth function with respect to w then using the Taylor series expansion $\hat{f}(y;w^*)$ can be written as

$$\begin{aligned} \hat{f}(y;w^*) &= \hat{f}(y;w-\tilde{w}) \\ &= \hat{f}(y;w) - \frac{\partial\hat{f}}{\partial w}(y;w)\tilde{w} - \mathcal{F}(y;w), \end{aligned} \tag{6.77}$$

where $\tilde{w} := w - w^*$ is the parameter estimation error and $\mathcal{F}(y;w)$ represents the higher-order terms of $\hat{f}$ with respect to w.

Before proceeding with the analysis using eqn. (6.77), let us examine the properties of the higher-order term $\mathcal{F}$. Using the Mean Value Theorem [64], it can be shown that

$$|\mathcal{F}(y;w)| \le p(y;w)\|\tilde{w}\|$$

where

$$p(y;w) := \sup_{\bar{w}\in[w,\ w^*]} \left\| \frac{\partial\hat{f}}{\partial w}(y;\bar{w}) - \frac{\partial\hat{f}}{\partial w}(y;w) \right\|$$

and $[w,\ w^*]$ is a line segment connecting w and w^*; i.e.,

$$[w,\ w^*] := \{\chi \mid \chi = \lambda w + (1-\lambda)w^*;\ 0 \le \lambda \le 1\}.$$

It is noted that based on the definition of $p(y;w)$, the following property holds:

$$\lim_{w\to w^*} p(y;w) = 0 \qquad \forall y \in \mathcal{D}.$$

Therefore,

$$\lim_{w\to w^*} \frac{|\mathcal{F}(y;w)|}{\|\tilde{w}\|} = 0.$$

The higher-order term $\mathcal{F}$ encapsulates the nonlinear parametrization structure of the approximator. In the special case of a linearly parameterized approximator, $\mathcal{F}$ is identically equal to zero.

By substituting (6.77) in (6.76) we obtain

$$f^*(y) = \hat{f}(y;w) - \frac{\partial \hat{f}}{\partial w}(y;w)\tilde{w} - \mathcal{F}(y;w) + \bar{e}_f(y).$$

This can be written as

$$f^*(y) - \hat{f}(y;w) = -\frac{\partial \hat{f}}{\partial w}(y;w)\tilde{w} + \delta(y;w), \tag{6.78}$$

where $\delta(y;w) := \bar{e}_f(y) - \mathcal{F}(y;w)$. Now, let us consider the term

$$\frac{\partial \hat{f}}{\partial w}(y;w)\tilde{w}$$

for the case where $\hat{f}(y;w) = \phi(y;\sigma)^\top \theta$. In this case we have

$$\begin{aligned} \frac{\partial \hat{f}}{\partial w}(y;w)\tilde{w} &= \phi(y;\sigma)^\top \tilde{\theta} + \theta^\top \frac{\partial \phi}{\partial \sigma}(y;\sigma)\tilde{\sigma} \\ &= \phi(y;\sigma)^\top \tilde{\theta} + \xi(y;\theta,\sigma)^\top \tilde{\sigma} \end{aligned} \tag{6.79}$$

where

$$\xi(y;\theta,\sigma) := \frac{\partial \phi}{\partial \sigma}(y;\sigma)^\top \theta.$$

Suppose θ is of dimension q_1, and σ is of dimension q_2. Then ξ will be a vector of dimension q_2, whose k-th element can be computer by

$$\xi_k = \sum_{i=1}^{q_1} \theta_i \frac{\partial \phi_i}{\partial \sigma_k}(y;\sigma).$$

By substituting (6.79) in (6.78), we obtain

$$f^*(y) - \phi(y;\sigma)^\top \theta = -\phi(y;\sigma)^\top \tilde{\theta} - \xi(y;\theta,\sigma)^\top \tilde{\sigma} + \delta(y;\theta,\sigma). \tag{6.80}$$

Now, let us consider the system described in Section 6.3.5 by eqn. (6.60) where the unknown functions $f^*(y)$ and $g^*(y)$ are represented by nonlinearly parameterized approximators:

$$\begin{aligned} f^*(y) &= \phi_f(y;\sigma_f^*)^\top \theta_f^* + \bar{e}_f(y) \\ g^*(y) &= \phi_g(y;\sigma_g^*)^\top \theta_g^* + \bar{e}_g(y). \end{aligned}$$

The feedback controller is given by

$$u = \frac{-a_m e_F - c(y - y_d) + \dot{y}_d - f_o(y) - \phi_f(y;\hat{\sigma}_f)^\top \hat{\theta}_f}{g_o(y) + \phi_g(y;\hat{\sigma}_g)^\top \hat{\theta}_g}, \tag{6.81}$$

where $\phi_f(y;\hat{\sigma}_f)^\top \hat{\theta}_f$ and $\phi_g(y;\hat{\sigma}_g)^\top \hat{\theta}_g$ are the online estimates of the unknown functions $f^*(y)$ and $g^*(y)$, respectively. If we substitute the control law (6.81) in (6.60) then after some algebraic manipulation it can be shown that the filtered tracking error dynamics satisfy

$$\dot{e}_F = -a_m e_F + \left(f^*(y) - \phi_f(y;\hat{\sigma}_f)^\top \hat{\theta}_f\right) + \left(g^*(y) - \phi_g(y;\hat{\sigma}_g)^\top \hat{\theta}_g\right) u.$$

Hence using (6.80) we obtain

$$\begin{aligned}\dot{e}_F &= -a_m e_F - \phi_f(y;\hat{\sigma}_f)^\top \tilde{\theta}_f - \xi_f(y;\hat{\theta}_f,\hat{\sigma}_f)^\top \tilde{\sigma}_f - \phi_g(y;\hat{\sigma}_g)^\top \tilde{\theta}_g u \\ &\quad - \xi_g(y;\hat{\theta}_g,\hat{\sigma}_g)^\top \tilde{\sigma}_g u + \delta_f(y;\hat{\theta}_f,\hat{\sigma}_f) + \delta_g(y;\hat{\theta}_g,\hat{\sigma}_g)u.\end{aligned}$$

Using a Lyapunov function of the form

$$V = \frac{1}{2}e_F^2 + \frac{1}{2}\tilde{\theta}_f^\top \Gamma_{\theta_f}^{-1}\tilde{\theta}_f + \frac{1}{2}\tilde{\sigma}_f^\top \Gamma_{\sigma_f}^{-1}\tilde{\sigma}_f + \frac{1}{2}\tilde{\theta}_g^\top \Gamma_{\theta_g}^{-1}\tilde{\theta}_g + \frac{1}{2}\tilde{\sigma}_g^\top \Gamma_{\sigma_g}^{-1}\tilde{\sigma}_g,$$

we obtain the following adaptive update algorithms for generating the parameter estimates $\hat{\theta}_f(t), \hat{\sigma}_f(t), \hat{\theta}_g(t), \hat{\sigma}_g(t)$:

$$\dot{\hat{\theta}}_f = \Gamma_{\theta_f}\phi_f(y;\hat{\sigma}_f)e_F \tag{6.82}$$

$$\dot{\hat{\sigma}}_f = \Gamma_{\sigma_f}\xi_f(y;\hat{\theta}_f,\hat{\sigma}_f)e_F \tag{6.83}$$

$$\dot{\hat{\theta}}_g = \Gamma_{\theta_g}\phi_g(y;\hat{\sigma}_g)e_F u \tag{6.84}$$

$$\dot{\hat{\sigma}}_g = \Gamma_{\sigma_g}\xi_g(y;\hat{\theta}_g,\hat{\sigma}_g)e_F u. \tag{6.85}$$

Based on the feedback control law and adaptive laws, assuming that $\delta_f = \delta_g = 0$, the derivative of the Lyapunov function satisfies $\dot{V} = -a_m e_f^2$, which implies that the closed-loop system is stable and the filtered tracking error converges to zero. Of course, for applications nonzero δ_f and δ_g, local minima, and other issues must be addressed.

6.5 CONCLUDING SUMMARY

Adaptive approximation based control can be viewed as one of the available tools that a control designer should have in her/his control toolbox. Therefore, it is desirable for the reader not only to be able to apply, for example, neural network techniques to a certain class of systems, but more importantly to gain enough intuition and understanding about adaptive approximation so that she/he knows when it is a useful tool to be used and how to make necessary modifications or how to combine it with other control tools, so that it can be applied to a system which has not be encountered before.

In this chapter we have learned various key aspects of approximation based control and, hopefully, we have acquired some useful intuition about this control tool. We have studied the problem of designing and analyzing adaptive approximation based control. The presentation of this chapter has been restricted to a class of simple scalar systems with unknown nonlinearities, which has allowed the thorough analysis of the closed-loop system without the complicating mathematics that are usually encountered in higher dimensional systems.

The first section of the chapter presented a general framework for modeling of a dynamical system, design of a feedback control system, and evaluation and testing of the overall, closed-loop system. This discussion has provided the reader with a general perspective for the application of adaptive approximation based control in terms of handling modeling errors.

We then studied the stabilization of a scalar system. Our study started with the case of a known nonlinearity, proceeded to the case where the nonlinearity is unknown but there is a known bound available, and finally we considered the case where the nonlinearity is unknown and is approximated online. We studied various aspects of the adaptive approximation based control problem, including the effect on closed-loop performance of the learning rate, feedback gain, and initial conditions. In order to make the design of adaptive approximation based control more robust with respect to residual approximation error and disturbances, we studied its combination with adaptive bounding techniques, and analyzed the stability properties of the closed-loop system.

We then considered the tracking problem of a scalar system with two unknown nonlinearities. We studied the synthesis of stable approximation based control schemes and investigated the stability and robustness properties of the closed-loop system. Finally, we discussed the case of nonlinearly parameterized approximators.

The results of this chapter are extended to higher-order systems in the next chapter, which provides a general theory for the synthesis and analysis of adaptive approximation based control systems.

6.6 EXERCISES AND DESIGN PROBLEMS

Exercise 6.1 Consider the simple example examined in Example 6.3, where there is a single basis function. In the analysis presented on page 246 for trajectories outside the approximation region, we derived some intuition by considering the problem where the approximation error satisfies $\bar{e}_f(y) = 0$ for $|y| \leq \alpha$, and $\bar{e}_f(y) \leq k_e$ for $|y| > \alpha$. Now consider the case where the approximation error increases incrementally as follows:

$$\begin{aligned} \bar{e}_f(y) &= 0 & &\text{for } |y| \leq \alpha \\ |\bar{e}_f(y)| &\leq k_1 & &\text{for } \alpha \leq |y| \leq \beta \\ |\bar{e}_f(y)| &\leq k_2 & &\text{for } |y| \geq \beta, \end{aligned}$$

where $k_1 < k_2$ and $\alpha < \beta$. Repeat the derivation of the stability regions analytically and show them diagrammatically.

Exercise 6.2 Show that eqn. (6.16) is valid.

Exercise 6.3 Show the stability analysis of the combined adaptive approximation and adaptive bounding method developed in Section 6.2.7.

Exercise 6.4 Consider the tracking problem formulated in Section 6.3.2. Show that in the case of linearizing around the desired trajectory y_d, the control law (6.37) results in the closed-loop dynamics described by (6.38).

Exercise 6.5 Simulate the second-order system of Example 6.3, which is described by

$$\begin{aligned} \dot{y} &= -a_m y - \tilde{\theta}\phi(y) + \bar{e}_f(y) & \qquad y(0) &= y_0 \\ \dot{\tilde{\theta}} &= \gamma\phi(y)y & \qquad \tilde{\theta}(0) &= \tilde{\theta}_0, \end{aligned}$$

where $a_m = 0.4$, $\gamma = 1$, $\phi(y) = e^{-y^2}$, $y_0 = 0.5$ and

$$\bar{e}_f(y) = \begin{cases} 0 & \text{for } |y| \leq 1 \\ 5 & \text{for } |y| > 1 \end{cases}$$

Let the initial condition $\tilde{\theta}_0$ vary between 0 and -2 in increments of 0.2. For $t \in [0,\ 50]$, plot on the same graph $y(t)$ versus $\tilde{\theta}(t)$ for the cases where $\tilde{\theta}_0 = 0,\ -0.2,\ -0.4,\ \ldots\ -2.0$.

Exercise 6.6 Repeat the simulation of Exercise 6.5 with $\tilde{\theta}_0$ fixed as $\tilde{\theta}_0 = -1.5$, while γ is allowed to vary between 0.1 and 1.5 in increments of 0.2; i.e., $\gamma = 0.1,\ 0.3,\ 0.5,\ \ldots 1.3,\ 1.5$.

Exercise 6.7 Consider the scalar system model of Example 6.4 on page 255. Simulate the linearizing control law for:

1. the case of linearizing around $y = 0$; i.e., the error dynamics given by eqn. (6.39);
2. the case of linearizing around $e = 0$ (i.e., the error dynamics given by eqn. (6.40).

Select various initial conditions between $e(0) \in [-2,\ 2]$ and compare the performance of the two linearizing control schemes.

Exercise 6.8 Consider the nonlinear system

$$\dot{y} = (y-2)^2 + 2y^3 + (4-y^2)u.$$

The objective is to design a control law for tracking such that the system follows the desired reference signal $y_d = \sin t$. Following the small-signal linearization procedure of Section 6.3.2, first linearize the system around $y = 0$ and come up with a linear control law (let $a_m = 2$). Then linearize the system around the desired trajectory y_d and again derive the corresponding linear control law. In both cases, derive the closed-loop tracking error dynamics.

Exercise 6.9 For the problem described in Exercise 6.8 simulate the case of linearizing around the desired trajectory y_d. Let $a_m = 2$ and consider the following cases

1. $y(0) = 0$
2. $y(0) = 0.2$
3. $y(0) = -0.2$
4. $y(0) = 0.5$

By trying different initial conditions y(0), estimate the region of attraction for the closed-loop system; in other words, find the largest values for α and β such that if $y(0)$ satisfies $-\alpha \leq y(0) \leq \beta$ then $y(t)$ is able to follow the desired trajectory $y_d(t)$.

Exercise 6.10 In Section 6.3.3, a feedback control algorithm was designed and analyzed for the case where the unknown nonlinearities are within certain bounds. The design and analysis procedure was based on the feedback control law (6.41)–(6.43), which is discontinuous at $e = 0$ and $u_a = 0$.

In this exercise, design a smooth approximation of the form described by (6.9) and then perform a stability analysis of the smooth control law, similar to the analysis carried out in Section 6.2.4.

Exercise 6.11 Consider Example 6.5 presented in Section 6.3.3. Simulate the example for three values of ε: (i) $\varepsilon = 0$; (ii) $\varepsilon = 0.1$; (iii) $\varepsilon = 0.5$. In your simulation, assume that the unknown functions f^*, g^* are given by

$$f^*(y) = \begin{cases} \frac{1}{2}y^2 & \text{if } y \geq 0 \\ -\frac{1}{2}y^2 & \text{if } y < 0 \end{cases}$$

$$g^*(y) = \begin{cases} \frac{1}{2}y^2 & \text{if } y \geq -1 \\ -\frac{1}{2} & \text{if } y < -1 \end{cases}$$

and the reference input $y_c(t)$ is given by

$$y_c(t) = \begin{cases} 1 & \text{if } \ 2m \leq t \leq 2m+1 \quad m = 0,\ 1,\ 2,\ .... \\ -1 & \text{if } \ 2m+1 \leq t \leq 2m+2 \quad m = 0,\ 1,\ 2,\ .... \end{cases}$$

Note that $y_c(t)$ is a signal of period $t = 2s$, which oscillates between 1 and -1. Assume that $y(0) = 0$ and the initial conditions for the prefilter are zero. Plot $e(t)$, $y(t)$, $y_c(t)$, $y_d(t)$ and $u(t)$. Discuss both the positive and negative aspects of $u(t)$.

Exercise 6.12 The analysis on p. 259 considered one of four possible cases. Complete the proof for one of the remaining cases.

CHAPTER 7

ADAPTIVE APPROXIMATION BASED CONTROL: GENERAL THEORY

Chapter 6 motivated the use of adaptive approximation based control methods and discussed some of the key issues involved in the use of such methods for feedback control. In order to allow the reader to focus on the crucial issues without the distraction of mathematical complexities that occur while considering high-order systems, the design and analysis of that chapter was carried out on a class of scalar nonlinear systems. In this chapter, the design and analysis is extended to higher-order systems.

The objective of this chapter is to illustrate the design of adaptive approximation based control schemes for certain classes of n-th order nonlinear systems and to provide a rigorous stability analysis of the resulting closed-loop system. Although the mathematics become more involved as compared to Chapter 6, several important aspects of adaptive approximation extend directly from that previous analysis. These issues — such as stability analysis, control robustness, ensuring that the state remains in the region $\mathcal{D}$, and robustness modifications in the adaptive laws — are highlighted so that the reader is able to extract useful intuition for why various components of the control design follow a certain structure. A key objective is to help the reader obtain a sufficiently deep understanding of the mathematical analysis and design so that the results herein can be extended to a larger class of nonlinear systems or to a specific application whose model does not exactly fit within a standard class of nonlinear systems.

The design and analysis of adaptive approximation based control in this chapter is applied to two general classes of nonlinear systems with unknown nonlinearities: (i) feedback linearizable systems (Section 7.2); and (ii) triangular nonlinear systems that allow the use of the backstepping control design procedure (Section 7.3). For each class of nonlinear

Adaptive Approximation Based Control: Unifying Neural, Fuzzy and Traditional Adaptive Approximation Approaches. By Jay A. Farrell and Marios M. Polycarpou

systems, we first consider the ideal case where the uncertainties can be approximated exactly by the selected approximation model within a certain operating region of interest (i.e., the Minimum Function Approximation Error (MFAE) is zero within a certain domain $\mathcal{D}$), and then we consider the case that includes the presence of residual approximation errors and disturbances. The latter case is referred to as *robust adaptive approximation based control.* As we will see, to achieve robustness, we utilize a modification in the adaptive laws for updating the weights of the adaptive approximation. This modification in the adaptive laws is based on a combination of projection and dead-zone— techniques that have been introduced in Chapter 4, and also used in Chapter 6.

It is important to note that this chapter follows a structure parallel to that of Chapter 5, where we introduced various design and analysis tools for nonlinear systems under the assumption that the nonlinearities were known. In this chapter, we revisit these techniques (e.g., feedback linearization, backstepping), with adaptive approximation models representing the unknown nonlinearities.

7.1 PROBLEM FORMULATION

This section presents some issues in the problem formulation for adaptive approximation based control. As we will see, certain notation, assumptions, and control law terms will be used repeatedly throughout this chapter. To decrease redundancy throughout the following sections, these items and their related discussion are collected here.

7.1.1 Trajectory Tracking

Throughout this chapter, the objective is to design tracking controllers such that the system output $y(t)$ converges to $y_d(t)$ as $t \to \infty$. The controller may use the derivatives $y_d^{(i)}(t)$ for $i = 1, \ldots, n$. As discussed in Appendix A.4, these signals are continuous, bounded, and available without the need for explicit differentiation of the tracking signal.

Associated with the tracking signal $y_d(t)$, there is a desired state $x_d(t)$ of the system, which is assumed to belong to a certain known compact set $\mathcal{D}$ for all $t > 0$. In feedback linearization control methods it will typically be the case that the i-th component of the desired state satisfies: $x_{d_i}(t) = y_d^{(i-1)}(t)$, where $y_d^{(0)}(t) = y_d(t)$. In backstepping control approaches, the desired state is defined by certain intermediate control variables, denoted by α_i. The tracking error between $x_{d_i}(t)$ and $x_i(t)$ will be denoted by $\tilde{x}_i(t) = x_i(t) - x_{d_i}(t)$. The vector of tracking errors is denoted as $\tilde{x} = [\tilde{x}_1, \ldots, \tilde{x}_n]^\top$.

7.1.2 System

Throughout this chapter, the dynamics of each state variable may contain unknown functions. For example, the dynamics of the i-th state variable may be represented as

$$\dot{x}_i \quad = \quad (f_{o_i}(x) + f_i^*(x)) + (g_{o_i}(x) + g_i^*(x))\, z_i, \tag{7.1}$$

where z_i can be the control variable u or the next state x_{i+1}. The functions $f_{o_i}(x)$ and $g_{o_i}(x)$ are the known components of the dynamics and $f_i^*(x)$ and $g_i^*(x)$ are the unknown parts of the dynamics. Both the known portion of the dynamics, $f_{o_i}(x)$ and $g_{o_i}(x)$, and the unknown functions $f_i^*(x)$ and $g_i^*(x)$ are assumed to be locally Lipschitz continuous in x.

The unknown portions of the model will be approximated over the compact region $\mathcal{D}$. This region is sometimes referred to as the safe operating envelope. For any system, the region $\mathcal{D}$ is physically determined at the design stage. For example, an electrical motor is designed to operate within certain voltage, current, torque, and speed constraints. If these constraints are violated, then the electrical or mechanical components of the motor may fail; therefore, the controller must ensure that the safe physical limits of the system represented by $\mathcal{D}$ are not violated. The majority of this chapter focuses on analysis within $\mathcal{D}$. The control law does include auxiliary control terms to ensure that initial conditions outside $\mathcal{D}$ will converge to and remain in $\mathcal{D}$. Section 7.2.4 discusses one method for designing such auxiliary control terms.

7.1.3 Approximator

The system dynamics for the i-th state may contain unknown nonlinear functions denoted by $f_i^*(x)$ and $g_i^*(x)$. These unknown nonlinearities will be approximated by smooth functions $\hat{f}_i(x, \hat{\theta}_{f_i})$ and $\hat{g}_i(x, \hat{\theta}_{g_i})$, respectively, where the vectors $\hat{\theta}_{f_i} \in \Re^{q_{f_i}}$ and $\hat{\theta}_{g_{f_i}} \in \Re^{q_{g_i}}$ denote the adjustable parameters (weights) of each approximating function.

The state eqn. (7.1) can be expressed as

$$\begin{aligned} \dot{x}_i &= \left(f_{o_i}(x) + \hat{f}_i(x, \theta^*_{f_i})\right) + \delta_{f_i}(x) \\ &\quad + \left(g_{o_i}(x) + \hat{g}_i(x, \theta^*_{g_i})\right) z_i + \delta_{g_i}(x) z_i \end{aligned} \tag{7.2}$$

where $\theta^*_{f_i}$ and $\theta^*_{g_i}$ are some unknown "optimal" weight vectors, and δ_{f_i} and δ_{g_i} represent the (minimal) approximation error given by:

$$\begin{aligned} \delta_{f_i}(x) &= f_i^*(x) - \hat{f}_i(x, \theta^*_{f_i}) \\ \delta_{g_i}(x) &= g_i^*(x) - \hat{g}_i(x, \theta^*_{g_i}). \end{aligned}$$

Here, the terms *optimal* and *minimal* are used in the sense of the infinity norm of the error over $\mathcal{D}$, see eqns. (7.3) and (7.4). This minimal approximation error is a critical quantity, representing the minimum possible deviation between the unknown function f_i^* and the input/output function of the adaptive approximator $\hat{f}_i(x, \hat{\theta}_{f_i})$. In general, increasing the number of adjustable weights (denoted by q_{f_i}) reduces the minimal approximation error. The universal approximation results discussed in Section 2.4.5 indicate that any specified approximation accuracy ϵ can be attained uniformly on the compact region $\mathcal{D}$ if q_{f_i} is sufficiently large.

The optimal weight vectors $\theta^*_{f_i}$ and $\theta^*_{g_{f_i}}$ are unknown quantities required only for analytical purposes. Typically, $\theta^*_{f_i}$ is chosen as the value of θ_{f_i} that minimizes the network approximation error uniformly for all $x \in \mathcal{D}$; i.e.,

$$\theta^*_{f_i} := \arg \min_{\hat{\theta}_{f_i} \in \mathcal{R}^{q_{f_i}}} \left\{ \sup_{x \in \mathcal{D}} \left| f_i^*(x) - \hat{f}_i(x, \hat{\theta}_{f_i}) \right| \right\}. \tag{7.3}$$

Similarly, the optimal weight vector $\theta^*_{g_i}$ is chosen as

$$\theta^*_{g_i} := \arg \min_{\hat{\theta}_{g_i} \in \mathcal{R}^{q_{g_i}}} \left\{ \sup_{x \in \mathcal{D}} \left| g_i^*(x) - \hat{g}_i(x, \hat{\theta}_{g_i}) \right| \right\}. \tag{7.4}$$

With these definitions of the "optimal" parameters, we define the parameter estimation error vectors as

$$\tilde{\theta}_{f_i} = \hat{\theta}_{f_i} - \theta^*_{f_i} \text{ and } \tilde{\theta}_{g_i} = \hat{\theta}_{g_i} - \theta^*_{g_i}.$$

As we saw in the previous chapters (see Chapters 4 and 6), it is often desirable in the update law of a parameter estimate vector $\hat{\theta}$ to incorporate a projection modification $\mathcal{P}$ in order to constrain the parameter estimate within a certain region. Typically, there are two objectives in using the projection modification in the update law: (a) to ensure the boundedness of the parameter estimate vector, e.g., to avoid parameter drift; (b) to ensure the stabilizability of the parameter estimate, e.g., to guarantee that the parameter estimate does not enter a region that would cause the approximation function $(g_{o_i} + \hat{g}_i)$ to become too close to zero, since that may create stabilizability problems. In some cases, it is desirable for the projection modification to achieve both boundedness and stabilizability. In order to distinguish the different cases of using the projection modification, in this chapter we use the following notation:

- $\mathcal{P}_B$: projection to ensure boundedness;
- $\mathcal{P}_S$: projection to ensure stabilizability;
- $\mathcal{P}_{SB}$: projection to ensure both stabilizability and boundedness.

7.1.4 Control Design

The control design is based on the concept of replacing the unknown nonlinearities in the feedback control law by adaptive approximators, whose weights are updated according to suitable adaptive laws. Therefore, the feedback control law is a feedback linearizing controller (or backstepping controller) combined with adaptive laws for updating the weights of the adaptive approximators. The adaptive laws are derived based on a Lyapunov synthesis approach, which guarantees certain stability criteria.

The main emphasis of the feedback control design and analysis in this chapter is for $x \in \mathcal{D}$. We discuss briefly the stability analysis of the closed-loop system for $x \in (\Re^n - \mathcal{D})$, which is based on the use of two robustifying terms, denoted by v_f and v_g. The design of the robustifying terms is based on a bounding control approach (see Chapter 5).

Even within $\mathcal{D}$, there will exist nonzero, bounded approximation errors. Therefore, the control analysis is broken up into two parts: (i) the ideal case where it is assumed that the approximation error is zero; and, (ii) the realistic case where the approximation error is nonzero and in addition there may be disturbance terms. In the latter case, the main difference in the control design is the use of a combined projection and dead-zone modification to the adaptive laws, which prevents the parameter estimates from going into an undesirable parameter estimation region.

7.2 APPROXIMATION BASED FEEDBACK LINEARIZATION

In this section we consider the design and analysis of adaptive approximation based control for feedback linearizable systems. The reader will recall from Chapter 5 that feedback linearization is one of the most commonly used techniques for controlling nonlinear systems. Feedback linearization is based on the concept of cancelling the nonlinearities by the combined use of feedback and change of coordinates. In Section 5.2, we developed the main framework for feedback linearization based on the key assumption that the nonlinearities are completely known. In Section 5.4, we developed a set of robust nonlinear control design tools for addressing special cases of uncertainty, mostly based on taking a worst-case scenario. In Chapter 6, we introduced adaptive approximation techniques for a simple scalar system, which is a first step in the design of feedback linearization. Specifically, in Section 6.3.5 we considered the tracking problem for a scalar system and investigated the key issues encountered in the use of adaptive approximation based control.

In this section we consider the feedback linearization problem with unknown nonlinearities, which are approximated online. We start in Section 7.2.1 with a scalar system (similar to Chapter 6) in order to examine carefully the ideal case, the need for projection and dead-zone techniques, and the robustness issues that are involved. In Section 7.2.2 we consider adaptive approximation based control of input-state feedback linearizable systems, and in Section 7.2.3 we consider input-output feedback linearizable systems.

7.2.1 Scalar System

The simple scalar system

$$\begin{aligned} \dot{x} &= (f_o(x) + f^*(x)) + (g_o(x) + g^*(x))u & (7.5)\\ y &= x & (7.6)\end{aligned}$$

has already been extensively discussed in Chapter 6. To achieve tracking of y_d by y, the approximation based feedback linearizing control law is summarized as

$$\begin{aligned} u_a &= -a_m\tilde{x} + \dot{y}_d - f_o(y) - \phi_f(y)^\top\hat{\theta}_f - v_f & (7.7)\\ u &= \frac{u_a}{g_o(y) + \phi_g(y)^\top\hat{\theta}_g + v_g}. & (7.8)\\ \dot{\hat{\theta}}_f &= \Gamma_f\phi_f\tilde{x}, \qquad \text{for } x \in \mathcal{D} & (7.9)\\ \dot{\hat{\theta}}_g &= \mathcal{P}_S\left(\Gamma_g\phi_g\tilde{x}u\right), \qquad \text{for } x \in \mathcal{D} & (7.10)\end{aligned}$$

where $\tilde{x} = x - y_d$ is the tracking error, $a_m > 0$ is a positive design constant, Γ_f and Γ_g are positive definite matrices, and $\mathcal{P}_S$ is the projection operator that will be used to ensure the stabilizability condition on $\hat{\theta}_g$. The auxiliary terms v_f and v_g are included to ensure that the state remains within a certain approximation region $\mathcal{D}$, see Section 7.2.4.

Theorem 7.2.1 summarizes the stability properties for the adaptive approximation based controller in the ideal case where the MFAE is zero and there are no disturbance terms.

Theorem 7.2.1 [Ideal Case] *Let v_f and v_g be zero for $x \in \mathcal{D}$ and assume that ϕ_f, ϕ_g are bounded. In the ideal case where the MFAE and disturbances are zero, the closed-loop system composed of the system model (7.5) with the control law (7.7)–(7.10) satisfies the following properties:*

- $\tilde{x}, x, \hat{\theta}_f, \hat{\theta}_g \in \mathcal{L}_\infty$
- $\tilde{x} \in \mathcal{L}_2$
- $\tilde{x}(t) \to 0$ *as* $t \to \infty$.

Proof: Outside the region $\mathcal{D}$, we assume that the terms v_f and v_g have been defined to ensure that the state will converge to and remain in $\mathcal{D}$ (i.e., the set $\mathcal{D}$ is positively invariant). Therefore, the proof is only concerned with $x \in \mathcal{D}$.

For $x \in \mathcal{D}$ with the stated control law, after some algebraic manipulation, the dynamics of the tracking error $\tilde{x} = y - y_d$ reduce to

$$\dot{\tilde{x}} = -a_m\tilde{x} - \tilde{\theta}_f^\top\phi_f - \tilde{\theta}_g^\top\phi_g u.$$

The derivative of the Lyapunov function

$$V = \frac{1}{2}\left(\tilde{x}^2 + \tilde{\theta}_f^\top \Gamma_f^{-1}\tilde{\theta}_f + \tilde{\theta}_g^\top \Gamma_g^{-1}\tilde{\theta}_g\right) \tag{7.11}$$

satisfies

$$\dot{V} = -a_m\tilde{x}^2 + \tilde{\theta}_f^\top \Gamma_f^{-1}\left(\dot{\tilde{\theta}}_f - \Gamma_f\phi_f\tilde{x}\right) + \tilde{\theta}_g^\top \Gamma_g^{-1}\left(\dot{\tilde{\theta}}_g - \Gamma_g\phi_g u\tilde{x}\right).$$

Therefore, with the adaptive laws (7.9), (7.10), the time derivative of the Lyapunov function V becomes

$$\dot{V} = -a_m\tilde{x}^2$$

when the projection operator is not enforcing the stabilizability condition. Note that as long as ϕ_f, ϕ_g are bounded, Lemma A.3.1 completes the proof. When the projection operator is active, as discussed in Theorem 4.6.1, the stability properties of the control algorithm are preserved. ∎

The previous theorem considered the ideal case. The following theorems analyze two possible approaches applicable to more realistic situations, where we consider the presence of disturbance terms and MFAE.

Consider again the system described by (7.5) with the addition of another term $d(t)$, which may represent disturbances:

$$\begin{aligned} \dot{x} &= f_o(x) + f^*(x) + (g_o(x) + g^*(x))u + d \\ &= f_o(x) + \phi_f^\top\hat{\theta}_f^* + (g_o(x) + \phi_g^\top\hat{\theta}_g^*)u + \delta_f + \delta_g u + d \\ &= f_o(x) + \phi_f^\top\hat{\theta}_f^* + (g_o(x) + \phi_g^\top\hat{\theta}_g^*)u + \delta \end{aligned} \tag{7.12}$$

where δ is given by

$$\delta(x,u,t) = \delta_f(x) + \delta_g(x)u + d.$$

As discussed earlier, the first two terms of δ represent the MFAE, which arise due to the fact that the corresponding adaptive approximator is not able to match exactly the unknown functions f^* and g^* within the region $\mathcal{D}$.

Theorem 7.2.2 [Projection] *Let v_f and v_g be zero for $x \in \mathcal{D}$ and assume that ϕ_f, ϕ_g are bounded. Let the parameter estimates be adjusted according to*

$$\dot{\hat{\theta}}_f = \mathcal{P}_B\left(\Gamma_f\phi_f\tilde{x}\right), \quad \text{for } x \in \mathcal{D} \tag{7.13}$$

$$\dot{\hat{\theta}}_g = \mathcal{P}_{SB}\left(\Gamma_g\phi_g\tilde{x}u\right), \quad \text{for } x \in \mathcal{D} \tag{7.14}$$

where $\mathcal{P}_B$ is a projection operator designed to keep $\hat{\theta}_f$ in the convex and compact set $\mathcal{S}_f$ and $\mathcal{P}_{SB}$ is a projection operator designed to keep $\hat{\theta}_g$ in the convex and compact set $\mathcal{S}_g$, which is designed to ensure the stabilizability condition, the condition that $\theta_g^* \in \mathcal{S}_g$, and the boundedness of $\hat{\theta}_g$.

1. *In the case where $\delta = 0$,*

   - $\tilde{x}, x, \hat{\theta}_f, \hat{\theta}_g \in \mathcal{L}_\infty$
   - $\tilde{x} \in \mathcal{L}_2$

- $\tilde{x}(t) \to 0$ *as* $t \to \infty$.

2. *In the case where* $\delta \neq 0$*, but* $|\delta| \leq \delta_0$ *where* δ_0 *is an unknown positive constant, then*

   - $x, \tilde{x}, \tilde{\theta}_f, \tilde{\theta}_g, \hat{\theta}_f, \hat{\theta}_g \in \mathcal{L}_\infty$.

Proof: The proof of the case where $\delta = 0$ is straightforward based on the proofs of Theorems 4.6.1 and 7.2.1, and therefore it is not included here.

For $\delta \neq 0$, for $x \in \mathcal{D}$, with the stated control law, the dynamics of the tracking error $\tilde{x} = y - y_d$ reduce to

$$\dot{\tilde{x}} = -a_m\tilde{x} - \tilde{\theta}_f^\top \phi_f - \tilde{\theta}_g^\top \phi_g u + \delta.$$

The time derivative of the Lyapunov function of eqn. (7.11) becomes

$$\dot{V} = -a_m\tilde{x}^2 + \tilde{\theta}_f^\top \Gamma_f^{-1}\left(\dot{\tilde{\theta}}_f - \Gamma_f\phi_f\tilde{x}\right) + \tilde{\theta}_g^\top \Gamma_g^{-1}\left(\dot{\tilde{\theta}}_g - \Gamma_g\phi_g u\tilde{x}\right) + \tilde{x}\delta.$$

Using (7.13) and (7.14) the time derivative of the Lyapunov function becomes

$$\dot{V} = -a_m\tilde{x}^2 + \tilde{x}\delta,$$

if the projection modification is not in effect. In the case that the projection is active, then, as shown in Theorem 4.6.1, the stability properties are retained (in the sense that the additional terms in the derivative of the Lyapunov function are negative) and in addition it is guaranteed that the parameter estimates $\hat{\theta}_f, \hat{\theta}_g$ remain within the desired regions $\mathcal{S}_f$ and $\mathcal{S}_g$, respectively. Therefore, with the projection modification, we have

$$\dot{V} \leq -a_m\tilde{x}^2 + \tilde{x}\delta.$$

We note that as $|\tilde{x}|$ increases, at some point the term $-a_m\tilde{x}^2 + \tilde{x}\delta$ becomes negative. Therefore, $\tilde{x} \in \mathcal{L}_\infty$. When $a_m\tilde{x}^2 < \delta\tilde{x}$, the time derivative of the Lyapunov function may become positive. In this case, the parameter errors $\tilde{\theta}_f$ and $\tilde{\theta}_g$ may increase (this is referred to as parameter drift); however, the projection on the parameter estimates will maintain $\hat{\theta}_f \in \mathcal{S}_f$ and $\hat{\theta}_g \in \mathcal{S}_g$. By the compactness of $\mathcal{S}_f$ and $\mathcal{S}_g$, we attain $\hat{\theta}_f, \hat{\theta}_g, \tilde{\theta}_f, \tilde{\theta}_g \in \mathcal{L}_\infty$.

So far we have established that $\tilde{x}$, $\tilde{\theta}_f$ $\tilde{\theta}_g$ are bounded; however, it is not yet clear what is an upper bound or the limit for $\tilde{x}$, which is a key performance issue. Consider the two cases:

1. If $a_m\tilde{x}^2 < \delta\tilde{x}$, then $\tilde{\theta}_f$ and $\tilde{\theta}_g$ may increase with either $\hat{\theta}_f \to \partial\mathcal{S}_f$ or $\hat{\theta}_g \to \partial\mathcal{S}_g$, where $\partial\mathcal{S}_f$ and $\partial\mathcal{S}_g$ denote the bounding surfaces for $\mathcal{S}_f$ and $\mathcal{S}_g$, respectively. While this case remains valid, we have $a_m|\tilde{x}| < |\delta| < \delta_0$; however, a change in ϕ_f or ϕ_g may cause the state to switch to Case 2 at any time.

2. With $a_m\tilde{x}^2 \geq \delta\tilde{x}$, the time derivative of the Lyapunov function is decreasing. Let the condition $a_m\tilde{x}^2 \geq \delta\tilde{x}$ be satisfied for $t \in [t_{s_i}, t_{f_i}]$ with $|\tilde{x}(t_{s_i})| = |\tilde{x}(t_{f_i})| = \frac{\delta}{a_m}$. For t in this interval,

$$\frac{1}{2}\tilde{x}(t)^2 \leq V(t_{s_i}) < \frac{1}{2}\left(\left(\frac{\delta_0}{a_m}\right)^2 + B\right) \tag{7.15}$$

where

$$B = \max_{\hat{\theta}_f \in \mathcal{S}_f,\, \hat{\theta}_g \in \mathcal{S}_g} \left(\tilde{\theta}_f^\top \Gamma_f^{-1}\tilde{\theta}_f + \tilde{\theta}_g^\top \Gamma_g^{-1}\tilde{\theta}_g\right) < \infty.$$

Since $\tilde{x}$ is bounded and x_d is bounded, then x is also bounded.

Note that there is no limit to the number of times that the system can switch between cases 1 and 2. The fact that $\tilde{x}(t)$ becomes small for an extended period of time (i.e., Case 1) does not guarantee that it will stay small. Parameter drift or changes in the reference input may cause the system to switch from Case 1 to Case 2. The bound B applicable in Case 2 may be quite large as it is determined by the maximum value achieved over the allowable parameter sets. The term "bursting" has been used in the literature [5, 114, 119, 158] to describe the phenomenon where the tracking error $\tilde{x}$ is small in Case 1, and while it appears to have reached a steady state behavior, there occurs a switch to Case 2, which results in $\tilde{x}$ increasing dramatically. In summary, the best guaranteed bound for this approach is given by eqn. (7.15) and this bound is not small. ■

The previous result and the proof highlights the fact that merely proving boundedness is not necessarily useful in practice. From a designer's viewpoint, it is important to be able to manipulate the design variables in a way that improves the level of performance. The bound provided by (7.15) is not useful from a designer's point of view since it cannot be made sufficiently small by an appropriate selection of certain design variables.

In the next design approach, we introduce a dead-zone on the error variable and investigate the closed-loop stability properties of this new scheme.

Theorem 7.2.3 [Projection with Dead-Zone] *Let v_f and v_g be zero for $x \in \mathcal{D}$ and assume that ϕ_f, ϕ_g are bounded. Let the parameter estimates be adjusted according to*

$$\dot{\hat{\theta}}_f = \mathcal{P}_B\left(\Gamma_f \phi_f d\left(\tilde{x}, \varepsilon\right)\right), \quad \text{for } x \in \mathcal{D} \tag{7.16}$$

$$\dot{\hat{\theta}}_g = \mathcal{P}_{SB}\left(\Gamma_g \phi_g d\left(\tilde{x}, \varepsilon\right) u\right), \quad \text{for } x \in \mathcal{D} \tag{7.17}$$

where

$$d(\tilde{x}, \varepsilon) = \begin{cases} 0 & \text{if } |\tilde{x}| \le \varepsilon, \\ \tilde{x} & \text{if } |\tilde{x}| > \varepsilon, \end{cases}$$

and

$$\varepsilon = \frac{\delta_0}{a_m} + \mu$$

for some $\mu > 0$. $\mathcal{P}_B$ is a projection operator designed to keep $\hat{\theta}_f$ in the convex and compact set $\mathcal{S}_f$ and $\mathcal{P}_{SB}$ is a projection operator designed to keep $\hat{\theta}_g$ in the convex and compact set $\mathcal{S}_g$, which is designed to ensure the stabilizability condition, the condition that $\theta_g^* \in \mathcal{S}_g$, and the boundedness of $\hat{\theta}_g$. In the case where $|\delta| < \delta_0$,

1. *$\tilde{x}$, x, $\hat{\theta}_f$, $\hat{\theta}_g \in \mathcal{L}_\infty$;*

2. *$\tilde{x}$ is small-in-the-mean-square sense, satisfying*

$$\int_t^{t+T} \tilde{x}^2(\tau)d\tau \le \frac{2}{a_m}V(t) + \varepsilon^2 T;$$

3. *$\tilde{x}(t)$ is uniformly ultimately bounded by ε; i.e., the total time such that $|\tilde{x}(t)| > \varepsilon$ is finite.*

Proof: Let the condition $|\tilde{x}(t)| > \varepsilon$ be satisfied for $t \in (t_{s_i}, t_{f_i})$, $i = 1, 2, 3, \ldots$, where $t_{s_i} < t_{f_i} \le t_{s_{i+1}}$, $|\tilde{x}(t)| \le \varepsilon$ for $t \in (t_{f_i}, t_{s_{i+1}})$, t_{s_0} is assumed to be zero without

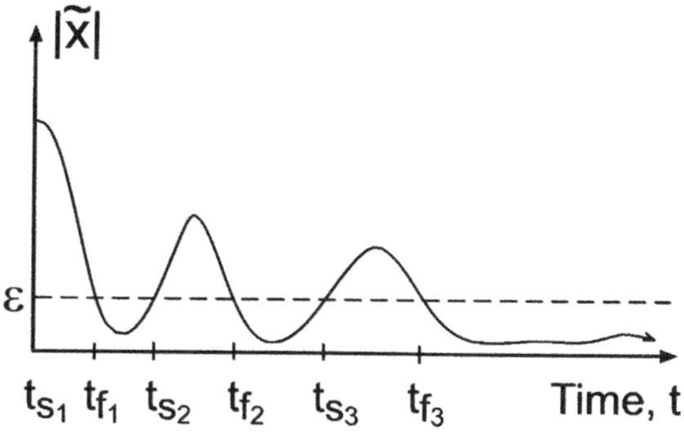

Figure 7.1: Illustration of the definitions of the time indices for the proof of Theorem 7.2.3.

loss of generality, and $t_{s_{i+1}}$ may be infinity for some i (see Figure 7.1). Following the same procedure as in the previous proof, for $t \in (t_{s_i}, t_{f_i})$ where $i = 1, 2, 3, \ldots$, the time derivative of the Lyapunov function (7.11) reduces to

$$\begin{aligned} \dot{V} &= -a_m \tilde{x}^2 + \tilde{x}\delta \\ &\leq -a_m|\tilde{x}|\left(|\tilde{x}| - \frac{|\delta|}{a_m}\right) \\ &\leq -a_m \varepsilon \mu. \end{aligned}$$

In this case, $\dot{V}$ is negative; therefore,

$$V(t_{f_i}) \leq V(t_{s_i}) - a_m \varepsilon \mu \left(t_{f_i} - t_{s_i}\right).$$

During the interval $t \in [t_{f_i}, t_{s_{i+1}}]$, parameter adaptation is off. In addition, $|\tilde{x}(t_{f_i})| = |\tilde{x}(t_{s_{i+1}})| = \varepsilon$; therefore, $V(t_{f_i}) = V(t_{s_{i+1}})$. This implies:

$$\begin{aligned} V(t_{f_i}) &\leq V(t_{s_i}) - a_m \varepsilon \mu \left(t_{f_i} - t_{s_i}\right) \\ &\leq V(t_{f_{i-1}}) - a_m \varepsilon \mu \left(t_{f_i} - t_{s_i}\right) \\ &\leq V(t_{f_{i-2}}) - a_m \varepsilon \mu \left(\left(t_{f_i} - t_{s_i}\right) + \left(t_{f_{i-1}} - t_{s_{i-1}}\right)\right) \\ &\vdots \\ V(t_{f_i}) &\leq V(0) - a_m \varepsilon \mu \sum_{k=1}^{i} \left(t_{f_k} - t_{s_k}\right). \end{aligned}$$

Therefore, since $V(t_{f_i}) \geq 0$,

$$\sum_{k=1}^{i} \left(t_{f_k} - t_{s_k}\right) \leq \frac{V(0)}{a_m \varepsilon \mu},$$

which shows that the total time spent with $|\tilde{x}| > \varepsilon$ must be finite. Note also that $V(t_{f_i})$ for $i = 1, 2, 3, \ldots$ is a positive decreasing sequence. This implies that either the sequence terminates with i, t_{f_i} and $V(t_{f_i})$ being finite or $\lim_{i\to\infty} V(t_{f_i}) = V_\infty$ exists and is finite. In addition, if $t > t_{f_i}$, then $V(t) \leq V(t_{f_i})$.

Using the inequality

$$xy \leq \alpha^2 x^2 + \frac{1}{4\alpha^2} y^2, \qquad \forall \alpha \neq 0,$$

we have that

$$\begin{aligned} -a_m \tilde{x}^2 + \tilde{x}\delta &\leq (\alpha^2 - a_m)\tilde{x}^2 + \frac{1}{4\alpha^2}\delta_0^2 \\ &= \frac{-a_m}{2}\tilde{x}^2 + \frac{1}{2a_m}\delta_0^2 \end{aligned}$$

where $\alpha^2 = a_m/2$. Therefore, we obtain

$$\dot{V} \leq \frac{-a_m}{2}\tilde{x}^2 + \frac{1}{2a_m}\delta_0^2. \tag{7.18}$$

Integrating both sides of (7.18) over the time interval $[t,\ t+T]$ yields

$$\begin{aligned} \int_t^{t+T} \dot{V}(\tau)d\tau &\leq \frac{-a_m}{2}\int_t^{t+T} \tilde{x}^2(\tau)d\tau + \frac{1}{2a_m}\int_t^{t+T} \delta_0^2 d\tau \\ \frac{2}{a_m}\left(V(t) - V(t+T) + \frac{T}{2a_m}\delta_0^2\right) &\geq \int_t^{t+T} \tilde{x}^2(\tau)d\tau \\ \left(\frac{2}{a_m}V(t) + \frac{\delta_0^2}{a_m^2}T\right) &\geq \int_t^{t+T} \tilde{x}^2(\tau)d\tau. \end{aligned}$$

Hence

$$\int_t^{t+T} \tilde{x}^2(\tau)d\tau \leq \frac{2}{a_m}V(t) + \varepsilon^2 T,$$

which completes the proof. ■

In this case, where $|\delta(t)| \leq \delta_0$ for all $t > 0$, then the ultimate bound for the tracking error is $|\tilde{x}(t)| \leq \varepsilon$. In practice, it is usually not desirable to decrease the bound by increasing a_m, since this parameter is directly related to the magnitude of the control signal (see eqn. (7.7)) and the rate of decay of transient errors. Instead, the designer can consider whether it is possible to decrease δ_0. If δ_0 was determined by disturbances, then not much can be done – unless the general structure of the disturbance is already known. If δ_0 was determined by unmodeled nonlinear effects, then the designer can enhance the structure of the adaptive approximator.

One of the possible disadvantages of the dead-zone is the need to know an upper bound δ_0 on the uncertainty. However, if the designer utilizes a smaller than necessary δ_0 which results in the inequality $|\delta(t)| \leq \delta_0$ not being valid for some $t > 0$, then the stability result is essentially the same as for Theorem 7.2.2. This is left as an exercise (see Problem 7.1).

7.2.2 Input-State

To illustrate some of the key concepts and to obtain some intuition regarding the control and robustness design, in Section 7.2.1, we considered the scalar case. In this subsection we consider the n-th order input-state feedback linearization case. In Section 7.2.2.1, we first consider the ideal case where the MFAE and disturbances are all zero. In Section 7.2.2.2, we consider an approach to achieve robustness with respect to these same issues.

7.2.2.1 *Ideal Case.* Consider nonlinear systems of the so-called companion form:

$$\dot{x}_1 = x_2 \tag{7.19}$$

$$\dot{x}_2 = x_3 \tag{7.20}$$

$$\vdots$$

$$\dot{x}_n = (f_o(x) + f^*(x)) + (g_o(x) + g^*(x))u, \tag{7.21}$$

where $x = [x_1\ x_2\ \cdots\ x_n]^\top$ is the state vector, f_o, g_o are known functions, while $f^*(x)$ and $g^*(x)$ are unknown functions, which are to be estimated using adaptive approximators. The tracking objective is satisfied if $y(t) = x_1(t)$ converges to a desirable tracking signal $y_d(t)$. The tracking error dynamics are

$$\begin{aligned}
\dot{\tilde{x}}_1 &= \tilde{x}_2 \\
\dot{\tilde{x}}_2 &= \tilde{x}_3 \\
&\vdots \\
\dot{\tilde{x}}_n &= (f_o(x) + f^*(x)) + (g_o(x) + g^*(x))u - y_d^{(n)}(t),
\end{aligned}$$

where $\tilde{x}_i(t) = x_i(t) - y_d^{(i-1)}(t)$. The tracking error dynamics can be written in matrix state space form as

$$\dot{\tilde{x}} = A\tilde{x} + B\left(f_o(x) + f^*(x) + (g_o(x) + g^*(x))u - y_d^{(n)}(t)\right) \tag{7.22}$$

where

$$A = \begin{bmatrix} 0 & 1 & 0 & \dots & 0 \\ 0 & 0 & 1 & \dots & 0 \\ \vdots & \vdots & \vdots & \ddots & \vdots \\ 0 & 0 & 0 & \dots & 1 \\ 0 & 0 & 0 & \dots & 0 \end{bmatrix}, \qquad B = \begin{bmatrix} 0 \\ 0 \\ \vdots \\ 0 \\ 1 \end{bmatrix}.$$

The approximation based feedback linearizing controller is selected as

$$u_a = v_l + y_d^{(n)} - f_o(x) - \phi_f(x)^\top \hat{\theta}_f - v_f \tag{7.23}$$

$$u = \frac{u_a}{g_o(x) + \phi_g(x)^\top \hat{\theta}_g + v_g} \tag{7.24}$$

$$\dot{\hat{\theta}}_f = \Gamma_f \phi_f e, \qquad \text{for } x \in \mathcal{D} \tag{7.25}$$

$$\dot{\hat{\theta}}_g = \mathcal{P}_S\left(\Gamma_g \phi_g e u\right), \qquad \text{for } x \in \mathcal{D} \tag{7.26}$$

where v_l is a stabilizing linear controller to be defined later, Γ_f and Γ_g are positive definite matrices representing the adaptive gain, $e(t)$ is the training error that will be defined subsequently, and $\mathcal{P}_S$ is the projection operator that will be used to ensure the stabilizability condition (i.e., $g_o(x) + \phi_g(x)^\top \hat{\theta}_g > 0$).

As previously, v_f and v_g are included to ensure that the states remain within the predefined approximation region $\mathcal{D}$. While approximation based controllers could be developed for a variety of stabilizing linear controllers v_l, for simplicity we will select the state feedback $v_l = -K^\top \tilde{x}$ with $K = [k_1\ \ k_2\ \cdots\ k_n]^\top \in \Re^n$, resulting in $A - BK^\top$ being a Hurwitz matrix; i.e., all the eigenvalues of $A - BK^\top$ are in the open left half complex plane.

In the ideal case where we assume that the MFAE is zero (i.e., there exists θ_f^* such that $f^*(x) = \phi_f(x)^\top \theta_f^*$ for all $x \in \mathcal{D}$, and correspondingly for θ_g^*), then the control law defined in (7.23) and (7.24) results in the following closed-loop tracking error dynamics

$$\dot{\tilde{x}} = \left(A - BK^\top\right)\tilde{x} - B\phi_f(x)^\top \tilde{\theta}_f - B\phi_g(x)^\top \tilde{\theta}_g u. \tag{7.27}$$

Since the feedback gain vector K is selected such that $A - BK^\top$ is Hurwitz, for any positive definite Q there exists a positive definite matrix P satisfying the Lyapunov equation

$$P(A - BK^\top) + (A - BK^\top)^\top P = -Q. \tag{7.28}$$

In the following, without any loss of generality we will select $Q = I$. Finally, based on the solution of the Lyapunov equation, we define the scalar training error $e(t)$ as follows:

$$e = B^\top P\tilde{x}.$$

The stability properties of this control law are summarized in Theorem 7.2.4.

Theorem 7.2.4 [Ideal Case] *Let v_f and v_g be zero for $x \in \mathcal{D}$ and assume that ϕ_f, ϕ_g are bounded. In the ideal case where the MFAE and disturbances are zero, the closed-loop system (7.19)-(7.21) with the control law (7.23)-(7.26) satisfies the following properties:*

- $\tilde{x}$, x, $\hat{\theta}_f$, $\hat{\theta}_g \in \mathcal{L}_\infty$
- $\tilde{x} \in \mathcal{L}_2$
- $\tilde{x}(t) \to 0$ *as* $t \to \infty$.

Proof: Outside the region $\mathcal{D}$, we assume that the terms v_f and v_g have been defined to ensure that the state will converge to and remain in $\mathcal{D}$. Therefore, the proof is only concerned with $x \in \mathcal{D}$.

For $x \in \mathcal{D}$, the time derivative of the Lyapunov function

$$V = \tilde{x}^\top P\tilde{x} + \tilde{\theta}_f^\top \Gamma_f^{-1}\tilde{\theta}_f + \tilde{\theta}_g^\top \Gamma_g^{-1}\tilde{\theta}_g \tag{7.29}$$

satisfies

$$\dot{V} = -\tilde{x}^\top \tilde{x} + 2\tilde{\theta}_f^\top \Gamma_f^{-1}\left(\dot{\tilde{\theta}}_f - \Gamma_f\phi_f e\right) + 2\tilde{\theta}_g^\top \Gamma_g^{-1}\left(\dot{\tilde{\theta}}_g - \Gamma_g\phi_g e u\right)$$

for $Q = I$. Therefore, with the adaptive laws (7.25)–(7.26), the Lyapunov time derivative $\dot{V}$ becomes

$$\dot{V} = -\tilde{x}^\top \tilde{x},$$

which is negative semidefinite. In the case that the projection operator $\mathcal{P}_S$ becomes active in order to ensure the stabilizability condition, as discussed earlier, the stability properties are preserved. Therefore, $\dot{V}$ satisfies $\dot{V} \leq -\tilde{x}^\top \tilde{x}$ for all $x \in \mathcal{D}$. Hence the application of Lemma A.3.1 completes the proof. ∎

The above design and analysis of approximation based input-state feedback linearization was developed for nonlinear systems of the companion form (7.19)–(7.21). This can be extended to a more general class of feedback linearizable systems of the form

$$\dot{x} = Ax + B\beta^{-1}(x)\left[u - \alpha(x)\right], \tag{7.30}$$

where u is a scalar control input, x is an n-dimensional state vector, A is an $n \times n$ matrix, B is an $n \times 1$ matrix, and the pair (A, B) is controllable. The unknown nonlinearities are contained in the continuous functions $\alpha : \Re^n \mapsto \Re^1$ and $\beta : \Re^n \mapsto \Re^1$, which are defined on an appropriate domain of interest $\mathcal{D}$, with the function $\beta(x)$ assumed to be nonzero for every $x \in \mathcal{D}$. It is noted that the case of systems of the form (7.30) with known nonlinearities was studied in Section 5.2. The tracking control objective is for $x(t)$ to track the desired state $x_d(t)$, where $x_d(t)$ is generated by the reference model

$$\dot{x}_d = Ax_d + Br, \tag{7.31}$$

and $r(t)$ denotes a certain command tracking signal (see Appendix Section A.4). Using the definition $\tilde{x} = x - x_d$, the tracking error dynamics are described by

$$\dot{\tilde{x}} = A\tilde{x} + B\left[\beta^{-1}(x)\left(u - \alpha(x)\right) - r\right].$$

These tracking error dynamics can be written in the form (similar to (7.22)):

$$\dot{\tilde{x}} = A\tilde{x} + B\left((f_o(x) + f^*(x)) + (g_o(x) + g^*(x))u - r(t)\right) \tag{7.32}$$

where

- $f_o(x)$ is the known component of $-\beta^{-1}(x)\alpha(x)$;
- $f^*(x)$ is the unknown component of $-\beta^{-1}(x)\alpha(x)$;
- $g_o(x)$ is the known component of $\beta^{-1}(x)$;
- $g^*(x)$ is the unknown component of $\beta^{-1}(x)$.

Note that the command signal $r(t)$ corresponds to the signal $y_d^{(n)}(t)$ that was used for the nonlinear system in companion form.

Once the tracking error dynamics are formulated as in eqn. (7.32), the approximation based feedback controller (7.23)–(7.26), with $y_d^{(n)}(t)$ being replaced by $r(t)$, can be used to achieve the tracking results described by Theorem 7.2.4.

In the next subsection, we consider the case where the adaptive approximators cannot match exactly the unknown nonlinearities within the domain of interest $\mathcal{D}$ (i.e., the MFAE is nonzero), and there may be disturbance terms.

7.2.2.2 Robustness Considerations. In a realistic situation, there will be modeling errors. If the modeling error, represented by δ, satisfies a matching condition, then the tracking error dynamics satisfy

$$\dot{\tilde{x}} = A\tilde{x} + B\left(f_o(x) + f^*(x) + (g_o(x) + g^*(x))u - y_d^{(n)}(t)\right) + B\delta.$$

As previously, the term $\delta(t)$ may contain disturbance terms, as well as residual approximation errors due to the MFAE, which was discussed earlier. The following theorem presents a projection and dead-zone modification in the adaptive laws of the parameter estimates that ensures some key robustness properties.

Theorem 7.2.5 [Projection with Dead-Zone] *Let v_f and v_g be zero for $x \in \mathcal{D}$ and assume that ϕ_f, ϕ_g are bounded. Let the parameter estimates be adjusted according to*

$$\dot{\hat{\theta}}_f = \mathcal{P}_B\left(\Gamma_f \phi_f d\left(e, \tilde{x}, \varepsilon\right)\right), \qquad \textit{for } x \in \mathcal{D} \tag{7.33}$$

$$\dot{\hat{\theta}}_g = \mathcal{P}_{SB}\left(\Gamma_g \phi_g d\left(e, \tilde{x}, \varepsilon\right) u\right), \qquad \textit{for } x \in \mathcal{D} \tag{7.34}$$

where for P satisfying eqn. (7.28)

$$d(e,\tilde{x},\varepsilon) = \begin{cases} 0 & \text{if } \tilde{x}^\top P\tilde{x} \leq \bar{\lambda}_P\varepsilon^2, \\ e & \text{if } \tilde{x}^\top P\tilde{x} > \bar{\lambda}_P\varepsilon^2, \end{cases} \tag{7.35}$$

$$\varepsilon = 2\|PB\|_2\delta_0 + \mu \tag{7.36}$$

where $\mu > 0$ is a positive constant, and $\bar{\lambda}_P$ and $\underline{\lambda}_P$ are the maximum and minimum eigenvalues of P respectively. $\mathcal{P}_B$ is a projection operator designed to keep $\hat{\theta}_f$ in the convex and compact set S_f and $\mathcal{P}_{SB}$ is a projection operator designed to keep $\hat{\theta}_g$ in the convex and compact set S_g, which is designed to ensure the stabilizability condition, the condition that $\theta_g^* \in S_g$, and the boundedness of $\hat{\theta}_g$. In the case where $|\delta| < \delta_0$,

1. *e, x, $\hat{\theta}_f$, $\hat{\theta}_g \in \mathcal{L}_\infty$;*

2. *$\tilde{x}$ is small-in-the-mean-square sense, satisfying*

$$\int_t^{t+T} \|\tilde{x}^2(\tau)\|_2^2 d\tau \leq 2V(t) + \frac{\bar{\lambda}_P}{\underline{\lambda}_P}\varepsilon^2 T;$$

3. *$\|\tilde{x}(t)\|_2$ is uniformly ultimately bounded by ε; i.e., the total time such that $\tilde{x}^\top P\tilde{x} > \bar{\lambda}_P\varepsilon^2$ is finite.*

Proof: Outside the region $\mathcal{D}$, we assume that the terms v_f and v_g have been defined to ensure that the state will return to and remain in $\mathcal{D}$. Therefore, the proof will only be concerned with $x \in \mathcal{D}$.

In the region $\mathcal{D}$, with P selected to solve the Lyapunov eqn. (7.28) with $Q = I$, the time derivative of the Lyapunov function (7.29) is

$$\dot{V} = -\tilde{x}^\top\tilde{x} + 2\tilde{\theta}_f^\top\Gamma_f^{-1}\left(\dot{\tilde{\theta}}_f - \Gamma_f\phi_f e\right) + 2\tilde{\theta}_g^\top\Gamma_g^{-1}\left(\dot{\tilde{\theta}}_g - \Gamma_g\phi_g eu\right) + 2\tilde{x}^\top PB\delta.$$

Suppose that the time intervals (t_{s_i}, t_{f_i}) are defined as discussed relative to Figure 7.1, so that the condition $\tilde{x}(t)^\top P\tilde{x}(t) > \bar{\lambda}_P\varepsilon^2$ is satisfied only for $t \in (t_{s_i}, t_{f_i})$, $i = 1, 2, 3, \ldots$, where $t_{s_i} < t_{f_i} \leq t_{s_{i+1}}$. Since $\tilde{x}(t_{f_i})^\top P\tilde{x}(t_{f_i}) = \tilde{x}(t_{s_{i+1}})^\top P\tilde{x}(t_{s_{i+1}}) = \bar{\lambda}_P\varepsilon^2$ and parameter estimation is off for $t \in [t_{f_i}, t_{s_{i+1}}]$, we have that $V(t_{f_i}) = V(t_{s_{i+1}})$. When $t \in (t_{s_i}, t_{f_i})$ for any i the fact that $\tilde{x}^\top P\tilde{x} > \bar{\lambda}_P(2\|PB\|_2\delta_0 + \mu)^2$ ensures that $\|\tilde{x}\|_2 > 2\|PB\|_2\delta_0 + \mu$; therefore, when projection is not in effect,

$$\begin{aligned} \dot{V} &= -\tilde{x}^\top\tilde{x} + 2\tilde{x}^\top PB\delta \\ &\leq -\|\tilde{x}\|_2^2 + 2\|\tilde{x}\|_2\|PB\|_2|\delta| \\ &\leq -\|\tilde{x}\|_2(\|\tilde{x}\|_2 - 2\|PB\|_2\delta_0) \\ &\leq -\varepsilon\mu. \end{aligned}$$

Therefore, by integrating both sides over (t_{s_i}, t_{f_i}),

$$\begin{aligned} V(t_{f_i}) &\leq V(t_{s_i}) - \varepsilon\mu\,(t_{f_i} - t_{s_i}) \\ &\leq V(t_{f_{i-1}}) - \varepsilon\mu\,(t_{f_i} - t_{s_i}) \\ &\leq V(t_{f_{i-2}}) - \varepsilon\mu\,((t_{f_i} - t_{s_i}) + (t_{f_{i-1}} - t_{s_{i-1}})) \\ &\;\;\vdots \\ &\leq V(0) - \varepsilon\mu\left(\sum_{k=1}^{i}(t_{f_k} - t_{s_k})\right). \end{aligned}$$

Hence, since $V(t_{f_i}) \geq 0$,

$$\sum_{k=1}^{i}(t_{f_k} - t_{s_k}) \leq \frac{V(0)}{\varepsilon\mu},$$

which shows that the total time spent with $\tilde{x}^\top P\tilde{x} > \bar{\lambda}_P\varepsilon^2$ is finite. In addition, $V(t_{f_i})$ $i = 1, 2, 3, \ldots$ is a positive decreasing sequence, either this is a finite sequence or $\lim_{i\to\infty} V(t_{f_i}) = V_\infty$ exists and is finite. In addition, if $t > t_{f_i}$, then $V(t) < V(t_{f_i})$.

Within the dead-zone, it is obvious that $\underline{\lambda}_P\|\tilde{x}\|_2^2 \leq \tilde{x}^\top P\tilde{x} \leq \bar{\lambda}_P\varepsilon^2$ implies

$$\int_t^{t+T} \|\tilde{x}(\tau)\|_2^2 d\tau \leq \frac{\bar{\lambda}_P}{\underline{\lambda}_P}\varepsilon^2 T.$$

Outside the dead-zone, using the inequality,

$$xy \;\leq\; \beta^2 x^2 + \frac{1}{4\beta^2}y^2, \qquad \forall\beta \neq 0,$$

we have that

$$\begin{aligned}
\dot{V} &\leq -\|\tilde{x}\|_2^2 + 2\|\tilde{x}\|_2\|PB\|_2\delta_0 \\
&\leq -\|\tilde{x}\|_2^2 + \beta^2\|\tilde{x}\|_2^2 + \frac{4\|PB\|_2^2}{4\beta^2}\delta_0^2 \\
&= (\beta^2 - 1)\,\|\tilde{x}\|_2^2 + \frac{\|PB\|_2^2}{\beta^2}\delta_0^2 \\
&= -\frac{1}{2}\|\tilde{x}\|_2^2 + 2\|PB\|_2^2\delta_0^2,
\end{aligned}$$

for $\beta^2 = 0.5$. Integrating both sides of the last equation over the time interval $[t,\ t+T]$ we obtain

$$\begin{aligned}
\int_t^{t+T} \dot{V}(\tau)d\tau &\leq -\frac{1}{2}\int_t^{t+T} \|\tilde{x}(\tau)\|_2^2\, d\tau + 2\|PB\|_2^2 \int_t^{t+T} \delta_0^2\, d\tau \\
\int_t^{t+T} \|\tilde{x}(\tau)\|_2^2\, d\tau &\leq 2\left(V(t) - V(t+T) + 2\|PB\|_2^2 T\delta_0^2\right) \\
\int_t^{t+T} \|\tilde{x}(\tau)\|_2^2\, d\tau &\leq 2V(t) + 4\|PB\|_2^2\delta_0^2 T.
\end{aligned}$$

Therefore,

$$\int_t^{t+T} \|\tilde{x}(\tau)\|_2^2\, d\tau \leq 2V(t) + \varepsilon^2 T \leq 2V(t) + \frac{\bar{\lambda}_P}{\underline{\lambda}_P}\varepsilon^2 T,$$

which completes the proof. ■

The mean-square and the ultimate bound ε are increasing functions of the bound δ_0 on the model error. When the model error is determined predominantly by the MFAE, the performance can be improved, independent of the control parameters K, by increasing the capabilities of the adaptive approximator, which decreases δ_0.

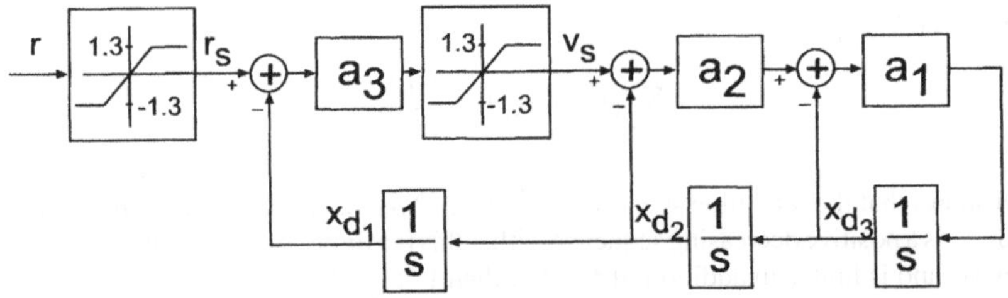

Figure 7.2: Block diagram implementation of the trajectory generation prefilter described in Section 7.2.2.3.

7.2.2.3 Detailed Example This subsection presents a simulation implementation of the control approach of Section 7.2.2 applied to the system

$$\begin{aligned} \dot{x}_1 &= x_2 \\ \dot{x}_2 &= x_3 \\ \dot{x}_3 &= f(x_1, x_2) + g(x_1, x_2)u \end{aligned}$$

which is of the form of eqns. (7.19)–(7.21). The only knowledge of f and g assumed at the design stage is that both are continuous with $-1 \leq f \leq 1$ and $0.05 \leq g$. Therefore, $f_o = g_o = 0$. We also assume that the system is designed to safely operate over the region $(x_1, x_2) \in \mathcal{D} = [-1.3, 1.3] \times [-1.3, 1.3]$.

The user of the system specifies a desired output $r(t)$ that will be used to generate a desired trajectory $x_d(t) = [x_{d_1}(t)\; x_{d_2}(t)\; x_{d_3}(t)]^\top$ and $\dot{x}_{d_3}(t)$ such that x_d is continuous; x_d and $\dot{x}_{d_3}(t)$ are bounded; $\dot{x}_{d_1} = x_{d_2}$ and $\dot{x}_{d_2} = x_{d_3}$; and, $(x_{d_1}(t), x_{d_2}(t)) \in \mathcal{D}$ for all $t > 0$. The trajectory generation system is defined by

$$\begin{aligned} \dot{x}_{d_1} &= x_{d_2} \\ \dot{x}_{d_2} &= x_{d_3} \\ \dot{x}_{d_3} &= a_1 \left(a_2 \left[\sigma\left(a_3(r_s - x_{d_1})\right) - x_{d_2}\right] - x_{d_3}\right) \end{aligned}$$

where $r_s = \sigma(r)$ and $\sigma(\cdot)$ is that saturation function

$$\sigma(x) = \begin{cases} 1.3 & \text{if} \quad x > 1.3 \\ x & \text{if} \quad |x| \leq 1.3 \\ -1.3 & \text{if} \quad x < -1.3. \end{cases}$$

Figure 7.2 shows this trajectory generation prefilter in block diagram form. The signal $r_s(t)$ is a magnitude limited version of $r(t)$ that is treated as the commanded value of x_{d_1}. The signal $v(t) = a_3(r_s(t) - x_{d_1}(t))$ has the correct sign to drive x_{d_1} toward r_s. The signal v_s has the same sign as v, but its magnitude is constrained to $[-1.3, 1.3]$, so that it can be interpreted as a desired value for x_{d_2}. By the design of the filter, $(r_s(t), v_s(t)) \in \mathcal{D}$ for all $t \geq 0$ and the filter is designed so that (x_{d_1}, x_{d_2}) track (r_s, v_s). However, due to the dynamics of the filter, tracking may not be perfect. If it is essential that the commanded trajectory always remain in $\mathcal{D}$, then the magnitude limits in the function $\sigma(\cdot)$ should be decreased from ± 1.3. We select the parameter vector $[a_1, a_2, a_3] = [9, 3, 1]$. Within the linear range of the trajectory generator, this choice yields the transfer functions

$$\frac{x_{d_1}}{r} = \frac{27}{s^3 + 9s^2 + 27s + 27}$$

$$\frac{x_{d_2}}{r} = \frac{27s}{s^3+9s^2+27s+27}$$
$$\frac{x_{d_3}}{r} = \frac{27s^2}{s^3+9s^2+27s+27}$$

which have three poles at $s = -3$. As long as r is bounded, the signal $x_d^{(3)}$ will be bounded, but it is not necessarily continuous. Issues related to the design of such trajectory generation prefilters are discussed in Appendix Section A.4.

For $x \in \mathcal{D}$, the adaptive approximation based controller is designed to satisfy the requirements of Theorem 7.2.5. The control gain is selected as $K = [1, 3, 3]$ which gives

$$A - BK = \begin{bmatrix} 0 & 1 & 0 \\ 0 & 0 & 1 \\ -1 & -3 & -3 \end{bmatrix}$$

with A and B defined as in eqn. (7.22). The matrix $A - BK$ is Hurwitz with all three eigenvalues equal to -1. The matrix solving Lyapunov eqn. (7.28) with $Q = I$ is

$$P = \begin{bmatrix} 2.3125 & 1.9375 & 0.5000 \\ 1.9375 & 3.2500 & 0.8125 \\ 0.5000 & 0.8125 & 0.4375 \end{bmatrix}$$

which has eigenvalues 0.2192, 0.8079, and 4.9728. The vector $L = B^\top P$ takes the value $L = [0.5000, 0.8125, 0.4375]$. Note that this choice of L ensures that the transfer function $L\left(sI - (A - BK)\right)^{-1} B$ is strictly positive real, according to the Kalman-Yakubovich Lemma (see page 392). Satisfaction of the SPR condition is critical to the design of a stable adaptive system.

For the approximators, a lattice network was designed with centers located on the grid defined by $b \times b$ with $b = [-1.300, -0.975, -0.650, -0.325, 0.000, 0.325, 0.650, 0.975, 1.300]$ to yield a set of 81 centers:

$$c = \begin{bmatrix} c_1 \\ c_2 \\ \vdots \\ c_{80} \\ c_{81} \end{bmatrix} = \begin{bmatrix} (-1.300, -1.300) \\ (-1.300, -0.975) \\ \vdots \\ (\ 1.300,\ \ 0.975) \\ (\ 1.300,\ \ 1.300) \end{bmatrix}.$$

We define the i-th regressor element by the biquadratic function

$$\phi_{f_i}(x_1, x_2) = \phi_{g_i}(x_1, x_2) = \left(1 - \left(\frac{\|v - c_i\|}{\mu}\right)^2\right)^2$$

where $v = (x_1, x_2)$. The value of μ was selected to be 0.66. The approximators are $\hat{f} = \theta_f^\top \phi_f$ and $\hat{g} = \theta_g^\top \phi_g$. The parameter vectors are estimated according to eqns. (7.33–7.34). The deadzone in the adaptation law was designed so that parameter estimation would occur for $x \in \mathcal{D}$ and $\tilde{x}^\top P \tilde{x} > 0.002$:

$$d(e, \tilde{x}) = \begin{cases} 0 & \text{if } \tilde{x}^\top P \tilde{x} \le 0.002, \\ e & \text{if } \tilde{x}^\top P \tilde{x} > 0.002. \end{cases} \tag{7.37}$$

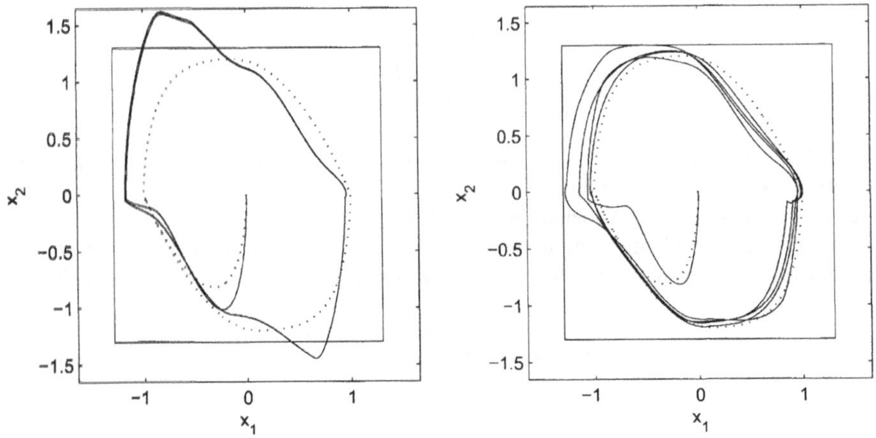

Figure 7.3: Phase plane plot of x_1 versus x_2 for $t \in [0, 100]$. The left plot shows the performance without adaptive approximation. The right plot shows the performance with adaptive approximation. In each plot, the dotted line is the desired trajectory for $t \in [0, 100)$ s. The thin solid line represents the actual trajectory. The domain of approximation $\mathcal{D} = [-1.3, 1.3] \times [-1.3, 1.3]$ is also shown.

The learning rate matrices were diagonal with all diagonal elements equal to $\frac{10}{\|L\|}$. A projection operator is included in the adaptation law for θ_g to ensure that each element of the vector θ_g remains larger than 0.05. All elements of θ_f are initialized to zero. All elements of θ_g are initialized to 0.5.

This paragraph focuses on the design of the control signal to ensure that states outside of $\mathcal{D}$ are returned to $\mathcal{D}$. Since $\mathcal{D}$ is defined only by the variables (x_1, x_2), the design focuses on forcing x_3 to take a value x_{3c} that is designed to cause (x_1, x_2) to return to $\mathcal{D}$. For $x \notin \mathcal{D}$, we define $x_{3c} = -x_1 - h(x_2)$ where $h(\cdot)$ can be selected from the class of functions such that $yh(y) > 0$ for all $y \neq 0$. Let $z = [x_1, x_2, (x_3 - x_{3c})]$. If $z_3 > 0$ we select

$$u = \begin{cases} 0 & \text{if} \quad 2x_2 + f_u + \frac{dh}{dx_2}x_3 \leq 0 \\ -\frac{1}{g_l}\left(2x_2 + f_u + \frac{dh}{dx_2}x_3\right) & \text{if} \quad 2x_2 + f_u + \frac{dh}{dx_2}x_3 > 0. \end{cases}$$

If $z_3 < 0$ we select

$$u = \begin{cases} 0 & \text{if} \quad 2x_2 + f_l + \frac{dh}{dx_2}x_3 \geq 0 \\ \frac{1}{g_l}\left|2x_2 + f_l + \frac{dh}{dx_2}x_3\right| & \text{if} \quad 2x_2 + f_l + \frac{dh}{dx_2}x_3 < 0. \end{cases}$$

We select $h(x_2) = signum(x_2)$ for which we define $\frac{dh}{dx_2} = 0$ even for $x_2 = 0$. If we select the Lyapunov function $V = \frac{1}{2}z^\top z$, then it is straightforward (see Problem 7.7) to show that this choice of u yields $\dot{V} \leq -z_2 h(z_2)$. The function V is decreasing outside of $\mathcal{D}$ except when $z_2 = 0$. Since $z_2 = 0$ is not a stationary point of the system, invariance theory shows that trajectories outside $\mathcal{D}$ will be forced into $\mathcal{D}$; however, because the boundary of $\mathcal{D}$ is not a portion of a level curve of V, we cannot show that $\mathcal{D}$ is a a positively invariant set.

The simulation results are shown in Figures 7.3–7.7. For implementation of the plant in the simulation, $f = cos\left(\frac{\pi}{1.2}R_2\right)$ and $g = (x_1 + x_2)^2 + 2e^{-R_2}$ where $R_2 = x_1^2 + x_2^2$.

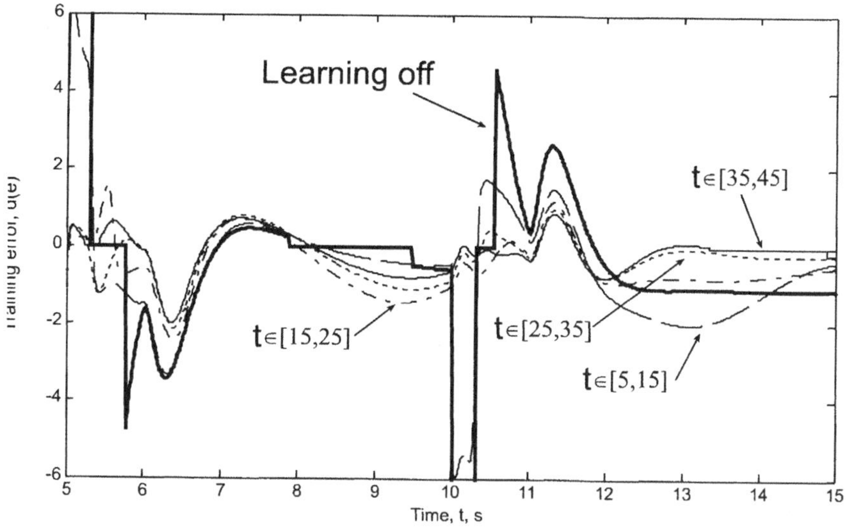

Figure 7.4: Training error $e = L\tilde{x}$ after processing through the deadzone operator $d(e, \tilde{x})$ for $t \in [5, 15]$ s (dashed), $t \in [15, 25]$ s (dash-dot), $t \in [25, 35]$ s (dotted), and $t \in [35, 45]$ s (thin solid). The wide solid line shows a portion of the error e for the simulation without adaptive approximation. The time axis of each plot has been shifted by a multiple of $T = 10$ s to increase the resolution of the time axis and to facilitate direct comparison across repetitions of the trajectory.

Both graphs in Figure 7.3 displays the desired trajectory as a dotted line. For $t \in [0, 100]$, the input $r(t)$ is a unit amplitude square wave with period $T = 10$ s. The state of the trajectory generation system starts at the origin. Therefore, for $t \in [0, 5)$ the effects of the initial condition of the state of x_d are dominant. For $t \in [5, 100]$s, the desired state has essentially converged to a repetitive trajectory pattern with period T. To analyze the performance improvement over the repetitive portion of the desired trajectory, the discussion of the next three paragraphs will focus on $t \in [5, 100]$s. Both graphs also display the square operating region $\mathcal{D}$.

The narrow solid curve of the left graph of Figure 7.3 is the plot of $x_1(t)$ versus $x_2(t)$ when the simulation is run with learning turned off. Note that the actual trajectory does leave $\mathcal{D}$ twice for every repetition of the desired trajectory, but is returned to $\mathcal{D}$ by the control law. Also, without learning, the tracking performance does not improve from one repetition of the pattern to the next.

The narrow solid curve of the right graph of Figure 7.3 is the plot of $x_1(t)$ versus $x_2(t)$ when the simulation is run with learning turned on. As the system operates, the tracking performance improves. This is shown more clearly in Figure 7.4, which displays the training error for the first four repetitions of the trajectory pattern. For graphical purposes the tracking error for each 10 s interval is shifted in time by a multiple of $T = 10$ s. This shifting enhances the resolution of the time axis and facilitates the comparison of the training errors at corresponding points in the repeating pattern.

For $t \in [5, 15]$ s the training error $e = L\tilde{x}$ is plotted as a dashed line. For $t \in [15, 25]$ s the training error is plotted as a dash-dot line. For $t \in [25, 35]$ s the training error is plotted

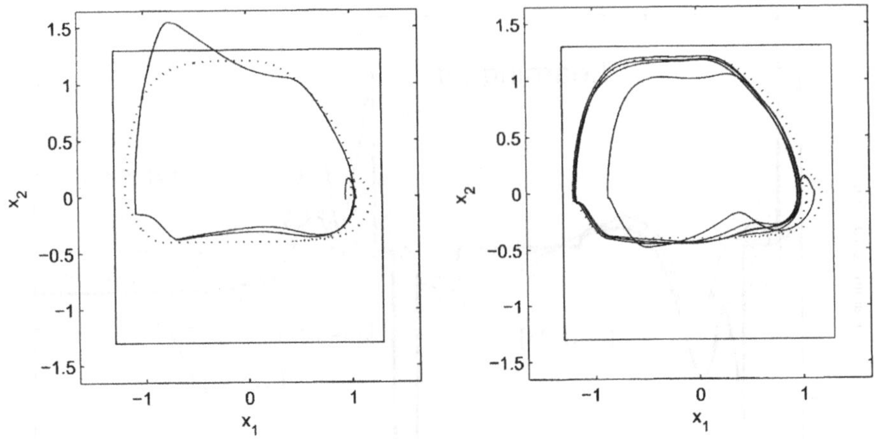

Figure 7.5: Phase plane plot of x_1 versus x_2 for $t \in [100, 200]$. The left plot shows the performance without online approximation. The right plot shows the performance with online approximation. In each plot, the dotted line is the desired trajectory for $t \in [100, 200]$ s. The thin solid line represents the actual trajectory. The domain of approximation $\mathcal{D} = [-1.3, 1.3] \times [-1.3, 1.3]$ is also shown.

as a dotted line. For $t \in [35, 45]$ s the training error is plotted as the thin solid line. Figure 7.4 plots $d(e, \tilde{x})$, not e. The effect of the deadzone is particularly evident in the plot for $t \in [35, 45]$ s. Note that with online approximation, the training error tends to decrease with each repetition of the trajectory. The wide solid line shows a clipped portion of the error e for the simulation with learning turned off. For the simulation without learning, the range of the training error was $(-13.5, 14)$.

At $t = 100$ s, the signal $r(t)$ is changed to a sawtooth wave with amplitude 2.0 and period $T = 10.0$. The first two components of the resulting desired trajectory x_d are again shown as the dotted line in both graphs of Figure 7.5. Note that for $x_2 > 0$ the trajectory lies in similar regions of $\mathcal{D}$ as did the previous trajectory. However, when $x_2 < 0$ the two trajectories pass through different portions of the operating envelope $\mathcal{D}$. Note that the system with learning maintains accurate tracking for $x_2 > 0$ where the functions had previously converged, but requires several repetitions before achieving accurate tracking on the new regions of $\mathcal{D}$. This demonstrates that learning is achieved as a function of the operating point, not as a function of a specific trajectory. The training errors for the sawtooth generated trajectory are displayed in Figure 7.6. Again the time axis is shifted to that corresponding portions of the repeating pattern line up vertically. The improvement in performance as the number of repetitions increases is easily observed. Again, the graph of the training error when learning is turned off (wide solid) is clipped from its maximum value of 15.5 to enhance the vertical resolution of the plot.

Define an indicator signal

$$p(\tilde{x}(t))) = \begin{cases} 1 & \text{if } \tilde{x}^\top P \tilde{x} > 0.002 \\ 0 & \text{otherwise.} \end{cases}$$

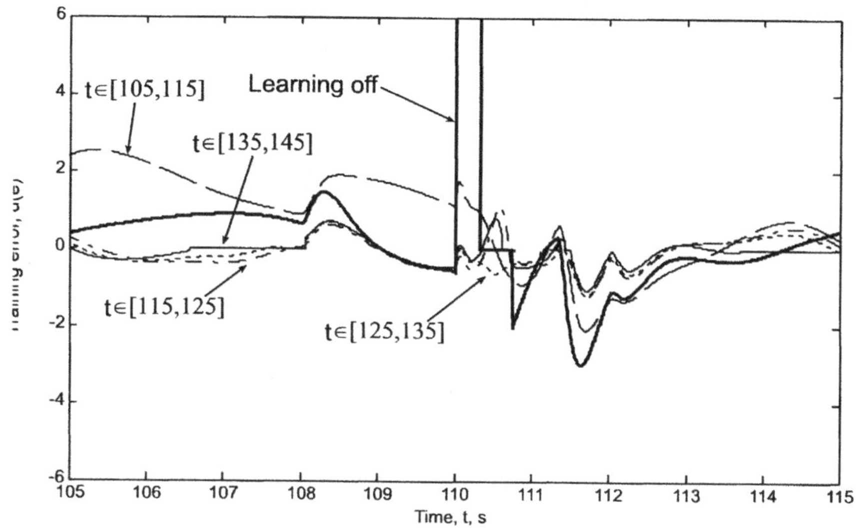

Figure 7.6: Training error $e = L\tilde{x}$ after processing through the deadzone operator $d(e, \tilde{x})$ for $t \in [105, 115]$ s (dashed), $t \in [115, 125]$ s (dash-dot), $t \in [125, 135]$ s (dotted), and $t \in [135, 145]$ s (thin solid). The wide dotted line shows a portion of the error e for the simulation without online approximation. The time axis of each plot has been shifted by a multiple of $T = 10$-s to increase the resolution of the time axis and to facilitate direct comparison across repetitions of the trajectory.

Also define the signal

$$y(t) = \int_{t-10}^{t} p(\tau) d\tau.$$

This signal represents the total time during the preceding 10 s interval that the tracking error was outside of the deadzone. The signal y is plotted in Figure 7.7, which shows that for each given trajectory the time outside the deadzone is decreasing, but not necessarily in a monotonic fashion. Also, changing the trajectory increases the time outside the deadzone temporarily when the new trajectory explores new regions of the operating envelope. Theorem 7.2.5 guarantees that, even with time variation of the desired trajectory, if the deadzone is sufficiently large, the total time outside the deadzone will be finite.

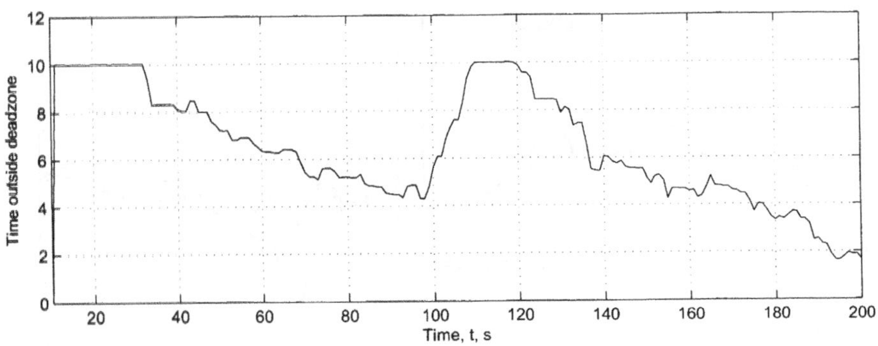

Figure 7.7: Time spent outside the deadzone $\tilde{x}P\tilde{x} = 0.002$ during the previous 10 s.

7.2.3 Input-Output

As discussed in Section 5.2 for the case of known nonlinearities, feedback linearization methods have been studied both in an input-state formulation as well as within an input-output framework. In the input-output formulation, a change of state coordinates is used to convert the system into a canonical form (normal form), where the nonlinear system is decomposed into two parts: the ζ-dynamics, which can be linearized by feedback; and the η-dynamics, which characterize the internal dynamics of the system. It is assumed that the internal dynamics are such that the the η-variables remain bounded as the ζ-variables are moving in the state-space following a tracking objective.

In this section, we consider nonlinear systems of the input-output linearizable canonical form

$$\dot{\eta} = \phi(\eta, \zeta) \tag{7.38}$$

$$\dot{\zeta} = A_0\zeta + B_0\beta_0^{-1}(\eta, \zeta)\left[u - \alpha_0(\eta, \zeta)\right] + B_0\delta \tag{7.39}$$

$$y = C_0\zeta \tag{7.40}$$

where $(A_0,\ B_0,\ C_0)$ is a canonical form representation of a system consisting of r integrators in series:

$$A_0 = \begin{bmatrix} 0 & 1 & 0 & \cdots & 0 \\ 0 & 0 & 1 & & 0 \\ \vdots & & & \ddots & \vdots \\ 0 & & & & 1 \\ 0 & 0 & & \cdots & 0 \end{bmatrix}, \quad B_0 = \begin{bmatrix} 0 \\ 0 \\ \vdots \\ 0 \\ 1 \end{bmatrix}, \quad C_0 = \begin{bmatrix} 1 & 0 & \cdots & 0 & 0 \end{bmatrix}. \tag{7.41}$$

In the above formulation the functions α_0 and β_0 are assumed unknown, while $(A_0,\ B_0,\ C_0)$ are known. The vector field ϕ does not necessarily need to be known, as long as it is such that it guarantees the boundedness of the internal states η for different values of ζ. As previously, $\delta(t)$ denotes the disturbance terms and the MFAE, which are assumed to satisfy a matching condition. The control objective is for $y(t)$ to track the signal $y_d(t)$, which is generated by

$$\dot{\zeta}_0 = A_0\zeta_0 + B_0 r \tag{7.42}$$

$$y_d = C_0\zeta_0, \tag{7.43}$$

where $r(t)$ denotes a certain command tracking signal (see Appendix Section A.4). Let $\tilde{\zeta}(t) = \zeta(t) - \zeta_0(t)$ denote the tracking error in the ζ-dynamics. Then, the tracking error dynamics can be written in the form

$$\dot{\tilde{\zeta}} = A_0\tilde{\zeta} + B_0\left(f_0(\eta,\zeta) + f^*(\eta,\zeta) + (g_0(\eta,\zeta) + g^*(\eta,\zeta))u - r\right) + B_0\delta \tag{7.44}$$

where

- $f_o(\eta,\zeta)$ is the known component of $-\beta^{-1}(\eta,\zeta)\alpha(\eta,\zeta)$;
- $f^*(\eta,\zeta)$ is the unknown component of $-\beta^{-1}(\eta,\zeta)\alpha(\eta,\zeta)$;
- $g_o(\eta,\zeta)$ is the known component of $\beta^{-1}(\eta,\zeta)$;
- $g^*(\eta,\zeta)$ is the unknown component of $\beta^{-1}(\eta,\zeta)$.

The reader will notice that once the input-output problem is formulated as shown above, then the control design and the adaptive laws for the weights of the adaptive approximator can proceed similar to the design shown in Section 7.2.2. One main difference is the presence of the internal dynamics variables η, which need to be guaranteed to remain bounded.

For completeness, we provide below the adaptive approximation based control design, which also incorporates the projection and dead-zone for robustness purposes:

$$u_a = -K^\top\tilde{\zeta} + r - f_o(\eta,\zeta) - \phi_f(\eta,\zeta)^\top\hat{\theta}_f - v_f \tag{7.45}$$

$$u = \frac{u_a}{g_o(\eta,\zeta) + \phi_g(\eta,\zeta)^\top\hat{\theta}_g + v_g} \tag{7.46}$$

$$\dot{\hat{\theta}}_f = \mathcal{P}_B\left(\Gamma_f\phi_f d(e,\tilde{\zeta},\varepsilon)\right), \quad \text{for } (\eta,\zeta) \in \mathcal{D} \tag{7.47}$$

$$\dot{\hat{\theta}}_g = \mathcal{P}_{SB}\left(\Gamma_g\phi_g d(e,\tilde{\zeta},\varepsilon)u\right), \quad \text{for } (\eta,\zeta) \in \mathcal{D}. \tag{7.48}$$

The training error $e(t)$ is defined as $e = B^\top P\tilde{\zeta}$, where P is the solution of the Lyapunov equation:

$$P(A - BK^\top) + (A - BK^\top)^\top P = -I.$$

As previously, the dead-zone is defined as

$$d(e, \tilde{\zeta}, \varepsilon) = \begin{cases} 0 & \text{if } \tilde{\zeta}^\top P \tilde{\zeta} \le \bar{\lambda}_P \varepsilon^2, \\ e & \text{if } \tilde{\zeta}^\top P \tilde{\zeta} > \bar{\lambda}_P \varepsilon^2, \end{cases} \tag{7.49}$$

$$\varepsilon = 2\|PB_0\|_2 \delta_0 + \mu \tag{7.50}$$

where $\mu > 0$ is a positive constant. Again, $\mathcal{P}_B$ is a projection operator designed to keep $\hat{\theta}_f$ in the convex and compact set $\mathcal{S}_f$ and $\mathcal{P}_{SB}$ is a projection operator designed to keep $\hat{\theta}_g$ in the convex and compact set $\mathcal{S}_g$, which is designed to ensure the stabilizability condition, the condition that $\theta_g^* \in \mathcal{S}_g$, and the boundedness of $\hat{\theta}_g$.

The analysis of the above approximation based feedback control scheme is left as an exercise (see Problem 7.2).

7.2.4 Control Design Outside the Approximation Region $\mathcal{D}$

So far in Section 7.2 we have considered the problem of approximation based feedback linearization under the assumption that the trajectory $x(t)$ remains within a predefined approximation region $\mathcal{D}$, which is a compact subset of $\mathcal{R}^n$. As discussed in the introduction to this chapter, the operating envelope $\mathcal{D}$ is a physically defined region over which it is safe and desirable for the system to operate. The trajectory generation system ensures that the desired state remains in $\mathcal{D}$. The control designer must ensure that the actual state converges to $\mathcal{D}$. Within $\mathcal{D}$ the objective is high accuracy trajectory tracking; therefore, the designer will select the approximator structure to provide confidence about the capability of the approximators $\hat{f}$ and $\hat{g}$ to approximate the unknown functions f^* and g^* accurately for $x \in \mathcal{D}$.

The techniques developed in Section 7.2.1 for scalar systems, in Section 7.2.2 for input-state feedback linearizable systems, and in Section 7.2.3 for input-output feedback linearizable systems have focused on the design, analysis and robustness of the closed-loop system under the key assumption that $x(t)$ remains in $\mathcal{D}$. Moreover, it was assumed that if $x(t)$ leaves the region $\mathcal{D}$, then the auxiliary control terms v_f and v_g are able to bring the state back within $\mathcal{D}$.

In this subsection, we show how to ensure that the design of the auxiliary terms v_f and v_g achieves the objective of bringing the trajectory within $\mathcal{D}$. Although the control design outside the approximation region $\mathcal{D}$ can be formulated and solved in a number of ways, such as sliding mode control, Lyapunov redesign method, etc., for simplicity we use the bounding method (see Section 5.4.1).

Let $\overline{\mathcal{D}} = \mathcal{R}^n - \mathcal{D}$; i.e., $\overline{\mathcal{D}}$ is the region outside of $\mathcal{D}$. Consider the class of nonlinear systems described by (7.19)-(7.21). We assume that outside of $\mathcal{D}$, the unknown functions $f^*(x)$ and $g^*(x)$ are bounded by known nonlinearities as follows:

$$\begin{aligned} f_L(x) \;&\le\; f^*(x) \;\le\; f_U(x) & x \in \overline{\mathcal{D}} \\ 0 < g_L(x) \;&\le\; g^*(x) \;\le\; g_U(x) & x \in \overline{\mathcal{D}}. \end{aligned}$$

The control design for $x \in \mathcal{D}$ has already been considered. For $x \in \overline{\mathcal{D}}$, the adaptation of the parameter estimates $\hat{\theta}_f$ and $\hat{\theta}_g$ is stopped and $\phi_f(x) = 0$, $\phi_g(x) = 0$; i.e., no basis functions are placed in $\overline{\mathcal{D}}$. Therefore, for $x \in \overline{\mathcal{D}}$, the feedback linearizing controller is given by

$$\begin{aligned} u_a &= -K^\top \tilde{x} + y_d^{(n)} - f_o(x) - v_f \\ u &= \frac{u_a}{g_o(x) + v_g}, \end{aligned}$$

where the design of the auxiliary terms v_f and v_g for $x \in \overline{\mathcal{D}}$ is as follows:

$$v_f = \begin{cases} f_U(x) & \text{if } e \geq 0 \\ f_L(x) & \text{if } e \leq 0 \end{cases} \tag{7.51}$$

$$v_g = \begin{cases} g_U(x) & \text{if } eu_a \geq 0 \\ g_L(x) & \text{if } eu_a \leq 0, \end{cases} \tag{7.52}$$

where $e = B^\top P\tilde{x}$.

The stability of the closed-loop system for $x \in \overline{\mathcal{D}}$ is obtained by considering the Lyapunov function

$$V_{\overline{\mathcal{D}}} = \tilde{x}^\top P\tilde{x}.$$

Note that for $x \in \overline{\mathcal{D}}$ adaptation is off, therefore, the parameter estimation error terms $\tilde{\theta}_f$, $\tilde{\theta}_g$ do not appear in the Lyapunov function. The time derivative of $V_{\overline{\mathcal{D}}}$ along the solutions of the closed-loop system is given by

$$\begin{aligned} \dot{V}_{\overline{\mathcal{D}}} &= \tilde{x}^\top \left(P(A - BK^\top) + (A - BK^\top)^\top P\right)\tilde{x} \\ &\quad + 2B^\top P\tilde{x}\left(f^*(x) - v_f + (g^*(x) - v_g)u\right) \\ &= -\tilde{x}^\top\tilde{x} + 2e\left(f^*(x) - v_f\right) + 2eu\left(g^*(x) - v_g\right) \\ &\leq -\|\tilde{x}\|_2^2. \end{aligned}$$

Since the desired state is strictly within $\mathcal{D}$, $\|\tilde{x}\|_2^2$ is positive for $x \in \overline{\mathcal{D}}$. Therefore, $\dot{V}_{\overline{\mathcal{D}}}$ is negative on $\overline{\mathcal{D}}$, which shows that $x(t)$ enters $\mathcal{D}$ in finite time.

The functions v_f and v_g defined in (7.51)–(7.52) are not Lipschitz functions. Their simplicity facilitates a clear discussion of methods to enforce convergence to $\mathcal{D}$. Usually these functions are smoothed across the boundary of $\mathcal{D}$ for practical implementations. For example, let $\mathcal{D}_0 \subset \mathcal{D}$, where the minimum distance between points on the boundaries of these sets is $\mu > 0$. Assume that all trajectories are defined such that $x_d(t) \in \mathcal{D}_0$ for all $t \geq 0$. Here we perform function approximation over the set $\mathcal{D}$, which is slightly larger than the region $\mathcal{D}_0$ containing all expected trajectories. Therefore, if $x \in \overline{\mathcal{D}}$, then $\|x - x_d\| \geq \mu$. The functions v_f and v_g can be defined to be zero on $\mathcal{D}_0$, as in the previous paragraphs of this section on $\overline{\mathcal{D}}$, and increasing from the former to the latter as x crosses $\mathcal{D} - \mathcal{D}_0$. This interpolation must be done carefully so that the terms including v_f and v_g are negative semidefinite on $\mathcal{D} - \mathcal{D}_0$ leaving the stability analysis on $\mathcal{D}$ effectively unchanged. An example of such a design is included in Section 8.3.2.3.

7.3 APPROXIMATION BASED BACKSTEPPING

In this section we consider the design and analysis of approximation based backstepping control. The control design procedure follows the same general formulation as in Section 5.3, with the adaptive approximators replacing the unknown nonlinearities. We start in Section 7.3.1 with a second-order system, which is extended to higher-order systems in Section 7.3.2. Finally, in Section 7.3.3 we present an alternative approximation based backstepping design, referred to as the command filtering approach.

7.3.1 Second Order Systems

In this section we consider second order systems of the form

$$\dot{x}_1 = f_{o_1}(x_1) + f_1^*(x_1) + \left(g_{o_1}(x_1) + g_1^*(x_1)\right)x_2 \tag{7.53}$$

$$\dot{x}_2 = f_{o_2}(x_1, x_2) + f_2^*(x_1, x_2) + (g_{o_2}(x_1, x_2) + g_2^*(x_1, x_2))\, u, \qquad (7.54)$$

where $x_1(t)$, $x_2(t)$ are the state variables and $u(t)$ is the control variable. The functions $f_{o_1}(x_1)$, $g_{o_1}(x_1)$, $f_{o_2}(x_1, x_2)$, $g_{o_2}(x_1, x_2)$ represent the known components of the system nonlinearities and $f_1^*(x_1)$, $g_1^*(x_1)$, $f_2^*(x_1, x_2)$, $g_2^*(x_1, x_2)$ represent the corresponding unknown components of the nonlinearities. The control objective is for $y(t) = x_1(t)$ to track a desired signal $y_d(t)$. We assume that $g_{o_1}(x_1) + g_1^*(x_1) > 0$ and $g_{o_2}(x_1, x_2) + g_2^*(x_1, x_2) > 0$ for all $(x_1, x_2) \in \mathcal{D}$, even though the results can easily be modified if these functions are entirely negative instead of positive — the important assumption is to ensure that these functions do not cross through zero, since that would imply loss of controllability.

7.3.1.1 Ideal Case. As discussed in Chapter 5, the main idea behind backstepping is to treat x_2 as a virtual control for the x_1-subsystem. Therefore, we introduce the virtual control variable α_1 which is now defined in terms of the adaptive approximators $\hat{f}_1(x, \hat{\theta}_{f_1})$, $\hat{g}_1(x, \hat{\theta}_{g_1})$ as follows:

$$\alpha_1(x_1, \hat{\theta}_{f_1}, \hat{\theta}_{g_1}, y_d, \dot{y}_d) = \frac{1}{g_{o_1}(x_1) + \hat{g}_1(x_1, \hat{\theta}_{g_1})} \left(-f_{o_1}(x_1) - \hat{f}_1(x_1, \hat{\theta}_{f_1}) - k_1 \tilde{x}_1 + \dot{y}_d \right),$$

where $k_1 > 0$ is a design constant. Following this definition of α_1, the x_1 tracking error dynamics, denoted as $\tilde{x}_1 = x_1 - y_d$, reduce to

$$\begin{aligned} \dot{\tilde{x}}_1 &= \dot{x}_1 - \dot{y}_d \\ &= (f_{o_1} + \hat{f}_1) + (g_{o_1} + \hat{g}_1)\,\alpha_1 + (g_{o_1} + \hat{g}_1)\,(x_2 - \alpha_1) \\ &\quad + (f_1^* - \hat{f}_1) + (g_1^* - \hat{g}_1)\, x_2 - \dot{y}_d \\ &= -k_1 \tilde{x}_1 + (g_{o_1} + \hat{g}_1)\, \tilde{x}_2 - \tilde{\theta}_{f_1}^{\top} \phi_{f_1} - \tilde{\theta}_{g_1}^{\top} \phi_{g_1} x_2, \end{aligned} \qquad (7.55)$$

where $\hat{f}_1(x_1, \hat{\theta}_{f_1}) = \hat{\theta}_{f_1}^{\top} \phi_{f_1}$, $\hat{g}_1(x_1, \hat{\theta}_{g_1}) = \hat{\theta}_{g_1}^{\top} \phi_{g_1}$ (i.e., $\hat{f}_1$, $\hat{g}_1$ are linearly parameterized approximators), and $\tilde{x}_2$ is defined as $\tilde{x}_2 = x_2 - \alpha_1$. Therefore, according to the definition of $\tilde{x}_2$, the signal α_1 is treated as the command signal for x_2. The dynamics of $\tilde{x}_2$ are described by

$$\begin{aligned} \dot{\tilde{x}}_2 &= (f_{o_2} + f_2^*) + (g_{o_2} + g_2^*)\, u - \dot{\alpha}_1 \\ &= (f_{o_2} + \hat{f}_2) + (g_{o_2} + \hat{g}_2)\, u - \tilde{\theta}_{f_2}^{\top} \phi_{f_2} - \tilde{\theta}_{g_2}^{\top} \phi_{g_2} u - \dot{\alpha}_1, \end{aligned} \qquad (7.56)$$

where for simplicity it is assumed that $\hat{f}_2$, $\hat{g}_2$ are also linearly parameterized. The time derivative $\dot{\alpha}_1$ is given by

$$\begin{aligned} \dot{\alpha}_1 &= \frac{\partial \alpha_1}{\partial x_1} \dot{x}_1 + \frac{\partial \alpha_1}{\partial \hat{\theta}_{f_1}} \dot{\hat{\theta}}_{f_1} + \frac{\partial \alpha_1}{\partial \hat{\theta}_{g_1}} \dot{\hat{\theta}}_{g_1} + \frac{\partial \alpha_1}{\partial y_d} \dot{y}_d + \frac{\partial \alpha_1}{\partial \dot{y}_d} \ddot{y}_d \\ &= -\frac{\partial \alpha_1}{\partial x_1} \left(\tilde{\theta}_{f_1}^{\top} \phi_{f_1} + \tilde{\theta}_{g_1}^{\top} \phi_{g_1} x_2 \right) + \frac{\partial \alpha_1}{\partial x_1} \left((f_{o_1} + \hat{f}_1) + (g_{o_1} + \hat{g}_1)\, x_2 \right) \\ &\quad + \frac{\partial \alpha_1}{\partial \hat{\theta}_{f_1}} \dot{\hat{\theta}}_{f_1} + \frac{\partial \alpha_1}{\partial \hat{\theta}_{g_1}} \dot{\hat{\theta}}_{g_1} + \frac{\partial \alpha_1}{\partial y_d} \dot{y}_d + \frac{\partial \alpha_1}{\partial \dot{y}_d} \ddot{y}_d \\ &= -\frac{\partial \alpha_1}{\partial x_1} \left(\tilde{\theta}_{f_1}^{\top} \phi_{f_1} + \tilde{\theta}_{g_1}^{\top} \phi_{g_1} x_2 \right) + \hat{\mu}_1(x_1, x_2, \hat{\theta}_{f_1}, \hat{\theta}_{g_1}, y_d, \dot{y}_d, \ddot{y}_d), \end{aligned} \qquad (7.57)$$

where

$$\hat{\mu}_1 = \frac{\partial \alpha_1}{\partial x_1} \left((f_{o_1} + \hat{f}_1) + (g_{o_1} + \hat{g}_1)\, x_2 \right) + \frac{\partial \alpha_1}{\partial \hat{\theta}_{f_1}} \dot{\hat{\theta}}_{f_1} + \frac{\partial \alpha_1}{\partial \hat{\theta}_{g_1}} \dot{\hat{\theta}}_{g_1} + \frac{\partial \alpha_1}{\partial y_d} \dot{y}_d + \frac{\partial \alpha_1}{\partial \dot{y}_d} \ddot{y}_d.$$

It is noted that $\dot{\alpha}_1$ is broken into two components:

- $\hat{\mu}_1$, which is available analytically in terms of known functions and measurable variables; and
- $\frac{\partial \alpha_1}{\partial x_1}\left(\tilde{\theta}_{f_1}^{\mathrm{T}}\phi_{f_1} + \tilde{\theta}_{g_1}^{\mathrm{T}}\phi_{g_1}x_2\right)$, which is not available analytically due to the fact that $\tilde{\theta}_{f_1}, \tilde{\theta}_{g_1}$ are unknown. As we will see, this second component will be carried through the backstepping procedure until the end, and eventually it will be handled by appropriately selecting the adaptive laws for $\hat{\theta}_{f_1}, \hat{\theta}_{g_1}$.

Now, define a Lyapunov function as

$$V = \frac{1}{2}\left(\tilde{x}_1^2 + \tilde{x}_2^2 + \sum_{i=1}^{2}\left(\tilde{\theta}_{f_i}^{\mathrm{T}}\Gamma_{f_i}^{-1}\tilde{\theta}_{f_i} + \tilde{\theta}_{g_i}^{\mathrm{T}}\Gamma_{g_i}^{-1}\tilde{\theta}_{g_i}\right)\right). \tag{7.58}$$

Using eqns. (7.55)–(7.57), the time derivative of the Lyapunov function is

$$\begin{aligned}
\dot{V} &= -k_1\tilde{x}_1^2 + \tilde{x}_1(g_{o_1} + \hat{g}_1)\tilde{x}_2 - \tilde{x}_1\tilde{\theta}_{f_1}^{\mathrm{T}}\phi_{f_1} - \tilde{x}_1\tilde{\theta}_{g_1}^{\mathrm{T}}\phi_{g_1}x_2 \\
&\quad + \tilde{x}_2\left((f_{o_2} + \hat{f}_2) + (g_{o_2} + \hat{g}_2)u - \dot{\alpha}_1\right) - \tilde{x}_2\tilde{\theta}_{f_2}^{\mathrm{T}}\phi_{f_2} - \tilde{x}_2\tilde{\theta}_{g_2}^{\mathrm{T}}\phi_{g_2}u \\
&\quad + \sum_{i=1}^{2}\left(\tilde{\theta}_{f_i}^{\mathrm{T}}\Gamma_{f_i}^{-1}\dot{\hat{\theta}}_{f_i} + \tilde{\theta}_{g_i}^{\mathrm{T}}\Gamma_{g_i}^{-1}\dot{\hat{\theta}}_{g_i}\right) \\
&= -k_1\tilde{x}_1^2 - k_2\tilde{x}_2^2 + \tilde{x}_2\left(k_2\tilde{x}_2 + \tilde{x}_1(g_{o_1} + \hat{g}_1) + (f_{o_2} + \hat{f}_2) - \hat{\mu}_1 + (g_{o_2} + \hat{g}_2)u\right) \\
&\quad + \tilde{\theta}_{f_1}^{\mathrm{T}}\Gamma_{f_1}^{-1}\left(\dot{\hat{\theta}}_{f_1} - \Gamma_{f_1}\phi_{f_1}\left(\tilde{x}_1 - \frac{\partial \alpha_1}{\partial x_1}\tilde{x}_2\right)\right) \\
&\quad + \tilde{\theta}_{g_1}^{\mathrm{T}}\Gamma_{g_1}^{-1}\left(\dot{\hat{\theta}}_{g_1} - \Gamma_{g_1}\phi_{g_1}x_2\left(\tilde{x}_1 - \frac{\partial \alpha_1}{\partial x_1}\tilde{x}_2\right)\right) \\
&\quad + \tilde{\theta}_{f_2}^{\mathrm{T}}\Gamma_{f_2}^{-1}\left(\dot{\hat{\theta}}_{f_2} - \Gamma_{f_2}\phi_{f_2}\tilde{x}_2\right) + \tilde{\theta}_{g_2}^{\mathrm{T}}\Gamma_{g_2}^{-1}\left(\dot{\hat{\theta}}_{g_2} - \Gamma_{g_2}\phi_{g_2}\tilde{x}_2u\right).
\end{aligned}$$

In order to make the derivative of the Lyapunov function negative semidefinite, we choose the control law and the adaptive laws as follows:

$$u = \frac{1}{g_{o_2} + \hat{g}_2}\left(-k_2\tilde{x}_2 - \tilde{x}_1(g_{o_1} + \hat{g}_1) - (f_{o_2} + \hat{f}_2) + \hat{\mu}_1\right) \tag{7.59}$$

$$\dot{\hat{\theta}}_{f_1} = \Gamma_{f_1}\phi_{f_1}\left(\tilde{x}_1 - \frac{\partial \alpha_1}{\partial x_1}\tilde{x}_2\right) \tag{7.60}$$

$$\dot{\hat{\theta}}_{g_1} = \mathcal{P}_S\left(\Gamma_{g_1}\phi_{g_1}x_2\left(\tilde{x}_1 - \frac{\partial \alpha_1}{\partial x_1}\tilde{x}_2\right)\right) \tag{7.61}$$

$$\dot{\hat{\theta}}_{f_2} = \Gamma_{f_2}\phi_{f_2}\tilde{x}_2 \tag{7.62}$$

$$\dot{\hat{\theta}}_{g_2} = \mathcal{P}_S\left(\Gamma_{g_2}\phi_{g_2}\tilde{x}_2u\right) \tag{7.63}$$

where $\mathcal{P}_S$ is the projection operator that is used to ensure the stabilizability conditions:

$$\begin{aligned}
g_{o_1}(x_1) + \hat{g}_1(x_1, \hat{\theta}_{g_1}) &> 0 \\
g_{o_2}(x_1, x_2) + \hat{g}_2(x_1, x_2, \hat{\theta}_{g_2}) &> 0.
\end{aligned}$$

Moreover, it is assumed that the state remains in the approximation region $\mathcal{D}$ via the use of some robustifying terms v_{f_1}, v_{g_1}, v_{f_2}, and v_{g_2}, whose design will be discussed later in this subsection.

The derivative of V along solutions of the closed-loop system when the projection is not active reduces to

$$\dot{V} = -k_1\tilde{x}_1^2 - k_2\tilde{x}_2^2, \tag{7.64}$$

which provides the required closed-loop stability result, summarized in Theorem 7.3.1, in the ideal case of no approximation error and no disturbances.

Theorem 7.3.1 [Ideal Case] *For the closed-loop system composed of the system model described by eqns. (7.53)–(7.54) and the feedback controller defined by eqns. (7.59)–(7.63), satisfies the following properties*

1. $\tilde{x}_i,\ x_i,\ \hat{\theta}_{f_i},\ \hat{\theta}_{g_i} \in \mathcal{L}_\infty \qquad i = 1,\ 2;$
2. $\tilde{x} \in \mathcal{L}_2;$
3. $x_1(t) \to y_d(t)$ *and* $x_2(t) \to \alpha_1$ *as* $t \to \infty$.

Proof: The proof follows trivially based on the design of the feedback control law (see Problem 7.3). As discussed earlier, in the case where the projection operator is active, the stability properties of the algorithm are preserved. ■

The controller specified in this section was successfully defined by deferring the choice of the parameter update laws until the second step of the backstepping recursion. As we will see, this approach to defining the approximation based backstepping controller becomes increasing complicated for higher order systems. Section 7.3.3 presents an alternative approach.

7.3.1.2 Robustness Considerations. In this subsection, we consider the case where there are residual modeling errors for $x \in \mathcal{D}$. We consider the following, more general class of second order systems:

$$\begin{aligned}
\dot{x}_1 &= f_{o_1}(x_1) + f_1^*(x_1) + (g_{o_1}(x_1) + g_1^*(x_1))\,x_2 + \delta_1(x) && (7.65)\\
\dot{x}_2 &= f_{o_2}(x_1,x_2) + f_2^*(x_1,x_2) + (g_{o_2}(x_1,x_2) + g_2^*(x_1,x_2))\,u + \delta_2(x), && (7.66)
\end{aligned}$$

where δ_1 and δ_2 may contain disturbance terms as well as residual approximation errors, referred to as MFAE. Let $\delta = [\delta_1 \ \ \delta_2]^\top$. As previously, the main idea is to modify the adaptive laws (7.60)–(7.63), using the dead-zone and projection modifications, such that the tracking error of the closed-loop system is small-in-the-mean-square sense and is uniformly ultimately bounded by a certain constant ε that depends on the size of the modeling error, denoted by δ_0. We assume that the modeling error term satisfies $\|\delta\|_2 \le \delta_0$ for all $x \in \mathcal{D}$.

In the presence of the modeling error terms δ_1, δ_2, the tracking error dynamics (7.55), (7.56), now become

$$\begin{aligned}
\dot{\tilde{x}}_1 &= -k_1\tilde{x}_1 + (g_{o_1} + \hat{g}_1)\,\tilde{x}_2 - \tilde{\theta}_{f_1}^\top\phi_{f_1} - \tilde{\theta}_{g_1}^\top\phi_{g_1}x_2 + \delta_1, && (7.67)\\
\dot{\tilde{x}}_2 &= (f_{o_2} + \hat{f}_2) + (g_{o_2} + \hat{g}_2)\,u - \tilde{\theta}_{f_2}^\top\phi_{f_2} - \tilde{\theta}_{g_2}^\top\phi_{g_2}u - \dot{\alpha}_1 + \delta_2. && (7.68)
\end{aligned}$$

In this case, $\dot{\alpha}_1$ has an additional term which cannot be obtained analytically. Therefore, we have

$$\dot{\alpha}_1 = -\frac{\partial\alpha_1}{\partial x_1}\left(\tilde{\theta}_{f_1}^\top\phi_{f_1} + \tilde{\theta}_{g_1}^\top\phi_{g_1}x_2\right) + \frac{\partial\alpha_1}{\partial x_1}\delta_1 + \hat{\mu}_1. \tag{7.69}$$

For notational convenience, let e_1, e_2 be defined as

$$e_1 = \tilde{x}_1 - \frac{\partial \alpha_1}{\partial x_1}\tilde{x}_2 \tag{7.70}$$

$$e_2 = \tilde{x}_2. \tag{7.71}$$

Computing the time derivative of the Lyapunov function (7.58) yields the following expression, which is the same as for the ideal case, except for some additional terms due to the presence of the modeling errors δ_1, δ_2:

$$\begin{aligned}
\dot{V} = & -k_1\tilde{x}_1^2 - k_2\tilde{x}_2^2 + \tilde{x}_2\left(k_2\tilde{x}_2 + \tilde{x}_1(g_{o_1} + \hat{g}_1) + (f_{o_2} + \hat{f}_2) - \dot{\hat{\mu}}_1 + (g_{o_2} + \hat{g}_2)u\right) \\
& + \tilde{\theta}_{f_1}^T \Gamma_{f_1}^{-1}\left(\dot{\hat{\theta}}_{f_1} - \Gamma_{f_1}\phi_{f_1}e_1\right) + \tilde{\theta}_{g_1}^T \Gamma_{g_1}^{-1}\left(\dot{\hat{\theta}}_{g_1} - \Gamma_{g_1}\phi_{g_1}x_2e_1\right) \\
& + \tilde{\theta}_{f_2}^T \Gamma_{f_2}^{-1}\left(\dot{\hat{\theta}}_{f_2} - \Gamma_{f_2}\phi_{f_2}e_2\right) + \tilde{\theta}_{g_2}^T \Gamma_{g_2}^{-1}\left(\dot{\hat{\theta}}_{g_2} - \Gamma_{g_2}\phi_{g_2}e_2u\right) \\
& + e_1\delta_1 + e_2\delta_2.
\end{aligned}$$

We are now ready to present the robustness theorem with the projection and dead-zone modification in the adaptive laws.

Theorem 7.3.2 [Projection with Dead-Zone] *Suppose there are some terms v_{f_1}, v_{g_1}, v_{f_2}, v_{g_2} which are zero for $x \in \mathcal{D}$ and are designed to ensure that the state will return to and remain in $\mathcal{D}$. Assume that ϕ_{f_1}, ϕ_{g_1}, ϕ_{f_2}, ϕ_{g_2} are bounded and let the parameter estimates be adjusted according to*

$$\dot{\hat{\theta}}_{f_1} = \mathcal{P}_B\left(\Gamma_{f_1}\phi_{f_1}d(e_1, \tilde{x}, \varepsilon)\right), \quad \textit{for } x \in \mathcal{D} \tag{7.72}$$

$$\dot{\hat{\theta}}_{g_1} = \mathcal{P}_{SB}\left(\Gamma_{g_1}\phi_{g_1}x_2\, d(e_1, \tilde{x}, \varepsilon)\right), \quad \textit{for } x \in \mathcal{D} \tag{7.73}$$

$$\dot{\hat{\theta}}_{f_2} = \mathcal{P}_B\left(\Gamma_{f_2}\phi_{f_2}d(e_2, \tilde{x}, \varepsilon)\right), \quad \textit{for } x \in \mathcal{D} \tag{7.74}$$

$$\dot{\hat{\theta}}_{g_2} = \mathcal{P}_{SB}\left(\Gamma_{g_2}\phi_{g_2}u\, d(e_2, \tilde{x}, \varepsilon)\right), \quad \textit{for } x \in \mathcal{D} \tag{7.75}$$

where

$$d(z, \tilde{x}, \varepsilon) = \begin{cases} 0 & \textit{if } \|\tilde{x}\|_2 \le \varepsilon, \\ z & \textit{if } \|\tilde{x}\|_2 > \varepsilon, \end{cases} \tag{7.76}$$

$$\varepsilon = c_e\delta_0 + \mu \tag{7.77}$$

where $\mu > 0$ is a positive constant, and $c_e > 0$ will be defined in the proof. $\mathcal{P}_B$ is a projection operator designed to keep $\hat{\theta}_{f_1}$, $\hat{\theta}_{f_2}$ in some convex and compact sets $\mathcal{S}_{f_1}$, $\mathcal{S}_{f_2}$ respectively, and $\mathcal{P}_{SB}$ is a projection operator designed to keep $\hat{\theta}_{g_1}$, $\hat{\theta}_{g_2}$ in the convex and compact sets $\mathcal{S}_{g_1}$, $\mathcal{S}_{g_2}$, which are designed to ensure the stabilizability condition, the condition that $\theta_{g_i}^* \in \mathcal{S}_{g_i}$, and the boundedness of $\hat{\theta}_{g_i}$. In the case where $\|\delta\|_2 < \delta_0$,

1. *$\tilde{x}_i$, x_i, $\hat{\theta}_{f_i}$, $\hat{\theta}_{g_i} \in \mathcal{L}_\infty$ for $i = 1,\ 2$;*

2. *$\tilde{x}$ is small-in-the-mean-square sense, satisfying*

$$\int_t^{t+T} \|\tilde{x}^2(\tau)\|_2^2\, d\tau \le \frac{2}{k}V(t) + \varepsilon^2 T;$$

3. $\|\tilde{x}(t)\|_2$ *is uniformly ultimately bounded by* ε.

Proof: Let $e = [e_1 \; e_2]^\top$. Based on the definition of e_1, e_2, given by (7.70), (7.71), there exists a finite constant $c > 0$ such that $\|e\|_2 \le c\|\tilde{x}\|_2$, where c is defined over all $x \in \mathcal{D}$. The time derivative of the Lyapunov function (7.58) for $x \in \mathcal{D}$ satisfies

$$\begin{aligned}\dot{V} \;\le\;& -k\|\tilde{x}\|_2^2 + e^\top \delta \\ &+ \tilde{\theta}_{f_1}^\top \Gamma_{f_1}^{-1}\left(\dot{\hat{\theta}}_{f_1} - \Gamma_{f_1}\phi_{f_1} e_1\right) + \tilde{\theta}_{g_1}^\top \Gamma_{g_1}^{-1}\left(\dot{\hat{\theta}}_{g_1} - \Gamma_{g_1}\phi_{g_1} x_2 e_1\right) \\ &+ \tilde{\theta}_{f_2}^\top \Gamma_{f_2}^{-1}\left(\dot{\hat{\theta}}_{f_2} - \Gamma_{f_2}\phi_{f_2} e_2\right) + \tilde{\theta}_{g_2}^\top \Gamma_{g_2}^{-1}\left(\dot{\hat{\theta}}_{g_2} - \Gamma_{g_2}\phi_{g_2} e_2 u\right),\end{aligned}$$

where $k = \min\{k_1, \; k_2\}$ is a positive constant. Suppose that the time intervals (t_{s_i}, t_{f_i}) are defined as discussed relative to Figure 7.1, so that the condition $\|\tilde{x}(t)\|_2 > \varepsilon$ is satisfied only for $t \in (t_{s_i}, t_{f_i})$, $i = 1, 2, 3, \ldots$, where $t_{s_i} < t_{f_i} \le t_{s_{i+1}}$. Since $\|\tilde{x}(t_{f_i})\|_2 = \|\tilde{x}(t_{s_{i+1}})\|_2 = \varepsilon$ and parameter estimation is off for $t \in [t_{f_i}, \; t_{s_{i+1}}]$, we have that $V(t_{f_i}) = V(t_{s_{i+1}})$. When $t \in (t_{s_i}, t_{f_i})$ for any i and projection is not in effect, then

$$\begin{aligned}\dot{V} \;&\le\; -k\|\tilde{x}\|_2^2 + \|e\|_2 \cdot \|\delta\|_2 \\ &\le\; -k\|\tilde{x}\|_2^2 + c\|\tilde{x}\|_2 \delta_0 \\ &\le\; -k\|\tilde{x}\|_2 \left(\|\tilde{x}\|_2 - c_e \delta_0\right) \\ &=\; -k\varepsilon\mu,\end{aligned} \tag{7.78}$$

where $c_e = c/k$. Therefore, by integrating both sides over (t_{s_i}, t_{f_i}),

$$\begin{aligned}V(t_{f_i}) \;&\le\; V(t_{s_i}) - k\varepsilon\mu\,(t_{f_i} - t_{s_i}) \\ &\le\; V(t_{f_{i-1}}) - k\varepsilon\mu\,(t_{f_i} - t_{s_i}) \\ &\le\; V(t_{f_{i-2}}) - k\varepsilon\mu\left((t_{f_i} - t_{s_i}) + (t_{f_{i-1}} - t_{s_{i-1}})\right) \\ &\;\;\vdots \\ &\le\; V(0) - k\varepsilon\mu\left(\sum_{k=1}^{i}(t_{f_k} - t_{s_k})\right).\end{aligned}$$

Hence, since $V(t_{f_i}) \ge 0$,

$$\sum_{k=1}^{i}(t_{f_k} - t_{s_k}) \le \frac{V(0)}{k\varepsilon\mu},$$

which shows that the total time spent with $\|\tilde{x}(t)\|_2 > \varepsilon$ is finite. In addition, $V(t_{f_i})$ $i = 1, 2, 3, \ldots$ is a positive decreasing sequence, either this is a finite sequence or $\lim_{i\to\infty} V(t_{f_i}) = V_\infty$ exists and is finite. In addition, if $t > t_{f_i}$, then $V(t) < V(t_{f_i})$.

Within the dead-zone, it is obvious that $\|\tilde{x}(t)\|_2 \le \varepsilon$ implies

$$\int_t^{t+T} \|\tilde{x}(\tau)\|_2^2 d\tau \le \varepsilon^2 T.$$

Outside the dead-zone, using the inequality,

$$xy \;\le\; \beta^2 x^2 + \frac{1}{4\beta^2} y^2,$$

with $\beta^2 = \frac{k}{2}$ it can be readily shown from eqn. (7.78) that

$$\dot{V} \leq -\frac{k}{2}\|\tilde{x}\|_2^2 + \frac{k}{2}c_e^2\delta_0^2.$$

Integrating both sides of this inequality over the time interval $[t,\ t+T]$ yields

$$\int_t^{t+T} \|\tilde{x}(\tau)\|_2^2\, d\tau \leq \frac{2}{k}V(t) + \varepsilon^2 T,$$

which completes the proof. ∎

So far, we have considered the ideal case where all the uncertainties can be represented exactly in the region $\mathcal{D}$ by the adaptive approximators, and the robust case, where we allow the presence of residual approximation errors, as well as disturbance terms. In the robust case, the adaptive laws are modified accordingly. In the next subsection, we consider the design of the control for x outside the approximation region $\mathcal{D}$.

7.3.1.3 Control Outside the Region $\mathcal{D}$.

In the previous design and analysis, it was assumed that if $x(t)$ starts outside the region $\mathcal{D}$, then the auxiliary control terms $v_{f_1}, v_{g_1}, v_{f_2}, v_{g_2}$, are able to bring the trajectory within $\mathcal{D}$. In this subsection, we show how to ensure that the design of the auxiliary terms $v_{f_1}, v_{g_1}, v_{f_2}, v_{g_2}$ achieves the desired objective.

Again, we consider the second-order system

$$\dot{x}_1 = f_{o_1}(x_1) + f_1^*(x_1) + (g_{o_1}(x_1) + g_1^*(x_1))\, x_2 \tag{7.79}$$
$$\dot{x}_2 = f_{o_2}(x_1,x_2) + f_2^*(x_1,x_2) + (g_{o_2}(x_1,x_2) + g_2^*(x_1,x_2))\, u, \tag{7.80}$$

where for $x \in \overline{\mathcal{D}}$ the unknown functions $f_1^*,\ g_1^*,\ f_2^*,\ g_2^*$ are assumed to be bounded by known nonlinearities as follows:

$$\begin{array}{rcccll} f_{1_L}(x_1) & \leq & f_1^*(x_1) & \leq & f_{1_U}(x_1) & x \in \overline{\mathcal{D}} \\ 0 < g_{1_L}(x_1) & \leq & g_1^*(x_1) & \leq & g_{1_U}(x_1) & x \in \overline{\mathcal{D}} \\ f_{2_L}(x_1,x_2) & \leq & f_2^*(x_1,x_2) & \leq & f_{2_U}(x_1,x_2) & x \in \overline{\mathcal{D}} \\ 0 < g_{2_L}(x_1,x_2) & \leq & g_2^*(x_1,x_2) & \leq & g_{2_U}(x_1,x_2) & x \in \overline{\mathcal{D}}. \end{array}$$

As discussed previously, for $x \in \overline{\mathcal{D}}$, the regressor vectors $\phi_{f_1}, \phi_{g_1}, \phi_{f_2}, \phi_{g_2}$ are all zero (i.e., no basis functions are placed in $\overline{\mathcal{D}}$) and the adaptation of the parameter estimates is stopped. The feedback control for $x \in \overline{\mathcal{D}}$ is derived as follows.

Let the virtual control variable α_1 be defined as

$$\alpha_1(x_1, y_d, \dot{y}_d) = \frac{1}{g_{o_1} + v_{g_1}}\left(-f_{o_1} - k_1\tilde{x}_1 + \dot{y}_d - v_{f_1}\right).$$

After some algebraic manipulation, the x_1 tracking error dynamics become

$$\dot{\tilde{x}}_1 = -k_1\tilde{x}_1 + (g_{o_1} + v_{g_1})\tilde{x}_2 + (f_1^* - v_{f_1}) + (g_1^* - v_{g_1})\, x_2 \tag{7.81}$$

where $\tilde{x}_2 = x_2 - \alpha_1$. The error dynamics for $\tilde{x}_2$ are described by

$$\begin{aligned} \dot{\tilde{x}}_2 &= (f_{o_2} + f_2^*) + (g_{o_2} + g_2^*)\, u - \dot{\alpha}_1 \\ &= (f_{o_2} + v_{f_2}) + (g_{o_2} + v_{g_2})\, u - \overline{\mu}_1 + (f_2^* - v_{f_2}) + (g_2^* - v_{g_2})\, u \\ &\quad + \frac{\partial \alpha_1}{\partial x_1}(v_{f_1} - f_1^*) + \frac{\partial \alpha_1}{\partial x_1}(v_{g_1} - g_1^*)\, x_2, \end{aligned} \tag{7.82}$$

where $\overline{\mu}_1$ is given by

$$\overline{\mu}_1 = \frac{\partial \alpha_1}{\partial x_1}\left((f_{o_1} + v_{f_1}) + (g_{o_1} + v_{g_1})x_2\right) + \frac{\partial \alpha_1}{\partial y_d}\dot{y}_d + \frac{\partial \alpha_1}{\partial \dot{y}_d}\ddot{y}_d.$$

The closed-loop stability for $x \in \overline{\mathcal{D}}$ is investigated by considering the Lyapunov function

$$V_{\overline{\mathcal{D}}} = \frac{1}{2}\tilde{x}_1^2 + \frac{1}{2}\tilde{x}_2^2.$$

The time derivative of $V_{\overline{\mathcal{D}}}$ along the solutions of (7.81), (7.82) is given by

$$\begin{aligned} \dot{V}_{\overline{\mathcal{D}}} &= -k_1\tilde{x}_1^2 - k_2\tilde{x}_2^2 + \tilde{x}_2\left((f_{o_2} + v_{f_2}) + (g_{o_1} + v_{g_1})\tilde{x}_1 + k_2\tilde{x}_2 - \overline{\mu}_1\right) \\ &\quad + \tilde{x}_2(g_{o_2} + v_{g_2})u + \Lambda \end{aligned}$$

where Λ is

$$\begin{aligned} \Lambda &= (f_1^* - v_{f_1})\left(\tilde{x}_1 - \frac{\partial \alpha_1}{\partial x_1}\tilde{x}_2\right) + (g_1^* - v_{g_1})\left(\tilde{x}_1 - \frac{\partial \alpha_1}{\partial x_1}\tilde{x}_2\right)x_2 \\ &\quad + (f_2^* - v_{f_2})\,\tilde{x}_2 + (g_2^* - v_{g_2})\,\tilde{x}_2 u \\ &= (f_1^* - v_{f_1})\,e_1 + (g_1^* - v_{g_1})\,x_2 e_1 + (f_2^* - v_{f_2})\,e_2 + (g_2^* - v_{g_2})\,e_2 u. \end{aligned}$$

The control law is selected as

$$u = \frac{1}{g_{o_2} + v_{g_2}}\left(-k_2\tilde{x}_2 - (f_{o_2} + v_{f_2}) - (g_{o_1} + v_{g_1})\tilde{x}_1 + \overline{\mu}_1\right),$$

which results in the following Lyapunov function derivative

$$\dot{V}_{\overline{\mathcal{D}}} = -k_1\tilde{x}_1^2 - k_2\tilde{x}_2^2 + \Lambda \le -\min(k_1, k_2)\|\tilde{x}\|^2 + \Lambda.$$

In order to ensure that $\Lambda \le 0$, the design of the auxiliary terms $v_{f_1}, v_{g_1}, v_{f_2}, v_{g_2}$ for $x \in \overline{\mathcal{D}}$ is chosen as follows:

$$v_{f_1} = \begin{cases} f_{1_U}(x_1) & \text{if } e_1 \ge 0 \\ f_{1_L}(x_1) & \text{if } e_1 \le 0 \end{cases} \tag{7.83}$$

$$v_{g_1} = \begin{cases} g_{1_U}(x_1) & \text{if } x_2 e_1 \ge 0 \\ g_{1_L}(x_1) & \text{if } x_2 e_1 \le 0 \end{cases} \tag{7.84}$$

$$v_{f_2} = \begin{cases} f_{2_U}(x) & \text{if } e_2 \ge 0 \\ f_{2_L}(x) & \text{if } e_2 \le 0 \end{cases} \tag{7.85}$$

$$v_{g_2} = \begin{cases} g_{2_U}(x) & \text{if } e_2 u \ge 0 \\ g_{2_L}(x) & \text{if } e_2 u \le 0. \end{cases} \tag{7.86}$$

Since the desired state is strictly within $\mathcal{D}$, $\|\tilde{x}\|_2^2$ is positive for $x \in \overline{\mathcal{D}}$. Therefore, $\dot{V}_{\overline{\mathcal{D}}}$ is negative on $\overline{\mathcal{D}}$, which shows that $x(t)$ enters $\mathcal{D}$ in finite time.

7.3.2 Higher Order Systems

In this subsection, we extend the results of Section (7.3.1) from second-order systems to higher-order systems. We consider n-order single-input single-output (SISO) systems described by

$$\begin{aligned}
\dot{x}_1 &= f_1(x_1) + g_1(x_1)x_2 + d_1(t) \\
\dot{x}_2 &= f_2(x_1, x_2) + g_2(x_1, x_2)x_3 + d_2(t) \\
&\vdots \\
\dot{x}_i &= f_i(x_1, \ldots, x_i) + g_i(x_1, \ldots, x_i)x_{i+1} + d_i(t) \\
&\vdots \\
\dot{x}_n &= f_n(x_1, \ldots, x_n) + g_n(x_1, \ldots, x_n)u + d_n(t),
\end{aligned}$$

where $d_i(t)$ denote unknown disturbance terms. If we define $\bar{x}_i = [x_1\ x_2\ \cdots\ x_i]^\top$, then the above system can be written in compact form as

$$\dot{x}_i = f_i(\bar{x}_i) + g_i(\bar{x}_i)x_{i+1} + d_i(t) \qquad \text{for } i = 1,\ 2\ \ldots\ n-1 \tag{7.87}$$

$$\dot{x}_n = f_n(\bar{x}_n) + g_n(\bar{x}_n)u + d_n(t). \tag{7.88}$$

Each function $f_i(\bar{x}_i)$ and $g_i(\bar{x}_i)$ is assumed to consist of two parts: (i) the known part, or nominal model, which is denoted by $f_{o_i}(\bar{x}_i)$; and (ii) the unknown part, or the model uncertainty, which is denoted by $f_i^*(\bar{x}_i)$ (correspondingly for $g_i(\bar{x}_i)$). Each unknown nonlinearity $f_i^*(\bar{x}_i)$ will be represented by a linearly parameterized approximator of the form ${\theta_{f_i}^*}^\top \phi_{f_i}$, where $\theta_{f_i}^*$ is an unknown vector of network weights, referred to as the optimal weights of the approximator. As previously, the residual approximation error $\delta_{f_i} = f_i^*(\bar{x}_i) - {\theta_{f_i}^*}^\top \phi_{f_i}(\bar{x}_i)$ is referred to as the MFAE.

Therefore, (7.87), (7.88) can be rewritten as

$$\begin{aligned}
\dot{x}_i &= f_{o_i}(\bar{x}_i) + {\theta_{f_i}^*}^\top \phi_{f_i}(\bar{x}_i) + \left(g_{o_i}(\bar{x}_i) + {\theta_{g_i}^*}^\top \phi_{g_i}(\bar{x}_i)\right) x_{i+1} + \delta_i \\
\dot{x}_n &= f_{o_n}(\bar{x}_n) + {\theta_{f_n}^*}^\top \phi_{f_n}(\bar{x}_n) + \left(g_{o_n}(\bar{x}_n) + {\theta_{g_n}^*}^\top \phi_{g_n}(\bar{x}_n)\right) u + \delta_n,
\end{aligned}$$

where $i = 1,\ 2\ \ldots\ n-1$ and δ_i is defined as

$$\delta_i = \begin{cases} \delta_{f_i}(\bar{x}_i) + \delta_{g_i}(\bar{x}_i)x_{i+1} + d_i & \text{if } i = 1,\ 2\ \ldots\ n-1 \\ \delta_{f_i}(\bar{x}_i) + \delta_{g_i}(\bar{x}_i)u + d_i & \text{if } i = n. \end{cases}$$

In the subsequent analysis, we will assume that a known bound is available for δ_i. We denote the bound by $\bar{\delta}_i$; i.e.,

$$|\delta_i(x)| \leq \bar{\delta}_i, \qquad \forall x \in \mathcal{D}.$$

If such a bound is not available, then the adaptive bounding methodology (see Chapter 5) can be employed.

It is assumed that each $g_i(\bar{x}_i) > 0$ for all $x \in \mathcal{D}$, which allows controllability through the backstepping procedure. The control objective is for $y(t) = x_1(t)$ to track some desired reference signal $y_d(t)$. It is assumed that $y_d, \dot{y}_d, \ldots y_d^{(n)}$ are known and uniformly bounded.

Let

$$\tilde{x}_i = x_i - \alpha_{i-1} \qquad 1 \leq i \leq n, \tag{7.89}$$

where α_i are virtual control inputs or intermediate control variables. For notational convenience we let $\alpha_0 = y_d$. The design of the adaptive controller is recursive in the sense that computation of α_i relies on first computing α_{i-1}. The overall design procedure yields a dynamic controller u that depends on the adaptive parameters $\hat{\theta}_{f_k}$, $\hat{\theta}_{g_k}$, whose right-hand side of the adaptation is also computed recursively:

$$\dot{\hat{\theta}}_{f_k} = \tau_{f_{kn}} \qquad 1 \le k \le n \tag{7.90}$$

$$\dot{\hat{\theta}}_{g_k} = \tau_{g_{kn}} \qquad 1 \le k \le n. \tag{7.91}$$

The recursive steps of the backstepping procedure are described next. For notational simplicity we drop the functional dependence on the state.

Step 1 : Using (7.87) and the change of coordinates (7.89) we obtain

$$\begin{aligned}\dot{\tilde{x}}_1 &= f_{o_1} + \hat{\theta}_{f_1}^\top \phi_{f_1} + \left(g_{o_1} + \hat{\theta}_{g_1}^\top \phi_{g_1}\right)\alpha_1 + \left(g_{o_1} + \hat{\theta}_{g_1}^\top \phi_{g_1}\right)(x_2 - \alpha_1) \\ &\quad - \tilde{\theta}_{f_1}^\top \phi_{f_1} - \tilde{\theta}_{g_1}^\top \phi_{g_1} x_2 - \dot{y}_d + \delta_1 .\end{aligned} \tag{7.92}$$

Now consider the intermediate Lyapunov function

$$V_1 = \frac{1}{2}\tilde{x}_1^2 + \frac{1}{2}\tilde{\theta}_{f_1}^\top \Gamma_{f_1}^{-1}\tilde{\theta}_{f_1} + \frac{1}{2}\tilde{\theta}_{g_1}^\top \Gamma_{g_1}^{-1}\tilde{\theta}_{g_1},$$

whose time derivative along (7.92) is given by

$$\begin{aligned}\dot{V}_1 &= \tilde{x}_1\left(f_{o_1} + \hat{\theta}_{f_1}^\top \phi_{f_1} - \dot{y}_d + \left(g_{o_1} + \hat{\theta}_{g_1}^\top \phi_{g_1}\right)\alpha_1\right) \\ &\quad + \left(g_{o_1} + \hat{\theta}_{g_1}^\top \phi_{g_1}\right)\tilde{x}_1\tilde{x}_2 + \tilde{x}_1\delta_1 \\ &\quad + \tilde{\theta}_{f_1}^\top \Gamma_{f_1}^{-1}\left(\dot{\hat{\theta}}_{f_1} - \Gamma_{f_1}\phi_{f_1}\tilde{x}_1\right) + \tilde{\theta}_{g_1}^\top \Gamma_{g_1}^{-1}\left(\dot{\hat{\theta}}_{g_1} - \Gamma_{g_1}\phi_{g_1}x_2\tilde{x}_1\right)\end{aligned} \tag{7.93}$$

We let

$$\alpha_1(\tilde{x}_1, \hat{\theta}_{f_1}, \hat{\theta}_{f_1}, y_d, \dot{y}_d) = \frac{-k_1\tilde{x}_1 - f_{o_1} - \hat{\theta}_{f_1}^\top \phi_{f_1} + \dot{y}_d}{g_{o_1} + \hat{\theta}_{g_1}^\top \phi_{g_1}} \tag{7.94}$$

$$\tau_{f_{11}} = \Gamma_{f_1}\phi_{f_1}\tilde{x}_1 \tag{7.95}$$

$$\tau_{g_{11}} = \Gamma_{g_1}\phi_{g_1}x_2\tilde{x}_1, \tag{7.96}$$

where $k_1 > 0$ is a design constant. By substituting (7.94)–(7.96) in (7.93) we obtain

$$\begin{aligned}\dot{V}_1 &= -k_1\tilde{x}_1^2 + \left(g_{o_1} + \hat{\theta}_{g_1}^\top \phi_{g_1}\right)\tilde{x}_1\tilde{x}_2 + \Lambda_1 \\ &\quad + \tilde{\theta}_{f_1}^\top \Gamma_{f_1}^{-1}\left(\dot{\hat{\theta}}_{f_1} - \tau_{f_{11}}\right) + \tilde{\theta}_{g_1}^\top \Gamma_{g_1}^{-1}\left(\dot{\hat{\theta}}_{g_1} - \tau_{g_{11}}\right),\end{aligned} \tag{7.97}$$

where $\Lambda_1 = \tilde{x}_1\delta_1$.

Step 2 : The $\tilde{x}_2$ tracking error dynamics satisfy

$$\dot{\tilde{x}}_2 = \dot{x}_2 - \dot{\alpha}_1 \tag{7.98}$$

$$\begin{aligned}&= f_{o_2} + \hat{\theta}_{f_2}^\top \phi_{f_2} + \left(g_{o_2} + \hat{\theta}_{g_2}^\top \phi_{g_2}\right)\alpha_2 + \left(g_{o_2} + \hat{\theta}_{g_2}^\top \phi_{g_2}\right)(x_3 - \alpha_2) + \delta_2 \\ &\quad - \tilde{\theta}_{f_2}^\top \phi_{f_2} - \tilde{\theta}_{g_2}^\top \phi_{g_2} x_3 + \frac{\partial \alpha_1}{\partial x_1}\left(\tilde{\theta}_{f_1}^\top \phi_{f_1} + \tilde{\theta}_{g_1}^\top \phi_{g_1} x_2 - \delta_1\right) - \hat{\mu}_1,\end{aligned} \tag{7.99}$$

where $\hat{\mu}_1$ is given by

$$\hat{\mu}_1 = \frac{\partial \alpha_1}{\partial x_1}\left(f_{o_1} + \hat{\theta}_{f_1}^\top \phi_{f_1} + \left(g_{o_1} + \hat{\theta}_{g_1}^\top \phi_{g_1}\right) x_2\right) + \frac{\partial \alpha_1}{\partial \hat{\theta}_{f_1}}\dot{\hat{\theta}}_{f_1} + \frac{\partial \alpha_1}{\partial \hat{\theta}_{g_1}}\dot{\hat{\theta}}_{g_1} + \frac{\partial \alpha_1}{\partial y_d}\dot{y}_d + \frac{\partial \alpha_1}{\partial \dot{y}_d}\ddot{y}_d.$$

Let

$$V_2 = V_1 + \frac{1}{2}\tilde{x}_2^2 + \frac{1}{2}\tilde{\theta}_{f_2}^\top \Gamma_{f_2}^{-1}\tilde{\theta}_{f_2} + \frac{1}{2}\tilde{\theta}_{g_2}^\top \Gamma_{g_2}^{-1}\tilde{\theta}_{g_2}.$$

From (7.97) and (7.99), the time derivative of V_2 satisfies

$$\begin{aligned}
\dot{V}_2 &= -k_1\tilde{x}_1^2 + \left(g_{o_1} + \hat{\theta}_{g_1}^\top \phi_{g_1}\right)\tilde{x}_1\tilde{x}_2 + \Lambda_1 \\
&\quad + \tilde{\theta}_{f_1}^\top \Gamma_{f_1}^{-1}\left(\dot{\hat{\theta}}_{f_1} - \tau_{f_{11}}\right) + \tilde{\theta}_{g_1}^\top \Gamma_{g_1}^{-1}\left(\dot{\hat{\theta}}_{g_1} - \tau_{g_{11}}\right) \\
&\quad + \tilde{x}_2\left(f_{o_2} + \hat{\theta}_{f_2}^\top \phi_{f_2} + \left(g_{o_2} + \hat{\theta}_{g_2}^\top \phi_{g_2}\right)\alpha_2 + \left(g_{o_2} + \hat{\theta}_{g_2}^\top \phi_{g_2}\right)\tilde{x}_3 - \hat{\mu}_1\right) \\
&\quad + \tilde{x}_2\left(-\tilde{\theta}_{f_2}^\top \phi_{f_2} - \tilde{\theta}_{g_2}^\top \phi_{g_2}x_3 + \frac{\partial \alpha_1}{\partial x_1}\left(\tilde{\theta}_{f_1}^\top \phi_{f_1} + \tilde{\theta}_{g_1}^\top \phi_{g_1}x_2 - \delta_1\right) + \delta_2\right) \\
&\quad + \tilde{\theta}_{f_2}^\top \Gamma_{f_2}^{-1}\dot{\hat{\theta}}_{f_2} + \tilde{\theta}_{g_2}^\top \Gamma_{g_2}^{-1}\dot{\hat{\theta}}_{g_2}
\end{aligned} \tag{7.100}$$

We let

$$\alpha_2 = \frac{1}{g_{o_2} + \hat{\theta}_{g_2}^\top \phi_{g_2}}\left(-k_2\tilde{x}_2 - f_{o_2} - \hat{\theta}_{f_2}^\top \phi_{f_2} - \left(g_{o_1} + \hat{\theta}_{g_1}^\top \phi_{g_1}\right)\tilde{x}_1 + \hat{\mu}_1\right) \tag{7.101}$$

$$\tau_{f_{12}} = \tau_{f_{11}} - \Gamma_{f_1}\phi_{f_1}\frac{\partial \alpha_1}{\partial x_1}\tilde{x}_2 = \Gamma_{f_1}\phi_{f_1}\left(\tilde{x}_1 - \frac{\partial \alpha_1}{\partial x_1}\tilde{x}_2\right) \tag{7.102}$$

$$\tau_{g_{12}} = \tau_{g_{11}} - \Gamma_{g_1}\phi_{g_1}x_2\frac{\partial \alpha_1}{\partial x_1}\tilde{x}_2 = \Gamma_{g_1}\phi_{g_1}x_2\left(\tilde{x}_1 - \frac{\partial \alpha_1}{\partial x_1}\tilde{x}_2\right) \tag{7.103}$$

$$\tau_{f_{22}} = \Gamma_{f_2}\phi_{f_2}\tilde{x}_2 \tag{7.104}$$

$$\tau_{g_{22}} = \Gamma_{g_2}\phi_{g_2}x_3\tilde{x}_2 \tag{7.105}$$

$$\Lambda_2 = \Lambda_1 + \tilde{x}_2\left(\delta_2 - \frac{\partial \alpha_1}{\partial x_1}\delta_1\right) = \left(\tilde{x}_1 - \frac{\partial \alpha_1}{\partial x_1}\tilde{x}_2\right)\delta_1 + \tilde{x}_2\delta_2. \tag{7.106}$$

By substituting (7.101)–(7.106) in (7.100) we obtain

$$\begin{aligned}
\dot{V}_2 &= -k_1\tilde{x}_1^2 - k_2\tilde{x}_2^2 + \left(g_{o_2} + \hat{\theta}_{g_2}^\top \phi_{g_2}\right)\tilde{x}_2\tilde{x}_3 + \Lambda_2 \\
&\quad + \tilde{\theta}_{f_1}^\top \Gamma_{f_1}^{-1}\left(\dot{\hat{\theta}}_{f_1} - \tau_{f_{12}}\right) + \tilde{\theta}_{g_1}^\top \Gamma_{g_1}^{-1}\left(\dot{\hat{\theta}}_{g_1} - \tau_{g_{12}}\right) \\
&\quad + \tilde{\theta}_{f_2}^\top \Gamma_{f_2}^{-1}\left(\dot{\hat{\theta}}_{f_2} - \tau_{f_{22}}\right) + \tilde{\theta}_{g_2}^\top \Gamma_{g_2}^{-1}\left(\dot{\hat{\theta}}_{g_2} - \tau_{g_{22}}\right).
\end{aligned} \tag{7.107}$$

<u>*Step i*</u> ($3 \le i \le n-1$) : A similar procedure is employed recursively at each step $i = 3, 4, \ldots n-1$. The tracking error dynamics $\tilde{x}_i$ satisfy

$$\dot{\tilde{x}}_i = \dot{x}_i - \dot{\alpha}_{i-1} \tag{7.108}$$

$$\begin{aligned}
&= f_{o_i} + \hat{\theta}_{f_i}^\top \phi_{f_i} + \left(g_{o_i} + \hat{\theta}_{g_i}^\top \phi_{g_i}\right)\alpha_i + \left(g_{o_i} + \hat{\theta}_{g_i}^\top \phi_{g_i}\right)(x_{i+1} - \alpha_i) \\
&\quad - \tilde{\theta}_{f_i}^\top \phi_{f_i} - \tilde{\theta}_{g_i}^\top \phi_{g_i}x_{i+1} + \delta_i - \hat{\mu}_{i-1} \\
&\quad + \sum_{k=1}^{i-1}\frac{\partial \alpha_{i-1}}{\partial x_k}\left(\tilde{\theta}_{f_k}^\top \phi_{f_k} + \tilde{\theta}_{g_k}^\top \phi_{g_k}x_{k+1} - \delta_k\right),
\end{aligned} \tag{7.109}$$

where $\hat{\mu}_{i-1}$ is given by

$$\begin{aligned}\hat{\mu}_{i-1} &= -\sum_{k=1}^{i-1} \frac{\partial \alpha_{i-1}}{\partial x_k}\left(f_{o_k} + \hat{\theta}_{f_k}^\top \phi_{f_k} + \left(g_{o_k} + \hat{\theta}_{g_k}^\top \phi_{g_k}\right) x_{k+1}\right) \\ &\quad + \sum_{k=1}^{i-1}\left(\frac{\partial \alpha_{i-1}}{\partial \hat{\theta}_{f_k}} \dot{\hat{\theta}}_{f_k} + \frac{\partial \alpha_{i-1}}{\partial \hat{\theta}_{g_k}} \dot{\hat{\theta}}_{g_k}\right) + \sum_{k=1}^{i} \frac{\partial \alpha_{i-1}}{\partial y_d^{(k-1)}} y_d^{(k)}.\end{aligned}$$

We let

$$V_i = V_{i-1} + \frac{1}{2}\tilde{x}_i^2 + \frac{1}{2}\tilde{\theta}_{f_i}^\top \Gamma_{f_i}^{-1} \tilde{\theta}_{f_i} + \frac{1}{2}\tilde{\theta}_{g_i}^\top \Gamma_{g_i}^{-1} \tilde{\theta}_{g_i}.$$

From (7.107) and (7.109), the time derivative of V_i satisfies

$$\begin{aligned}\dot{V}_i &= -\sum_{j=1}^{i-1} k_j \tilde{x}_j^2 + \left(g_{o_{i-1}} + \hat{\theta}_{g_{i-1}}^\top \phi_{g_{i-1}}\right) \tilde{x}_{i-1}\tilde{x}_i + \Lambda_{i-1} \\ &\quad + \sum_{k=1}^{i-1} \tilde{\theta}_{f_k}^\top \Gamma_{f_k}^{-1}\left(\dot{\hat{\theta}}_{f_k} - \tau_{f_{k(i-1)}}\right) + \sum_{k=1}^{i-1} \tilde{\theta}_{g_k}^\top \Gamma_{g_k}^{-1}\left(\dot{\hat{\theta}}_{g_k} - \tau_{g_{k(i-1)}}\right) \\ &\quad + \tilde{x}_i\left(f_{o_i} + \hat{\theta}_{f_i}^\top \phi_{f_i} + \left(g_{o_i} + \hat{\theta}_{g_i}^\top \phi_{g_i}\right)\alpha_i\right) \\ &\quad + \tilde{x}_i\left((g_{o_i} + \hat{\theta}_{g_i}^\top \phi_{g_i})\tilde{x}_{i+1} - \tilde{\theta}_{f_i}^\top \phi_{f_i} - \tilde{\theta}_{g_i}^\top \phi_{g_i} x_{i+1} + \delta_i - \hat{\mu}_{i-1}\right) \\ &\quad + \tilde{x}_i \sum_{k=1}^{i-1} \frac{\partial \alpha_{i-1}}{\partial x_k}\left(\tilde{\theta}_{f_k}^\top \phi_{f_k} + \tilde{\theta}_{g_k}^\top \phi_{g_k} x_{k+1} - \delta_k\right) \\ &\quad + \tilde{\theta}_{f_i}^\top \Gamma_{f_i}^{-1} \dot{\hat{\theta}}_{f_i} + \tilde{\theta}_{g_i}^\top \Gamma_{g_i}^{-1} \dot{\hat{\theta}}_{g_i}. \end{aligned} \qquad (7.110)$$

Next we select the intermediate control variable α_i and update the intermediate adaptation functions:

$$\alpha_i = \frac{-k_i\tilde{x}_i - f_{o_i} - \hat{\theta}_{f_i}^\top \phi_{f_i} - \left(g_{o_{i-1}} + \hat{\theta}_{g_{i-1}}^\top \phi_{g_{i-1}}\right)\tilde{x}_{i-1} + \hat{\mu}_{i-1}}{g_{o_i} + \hat{\theta}_{g_i}^\top \phi_{g_i}} \qquad (7.111)$$

$$\begin{aligned}\tau_{f_{ki}} &= \tau_{f_{k(i-1)}} - \Gamma_{f_k}\phi_{f_k}\frac{\partial \alpha_{i-1}}{\partial x_k}\tilde{x}_i \qquad (7.112)\\ &= \Gamma_{f_k}\phi_{f_k}\left(\tilde{x}_k - \sum_{j=k}^{i-1}\frac{\partial \alpha_j}{\partial x_k}\tilde{x}_{j+1}\right)\end{aligned}$$

$$\begin{aligned}\tau_{g_{ki}} &= \tau_{g_{k(i-1)}} - \Gamma_{g_k}\phi_{g_k}x_{k+1}\frac{\partial \alpha_{i-1}}{\partial x_k}\tilde{x}_i \qquad (7.113)\\ &= \Gamma_{g_k}\phi_{g_k}x_{k+1}\left(\tilde{x}_k - \sum_{j=k}^{i-1}\frac{\partial \alpha_j}{\partial x_k}\tilde{x}_{j+1}\right)\end{aligned}$$

$$\tau_{f_{ii}} = \Gamma_{f_i}\phi_{f_i}\tilde{x}_i \qquad (7.114)$$

$$\tau_{g_{ii}} = \Gamma_{g_i}\phi_{g_i}x_{i+1}\tilde{x}_i \qquad (7.115)$$

for $k = 1, \ldots, i-1$. By substituting (7.111)–(7.115) in (7.110) we obtain

$$\dot{V}_i = -\sum_{j=1}^{i} k_j\tilde{x}_j^2 + \left(g_{o_i} + \hat{\theta}_{g_i}^\top \phi_{g_i}\right)\tilde{x}_{i+1}\tilde{x}_i + \Lambda_i$$

$$+\sum_{k=1}^{i}\tilde{\theta}_{f_k}^{\mathrm{T}}\Gamma_{f_k}^{-1}\left(\dot{\hat{\theta}}_{f_k}-\tau_{f_{ki}}\right)+\tilde{\theta}_{g_k}^{\mathrm{T}}\Gamma_{g_k}^{-1}\left(\dot{\hat{\theta}}_{g_k}-\tau_{g_{ki}}\right), \tag{7.116}$$

where

$$\begin{aligned}\Lambda_i &= \Lambda_{i-1}+\tilde{x}_i\left(\delta_i-\sum_{k=1}^{i-1}\frac{\partial\alpha_{i-1}}{\partial x_k}\delta_k\right)\\ &= \sum_{k=1}^{i}\left(\tilde{x}_k-\sum_{j=k}^{i-1}\frac{\partial\alpha_j}{\partial x_k}\tilde{x}_{j+1}\right)\delta_k.\end{aligned}$$

Step n : In the final design step, the actual control input u appears. We consider the overall Lyapunov function

$$\begin{aligned}V=V_n &= V_{n-1}+\frac{1}{2}\tilde{x}_n^2+\frac{1}{2}\tilde{\theta}_{f_n}^{\mathrm{T}}\Gamma_{f_n}^{-1}\tilde{\theta}_{f_n}+\frac{1}{2}\tilde{\theta}_{g_n}^{\mathrm{T}}\Gamma_{g_n}^{-1}\tilde{\theta}_{g_n}\\ &= \frac{1}{2}\sum_{k=1}^{n}\left(\tilde{x}_k^2+\tilde{\theta}_{f_k}^{\mathrm{T}}\Gamma_{f_k}^{-1}\tilde{\theta}_{f_k}+\tilde{\theta}_{g_k}^{\mathrm{T}}\Gamma_{g_k}^{-1}\tilde{\theta}_{g_k}\right).\end{aligned}$$

The time derivative of the Lyapunov function V becomes

$$\begin{aligned}\dot{V} &= -\sum_{j=1}^{n-1}k_j\tilde{x}_j^2+\left(g_{o_{n-1}}+\hat{\theta}_{g_{n-1}}^{\mathrm{T}}\phi_{g_{n-1}}\right)\tilde{x}_{n-1}\tilde{x}_n+\Lambda_{n-1}\\ &\quad+\sum_{k=1}^{n-1}\tilde{\theta}_{f_k}^{\mathrm{T}}\Gamma_{f_k}^{-1}\left(\dot{\hat{\theta}}_{f_k}-\tau_{f_{k(n-1)}}\right)+\sum_{k=1}^{n-1}\tilde{\theta}_{g_k}^{\mathrm{T}}\Gamma_{g_k}^{-1}\left(\dot{\hat{\theta}}_{g_k}-\tau_{g_{k(n-1)}}\right)\\ &\quad+\tilde{x}_n\left(f_{o_n}+\hat{\theta}_{f_n}^{\mathrm{T}}\phi_{f_n}+\left(g_{o_n}+\hat{\theta}_{g_n}^{\mathrm{T}}\phi_{g_n}\right)u\right)\\ &\quad+\tilde{x}_n\left(-\tilde{\theta}_{f_n}^{\mathrm{T}}\phi_{f_n}-\tilde{\theta}_{g_n}^{\mathrm{T}}\phi_{g_n}u+\delta_n-\hat{\mu}_{n-1}\right)\\ &\quad+\tilde{x}_n\sum_{k=1}^{n-1}\frac{\partial\alpha_{n-1}}{\partial x_k}\left(\tilde{\theta}_{f_k}^{\mathrm{T}}\phi_{f_k}+\tilde{\theta}_{g_k}^{\mathrm{T}}\phi_{g_k}x_{k+1}-\delta_k\right)\\ &\quad+\tilde{\theta}_{f_n}^{\mathrm{T}}\Gamma_{f_n}^{-1}\dot{\hat{\theta}}_{f_n}+\tilde{\theta}_{g_n}^{\mathrm{T}}\Gamma_{g_n}^{-1}\dot{\hat{\theta}}_{g_n}.\end{aligned} \tag{7.117}$$

Since this is the last step, we choose the control law and the adaptive laws for generating $\hat{\theta}_{f_k}(t)$, $\hat{\theta}_{g_k}(t)$, $k=1,\ 2,\ldots,n$:

$$u = \frac{-k_n\tilde{x}_n-f_{o_n}-\hat{\theta}_{f_n}^{\mathrm{T}}\phi_{f_n}-\left(g_{o_{n-1}}+\hat{\theta}_{g_{n-1}}^{\mathrm{T}}\phi_{g_{n-1}}\right)\tilde{x}_{n-1}+\hat{\mu}_{n-1}}{g_{o_n}+\hat{\theta}_{g_n}^{\mathrm{T}}\phi_{g_n}} \tag{7.118}$$

$$\dot{\hat{\theta}}_{f_k} = \tau_{f_{k(n-1)}}-\Gamma_{f_k}\phi_{f_k}\frac{\partial\alpha_{n-1}}{\partial x_k}\tilde{x}_n \tag{7.119}$$

$$= \Gamma_{f_k}\phi_{f_k}\left(\tilde{x}_k-\sum_{j=k}^{n-1}\frac{\partial\alpha_j}{\partial x_k}\tilde{x}_{j+1}\right)$$

$$\dot{\hat{\theta}}_{g_k} = \tau_{g_{k(n-1)}}-\Gamma_{g_k}\phi_{g_k}x_{k+1}\frac{\partial\alpha_{n-1}}{\partial x_k}\tilde{x}_n \tag{7.120}$$

$$
\begin{aligned}
&= \Gamma_{g_k}\phi_{g_k}x_{k+1}\left(\tilde{x}_k - \sum_{j=k}^{n-1}\frac{\partial \alpha_j}{\partial x_k}\tilde{x}_{j+1}\right) \\
\dot{\hat{\theta}}_{f_n} &= \Gamma_{f_n}\phi_{f_n}\tilde{x}_n && (7.121)\\
\dot{\hat{\theta}}_{g_k} &= \Gamma_{g_n}\phi_{g_n}u\tilde{x}_n. && (7.122)
\end{aligned}
$$

For notational convenience, we define

$$
\begin{aligned}
e_k &= \tilde{x}_k - \sum_{j=k}^{n-1}\frac{\partial \alpha_j}{\partial x_k}\tilde{x}_{j+1}, \qquad k = 1, \ldots n-1\\
e_n &= \tilde{x}_n.
\end{aligned}
$$

Therefore the update laws (7.119)–(7.122) can be rewritten in compact form as

$$
\begin{aligned}
\dot{\hat{\theta}}_{f_k} &= \Gamma_{f_k}\phi_{f_k}e_k, \qquad k = 1, \ldots n && (7.123)\\
\dot{\hat{\theta}}_{g_k} &= \mathcal{P}_S\left(\Gamma_{g_k}\phi_{g_k}x_{k+1}e_k\right), \qquad k = 1, \ldots n-1 && (7.124)\\
\dot{\hat{\theta}}_{g_n} &= \mathcal{P}_S\left(\Gamma_{g_n}\phi_{g_n}ue_n\right), && (7.125)
\end{aligned}
$$

where the projector operator $\mathcal{P}_S$ has been added to ensure the stabilizability property.

By substituting (7.118)–(7.122) in (7.117) we obtain

$$
\dot{V} = -\sum_{j=1}^{n} k_j\tilde{x}_j^2 + \Lambda_n \tag{7.126}
$$

where

$$
\begin{aligned}
\Lambda_n &= \Lambda_{n-1} + \tilde{x}_n\left(\delta_n - \sum_{k=1}^{n-1}\frac{\partial \alpha_{n-1}}{\partial x_k}\delta_k\right)\\
&= \sum_{k=1}^{n}\left(\tilde{x}_k - \sum_{j=k}^{n-1}\frac{\partial \alpha_j}{\partial x_k}\tilde{x}_{j+1}\right)\delta_k\\
&= \sum_{k=1}^{n} e_k\delta_k\\
&= e^\top\delta
\end{aligned}
$$

where $e = [e_1 \ \cdots \ e_n]^\top$ and $\delta = [\delta_1 \ \cdots \ \delta_n]^\top$.

First we consider the ideal case where each $\delta_i = 0$, for $i = 1, 2 \ldots n$. In this case, $\Lambda_n = 0$; therefore

$$
\dot{V} = -\sum_{j=1}^{n} k_j\tilde{x}_j^2. \tag{7.127}
$$

The following closed-loop stability result follows directly from the backstepping design procedure.

Theorem 7.3.3 [Ideal Case] *The closed-loop composed of the system described by (7.87), (7.88) with the approximation-based backstepping controller defined by (7.118)–(7.122), guarantees the following properties:*

1. $\tilde{x}_i, x_i, \hat{\theta}_{f_i}, \hat{\theta}_{g_i} \in \mathcal{L}_\infty \qquad i = 1, 2, \ldots, n$

2. $\tilde{x} \in \mathcal{L}_2$

3. $\tilde{x}(t) \to 0$ *as* $t \to \infty$.

Proof: The proof follows trivially based on the design of the feedback control law that results in eqn. (7.127). ∎

Next, we consider the robustness issues. In the presence of modeling errors δ_i, the time derivative of the Lyapunov function satisfies

$$\dot{V} = -\sum_{j=1}^{n} k_j \tilde{x}_j^2 + e^\top \delta.$$

In order to deal with modeling errors, the adaptive laws (7.123)–(7.125) are modified with the incorporation of projection and dead-zone as follows:

$$\dot{\hat{\theta}}_{f_k} = \mathcal{P}_B \left(\Gamma_{f_k} \phi_{f_k} d(e_k, \tilde{x}, \varepsilon) \right), \qquad k = 1, \ldots n \tag{7.128}$$

$$\dot{\hat{\theta}}_{g_k} = \mathcal{P}_{SB} \left(\Gamma_{g_k} \phi_{g_k} x_{k+1} d(e_k, \tilde{x}, \varepsilon) \right), \qquad k = 1, \ldots n-1 \tag{7.129}$$

$$\dot{\hat{\theta}}_{g_n} = \mathcal{P}_{SB} \left(\Gamma_{g_n} \phi_{g_n} u \, d(e_n, \tilde{x}, \varepsilon) \right), \tag{7.130}$$

where

$$d(e_k, \tilde{x}, \varepsilon) = \begin{cases} 0 & \text{if } \|\tilde{x}\|_2 \le \varepsilon \\ e_k & \text{if } \|\tilde{x}\|_2 > \varepsilon \end{cases}$$

$$\varepsilon = c_e \delta_0 + \mu,$$

where $\mu > 0$ and $c_e > 0$ are positive constants. $\mathcal{P}_B$ is a projection operator designed to keep $\hat{\theta}_{f_k}$ in some convex and compact set $\mathcal{S}_{f_k}$, and $\mathcal{P}_{SB}$ is a projection operator designed to keep $\hat{\theta}_{g_k}$ in the convex and compact set $\mathcal{S}_{g_k}$, which is designed to ensure the stabilizability condition, the condition that $\theta^*_{g_k} \in \mathcal{S}_{g_k}$, and the boundedness of $\hat{\theta}_{g_k}$. The proof of this result is similar to previous proofs in this Chapter using the projection with a dead-zone to obtain robustness — therefore it is left as an exercise (see Problem 7.4).

7.3.3 Command Filtering Approach

Due to the recursive nature of the approach of Section 7.3.2, the derivation and implementation of the feedback control algorithm becomes quite tedious for $n > 3$. This section presents an alternative approach that decouples the design of each pseudo-control using command filters.

Consider the system

$$\dot{x}_i = f_i(x) + g_i(x) x_{i+1}, \qquad \text{for } i = 1, \ldots, n-1 \tag{7.131}$$

$$\dot{x}_n = f_n(x) + g_n(x) u \tag{7.132}$$

where $x = [x_1, \ldots, x_n]^\top \in \Re^n$ is the state, $x_i \in \Re^1$, and u is the scalar control signal. The system is not assumed to be triangular, but is assumed to be feedback passive [139]. The functions f_i, g_i for $i = 1, \ldots, n$ are locally Lipschitz functions that are unknown. For each

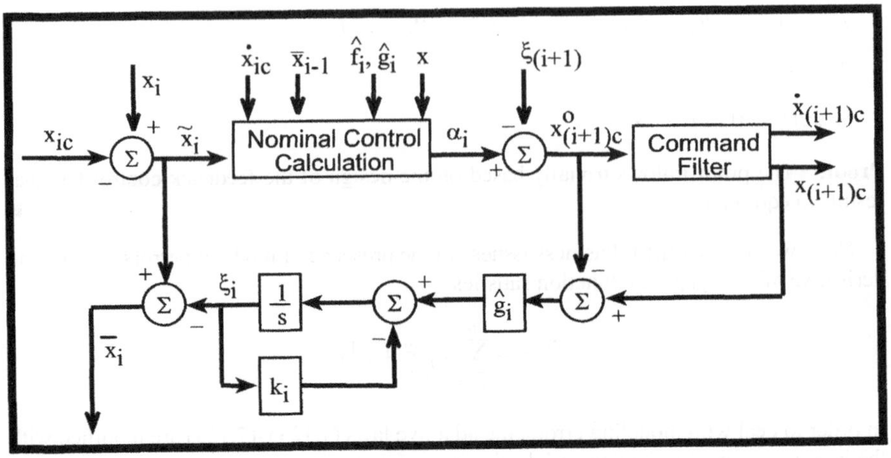

Figure 7.8: Block diagram of command filtered approximation based backstepping implementation for $i \in [2, n-1]$. The inputs to the block diagram are x from the plant; x_{i_c}, $\dot{x}_{i_c}$, and $\bar{x}_{i-1}$ from a previous block of the controller; and $\hat{f}_i$ and $\hat{g}_i$ from the approximation block (not shown). The outputs are the commands $x_{(i+1)_c}$ and $\dot{x}_{(i+1)_c}$ to the next block and $\bar{x}_i$ to the approximation block.

i, the sign of $g_i(x)$ is known and $g_i(x) \neq 0$ for any $x \in \mathcal{D}$. There is a desired trajectory $x_{1c}(t)$, with derivative $\dot{x}_{1c}(t)$, both of which lie in a region $\mathcal{D}$ for $t \geq 0$ and both signals are known.

The control law is defined by

$$\alpha_i = \begin{cases} \frac{1}{\hat{g}_1}\left(-\hat{f}_1 - k_1\tilde{x}_1 + \dot{x}_{1_c}\right) & \text{for } i = 1 \\ \frac{1}{\hat{g}_i}\left(-\hat{f}_i - k_i\tilde{x}_i + \dot{x}_{i_c} - \bar{x}_{i-1}\hat{g}_{i-1}\right), & \text{for } i = 2, \ldots, n \end{cases} \tag{7.133}$$

$$x_{i_c}^o = \alpha_{i-1} - \xi_i, \text{ for } i = 2, \ldots, n \tag{7.134}$$

$$\dot{\xi}_i = \begin{cases} -k_i\xi_i + \hat{g}_i\left(x_{(i+1)_c} - x_{(i+1)_c}^o\right), & \text{for } i = 1, \ldots, (n-1) \\ 0, & \text{for } i = n \end{cases} \tag{7.135}$$

$$\bar{x}_i = \tilde{x}_i - \xi_i, \text{ for } i = 1, \ldots, n \tag{7.136}$$

with $u = \alpha_n$ where each $k_i > 0$ for $i = 1, \ldots, n$. For each $i = 1, \ldots, n$, the signal x_{i_c} and its derivative $\dot{x}_{i_c}$ are produced without differentiation by using a command filter such as that defined in Figure A.4 with the input $x_{i_c}^o$. The tracking error is defined for $i = 1, \ldots, n$ as $\tilde{x}_i = x_i - x_{i_c}$. The variable ξ_i is a filtered version of the error $(x_{(i+1)_c} - x_{(i+1)_c}^o)$ imposed by the command filter. The variable $\bar{x}_i$ is referred to as the compensated tracking error as it is the tracking error after removal of ξ_i. A block diagram of this control calculation for one value of $i \in [2, n-1]$ is shown in Figure 7.8.

Given eqns. (7.133)–(7.136), the dynamics of the tracking errors and the compensated tracking errors can be derived. We present the derivations only for $i = 2, \ldots, (n-1)$ and the final results for all cases. The derivations for the $i = 1$ and $i = n$ cases are left as an exercise (see Problem 7.5). For $i = 2, \ldots, (n-1)$, the tracking error dynamics simplify as follows:

$$\begin{aligned}
\dot{\tilde{x}}_i &= \hat{f}_i + \hat{g}_i\, x^o_{(i+1)_c} - \dot{x}_{i_c} + \hat{g}_i \left(x_{(i+1)_c} - x^o_{(i+1)_c} \right) + \hat{g}_i \left(x_{(i+1)} - x_{(i+1)_c} \right) \\
&\quad + (f_i - \hat{f}_i) + (g_i - \hat{g}_i) x_{(i+1)} \\
&= \hat{f}_i + \hat{g}_i\, \alpha_i - \dot{x}_{i_c} - \hat{g}_i\, \xi_{(i+1)} + \hat{g}_i \left(x_{(i+1)_c} - x^o_{(i+1)_c} \right) + \hat{g}_i \left(x_{(i+1)} - x_{(i+1)_c} \right) \\
&\quad - \tilde{\theta}_{f_i}^\top \phi - \tilde{\theta}_{g_i}^\top \phi x_{(i+1)} \\
&= -k_i \tilde{x}_i - \bar{x}_{(i-1)} \hat{g}_{(i-1)} + \hat{g}_i \left(x_{(i+1)_c} - x^o_{(i+1)_c} \right) + \hat{g}_i \left(\tilde{x}_{(i+1)} - \xi_{(i+1)} \right) \\
&\quad - \tilde{\theta}_{f_i}^\top \phi - \tilde{\theta}_{g_i}^\top \phi x_{(i+1)} \\
&= -k_i \tilde{x}_i - \bar{x}_{(i-1)} \hat{g}_{(i-1)} + \hat{g}_i \left(x_{(i+1)_c} - x^o_{(i+1)_c} \right) + \hat{g}_i \bar{x}_{(i+1)} \\
&\quad - \tilde{\theta}_{f_i}^\top \phi - \tilde{\theta}_{g_i}^\top \phi x_{(i+1)}.
\end{aligned}$$

For $i = 2, \ldots, (n-1)$, the compensated tracking error dynamics simplify as follows:

$$\begin{aligned}
\dot{\bar{x}}_i &= \dot{\tilde{x}}_i - \dot{\xi}_i \\
&= -k_i \bar{x}_i - \bar{x}_{(i-1)} \hat{g}_{(i-1)} + \hat{g}_i \bar{x}_{(i+1)} - \tilde{\theta}_{f_i}^\top \phi - \tilde{\theta}_{g_i}^\top \phi x_{(i+1)}. \qquad (7.137)
\end{aligned}$$

For $i = 1$ the tracking error and compensated tracking error dynamics are

$$\begin{aligned}
\dot{\tilde{x}}_1 &= -k_1 \tilde{x}_1 + \hat{g}_1 \tilde{x}_2 + \hat{g}_1 \left(x_{2_c} - x^o_{2_c} \right) - \tilde{\theta}_{f_1}^\top \phi - \tilde{\theta}_{g_1}^\top \phi x_2, \\
\dot{\bar{x}}_1 &= -k_1 \bar{x}_1 + \hat{g}_1 \bar{x}_2 - \tilde{\theta}_{f_1}^\top \phi - \tilde{\theta}_{g_1}^\top \phi x_2. \qquad (7.138)
\end{aligned}$$

For $i = n$, we have that $\bar{x}_n = \tilde{x}_n$; therefore, the tracking error and compensated tracking error dynamics are

$$\dot{\tilde{x}}_n = \dot{\bar{x}}_n = -k_n \bar{x}_n - \bar{x}_{(n-1)} \hat{g}_{(n-1)} - \tilde{\theta}_{f_n}^\top \phi - \tilde{\theta}_{g_n}^\top \phi u. \qquad (7.139)$$

Consider the following Lyapunov function candidate

$$\mathcal{V} = \frac{1}{2} \sum_{i=1}^{n} \left(\bar{x}_i^2 + \tilde{\theta}_{f_i}^\top \Gamma_{f_i}^{-1} \tilde{\theta}_{f_i} + \tilde{\theta}_{g_i}^\top \Gamma_{g_i}^{-1} \tilde{\theta}_{g_i} \right). \qquad (7.140)$$

The time derivative of $\mathcal{V}$ along solutions of eqns. (7.137)–(7.139) satisfies

$$\begin{aligned}
\dot{\mathcal{V}} &= \bar{x}_1 \left(-k_1 \bar{x}_1 + \hat{g}_1 \bar{x}_2 - \tilde{\theta}_{f_1}^\top \phi - \tilde{\theta}_{g_1}^\top \phi x_2 \right) + \sum_{i=1}^{n} \left(\tilde{\theta}_{f_i}^\top \Gamma_{f_i}^{-1} \dot{\tilde{\theta}}_{f_i} + \tilde{\theta}_{g_i}^\top \Gamma_{g_i}^{-1} \dot{\tilde{\theta}}_{g_i} \right) \\
&\quad + \sum_{i=2}^{n-1} \bar{x}_i \left(-k_i \bar{x}_i - \bar{x}_{(i-1)} \hat{g}_{(i-1)} + \hat{g}_i \bar{x}_{(i+1)} - \tilde{\theta}_{f_i}^\top \phi - \tilde{\theta}_{g_i}^\top \phi x_{(i+1)} \right) \\
&\quad + \bar{x}_n \left(-k_n \bar{x}_n - \bar{x}_{(n-1)} \hat{g}_{(n-1)} - \tilde{\theta}_{f_n}^\top \phi - \tilde{\theta}_{g_n}^\top \phi u \right) \\
&= \sum_{i=1}^{n} k_i \bar{x}_i^2 + \sum_{i=1}^{n-1} \left(\tilde{\theta}_{f_i}^\top \Gamma_{f_i}^{-1} \left(\dot{\tilde{\theta}}_{f_i} - \Gamma_{f_i} \phi \bar{x}_i \right) + \tilde{\theta}_{g_i}^\top \Gamma_{g_i}^{-1} \left(\dot{\tilde{\theta}}_{g_i} - \Gamma_{g_i} \phi \bar{x}_i x_{(i+1)} \right) \right) \\
&\quad + \left(\tilde{\theta}_{f_n}^\top \Gamma_{f_n}^{-1} \left(\dot{\tilde{\theta}}_{f_n} - \Gamma_{f_n} \phi \bar{x}_n \right) + \tilde{\theta}_{g_n}^\top \Gamma_{g_n}^{-1} \left(\dot{\tilde{\theta}}_{g_n} - \Gamma_{g_n} \phi \bar{x}_n u \right) \right). \qquad (7.141)
\end{aligned}$$

Therefore, we select the parameter adaptation laws as

$$\dot{\hat{\theta}}_{f_i} = \Gamma_{f_i}\phi\bar{x}_i \qquad \text{for } i = 1,\ldots,n \tag{7.142}$$

$$\dot{\hat{\theta}}_{g_i} = \mathcal{P}_{S_i}\left(\Gamma_{g_i}\phi\bar{x}_i x_{(i+1)}\right) \qquad \text{for } i = 1,\ldots,n-1 \tag{7.143}$$

$$\dot{\hat{\theta}}_{g_n} = \mathcal{P}_{S_n}\left(\Gamma_{g_n}\phi\bar{x}_n u\right), \tag{7.144}$$

where $\mathcal{P}_{S_i}$ for $i = 1,\ldots,n$ are projections operators designed to maintain $\hat{\theta}_{g_i}$ in $\mathcal{S}_{g_i}$, where $\mathcal{S}_{g_i}$ is specified to ensure the stabilizability condition and possibly the boundedness of $\hat{\theta}_{g_i}$. When the projection operators are not in effect, the derivative of the Lyapunov function reduces to

$$\dot{\mathcal{V}} = -\sum_{i=1}^{n} k_i\bar{x}_i^2. \tag{7.145}$$

Therefore, we can summarize these results in the following theorem, which applies for the ideal case (i.e., $\delta = 0$).

Theorem 7.3.4 *Consider the closed-loop system composed of the plant described in eqns. (7.131)–(7.132) with the controller of eqns. (7.133)–(7.136) with parameter adaptation defined by eqns. (7.142)–(7.144). This system solves the tracking problem with the following properties:*

1. x, $\hat{\theta}_f$, $\hat{\theta}_g \in \mathcal{L}_\infty$,
2. $\bar{x}_i \in \mathcal{L}_2$, *and*
3. $\bar{x}_i \to 0$ *as* $t \to \infty$.

Proof: Outside the region $\mathcal{D}$, we assume that the terms v_{f_i} and v_{g_i} for $i = 1,\ldots,n$ have been defined to ensure that the state will converge to $\mathcal{D}$. Therefore, the proof will only be concerned with $x \in \mathcal{D}$.

For $x \in \mathcal{D}$ with the stated control law, along solutions of the closed-loop system, the Lyapunov function of eqn. (7.140) has the time derivative $\dot{\mathcal{V}} = -\sum_{i=1}^{n} k_i\bar{x}_i^2$, which is negative semidefinite. Note that as long as ϕ is bounded, Lemma A.3.1 completes the proof.

When the projection operator is active, as discussed in Theorem 4.6.1, the stability properties of the control algorithm are preserved. ■

Theorem 7.3.4 guarantees desirable properties for the compensated tracking errors $\bar{x}_i$, not the actual tracking errors $\tilde{x}_i$. The difference between these two quantities is ξ_i, which is the output of the stable filter

$$\dot{\xi}_i = -k_i\,\xi_i + \hat{g}_i\left(x_{(i+1)c} - x^o_{(i+1)c}\right).$$

The magnitude of the input $\left(x_{(i+1)c} - x^o_{(i+1)c}\right)$ to this filter is determined by the design of the $(i+1)$-st command filter. For a well-designed command filter, this error will be small. The continuous function $\hat{g}_i$ is bounded on the compact set $\mathcal{D}$. Therefore, ξ_i is expected to be small during transients and zero under steady state conditions.

The goal of the command filtered approach summarized in this theorem was to avoid tedious algebraic manipulations involved in the computation of the backstepping control signal. In addition to achieving the desired goal, the above command filtering approach allows parameter estimation to continue in the presence of — and can be used to enforce on the virtual control variables — any magnitude, rate, and bandwidth limitations of the actual physical system. This is achieved by the design of the command filters; however, when a physical limitation is imposed on the i-th state, then tracking of the filtered commands will not be achieved by states x_j for $j = 1, \ldots, i$. Once the physical constraint is no longer in effect, $\tilde{x}_i \to \bar{x}_i$ for all i. The following example is designed to clarify this issue.

EXAMPLE 7.1

The main issue of this example is the accommodation of constraints on the state variables in the backstepping control approach. In fact, we take this one step further, by also accommodating such constraints in the parameter adaptation process. To clearly present the issues, we focus in this example on a very simple system that contains a single unknown parameter.

Consider the system

$$\begin{aligned} \dot{x}_1 &= -x_1^3|x_2| + bx_2 \\ \dot{x}_2 &= u \end{aligned}$$

where the parameter b is not known, x_{1_c} is in $[-1, 1]$, and x_2 is constrained to be within $[-2,\ 2]$. In the notation of this section, $f_{1_o} = -x_1^3|x_2|$, $f_1^* = g_{1_o} = f_{2_o} = f_2^* = g_2^* = 0$, $g_{2_o} = 1$, and $g_1^* = b$. Note that this system is not triangular; therefore, the standard backstepping approach does not apply.

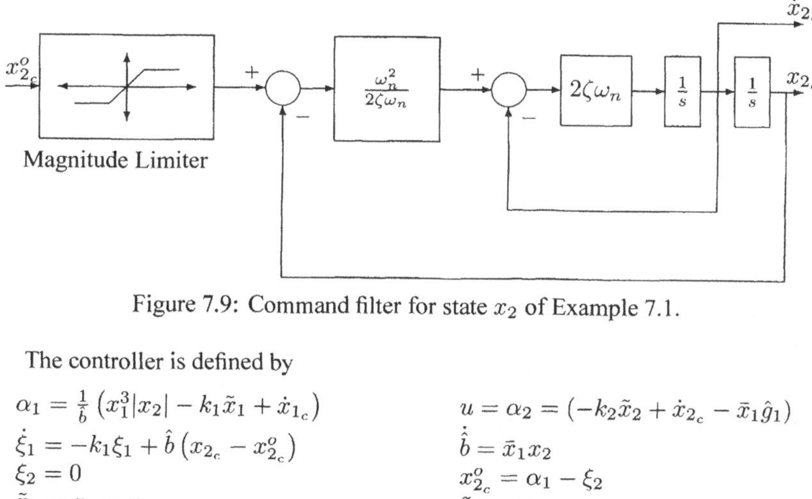

Figure 7.9: Command filter for state x_2 of Example 7.1.

The controller is defined by

$$\begin{aligned} &\alpha_1 = \tfrac{1}{\hat{b}}\left(x_1^3|x_2| - k_1\tilde{x}_1 + \dot{x}_{1_c}\right) && u = \alpha_2 = \left(-k_2\tilde{x}_2 + \dot{x}_{2_c} - \bar{x}_1\hat{g}_1\right) \\ &\dot{\xi}_1 = -k_1\xi_1 + \hat{b}\left(x_{2_c} - x_{2_c}^o\right) && \dot{\hat{b}} = \bar{x}_1 x_2 \\ &\xi_2 = 0 && x_{2_c}^o = \alpha_1 - \xi_2 \\ &\tilde{x}_1 = x_1 - x_{1_c} && \tilde{x}_2 = x_2 - x_{2_c} \\ &\bar{x}_1 = \tilde{x}_1 - \xi_1 && \bar{x}_2 = \tilde{x}_2 \end{aligned} \tag{7.146}$$

where $k_1 = 1$, and $k_2 = 2$. The signals x_{2_c} and $\dot{x}_{2_c}$ are outputs of the command filter shown in Figure 7.9 with magnitude limits of ± 2, $\omega_n = 100$ and $\zeta = 0.8$. For simulation purposes, $b = 3.0$. The estimated value of $\hat{b}$ is initialized as $\hat{b}(0) = 1.0$ and projection is used to ensure that $\hat{b}(t) > 0.5$ for all t.

Simulation results are shown in Figures 7.10, 7.11 and 7.12. The $x_{2_c}^o$ plot of Figure 7.10 shows that early in the simulation, the value of $x_{2_c}^o$ exceeds the ± 2 magnitude

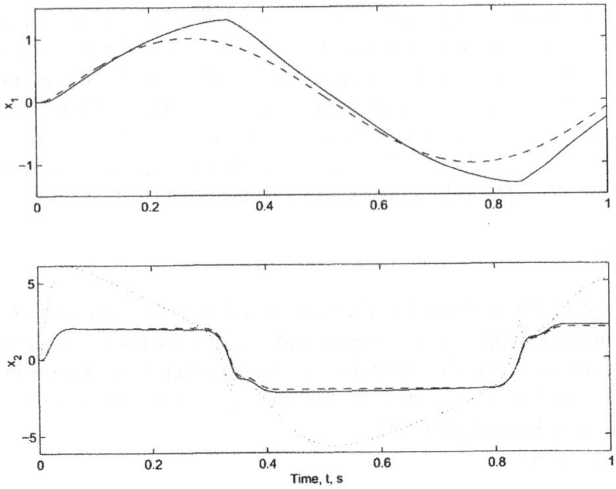

Figure 7.10: Simulated states and commands from the first simulated second for Example 7.1. *Top* — x_1 is solid, x_{1_c} is dashed; *Bottom* — x_2 is solid, x_{2_c} is dashed $x^o_{2_c}$ is dotted.

limit. The command filter ensures that x_{2_c} satisfies the ± 2 magnitude limit. Note that x_2 accurately tracks x_{2_c}. By the end of the 50-s simulation, see Figure 7.11, both $x^o_{2_c}$ and x_{2_c} satisfy the ± 2 magnitude limit. Throughout the entire simulation, even when $x^o_{2_c}$ is not achievable by the system, the Lyapunov function is decreasing, as shown in the top curve of Figure 7.12. The bottom curve of Figure 7.12 shows $\tilde{b}(t)$. If parameter adaptation is implemented using $\tilde{x}_1$ instead of $\bar{x}_1$ (i.e., $\dot{\hat{b}} = \tilde{x}_1 x_2$), the system does not converge. △

7.3.4 Robustness Considerations

Assume that perfect approximation is not possible, but instead, bounded model errors occur in each of the tracking error equations:

$$\begin{aligned}
\dot{\tilde{x}}_1 &= -k_1\tilde{x}_1 + \hat{g}_1\tilde{x}_2 + \hat{g}_1\left(x_{2_c} - x^o_{2_c}\right) - \tilde{\theta}^T_{f_1}\phi - \tilde{\theta}^T_{g_1}\phi x_2 + \delta_1(x,u), && (7.147)\\
\dot{\tilde{x}}_i &= -k_i\tilde{x}_i - \tilde{x}_{(i-1)}\hat{g}_{(i-1)} + \hat{g}_i\left(x_{(i+1)_c} - x^o_{(i+1)_c}\right) + \hat{g}_i\tilde{x}_{(i+1)} \\
&\quad -\tilde{\theta}^T_{f_i}\phi - \tilde{\theta}^T_{g_i}\phi x_{(i+1)} + \delta_i(x,u) \quad \text{for } i = 2,\ldots,(n-1), && (7.148)\\
\dot{\tilde{x}}_n &= -k_n\tilde{x}_n - \tilde{x}_{(n-1)}\hat{g}_{(n-1)} - \tilde{\theta}^T_{f_n}\phi - \tilde{\theta}^T_{g_n}\phi u + \delta_n(x,u). && (7.149)
\end{aligned}$$

The compensated tracking error dynamics simplify as follows:

$$\begin{aligned}
\dot{\bar{x}}_1 &= -k_1\bar{x}_1 + \hat{g}_1\bar{x}_2 - \tilde{\theta}^T_{f_1}\phi - \tilde{\theta}^T_{g_1}\phi x_2 + \delta_1(x,u) && (7.150)\\
\dot{\bar{x}}_i &= -k_i\bar{x}_i - \bar{x}_{(i-1)}\hat{g}_{(i-1)} + \hat{g}_i\bar{x}_{(i+1)} \\
&\quad -\tilde{\theta}^T_{f_i}\phi - \tilde{\theta}^T_{g_i}\phi x_{(i+1)} + \delta_i(x,u) \quad \text{for } i = 2,\ldots,(n-1) && (7.151)\\
\dot{\bar{x}}_n &= -k_n\bar{x}_n - \bar{x}_{(n-1)}\hat{g}_{(n-1)} - \tilde{\theta}^T_{f_n}\phi - \tilde{\theta}^T_{g_n}\phi u + \delta_n(x,u). && (7.152)
\end{aligned}$$

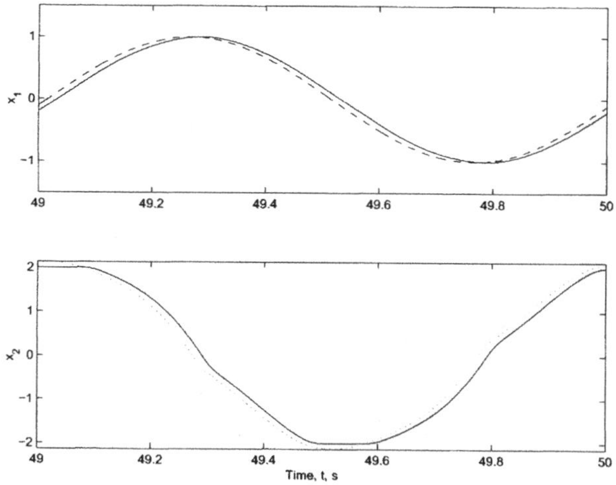

Figure 7.11: Simulated states and commands from the 49th (last) simulated second for Example 7.1. *Top* — x_1 is solid, x_{1_c} is dashed; *Bottom* — x_2 is solid, x_{2_c} is dashed $x^o_{2_c}$ is dotted.

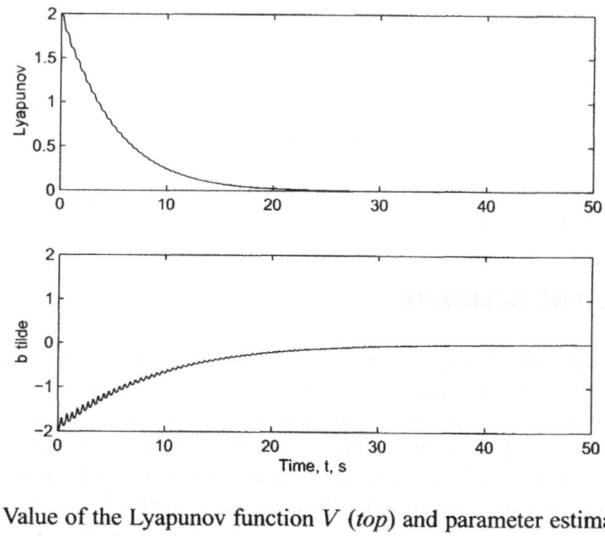

Figure 7.12: Value of the Lyapunov function V (*top*) and parameter estimation error $\tilde{b}(t)$ (*bottom*) versus time during the simulation of Example 7.1.

When the projection operators are not in effect, the derivative of the Lyapunov function reduces to

$$\dot{\mathcal{V}} = -\sum_{i=1}^{n} \left(k_i \bar{x}_i^2 - \bar{x}_i \delta_i\right) \tag{7.153}$$

which is negative for $\sum_{i=1}^{n} k_i \bar{x}_i^2 > \sum_{i=1}^{n} \bar{x}_i \delta_i$. Therefore, we can prove the following theorem.

Theorem 7.3.5 *Consider the closed-loop system composed of the plant described in eqns. (7.131)–(7.132) with the controller of eqns. (7.133)–(7.136) with parameter adaptation defined by*

$$\dot{\tilde{\theta}}_{f_i} = \mathcal{P}_{B_i}\left(\Gamma_{f_i} \phi d_i(\bar{x}, \delta_0)\right) \quad \text{for } i = 1, \ldots, n \tag{7.154}$$

$$\dot{\tilde{\theta}}_{g_i} = \mathcal{P}_{SB_i}\left(\Gamma_{g_i} \phi d_i(\bar{x}, \delta_0) x_{(i+1)}\right) \text{ for } i = 1, \ldots, n-1 \tag{7.155}$$

$$\dot{\tilde{\theta}}_{g_n} = \mathcal{P}_{SB_n}\left(\Gamma_{g_n} \phi d_n(\bar{x}, \delta_0) u\right) \tag{7.156}$$

where

$$d_i(\bar{x}, \varepsilon) = \begin{cases} 0 & \text{if } \sum_{i=1}^{n} \bar{x}_i^2 \leq \varepsilon, \\ \bar{x}_i & \text{if } \sum_{i=1}^{n} \bar{x}_i^2 > \varepsilon, \end{cases}$$

where $\varepsilon = \frac{1}{\kappa^2} \sum_{i=1}^{n} \delta_{i_0}^2 + \mu$ *for some* $\mu > 0$ *and* $\kappa = \min_i(k_i)$. *Assuming that* $\delta_i \leq \delta_{i_0}$ *where* $\delta_0 = [\delta_{1_0}, \ldots, \delta_{n_0}]$, *this system solves the tracking problem with the following properties:*

1. $\bar{x}_i$, x, $\tilde{\theta}_f$, $\tilde{\theta}_g$, $\hat{\theta}_f$, $\hat{\theta}_g \in \mathcal{L}_\infty$;
2. $\bar{x}_i$ *is small-in-the-mean-square sense, satisfying*

$$\int_t^{t+T} \|\bar{x}_i\|_2^2 d\tau \leq \frac{2}{\kappa} V(t) + T\varepsilon;$$

3. *as* $t \to \infty$, $\bar{x}_i$ *is ultimately bounded by* $\|\bar{x}\|^2 \leq \varepsilon$.

The proof follows the same lines as those of Theorems 7.2.3 and 7.2.5. Therefore, it is left as an exercise. For hints, see Problem 7.6

7.4 CONCLUDING SUMMARY

This chapter has presented a general theoretical framework for adaptive approximation based control. The main emphasis has been the derivation of provably stable feedback algorithms for some general classes of nonlinear systems, where the unknown nonlinearities are represented by adaptive approximation models. Two general classes of nonlinear systems have been considered: (i) feedback linearizable systems with unknown nonlinearities; (ii) triangular nonlinear systems that allow the use of the backstepping control design procedure.

Overall, this chapter has followed a similar development as Chapter 5, with the unknown nonlinearities being replaced by adaptive approximators. In some cases the mathematics get rather involved, especially when using the backstepping procedure. In this chapter, as well as in Chapter 6, there was a focus on understanding some of the key underlying concepts of adaptive approximation.

The development of a general theory for designing and analyzing adaptive approximation based control systems started in the early 1990s [40, 181, 208, 211, 212, 229, 232, 273]. In the beginning, most of the techniques dealt with the use of neural networks as approximators of unknown nonlinearities and they considered, in general, the ideal case of no approximation error. These works generated significant interest in the use of adaptive approximation methods for feedback control. One direction of research dealt with the design and analysis of robust adaptive approximation based control schemes [111, 191, 192, 209, 224]. There is also considerable research work focused on nonlinearly parameterized approximators [149, 216] and output-based adaptive approximation based control schemes [3, 90, 112, 144]. In addition to the continuous-time framework, several researchers have investigated the issue of designing and analyzing discrete-time adaptive approximation based control systems [41, 93, 95, 123, 210], as well as the multivariable case [94, 162]. Several researchers have investigated adaptive fuzzy control schemes and adaptive neuro-fuzzy control schemes [255, 282], as well as wavelet approximation models [25, 37, 199, 306]. A significant amount of research work has focused on adaptive approximation based control of specific applications, such as robotic systems [92, 241, 278, 275, 277] and aircraft systems [36, 77, 78]. In addition to feedback control, there has also been a lot of interest in the application of adaptive approximation methods to fault diagnosis [50, 207, 271, 276, 307, 308]. system identification [24, 44, 214, 137, 228], and adaptive critics [220, 247]. Finally, it is noted that several books have also appeared in the topics related to this chapter [15, 23, 32, 63, 87, 91, 101, 115, 129, 147, 148, 151, 168, 189, 197, 198, 254, 264, 283, 296].

7.5 EXERCISES AND DESIGN PROBLEMS

Exercise 7.1 Theorem 7.2.3 stability results for a scalar approximation based feedback linearization approach using a dead-zone. Discuss why violation of the inequality $|d| < \delta_o$ causes the proof of that theorem to break down. Show that even with the dead-zone, if the inequality $|d| < \delta_o$ does not always hold, then the method yields performance similar to that stated in Theorem 7.2.2.

Exercise 7.2 Complete the stability analysis for the closed-loop systems described in Section 7.2.3.

Exercise 7.3 Starting from eqn. (7.64) prove the properties of Theorem 7.3.1.

Exercise 7.4 Complete the stability analysis for the closed-loop systems described in Section 7.3.2 with $\delta \neq 0$ (discussed after Theorem 7.3.3).

Exercise 7.5 For the approach derived in Section 7.3.3:

1. derive the dynamic equations for the tracking errors $\tilde{x}_1$ and $\tilde{x}_n$;
2. derive eqns. (7.138) and (7.139).

Exercise 7.6 Complete the proof of Theorem 7.3.5.

Hint: Use Young's inequality in the form $\bar{x}_i\delta_i < \frac{k_i}{2}\bar{x}_i^2 + \frac{1}{2k_i}\delta_i^2$. Show that

$$\dot{\mathcal{V}} \leq \frac{\kappa}{2}\sum_{i=1}^{n}\left(-\bar{x}_i^2 + \left(\frac{\delta_{io}}{\kappa}\right)^2\right).$$

Use this expression to show that the time outside the dead-zone is finite. Integrate both sides of the inequality to derive the mean-squared error bound.

Exercise 7.7 This problem considers the design of u for $x \notin \mathcal{D}$, for the example of Section 7.2.2.3.

1. For the definition of z on page 302, find the equations for $\dot{z}$.
2. Evaluate $\dot{V}$ for $V = \frac{1}{2}z^\top z$.
3. Show that $\dot{V} \leq -z_2 h(z_2)$ for the specified u.
4. Discuss how this fact justifies the claim that initial conditions outside $\mathcal{D} = [-1.3, 1.3] \times [-1.3, 1.3]$ ultimately converge to $\mathcal{D}$, but $\mathcal{D}$ is not positively invariant. Consider the initial condition $x = [1.3, 1.3, 0]$.
5. If $\mathcal{D}$ were redefined to be $\mathcal{D} = \left\{(x_1, x_2) \middle| \, \|(x_1, x_2)\|_2 \leq 1.3\right\}$, can you show that the new $\mathcal{D}$ is positively invariant.

Exercise 7.8 *For the detailed example of Section 7.2.2.3, design and simulated a controller using the backstepping approach discussed in Section 7.3.2.*

Exercise 7.9 *For the detailed example of Section 7.2.2.3, design and simulated a controller using the command filtered backstepping approach discussed in Section 7.3.3.*

CHAPTER 8

ADAPTIVE APPROXIMATION BASED CONTROL FOR FIXED-WING AIRCRAFT

Various authors have investigated the applicability of nonlinear control methodologies to advanced flight vehicles. These methods offer both increases in aircraft performance as well as reduction of development times by dealing with the complete dynamics of the vehicle rather than local operating point designs (see Section 5.1.3). Feedback linearization, in its various forms, is perhaps the most commonly employed nonlinear control method in flight control [14, 34, 143, 165, 166, 250]. Backstepping-based approaches are discussed for example in [77, 98, 106, 107, 245]. Reference [135] presents a nonlinear model predictive control approach that relies on a Taylor series approximation to the system's differential equations. Optimal control techniques are applied to control load-factor in [96]. Prelinearization theory and singular perturbation theory are applied for the derivation of inner and outer loop controllers in [165]. The main drawback to the nonlinear control approaches mentioned above is that, as model-based control methods, they require accurate knowledge of the plant dynamics. This is of significance in flight control since aerodynamic parameters always contain some degree of uncertainty. Although, some of these approaches are robust to small modeling errors, they are not intended to accommodate significant unanticipated errors that can occur, for example, in the event of failure or battle damage. In such an event, the aerodynamics can change rapidly and deviate significantly from the model used for control design. Uninhabited Air Vehicles (UAVs) are particularly susceptible to such events since there is no pilot onboard. For high performance aircraft and UAVs, improved control may be achievable if the unknown nonlinearities are approximated adaptively.

Adaptive Approximation Based Control: Unifying Neural, Fuzzy and Traditional Adaptive Approximation Approaches. By Jay A. Farrell and Marios M. Polycarpou

This chapter presents detailed design and analysis of adaptive approximation based controllers applied to fixed-wing aircraft.[1] Therefore, we begin the chapter in Section 8.1 with a brief introduction to aircraft dynamics and the industry standard method for representing the aerodynamic forces and moments that act on the vehicle. The dynamic model for an aircraft is presented in Subsection 8.1.1. Subsection 8.1.2 introduces the nondimensional coefficient representation for the aerodynamic forces and moments in the dynamic model. For ease of reference, tables summarizing aircraft notation are included at the end of the chapter in Section 8.4.

Two control situations are considered. In Section 8.2, an angular rate controller is designed and analyzed. That controller is applicable in piloted aircraft applications where the stick motion of the pilot is processed into body-frame angular rate commands. That section will also discuss issues such as the effect of actuator distribution. In Section 8.3, we develop a full vehicle controller suitable for UAVs. The controller inputs are commands for climb rate γ, ground track χ, and airspeed V. An adaptive approximation based backstepping approach is used.

8.1 AIRCRAFT MODEL INTRODUCTION

Since entire books are written on aircraft dynamics and control, this section cannot completely cover the topic. The goal of this section is to briefly provide enough of an introduction so that readers unfamiliar with aircraft dynamics and control can understand the derivations that follow.

8.1.1 Aircraft Dynamics

Aircraft dynamics are derived and discussed in e.g. [7, 258]. Various choices are possible for the definition of the state variables. We will define the state vector using the standard choice $x = [\chi, \gamma, V, \mu, \alpha, \beta, P, Q, R]$. The subvector $[P, Q, R]$ is the body-frame angular rate vector. The components are the roll, pitch, and yaw rates, respectively. The subvector $[\mu, \alpha, \beta]$ will be referred to as the wind-axes angle vector. The bank angle of the vehicle is denoted by μ. The angle-of-attack α and sideslip β define the rotation between the body and wind frames-of-reference. The variables χ and γ are the ground-track angle and the climb angle. Finally, V is the airspeed. For convenience of the reader, the dynamics of this state vector are summarized here:

$$\begin{aligned}
\dot{\chi} &= \frac{1}{mV\cos\gamma}\left[D\sin\beta\cos\mu + Y\cos\beta\cos\mu + L\sin\mu\right. \\
&\quad \left. + T\left(\sin\alpha\sin\mu - \cos\alpha\sin\beta\cos\mu\right)\right] && (8.1) \\
\dot{\gamma} &= \frac{1}{mV}\left[-D\sin\beta\sin\mu - Y\cos\beta\sin\mu + L\cos\mu\right. \\
&\quad \left. + T\left(\sin\alpha\cos\mu + \cos\alpha\sin\beta\sin\mu\right)\right] - \frac{g}{V}\cos\gamma && (8.2) \\
\dot{V} &= \frac{1}{m}\left(T\cos\alpha\cos\beta - D\cos\beta + Y\sin\beta\right) - g\sin\gamma && (8.3) \\
\dot{\mu} &= \frac{1}{mV}\left[D\sin\beta\tan\gamma\cos\mu + Y\cos\beta\tan\gamma\cos\mu + L\left(\tan\beta + \tan\gamma\sin\mu\right) + \right.
\end{aligned}$$

[1] This research was performed in collaboration with by Barron Associates Inc. and builds on the ideas published in [77, 78] and the citations therein. The authors gratefully acknowledge the contributions of Manu Sharma and Nathan Richards to the theoretical development and for the implementation of the control algorithm software.

$$
\begin{aligned}
&+ T\left(\sin\alpha\tan\gamma\sin\mu + \sin\alpha\tan\beta - \cos\alpha\sin\beta\tan\gamma\cos\mu\right)\big] \\
&- \frac{g\tan\beta\cos\gamma\cos\mu}{V} + \frac{P_s}{\cos\beta} \qquad (8.4)
\end{aligned}
$$

$$
\begin{aligned}
\dot{\alpha} &= \frac{-1}{mV\cos\beta}\left[L + T\sin\alpha\right] + \frac{g\cos\gamma\cos\mu}{V\cos\beta} + Q - P_s\tan\beta \qquad (8.5) \\
\dot{\beta} &= \frac{1}{mV}\left[D\sin\beta + Y\cos\beta - T\cos\alpha\sin\beta\right] + \frac{g\cos\gamma\sin\mu}{V} - R_s \qquad (8.6) \\
\dot{P} &= \left(c_1 R + c_2 P\right) Q + c_3\bar{L} + c_4\bar{N} \qquad (8.7) \\
\dot{Q} &= c_5 PR - c_6\left(P^2 - R^2\right) + c_7\bar{M} \qquad (8.8) \\
\dot{R} &= \left(c_8 P - c_2 R\right) Q + c_4\bar{L} + c_9\bar{N}. \qquad (8.9)
\end{aligned}
$$

In these equations, m is the mass, g denotes gravity, and the c_i coefficients for $i = 1, ..., 9$ are defined on page 80 in [258]. The variables P_s and R_s are the stability axes roll and yaw rates:

$$
\begin{bmatrix} P_s \\ R_s \end{bmatrix} = \begin{bmatrix} \cos\alpha & \sin\alpha \\ -\sin\alpha & \cos\alpha \end{bmatrix} \begin{bmatrix} P \\ R \end{bmatrix}. \qquad (8.10)
$$

The symbols $[D, Y, L]$ denote the drag, side, and lift aerodynamic forces and the symbols $[\bar{L}, \bar{M}, \bar{N}]$ denote the aerodynamic moments about the body-frame x, y, and z axes, respectively. The aerodynamic forces and moments are functions of the aircraft state and of the control variables. The control variables are the engine thrust T and the angular deflection of each of the control surfaces denoted by the vector $\delta = [\delta_1, \ldots, \delta_m]$. The control signal δ does not appear explicitly in the above equations, but may affect the magnitude and sign of the aerodynamic forces and moments. See Section 8.1.2 for further discussion.

Tables 8.2, 8.3, and 8.4 at the end of this chapter define the constants, variable, and functions used in the above equations. For the discussion to follow, we will assume the (nominal) aircraft is tailless and configured with $p = 6$ control surfaces.

8.1.2 Nondimensional Coefficients

In the aircraft literature, these aerodynamic force and moments functions are represented by nondimensional coefficient functions. In this approach, the basic structure of the model and the major effects of speed, air density, etc. are accounted for explicitly for a general class of air vehicles. Nondimensional coefficient functions relate the general model to a specific vehicle in the class.

For example, the aerodynamic forces might be represented as

$$
\begin{aligned}
D(V, \rho, \alpha, \delta) &= \bar{q}S\left(C_{D_o} + \sum_{i=1}^{m} C_{D_{\delta_i}}\delta_i\right) \qquad (8.11) \\
Y(V, \rho, \alpha, \beta, P, \delta) &= \bar{q}S\left(C_{Y_o} + C_{Y_\beta}\beta + C_{Y_P}\frac{bP}{2V} + \sum_{i=1}^{m} C_{Y_{\delta_i}}\delta_i\right) \qquad (8.12) \\
L(V, \rho, \alpha, \delta) &= \bar{q}S\left(C_{L_o} + C_{L_\alpha}\alpha + \sum_{i=1}^{m} C_{L_{\delta_i}}\delta_i\right) \qquad (8.13)
\end{aligned}
$$

where $\bar{q} = \frac{\rho V^2}{2}$ is the aerodynamic pressure, S is the wing reference area, b is the reference wing span, and ρ is the air density. The subscripted 'C' symbols are the nondimensional aerodynamic coefficient functions, i.e., C_{D_o}, $C_{D_{\delta_i}}$, C_{Y_o} .... Different aerodynamic coefficient functions are dominant for different vehicles. The force and moment equations shown in this section include the dominant coefficient functions for the vehicle that we utilize in the simulations to follow. For the methods to follow, it will be clear how to extend the approach to use additional coefficient functions that may be applicable to other classes of vehicles. Typically, the nondimensional coefficients are functions of only one or two arguments. In the simulation examples to follow, the nondimensional coefficients will only be functions of angle-of-attack α and Mach number M. Similar to the above, the aerodynamic moments are represented as

$$\begin{aligned}
\bar{L}(V,\rho,\alpha,\beta,P,R,\delta) &= \bar{q}Sb\left(C_{\bar{L}_o} + C_{\bar{L}P}\frac{bP}{2V} + C_{\bar{L}R}\frac{bR}{2V} + C_{\bar{L}_\beta}\beta + \sum_{i=1}^{m} C_{\bar{L}_{\delta_i}}\delta_i\right)\\
\bar{M}(V,\rho,\alpha,\beta,Q,\delta) &= \bar{q}S\bar{c}\left(C_{\bar{M}_o} + C_{\bar{M}Q}\frac{\bar{c}Q}{2V} + \sum_{i=1}^{m} C_{\bar{M}_{\delta_i}}\delta_i\right)\\
\bar{N}(V,\rho,\alpha,\beta,P,R,\delta) &= \bar{q}Sb\left(C_{\bar{N}_o} + C_{\bar{N}P}\frac{bP}{2V} + C_{\bar{N}R}\frac{bR}{2V} + C_{\bar{N}_\beta}\beta + \sum_{i=1}^{m} C_{\bar{N}_{\delta_i}}\delta_i\right).
\end{aligned}$$

Whereas the aerodynamic forces and moments are functions of several variables and may change rapidly as a function of the vehicle state over the desired flight envelope, the nondimensional coefficients are continuous functions of only a few states (e.g., α and M in this case study).

In the control derivations that follow, for the convenience of representation of the control surface effectiveness matrix, the moment functions are decomposed as:

$$\bar{L} = \bar{L}' + \sum_{i=1}^{6} \bar{L}_{\delta_i}\delta_i, \quad \bar{M} = \bar{M}' + \sum_{i=1}^{6} \bar{M}_{\delta_i}\delta_i, \quad \text{and } \bar{N} = \bar{N}' + \sum_{i=1}^{6} \bar{N}_{\delta_i}\delta_i.$$

8.2 ANGULAR RATE CONTROL FOR PILOTED VEHICLES

This section considers the design of an angular rate controller where the pilot stick inputs are processed to generate angular rate commands (P_c, Q_c, R_c) and rate command derivatives $(\dot{P}_c, \dot{Q}_c, \dot{R}_c)$. Note that this does not suggest that the pilot is analytically computing derivatives while flying the plane. Instead, the pilot maneuvers the stick. The stick motion is processed to produce the (continuous and bounded) angular rate commands (P_c^o, Q_c^o, R_c^o). These signals are filtered (see Appendix A.4) to produce (P_c, Q_c, R_c) and $(\dot{P}_c, \dot{Q}_c, \dot{R}_c)$. Such filters are referred to herein as *command filters*.

The objective of a command filter with bounded input z_c^o is to produce two continuous and bounded output signals z_c and $\dot{z}_c$. The error between z_c^o and z_c should be small. This is achieved by designing the command filter to have a bandwidth larger than the bandwidth of z_c^o. Ensuring the fact that the signal z_c is the integral of $\dot{z}_c$, is a design constraint on the command filter. We present the design of one such prefilter here, and will refer back to it several times throughout the remainder of this chapter.

Consider the filtering of P_c^o by

$$\begin{bmatrix} \dot{x}_1 \\ \dot{x}_2 \end{bmatrix} = \begin{bmatrix} 0 & 1 \\ -\omega_n^2 & -2\zeta\omega_n \end{bmatrix}\begin{bmatrix} x_1 \\ x_2 \end{bmatrix} + \begin{bmatrix} 0 \\ \omega_n^2 \end{bmatrix} P_c^o$$

$$\begin{bmatrix} P_c \\ \dot{P}_c \end{bmatrix} = \begin{bmatrix} 1 & 0 \\ 0 & 1 \end{bmatrix} \begin{bmatrix} x_1 \\ x_2 \end{bmatrix}.$$

The transfer function from P_c^o to P_c is given by

$$\frac{P_c(s)}{P_c^o(s)} = \frac{\omega_n^2}{s^2 + 2\zeta\omega_n s + \omega_n^2}, \tag{8.14}$$

which has unity gain at low frequencies, damping specified by ζ, and undamped natural frequency equal to ω_n. As long as ω_n is selected to be large relative to the bandwidth of $P_c^o(t)$, the error $P_c^o(t) - P_c(t)$ will be small. Also, by the design of the filter, the output $P_c(t)$ is the integral of the output $\dot{P}_c(t)$. In the analysis of the control law, we will prove that $P(t)$ converges to and tracks $P_c(t)$. Therefore, the response of $P(t)$ to the pilot command $P_c^o(t)$ is determined by this prefilter; therefore, the prefilter determines the aircraft handling qualities. Similar prefilters are designed for Q_c^o and R_c^o.

8.2.1 Model Representation

The angular rate dynamics of eqns. (8.7)–(8.9) can be written as

$$\dot{z} = A(f_0 + f^*) + F(x) + B(G_0 + G^*)\delta \tag{8.15}$$

where $z = [P, Q, R]^\top$ and $A = B = \begin{bmatrix} c_3 & 0 & c_4 \\ 0 & c_7 & 0 \\ c_4 & 0 & c_9 \end{bmatrix}$ are known matrices. The inertial terms represented by

$$F = \begin{bmatrix} (c_1 R + c_2 P) Q \\ c_5 PR - c_6 \left(P^2 - R^2\right) \\ (c_8 P - c_2 R) Q \end{bmatrix}$$

are assumed to be known. The aerodynamic moments are represented as

$$(f_0 + f^*) = \begin{bmatrix} \bar{L}' \\ \bar{M}' \\ \bar{N}' \end{bmatrix} \quad \text{and} \quad (G_0 + G^*) = \begin{bmatrix} \bar{L}_{\delta_1} & \cdots & \bar{L}_{\delta_6} \\ \bar{M}_{\delta_1} & \cdots & \bar{M}_{\delta_6} \\ \bar{N}_{\delta_1} & \cdots & \bar{N}_{\delta_6} \end{bmatrix}.$$

In this representation, f_0 and G_0 represent the baseline or design model, while f^* and G^* represent model error. The model error may represent error between the actual dynamics and the baseline model or model error due to in-flight events.

The control signal will be implemented through the surface deflection vector $\delta = [\delta_1, \ldots, \delta_6]^\top$. The objective of the control design is to select δ to force $[P(t), Q(t), R(t)]$ to track $[P_c(t), Q_c(t), R_c(t)]$ in the presence of the nonlinear model errors f^* and G^*.

8.2.2 Baseline Controller

This subsection considers the design of an angular rate controller based on the design model without function approximation. The objective is to analyze the affect of model error and to illustrate that the approximation based controller can be considered as a straightforward addition to the baseline controller that enhances stability and performance in the presence of errors between the baseline model and the actual aircraft dynamics.

Since the functions f^* and G^* are unknown, the baseline controller design is developed using the following design model.

$$\dot{z} \;=\; Af_0 + F(x) + BG_0\delta.$$

Therefore, we select a continuous signal δ such that

$$BG_0\delta = -Af_0 - F - K\tilde{z} + \dot{z}_c, \tag{8.16}$$

where K is a positive definite matrix, $\tilde{z} = z - z_c$, $z_c = [P_c, Q_c, R_c]^\top$ and $\dot{z}_c = [\dot{P}_c, \dot{Q}_c, \dot{R}_c]^\top$.

Since the aircraft is over-actuated (i.e., $G_0 \in \Re^{3\times 6}$), the matrix BG_0 will have more columns than rows and will have full row rank. Therefore, many solutions to eqn. (8.16) exist. Some form of actuator distribution [26, 68, 70] is required to select a specific δ. For example, the surface deflections could be defined according to

$$\delta = d + W^{-1}G_0^\top B^\top \left[BG_0W^{-1}G_0^\top B^\top\right]^{-1}(u_c - BG_0\, d), \tag{8.17}$$

where W is a positive definite matrix, $d \in \Re^6$ is a possibly time-varying vector, and $u_c = -Af_0 - F - K\tilde{z} + \dot{z}_c$. This actuator distribution approach minimizes $(\delta - d)^\top W(\delta - d)$ subject to the constraint that $u_c = BG_0\delta$. It is straightforward to simply let d be the zero vector; however, it is also possible to define d to decrease the magnitude and rate of change of δ.

When a surface deflection vector δ satisfying eqn. (8.16) is applied to the actual dynamics of eqn. (8.15), the resulting closed-loop tracking error dynamics reduce as follows:

$$\begin{aligned}
\dot{z} &= Af_0 + F(x) + BG_0\delta + Af^* + BG^*\delta \\
&= -K\tilde{z} + \dot{z}_c + Af^* + BG^*\delta \\
\dot{\tilde{z}} &= -K\tilde{z} + Af^* + BG^*\delta.
\end{aligned} \tag{8.18}$$

If the design model were perfect (i.e., $f^* = 0$ and $G^* = 0$), then we would analyze the Lyapunov function $V = \frac{1}{2}\tilde{z}^\top\tilde{z}$. The time derivative of V along solutions of eqn. (8.18) with $f^* = 0$ and $G^* = 0$ is $\dot{V} = -\tilde{z}^\top K\tilde{z}$ which is negative definite. Therefore, relative to the design model, the closed-loop system is exponentially stable (by item 5 of Theorem A.2.1).

Relative to the actual aircraft dynamics, the derivative of the Lyapunov function is $\dot{V} = -\tilde{z}^\top K\tilde{z} + \tilde{z}^\top(Af^* + BG^*\delta)$. Nothing can be said about the definiteness properties of this time derivative without further assumptions about the modeling errors f^* and G^*. If f^* and G^* satisfy certain growth conditions (e.g., see the topic of "vanishing perturbations" in [134]), then the system is still locally exponentially stable. Note that such vanishing perturbation conditions are difficult to apply in tracking applications. As the modeling errors f^* and G^* increase, the closed-loop system may have bounded tracking errors or be unstable. However, nothing specific can be said without more explicit knowledge of the model error.

8.2.3 Approximation Based Controller

The approximation based controller will select a continuous signal δ such that

$$B\left(G_0 + \hat{G}\right)\delta = -A\left(f_0 + \hat{f}\right) - F - K\tilde{z} + \dot{z}_c, \tag{8.19}$$

where the only differences relative to the definition from (8.16) are the inclusion of the approximations $\hat{f}$ and $\hat{G}$ to the model errors f^* and G^*. The approximator structure and the parameter adaptation will be defined in the next two subsections. The parameter adaptation process must ensure that $G = \left(G_0 + \hat{G}\right)$ maintains full row rank to ensure that a solution to eqn. (8.19) exists. The solution vector δ can again be found by some form of actuator distribution, e.g., eqn. (8.17) with G_0 replaced by G and with $u_c = -A\left(f_0 + \hat{f}\right) - F - K\tilde{z} + \dot{z}_c$. When the surface deflection vector δ satisfying eqn. (8.19) is applied to the actual dynamics of eqn. (8.15), the resulting closed-loop tracking error dynamics reduce as follows:

$$\begin{aligned}
\dot{z} &= A\left(f_0 + \hat{f}\right) + F(x) + B\left(G_0 + \hat{G}\right)\delta + A\left(f^* - \hat{f}\right) + B\left(G^* - \hat{G}\right)\delta \\
&= -K\tilde{z} + \dot{z}_c + A\left(f^* - \hat{f}\right) + B\left(G^* - \hat{G}\right)\delta \\
\dot{\tilde{z}} &= -K\tilde{z} + A\left(f^* - \hat{f}\right) + B\left(G^* - \hat{G}\right)\delta \\
\dot{\tilde{z}} &= -K\tilde{z} - A\tilde{f} - B\tilde{G}\delta,
\end{aligned} \tag{8.20}$$

where $\tilde{f} = \left(\hat{f} - f^*\right)$ and $\tilde{G} = \left(\hat{G} - G^*\right)$. Completion of the design of the adaptive approximation based controller requires specification of the approximators, specification of the parameter adaptation laws, and analysis of the stability of the resulting closed-loop systems. These items are addressed in the following three subsections.

8.2.3.1 Approximator Definition. The aircraft angular rate dynamics involve three moments $(\bar{L}, \bar{M}, \bar{N})$. The unknown portion of these moment functions determines the vector and matrix functions f^* and G^* that we wish to approximate.

The designer could choose to approximate directly the three functions

$$\bar{L}(V, \rho, \alpha, \beta, P, R, \delta), \;\; \bar{M}(V, \rho, \alpha, \beta, Q, \delta), \text{ and } \bar{N}(V, \rho, \alpha, \beta, P, R, \delta).$$

Since each of these functions has several arguments, useful generalization would be difficult to achieve and the curse of dimensionality would be an issue. Alternatively, a designer wishing to take advantage of the known model structure could chose to approximate the 28 nondimensional coefficient functions each as a function of only α and M. We choose this latter approach. In doing so, we realize that the 28 nondimensional coefficient functions will likely not converge to the actual coefficient functions; instead, the approximated coefficient functions will only converge to the extent sufficient to ensure accurate command tracking. If guaranteed convergence of the approximated functions is desired, then persistence of excitation conditions would need to be analyzed and ensured.

Let each nondimensional coefficient function be represented as the sum of a known portion denoted with a superscript "o" and an unknown portion indicated with a lower case "c". For example,

$$C_{\bar{L}_P} = C^o_{\bar{L}_P} + c_{\bar{L}_P},$$

where $C^o_{\bar{L}_P}$ is the known portion used in the baseline design model and $c_{\bar{L}_P}$ is the unknown portion to be approximated online. Then, the baseline model is described by

$$f_0 = \bar{q}S \begin{bmatrix} b & 0 & 0 \\ 0 & \bar{c} & 0 \\ 0 & 0 & b \end{bmatrix} \begin{bmatrix} \left(C^o_{\bar{L}_o} + C^o_{\bar{L}_P}\frac{bP}{2V} + C^o_{\bar{L}_R}\frac{bR}{2V} + C^o_{\bar{L}_\beta}\beta\right) \\ \left(C^o_{\bar{M}_o} + C^o_{\bar{M}_Q}\frac{\bar{c}Q}{2V}\right) \\ \left(C^o_{\bar{N}_o} + C^o_{\bar{N}_P}\frac{bP}{2V} + C^o_{\bar{N}_R}\frac{bR}{2V} + C^o_{\bar{N}_\beta}\beta\right) \end{bmatrix} \tag{8.21}$$

and

$$G_0 = \bar{q}S \begin{bmatrix} b & 0 & 0 \\ 0 & \bar{c} & 0 \\ 0 & 0 & b \end{bmatrix} \begin{bmatrix} C^o_{\bar{L}_{\delta_1}} & C^o_{\bar{L}_{\delta_2}} & \cdots & C^o_{\bar{L}_{\delta_m}} \\ C^o_{\bar{M}_{\delta_1}} & C^o_{\bar{M}_{\delta_2}} & \cdots & C^o_{\bar{M}_{\delta_m}} \\ C^o_{\bar{N}_{\delta_1}} & C^o_{\bar{N}_{\delta_2}} & \cdots & C^o_{\bar{N}_{\delta_m}} \end{bmatrix}. \tag{8.22}$$

The functions f^* and G^* are defined similarly as

$$f^* = \bar{q}S \begin{bmatrix} b & 0 & 0 \\ 0 & \bar{c} & 0 \\ 0 & 0 & b \end{bmatrix} \begin{bmatrix} \left(c_{\bar{L}_o} + c_{\bar{L}_P}\frac{bP}{2V} + C_{\bar{L}_R}\frac{bR}{2V} + c_{\bar{L}_\beta}\beta\right) \\ \left(c_{\bar{M}_o} + c_{\bar{M}_Q}\frac{\bar{c}Q}{2V}\right) \\ \left(c_{\bar{N}_o} + c_{\bar{N}_P}\frac{bP}{2V} + c_{\bar{N}_R}\frac{bR}{2V} + c_{\bar{N}_\beta}\beta\right) \end{bmatrix}$$

and

$$G^* = \bar{q}S \begin{bmatrix} b & 0 & 0 \\ 0 & \bar{c} & 0 \\ 0 & 0 & b \end{bmatrix} \begin{bmatrix} c_{\bar{L}_{\delta_1}} & c_{\bar{L}_{\delta_2}} & \cdots & c_{\bar{L}_{\delta_m}} \\ c_{\bar{M}_{\delta_1}} & c_{\bar{M}_{\delta_2}} & \cdots & c_{\bar{M}_{\delta_m}} \\ c_{\bar{N}_{\delta_1}} & c_{\bar{N}_{\delta_2}} & \cdots & c_{\bar{N}_{\delta_m}} \end{bmatrix}. \tag{8.23}$$

The unknown portion of each nondimensional coefficient function will be approximated during aircraft operation. The coefficient $c_{\bar{L}_o}$ will be approximated as $\hat{c}_{\bar{L}_o}(\alpha, M) = \theta^\top_{\bar{L}_o}\phi_{\bar{L}_o}(\alpha, M)$ where $\phi_{\bar{L}_o}(\alpha, M) : \Re^2 \mapsto \Re^N$ is a regressor vector that is selected by the designer. The coefficient $c_{\bar{N}_P}$ will be approximated as $\hat{c}_{\bar{N}_P}(\alpha, M) = \theta^\top_{\bar{N}_P}\phi_{\bar{N}_P}(\alpha, M)$. The approximations to the other 28 coefficient functions are defined similarly. While it is reasonable to use different regressor vectors such as $\phi_{\bar{N}_P}(\alpha, M)$ for each coefficient functions, in this case study we uses a single regressor vector for all the approximated coefficient functions for notational simplicity: $\phi(\alpha, M) = \phi_{\bar{L}_o} = \phi_{\bar{N}_P} = \ldots$. The regressor vector $\phi(\alpha, M)$ will be defined so that it is a partition of unity for every $(\alpha, M) \in \mathcal{D}$ where $\mathcal{D} = \mathcal{D}_\alpha \times \mathcal{D}_M$ is compact with $\mathcal{D}_\alpha = [-7, 15]$ degrees and $\mathcal{D}_M = [0.2, 1.0]$. The variables α and M are outside the control loop, but are affected by the angular rates. It is assumed that the pilot issues commands (P^o_c, Q^o_c, R^o_c) and controls the engine thrust to ensure that (α, M) remains in $\mathcal{D}$.

An alternative way of stating the ideas at the end of the previous paragraph is that the aircraft designers specify an operating envelope $\mathcal{D} = \mathcal{D}_\alpha \times \mathcal{D}_M$. The control designers develop a controller with guaranteed performance over $\mathcal{D}$. The pilot must ensure that the angular rate and thrust commands maintain $(\alpha(t), M(t)) \in \mathcal{D}$ for all t.

The functions $\hat{f}$ and $\hat{G}$ can be reconstructed from the approximated coefficient functions as

$$\hat{f} = \bar{q}S \begin{bmatrix} b & 0 & 0 \\ 0 & \bar{c} & 0 \\ 0 & 0 & b \end{bmatrix} \begin{bmatrix} \left(\hat{c}_{\bar{L}_o} + \hat{c}_{\bar{L}_P}\frac{bP}{2V} + \hat{c}_{\bar{L}_R}\frac{bR}{2V} + \hat{c}_{\bar{L}_\beta}\beta\right) \\ \left(\hat{c}_{\bar{M}_o} + \hat{c}_{\bar{M}_Q}\frac{\bar{c}Q}{2V}\right) \\ \left(\hat{c}_{\bar{N}_o} + \hat{c}_{\bar{N}_P}\frac{bP}{2V} + \hat{c}_{\bar{N}_R}\frac{bR}{2V} + \hat{c}_{\bar{N}_\beta}\beta\right) \end{bmatrix} \tag{8.24}$$

and

$$\hat{G} = \bar{q}S \begin{bmatrix} b & 0 & 0 \\ 0 & \bar{c} & 0 \\ 0 & 0 & b \end{bmatrix} \begin{bmatrix} \hat{c}_{\bar{L}_{\delta_1}} & \hat{c}_{\bar{L}_{\delta_2}} & \cdots & \hat{c}_{\bar{L}_{\delta_m}} \\ \hat{c}_{\bar{M}_{\delta_1}} & \hat{c}_{\bar{M}_{\delta_2}} & \cdots & \hat{c}_{\bar{M}_{\delta_m}} \\ \hat{c}_{\bar{N}_{\delta_1}} & \hat{c}_{\bar{N}_{\delta_2}} & \cdots & \hat{c}_{\bar{N}_{\delta_m}} \end{bmatrix}, \tag{8.25}$$

where the arguments to the functions have been dropped to simplify the notation. For the analysis that follows, it is useful to note that $\hat{f}$ can be manipulated into the standard Linear-In-the-Parameter (LIP) form $\hat{f} = \Phi^\top_f \Theta_f$ where

$$\Theta^\top_f = [\theta^\top_{\bar{L}_o}, \ldots, \theta^\top_{\bar{N}_\beta}] \in \Re^{10N \times 1}$$

and $\Phi_f \in \Re^{10N\times 3}$. This representation is not computationally efficient, since Φ_f is sparse, but simplifies the notation of the analysis. Similarly, the j-th column of the matrix $\hat{G}$ can be represented as $\hat{G}_j = \Phi_{G_j}^\top \Theta_{G_j}$ where

$$\Theta_{G_j}^\top = [\theta_{\bar{L}\delta j}^\top, \theta_{\bar{M}\delta_j}^\top, \theta_{\bar{N}_{\delta j}}^\top] \in \Re^{1\times 3N}$$

and $\Phi_{G_j} \in \Re^{3N\times 3}$ for $j = 1,\ldots,6$.

Finally, over a compact region $\mathcal{D}$, which represents the operating envelope, from Section 3.1.3, we know that there exists optimal Θ_f^* and $\Theta_{G_j}^*$ such that

$$f^* = \Phi_f^\top \Theta_f^* + e_f \tag{8.26}$$

$$G_j^* = \Phi_{G_j}^\top \Theta_{G_j}^* + e_{G_j} \quad \text{for } j = 1,\ldots,m \tag{8.27}$$

where e_f and e_{G_j} are bounded with the bound determined by $\mathcal{D}$ and the choice of ϕ. The approximation parameter errors are defined as

$$\tilde{\Theta}_f = \Theta_f - \Theta_f^* \tag{8.28}$$

$$\tilde{\Theta}_{G_j} = \Theta_{G_j} - \Theta_{G_j}^* \quad \text{for } j = 1,\ldots,m. \tag{8.29}$$

With the approximator defined as in this subsection, the tracking error dynamics of eqn. (8.20) reduce to

$$\dot{\tilde{z}} = -K\tilde{z} - A\left(\Phi_f^\top \tilde{\Theta}_f - e_f\right) - B\sum_{j=1}^{m}\left(\Phi_{G_j}^\top \tilde{\Theta}_{G_j} - e_{G_j}\right)\delta_j. \tag{8.30}$$

8.2.3.2 Parameter Adaptation. We select the parameter adaptation laws as

$$\dot{\tilde{\Theta}}_f = \dot{\Theta}_f = \mathcal{P}_f\left(\Gamma_f \Phi_f A^\top \underline{z}\right) \tag{8.31}$$

$$\dot{\tilde{\Theta}}_{G_j} = \dot{\Theta}_{G_j} = \mathcal{P}_{G_j}\left(\Gamma_{G_j}\Phi_{G_j} B^\top \underline{z}\delta_j\right), \tag{8.32}$$

where Γ_f and Γ_{G_j} are positive definite matrices. The signal

$$\underline{z} = \begin{cases} \tilde{z} & \text{if } \underline{\lambda}_K \|\tilde{z}\|_2 > \varepsilon \\ 0 & \text{otherwise} \end{cases}$$

where $\underline{\lambda}_K$ is the minimum eigenvalue of K. This adaptation law includes dead-zone and projection operators. The projection operator $\mathcal{P}_f$ ensures that each element of Θ_f remains within known upper and lower bounds: $\underline{\Theta}_{f_i} \leq \Theta_{f_i} \leq \bar{\Theta}_{f_i}$. Therefore, the $\mathcal{P}_f$ projection operator acts componentwise according to

$$\mathcal{P}_{f_i}(\tau_i) = \begin{cases} \tau_i & \text{if } \underline{\Theta}_{f_i} \leq \Theta_{f_i} \leq \bar{\Theta}_{f_i} \\ 0 & \text{otherwise} \end{cases}$$

where $\tau = \Gamma_f \Phi_f A^\top \underline{z}$. The projection operators $\mathcal{P}_{G_j}$ for $j = 1,\ldots,m$ must maintain boundedness of the elements of Θ_{G_j} and full row rank of the matrix G. The row rank of G is determined by the row rank of the matrix

$$C = \begin{bmatrix} C^o_{\bar{L}_{\delta_1}} + \hat{c}_{\bar{L}_{\delta_1}} & C^o_{\bar{L}_{\delta_2}} + \hat{c}_{\bar{L}_{\delta_2}} & \cdots & C^o_{\bar{L}_{\delta_m}} + \hat{c}_{\bar{L}_{\delta_m}} \\ C^o_{\bar{M}_{\delta_1}} + \hat{c}_{\bar{M}_{\delta_1}} & C^o_{\bar{M}_{\delta_2}} + \hat{c}_{\bar{M}_{\delta_2}} & \cdots & C^o_{\bar{M}_{\delta_m}} + \hat{c}_{\bar{M}_{\delta_m}} \\ C^o_{\bar{N}_{\delta_1}} + \hat{c}_{\bar{N}_{\delta_1}} & C^o_{\bar{N}_{\delta_2}} + \hat{c}_{\bar{N}_{\delta_2}} & \cdots & C^o_{\bar{N}_{\delta_m}} + \hat{c}_{\bar{N}_{\delta_m}} \end{bmatrix}$$

defined in eqns. (8.19), (8.22), and (8.25). Based on physical principals, each element of the C matrix has a known sign. If the sign structure of the matrix is maintained, then the full rank condition is also maintained. Therefore, with the fact that ϕ is a partition of unity on $\mathcal{D}$, it is straightforward to find upper and lower bounds on each element of Θ_{G_j} such that $\underline{\Theta}_{G_{ji}} \leq \Theta_{G_{ji}} \leq \bar{\Theta}_{G_{ji}}$ ensures both the boundedness of Θ_{G_j} and the full row rank of G. Therefore, the $\mathcal{P}_{G_j}$ projection operator acts componentwise according to

$$\mathcal{P}_{G_ji}(\tau_i) = \begin{cases} \tau_i & \text{if } \underline{\Theta}_{G_{ji}} \leq \Theta_{G_{ji}} \leq \bar{\Theta}_{G_{ji}} \\ 0 & \text{otherwise} \end{cases}$$

where $\tau = \Gamma_{G_j} \Phi_{G_j} B^\top \tilde{z} \delta_j$.

8.2.3.3 *Stability Analysis.* Define the Lyapunov function

$$\mathcal{V} \;=\; \frac{1}{2}\left(\tilde{z}^\top \tilde{z} + \tilde{\Theta}_f^\top \Gamma_f^{-1} \tilde{\Theta}_f + \sum_{j=1}^{m}\left(\tilde{\Theta}_{G_j}^\top \Gamma_{G_j}^{-1} \tilde{\Theta}_{G_j}\right)\right). \tag{8.33}$$

When neither the projection nor the dead-zone is in effect, the time derivative of $\mathcal{V}$ is

$$\begin{aligned}
\dot{\mathcal{V}} &= \tilde{z}^\top \dot{\tilde{z}} + \dot{\tilde{\Theta}}_f^\top \Gamma_f^{-1} \tilde{\Theta}_f + \sum_{j=1}^{m}\left(\dot{\tilde{\Theta}}_{G_j}^\top \Gamma_{G_j}^{-1} \tilde{\Theta}_{G_j}\right) \\
&= \tilde{z}^\top\left(-K\tilde{z} - A\left(\Phi_f^\top \tilde{\Theta}_f - e_f\right) - B\sum_{j=1}^{m}\left(\Phi_{G_j}^\top \tilde{\Theta}_{G_j} - e_{G_j}\right)\delta_j\right) \\
&\quad + \dot{\tilde{\Theta}}_f^\top \Gamma_f^{-1} \tilde{\Theta}_f + \sum_{j=1}^{m}\left(\dot{\tilde{\Theta}}_{G_j}^\top \Gamma_{G_j}^{-1} \tilde{\Theta}_{G_j}\right) \\
&= -\tilde{z}^\top K \tilde{z} + \left(\dot{\tilde{\Theta}}_f^\top - \tilde{z}^\top A \Phi_f^\top \Gamma_f\right)\Gamma_f^{-1}\tilde{\Theta}_f + \tilde{z}^\top\left(Ae_f + \sum_{j=1}^{m} Be_{G_j}\delta_j\right) \\
&\quad + \sum_{j=1}^{m}\left(\left(\dot{\tilde{\Theta}}_{G_j}^\top - \tilde{z}^\top B \Phi_{G_j}^\top \Gamma_{G_j}\delta_j\right)\Gamma_{G_j}^{-1}\tilde{\Theta}_{G_j}\right) \\
&= -\tilde{z}^\top K \tilde{z} + \tilde{z}^\top\left(Ae_f + \sum_{j=1}^{m} Be_{G_j}\delta_j\right) \\
\dot{\mathcal{V}} &\leq -\|\tilde{z}\|_2\left(\lambda_K \|\tilde{z}\|_2 - \|\rho(\delta)\|_2\right),
\end{aligned} \tag{8.34}$$

where $\rho(\delta) = Ae_f + \sum_{j=1}^{m} Be_{G_j}\delta_j$. Therefore, the Lyapunov function is decreasing for $\|\tilde{z}\|_2 > \frac{\|\rho(\delta)\|_2}{\lambda_K}$.

Since the surface deflection vector δ has bounded components, the quantity $\rho(\delta)$ is bounded. Unfortunately, since e_f and e_{G_j} are unknown, the bound on $\rho(\delta)$ is unknown. When $\rho(\delta) \leq \varepsilon$, then the dead-zone in the parameter update law prevents parameter drift when $\|\tilde{z}\|_2 < \frac{\|\rho(\delta)\|_2}{\lambda_K} < \frac{\varepsilon}{\lambda_K}$. As shown in Chapter 7, the error state $\tilde{z}$ will only spend a finite time outside the dead-zone with $\|\tilde{z}\|_2 > \varepsilon$. During periods of time when $\rho(\delta) > \varepsilon$ and $\frac{\varepsilon}{\lambda_K} < \|\tilde{z}\|_2 < \frac{\|\rho(\delta)\|_2}{\lambda_K}$, the parameter vector may wander; however, projection will maintain its boundedness.

8.2.3.4 Control Law and Stability Properties. This subsection summarizes the stability results of the closed-loop system composed of the aircraft angular rate dynamics of eqns. (8.7)–(8.9) with the control law of (8.19) and parameter adaptation defined by (8.31)–(8.32). The summary is phrased in terms of three theorems. The theorems differ in the assumptions applicable to the modeling error term $\rho(\delta)$. The proof of each theorem proceeds from eqn. (8.34) using the methods described in Chapter 7.

In each of the theorems of this subsection, we implicitly assume that the pilot issues continuous and bounded commands (P_c^o, Q_c^o, R_c^o) and adjusts the thrust so that (α, M) remain in $\mathcal{D}$. For the purpose of the design of the (P, Q, R) tracking controller, the variables (α, M) are considered as exogenous variables. The controller cannot simultaneously track the pilot specified (P_c^o, Q_c^o, R_c^o) signals and independently alter (P, Q, R) to maintain (α, M) in $\mathcal{D}$. In Section 8.3, we will consider the design of a full vehicle controller for unpiloted vehicles.

Theorem 8.2.1 *In the ideal situation where $\rho(\delta) = 0$, the approximation based controller defined above solves the tracking problem with the following properties:*

1. $\tilde{z}_i$, z, $\tilde{\Theta}_f$, $\tilde{\Theta}_g$, $\hat{\Theta}_f$, $\hat{\Theta}_g \in \mathcal{L}_\infty$;

2. $\tilde{z} \in \mathcal{L}_2$; *and,*

3. *the total time $\tilde{z}(t)$ spends outside the dead-zone (i.e., that $\underline{\lambda}_K \|\tilde{z}(t)\|_2 \geq \varepsilon$) is finite.*

Theorem 8.2.1 is idealized, since it is not reasonable to expect perfect approximation of unknown functions. The following theorem is much more reasonable as it assumes that the approximators can be defined such that the approximation error is less than a known bound ε. This assumption is more reasonable since it can often be satisfied, based on available knowledge about the application, simply by increasing the dimension of the regressor vector.

Theorem 8.2.2 *In the situation where $\|\rho(\delta)\|_2 < \varepsilon$, the approximation based controller defined above solves the tracking problem with the following properties:*

1. $\tilde{z}_i$, z, $\tilde{\Theta}_f$, $\tilde{\Theta}_g$, $\hat{\Theta}_f$, $\hat{\Theta}_g \in \mathcal{L}_\infty$;

2. $\tilde{z}$ *is small-in-the-mean-squared sense, satisfying*

$$\int_t^{t+T} \|\tilde{z}(\tau)\|_2^2 d\tau \leq \frac{2}{\underline{\lambda}_K} V(t) + \frac{\varepsilon^2}{\underline{\lambda}_K^2} T;$$

3. *as* $t \to \infty$, $\tilde{z}(t)$ *is ultimately bounded by* $\|\tilde{z}\|_2 \leq \frac{\varepsilon}{\underline{\lambda}_K}$; *and,*

4. *if* $\|\rho(\delta)\|_2 < \varepsilon_o < \varepsilon$, *then the total time* $\tilde{z}(t)$ *spends outside the dead-zone is finite.*

Proof: We will only prove item 2. Starting from eqn. (8.34), completing the square yields

$$\begin{aligned} \dot{\mathcal{V}} &\leq -\frac{\underline{\lambda}_K}{2}\|\tilde{z}\|_2^2 + \frac{1}{2\underline{\lambda}_K}\|\rho(\delta)\|_2^2 - \left(\sqrt{\frac{\underline{\lambda}_K}{2}}\|\tilde{z}\|_2 - \sqrt{\frac{1}{2\underline{\lambda}_K}}\|\rho(\delta)\|_2\right)^2 \\ &\leq -\frac{\underline{\lambda}_K}{2}\|\tilde{z}\|_2^2 + \frac{\varepsilon^2}{2\underline{\lambda}_K}. \end{aligned}$$

Integrating both sides and rearranging yields

$$\begin{aligned}\int_t^{t+T} \frac{\lambda_K}{2}\|\tilde{z}(\tau)\|_2^2 d\tau &\leq \mathcal{V}(t) - \mathcal{V}(t+T) + \frac{\varepsilon^2}{2\lambda_K}T \\ \int_t^{t+T} \|\tilde{z}(\tau)\|_2^2 d\tau &\leq \frac{2}{\lambda_K}\mathcal{V}(t) + \frac{\varepsilon^2}{\lambda_K^2}T\end{aligned}$$

■

Whereas the previous theorem is valid under reasonable conditions, the following theorem is a worst case result. Theorem 8.2.3 is applicable when the dead-zone of the parameter adaptation law was selected to be too small relative to the size of the inherent approximation error.

Theorem 8.2.3 *In the situation where $\|\rho(\delta)\|_2$ may exceed ε in certain regions of $\mathcal{D}$, the approximation based controller defined above solves the tracking problem with the following properties:*

1. $\tilde{z}_i$, z, $\tilde{\Theta}_f$, $\tilde{\Theta}_g$, $\hat{\Theta}_f$, $\hat{\Theta}_g \in \mathcal{L}_\infty$; *and,*

2. $\tilde{z}_i$ *is small-in-the-mean-squared sense, satisfying*

$$\int_t^{t+T} \|\tilde{z}_i(\tau)\|_2^2 d\tau \leq \frac{2}{\lambda_K}\mathcal{V}(t) + \frac{1}{\lambda_K^2}\int_t^{t+T} \|\rho\|_2^2 d\tau.$$

Note that $\|\rho(\delta)\|_2$ is bounded, but its bound exceeds ε.

The proof of Theorem 8.2.3 is not included due to its similarity with the previous theorems of this section.

The interpretation of Theorem 8.2.3 deserves additional comment. In this worst case scenario, we cannot guarantee that the tracking error is ultimately bounded by a known bound. There are two major issues. First, the structure of the approximator was not defined sufficiently well to ensure that $\|\rho(\delta)\|_2 < \varepsilon$; however, since f^* and G^* are unknown this situation may sometimes occur in practice. The second issue is one that requires interpretation. The optimal parameter vectors Θ_f^* and Θ_G^* are defined to minimize the L_∞ approximation error over the entire region $\mathcal{D}$; however, the parameter adaptation is using the tracking error $\tilde{z}$ at the present operating point to estimate the parameter vector. The infinity norm of the approximation error on $\mathcal{D}$ decreases as the size of $\mathcal{D}$ (i.e., radius of the largest ball containing $\mathcal{D}$) decreases. Note that if the region $\mathcal{D}$ were redefined to be a small neighborhood of the present operating point, the entire analysis would still go through. Also, the optimal parameter vectors would change to those applicable to the new $\mathcal{D}$ around the present operating point; however, the parameter update of eqn. (8.31)–(8.32) would not change. To summarize, in situations where the condition $\|\rho(\delta)\|_2 < \varepsilon$ is not satisfied over the entire region $\mathcal{D}$, the parameter adaptation law can drive the parameter estimates to values that do satisfy this condition at least in some neighborhood of the present operating point. These locally satisfactory parameter values change with the operating point and are different from the Θ_f^* and Θ_G^* used in the definition of the Lyapunov function of (8.33). Therefore, when $\frac{\varepsilon}{\lambda_K} < \|\tilde{z}\|_2 < \frac{\|\rho(\delta)\|_2}{\lambda_K}$ the Lyapunov function may increase, since we cannot prove anything about the negative definiteness of its derivative; however, the increase in the Lyapunov function may only be the result of the parameter estimates converging to the parameters that result in a locally accurate fit to the functions f^* and G^*. This would be an example of the approach adapting the parameters to the local situation when it is not capable of learning the parameters that would be globally satisfactory over $\mathcal{D}$. A very simple example illustrating this issue is described in Exercise 8.1.

8.2.4 Simulation Results

This section presents simulation results from the control algorithms developed in this section when applied to the Barron Associates Nonlinear Tailless Aircraft Model (BANTAM), which is a nonlinear 6-DOF model of a flying-wing aircraft. BANTAM was developed primarily using the technical memorandum [80], which contains aerodynamic data from wind-tunnel testing of several flying-wing planforms, but also using analytical estimates of dynamic stability derivatives from DATCOM and HASC-95. The flying wing airframe is particularly challenging to control as it is statically unstable at low angles-of-attack and possesses a restricted set of control effectors that provide less yaw authority than the traditional set used on tailed aircraft. The control surfaces consist of two pairs of body flaps mounted on the trailing edge of the wing. Additionally, a pair of spoilers are mounted upstream of the flaps. This configuration generally relies upon the flaps for pitch and roll authority and the spoilers for yaw and drag. The simulation model also contains realistic actuator models for the control effectors with second order dynamics and both position and rate limits. The body flap actuators have 40 rad/sec bandwidth with ± 30 deg position limits and ± 90 deg/sec rate limits. The spoiler actuators are identical except that they can only be deflected upwards and their motion is limited to 60 deg.

Simulation results are shown in Figures 8.1–8.3. The simulation time is 100 s, with a simulated pilot generating the signals (P_c^o, Q_c^o, R_c^o). Each (P_c^o, Q_c^o, R_c^o)-command filter is of the form of eqn. (8.14). Each uses a damping factor of 1.0 and undamped natural frequencies of 20, 20, and 10 $\frac{rad}{s}$, respectively.

At $t = 0$, the known portion of the model described in (8.21)–(8.22) is defined using constant values for each of the nondimensional coefficient functions C_*^o. The constant values were selected so that the coefficient functions were approximately accurate near $\alpha = 0°$ and $M = 0.46$. The approximated functions $\hat{f}$ and $\hat{G}$ were initialized to be zero by defining each approximated coefficient function $\hat{c}_*$ in (8.24)–(8.25) to be zero. The control law is specified by (8.19). The control gain matrix is $K = diag(20, 20, 10)$. Parameter adaptation is specified by (8.31)–(8.32) with the dead-zone defined by $\varepsilon = 1$, which implies that parameter adaptation will stop if $\|\tilde{z}\|_2 < \frac{\varepsilon}{\lambda_K} = 0.1\frac{deg}{s}$. As the simulation progresses, the tracking of the (filtered) pilot specified command trajectory should improve as the functions $\hat{f}$ and $\hat{G}$ are increasingly well-approximated.

During the first 5 s of the simulation, the P_c^o and R_c^o signals are zero. The pilot adjusts the Q_c^o signal to attain stable flight near the initial flight condition with an airspeed of 500 fps, altitude of 4940 ft, and angle-of-attack of 3.6°. For $t \in [5, 50]s$, the pilot issues (P_c^o, Q_c^o, R_c^o) to perform aircraft maneuvering along a nominal trajectory. At $t = 50$s, the right midflap fails to the zero position. Throughout the simulation, the approximated model must be adjusted to maintain stability of the closed-loop and to attain the desired level of tracking performance.

The simulation was ran twice, once with learning off and once with learning on. The learning off simulation corresponds to the baseline controller. Other than turning learning on or off, all parameters were identical for the two simulations. The results in the left column of each figure correspond to the simulation with learning off. The results in the right column of each figure correspond to the simulation with learning off.

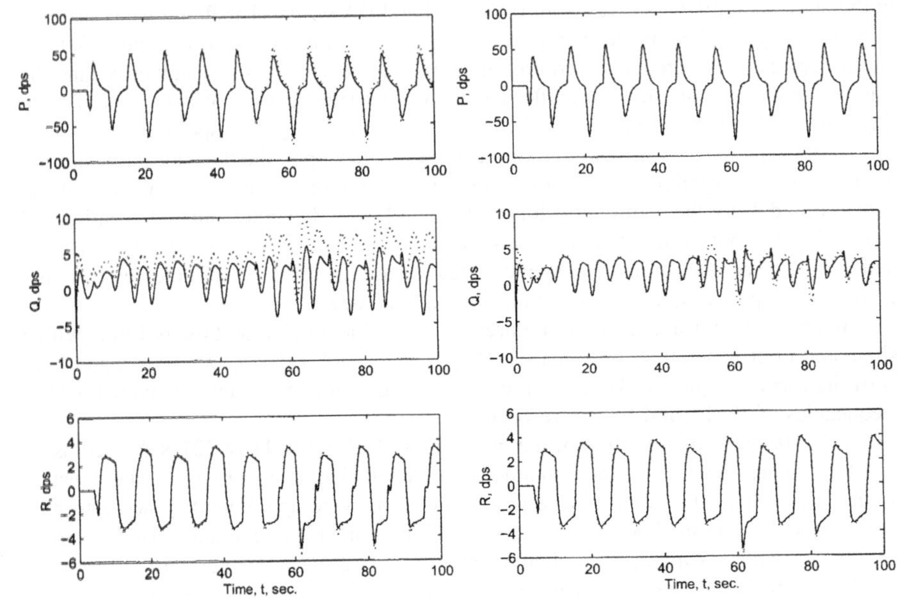

(a) *Response of* (P, Q, R) ***without learning.***

(b) *Response of* (P, Q, R) ***with learning.***

Figure 8.1: Response of the aircraft angular rate vector for the cases: (a) without learning and (b) with learning. At $t = 50$ s, the right midflap fails to zero. The solid lines are the state variables. The dotted lines are the commanded values of the state variables.

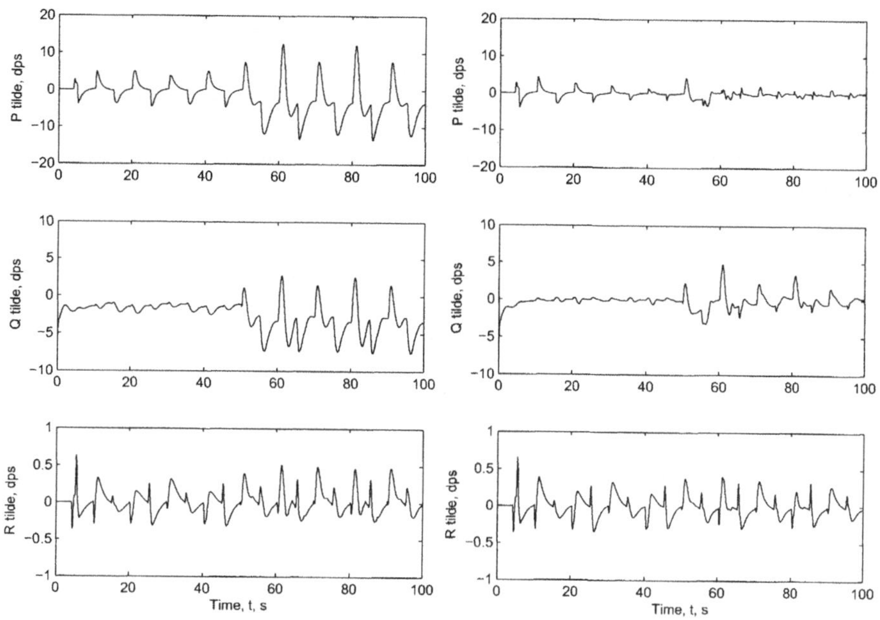

(a) ***Tracking error vector $\tilde{z}$ without learning.*** **(b)** ***Tracking error vector $\tilde{z}$ with learning.***

Figure 8.2: Tracking error vector $\tilde{z} = (P - P_c, Q - Q_c, R - R_c)$ for the cases: (a) without learning and (b) with learning. At $t = 50$ s, the right midflap fails to zero.

Figure 8.1 plots the variables (P, Q, R) as a solid line and (P_c, Q_c, R_c) as dashed lines. The units are degrees per second. Note that the pilot serves as an outer loop controller who adjusts the commands to maintain the nominal vehicle trajectory based on the response of the aircraft. Therefore, the nominal commands (P_c^o, Q_c^o, R_c^o) with and without learning are slightly different. Without the feedback action of the pilot, the trajectory tracking errors would accumulate resulting in the aircraft in the two simulations ultimately following very distinct trajectories. With the pilot feedback the operating point maintains $M \in [0.44, .47]$ throughout both simulations; and maintains $\alpha \in [1.8, 5.1]$ deg. throughout the simulation with learning and $\alpha \in [0.6, 5.1]$ deg. throughout the simulation without learning.

Due to the scale of Figure 8.1, the differences in tracking error between the two simulations are not easily observed; therefore, Figure 8.2 directly plots the tracking error vector $\tilde{z} = (P - P_c, Q - Q_c, R - R_c)$. The $\tilde{P}$ and $\tilde{Q}$ variables show clear improvements as a result of the adaptive function approximation. Note that, in the case that learning is used, as experience is accumulated, first for $t \in [0, 50]s$ and then for $t \in [50, 100]$, the tracking error decreases toward the point where it will be within the adaptation dead-zone. The change in performance in the $\tilde{R}$ variable is minor for a few reasons. First, the control authority for the R state is limited. Second, the rate of learning is related to the size of the tracking error. Since the magnitude of the R tracking errors are initially small, so are the changes to the functions affecting $\tilde{R}$.

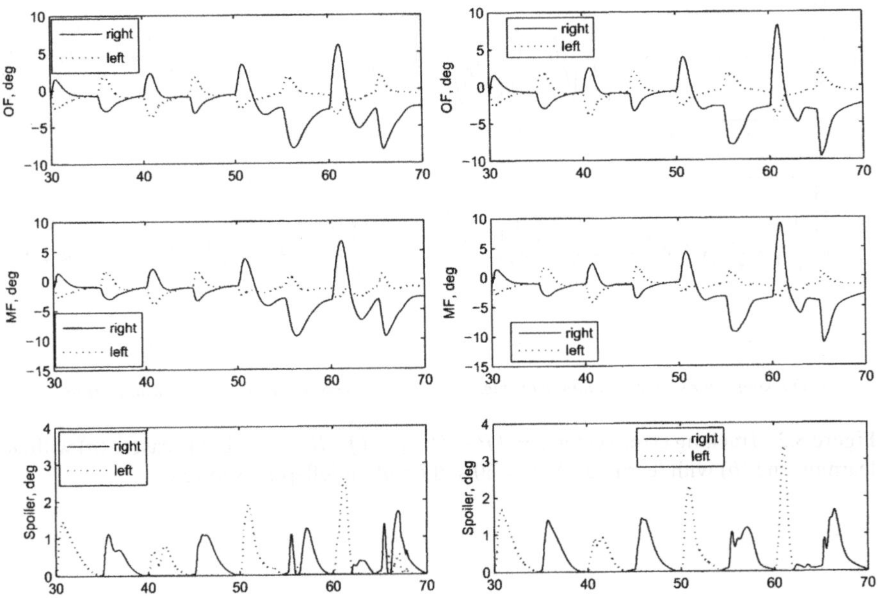

(a) ***Commanded surface deflections without learning.*** **(b)** ***Commanded surface deflections with learning.***

Figure 8.3: Commanded surface deflections for $t \in [30, 70]$ s. At $t = 50$ s, the right midflap fails to zero instead of tracking the command shown in this figure.

Figure 8.3 displays a portion of the time series of the surface position commands. Only a portion of the time series is shown so that the time axis can be expanded to a degree which allows the reader to clearly observe the signals. The selected time period includes 20 s before and after the actuator fault at $t = 50$ s. The previous two graphs indicate robustness to initial model error and to changes to the vehicle dynamics while in-flight. The main purpose of Figure 8.3 is to show that the robustness was achieved without using high gain or switching control. The actuator signals are very reasonable in magnitude and frequency content. In fact, the nature of the control signal does not change drastically after the fault (i.e., $t \geq 50$ s).

8.3 FULL CONTROL FOR AUTONOMOUS AIRCRAFT

This section presents an adaptive approximation based approach to the control of advanced flight vehicles. The controller is designed using three loops as illustrated in Figure 8.4 and the command filtered approximation based backstepping method described in Sections 5.3.3 and 7.3.3. The state of the vehicle x is subdivided into three subvectors: $z_1 = [\chi, \gamma, V]^\top$, $z_2 = ([\mu, \alpha, \beta]^\top$, and $z_3 = [P, Q, R]^\top$. The airspeed and flight path angle controller is the outermost loop. That controller receives a reference input command vector $z_{1c}(t)$ and its derivative $\dot{z}_{1c}(t)$ from an external system such as a mission planner. The airspeed and flight path angle controller is described in Section 8.3.1. It generates a command vector $z_{2c}(t)$ and its derivative $\dot{z}_{2c}(t)$, which are command inputs to the wind-axes angle controller that is described in Section 8.3.2. The wind-axes angle controller generates a command vector $z_{3c}(t)$ and its derivative $\dot{z}_{3c}(t)$, which are command inputs to the (body-axis) angular rate controller that is described in Section 8.3.3. Each of the blocks in Figure 8.4 is expanded in a later figure, in the same section in which the equations of the block are analyzed.

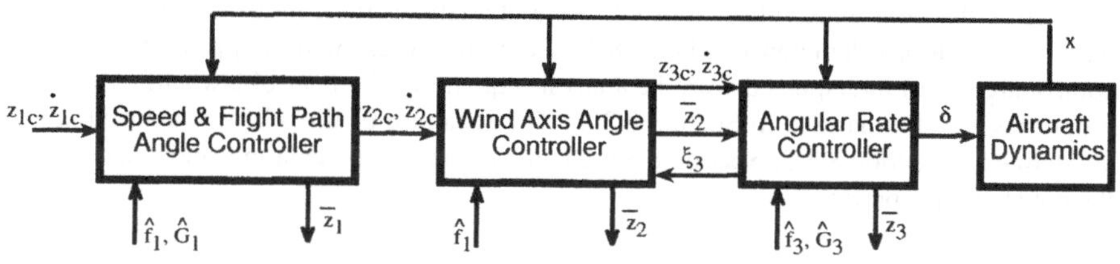

Figure 8.4: Block diagram of the full aircraft controller. The signals $x(t)$ and $\bar{z}_i(t)$ for $i = 1, 2, 3$ are inputs to the adaptive function approximation process (not shown) that develops $\hat{f}_1$, $\hat{G}_1$, $\hat{f}_3$, and $\hat{G}_3$.

The control approach includes adaptive approximation of the aerodynamic force and moment coefficient functions, as discussed in Section 8.1.2. The approach presented herein attains stability (in the sense of Lyapunov) of the aircraft state and of the adaptive function approximation process in the presence of unmodeled nonlinear effects. In Figure 8.4, $\hat{f}_1$, $\hat{f}_3$, $\hat{G}_1$, and $\hat{G}_3$ are approximated functions. The signals $\bar{z}_1$, $\bar{z}_2$, and $\bar{z}_3$ are signals used to implement the parameter estimation in the function approximation process.

The main advantages of the approach presented herein are the following: the aerodynamic force and moment models are automatically adjusted to accommodate changes to the aerodynamic properties of the vehicle and the Lyapunov stability results are provable. The main motivations for this work were to produce a simplified control design that is also more robust to model error without resorting to high gain or switching control, to accommodate large changes in the vehicle dynamics (e.g., damage) adaptively during operation, and to *learn* the aerodynamic coefficient functions for the vehicle. An anticipated benefit from these properties is that the controller could be applied to an aircraft for which it was not explicitly designed, e.g., an aircraft of the same family but different configuration. Additionally, the controller could be developed using a lower fidelity model than required by current methods, thereby offering a cost savings. This control method is expected to provide significant reduction in design time since the control system design does not depend on a conglomeration of point designs.

The functions that are approximated adaptively will use a basis set defined as a function of angle-of-attack α and Mach M. Successful implementation of the approach assumes that we can define a set $\mathcal{D}_\alpha = [\underline{\alpha}, \bar{\alpha}]$ with $\underline{\alpha} < 0 < \bar{\alpha}$ and $L(\underline{\alpha} - \varepsilon_\alpha) < 0 < L(\bar{\alpha} + \varepsilon_\alpha)$ where $L(x)$ denotes the lift force evaluated at x, and $\varepsilon_\alpha > 0$ is a designer-specified small constant. The approximated functions will be designed assuming that $M \in [0.2, 1.0]$ and $\alpha \in \mathcal{D}_\alpha^+ = [\underline{\alpha} - \varepsilon_\alpha, \bar{\alpha} + \varepsilon_\alpha]$. We assume that the region $\mathcal{D}_\alpha^+$ has been defined so that stall will not occur for $\alpha \in \mathcal{D}_\alpha^+$. Finally, we assume that $\alpha(0) \in \mathcal{D}_\alpha^+$ and that $\alpha_c(t) \in \mathcal{D}_\alpha$ for all $t \geq 0$, where α_c is an angle-of-attack command defined following eqn. (8.41) in Section 8.3.1.

Most of the *assumptions* stated in the previous paragraph are, in fact, *operating envelope* design constraints that the planner can enforce by monitoring and altering the $z_{1_c} = [\chi_{1_c}, \gamma_{1_c}, V_{1_c}]^\top$ commands that it issues. For example, as the control signal $\alpha_c(t)$ approaches $\bar{\alpha}$ from below, the planner can decrease γ_{1_c}, decrease the magnitude of χ_{1_c}, or increase V_{1_c}. Determining the combination of these options most appropriate for the current circumstance of the aircraft is straightforward within a planning framework. Given the above conditions, analysis showing that $\alpha(t) \in \mathcal{D}_\alpha^+, \forall t > 0$ is presented in Subsection 8.3.2.3.

Each of the next three subsections derives and analyzes the control law for one of the three control loops depicted in Figure 8.4. Since that presentation approach leaves the control algorithm interspersed with the analysis equations, the control law and its stability properties are summarized in Section 8.3.4. The structure of the adaptive approximators are defined in Section 8.3.5. Section 8.3.6 contains a simulation example and discussion of the controller properties.

8.3.1 Airspeed and Flight Path Angle Control

Let the state vector z_1 be defined by $z_1 = [\chi, \gamma, V]^\top$. To initiate the command-filtered backstepping process, we need a control law that stabilizes the z_1 dynamics in the presence of nonlinear model error. We assume that the command signal vector $z_{1c} = (\chi_c, \gamma_c, V_c)$ and its derivative $\dot{z}_{1c}$ is available, bounded, and continuous. The airspeed V will be controlled via thrust T. The flight path angles (χ, γ) will be controlled through the wind-axes angles (μ, α); therefore, $\mu_1 = [\mu, \alpha, T]^\top$ is the control signal for z_1. The block diagram of the controller derived in this subsection is shown in Figure 8.5.

The airspeed and flight path angle dynamics of eqns. (8.1)–(8.3) can be represented as

$$\dot{z}_1 = \bar{A}_1 f_1 + F_1 + G_1(\mu_1, x) \tag{8.35}$$

with $\bar{A}_1 = \frac{1}{mV}\begin{bmatrix} \sin\beta\cos\mu/\cos\gamma & \cos\beta\cos\mu/\cos\gamma & 0 \\ -\sin\beta\sin\mu & \cos\beta\sin\mu & 0 \\ -V\cos\beta & V\sin\beta & 0 \end{bmatrix}$,

$$F_1 = \begin{bmatrix} -T\cos\alpha\sin\beta\cos\mu\frac{1}{mV\cos\gamma} \\ (T\cos\alpha\sin\beta\sin\mu - mg\cos\gamma)\frac{1}{mV} \\ -g\sin\gamma \end{bmatrix}, \quad \mu_1 = \begin{bmatrix} \mu_{1_1} \\ \mu_{1_2} \\ \mu_{1_3} \end{bmatrix}, \quad f_1(x) = \begin{bmatrix} D(x) \\ Y(x) \\ L(x) \end{bmatrix},$$

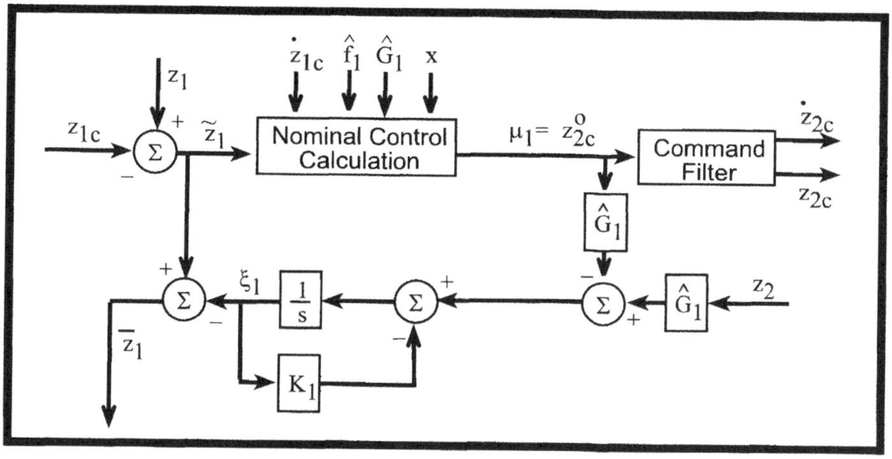

Figure 8.5: Block diagram of the airspeed and flight path angle controller described in Section 8.3.1. The signal $z_1(t)$ is a subvector of $x(t)$. The functions $\hat{f}_1$ and $\hat{G}_1$ are outputs of the adaptive function approximation process (not shown). The nominal control calculation refers to the solution of eqn. (8.39). The signals z_{1_c} and $\dot{z}_{1_c}$ are inputs from the mission planner. The signals z_{2_c} and $\dot{z}_{2_c}$ are outputs to the wind-axes angle controller described in Section 8.3.2. The signal $\bar{z}_1$ is a training signal output to the function approximation process.

and

$$G_1(\mu_1, x) = \begin{bmatrix} u_\chi \\ u_\gamma \\ u_V \end{bmatrix} = \begin{bmatrix} \dfrac{g(\mu_{1_2}, x)\sin\mu_{1_1}}{mV\cos\gamma} \\ \dfrac{g(\mu_{1_2}, x)\cos\mu_{1_1}}{mV} \\ \dfrac{\cos\beta\cos\mu_{1_2}}{m}T \end{bmatrix}, \tag{8.36}$$

where

$$\begin{aligned} g(\mu_{1_2}, x) &= (L(\mu_{1_2}, x) + T\sin\mu_{1_2}) && (8.37) \\ L(\mu_{1_2}, x) &= L_o(x) + L_\alpha(x)\mu_{1_2}. && (8.38) \end{aligned}$$

The drag, lift, and side force functions that are used in the definitions of f_1, $L_o(x)$, and $L_\alpha(x)$ are unknown. The function F_1 is known.

We select the control signal μ_1, with K_1 positive definite, so that the following equation is satisfied

$$\hat{G}_1(\mu_1, x) = -K_1\tilde{z}_1 + \dot{z}_{1_c} - \bar{A}_1\hat{f}_1 - F_1 \tag{8.39}$$

where $\hat{f}_1 = [\hat{D}(x), \hat{Y}(x), \hat{L}(x)]^\top$ and

$$\hat{G}_1 = \begin{bmatrix} \hat{u}_\chi \\ \hat{u}_\gamma \\ \hat{u}_V \end{bmatrix} = \begin{bmatrix} \dfrac{\hat{g}(\mu_{1_2}, x)\sin\mu_{1_1}}{mV\cos\gamma} \\ \dfrac{\hat{g}(\mu_{1_2}, x)\cos\mu_{1_1}}{mV} \\ \dfrac{\cos\beta\cos\mu_{1_2}}{m}T \end{bmatrix} \tag{8.40}$$

with $\hat{g}(\mu_{1_2}, x) = \hat{L}(\mu_{1_2}, x) + T \sin \mu_{1_2}$. The functions $[\hat{D}(x), \hat{Y}(x), \hat{L}(x)]$ are approximations to $[D(x), Y(x), L(x)]$. The effect of the error between these functions is considered in the analysis of Section 8.3.2.1. The solution of eqn. (8.39) for μ_1 is discussed in Section 8.3.1.1.

Assuming that the solution μ_1 to (8.39) has been found, let $z_{2_c}^o = [\mu_c^o, \alpha_c^o, \beta_c^o]^\top$. To produce the signals z_{2_c} and $\dot{z}_{2_c}$, which are the command inputs to the wind-axes angle controller, we pass $z_{2_c}^o$ through a command filter. The error between $z_{2_c}^o$ and z_{2_c} will be explicitly accounted for in the subsequent stability analysis. Define $\bar{z}_1 = \tilde{z}_1 - \xi_1$ where the variable ξ_1 is the output of the filter

$$\dot{\xi}_1 = -K_1 \xi_1 + \left(\hat{G}_1(z_2, x) - \hat{G}_1(z_{2_c}^o, x) \right). \tag{8.41}$$

The purpose of the command filter is to compute the command signal z_{2_c} and its derivative $\dot{z}_{2_c}$. This is accomplished without differentiation. The purpose of the ξ_1-filter is to compensate the tracking error $\tilde{z}_1$ for the effect of any differences between z_2 and $z_{2_c}^o$. In the analysis to follow, we will prove that $\hat{G}_1$ is a bounded function. By the design of the command filter, the difference between z_{2_c} and $z_{2_c}^o$ will be small. Finally, in the following subsections, we will design tracking controllers to ensure that the difference between z_2 and z_{2_c} is small. Therefore, ξ_1 will be bounded, because it is the output of a stable linear filter with a bounded input.

8.3.1.1 *Selection of α and μ Commands.*

The value of the vector μ_1 in the left-hand side of eqn. (8.39) must be derived, as it determines the command input to the wind-axes angle loop. Because all quantities in the right-hand side of eqn. (8.39) are known, the desired value of $\hat{G}_1(\mu_1, x)$ can be computed at any time instant. The purpose of this subsection is to discuss the solution of eqn. (8.40) for μ_1. Note that $\mu_{1_1} = \mu_c^o$ and $\mu_{1_2} = \alpha_c^o$ are the roll-angle and angle-of-attack commands. Also, to decrease the complexity of the notation, we will use the notation $\hat{g}(\alpha_c^o)$ instead of $\hat{g}(\mu_{1_2}, x)$. Finally, for complete specification of the desired wind-axes state, we will always specify β_c^o as zero.

Defining (X, Y) such that the first two rows of eqn. (8.40) can be written as

$$X \equiv \cos(\gamma) m V \hat{u}_\chi = \hat{g}(\alpha_c^o) \sin(\mu_c^o) \tag{8.42}$$
$$Y \equiv m V \hat{u}_\gamma = \hat{g}(\alpha_c^o) \cos(\mu_c^o), \tag{8.43}$$

we can interpret $(X, Y) = (\cos(\gamma) m V \hat{u}_\chi, m V \hat{u}_\gamma)$ as the known rectangular coordinates for a point with (signed) radius $\hat{g}(\alpha_c^o)$ and angle μ_c^o relative to the positive Y axis. Since the force $\hat{g}(\alpha_c^o)$ may be either positive or negative, there are always two possible solutions, as depicted in Figures 8.6a and 8.6b. Switching between the two possible solutions requires μ_c^o to change by 180 degrees as $\hat{g}(\alpha_c^o)$ reverses its sign. When $\hat{g}(\alpha_c^o)$ reverses its sign, the point (X, Y) passes through the origin. If $\hat{g}(\alpha_c^o)$ is selected to be positive for a sufficiently aggressive *diving* turn (i.e., $\dot{\chi}_c$ and $\dot{\gamma}_c$ both large), then the maneuver would be performed with the aircraft inverted (i.e., roll greater than 90 deg). When choosing (μ_c^o, α_c^o) to satisfy eqn. (8.40), the designer should only allow $\hat{g}(\alpha_c^o)$ to reverse its sign when $\hat{u}_\chi$ is near zero. If the sign of $g(\alpha)$ reversed while $\hat{u}_\chi$ was non-zero, then μ_c^o would also need to change so that $(\sin(\mu_c^o), \cos(\mu_c^o))$ would have the correct signs to attain the desired control signals. This change is a 180° roll reversal.

Once μ_c^o and α_c^o have been specified, the third equation of eqn. (8.40) can be directly solved for T.

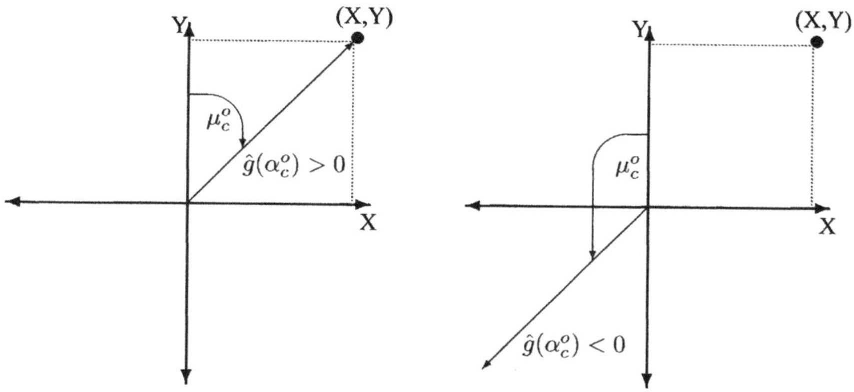

(a) ***The (α_c^o, μ_c^o) solution with positive lift.*** **(b)** ***The (α_c^o, μ_c^o) solution with negative lift.***

Figure 8.6: Two possible choices for α_c^o and μ_c^o to solve the (χ, γ) control.

EXAMPLE 8.1

This example illustrates, using Figures 8.7–8.9, the process of selecting the (μ_c^o, α_c^o) signals. The top row of graphs in Figure 8.7 shows the γ and χ signals during a sequence of dive and turn maneuvers. Various points are labelled to aid the following discussion. The bottom row of graphes in Figure 8.7 shows the α and μ signals selected to force the (χ, γ) response. Figure 8.8 is a plot of (X, Y) from eqns. (8.42)–(8.43) in a polar format. The radius is $\hat{g}(\alpha) = \|(X, Y)\|_2$ and the angle is $\mu = atan2(X, Y)$ where $atan2$ is a four-quadrant inverse tangent function. Figure 8.9 is a polar plot of the magnitude of α versus the angle μ. Figure 8.9 is included for comparison with Figure 8.8 to illustrate the fact that the main difference between the two is the distortion caused by inverting the nonlinear function $g(\alpha)$. At any point in time, the angles of the two contours are the same.

The time series begins at the point indicated by "A". At that time, the aircraft is diving and about to initiate a turn to 20°. At the time indicated by "C", the aircraft is nearly finished with its turn and also about to bring the dive rate γ back to zero. Between times "A" and "B", both μ and α are increased, even though the aircraft is still increasing its dive rate. While the aircraft is increasing α, it is also banking the aircraft (increasing μ) so that the increased lift is directed appropriately to turn the vehicle while still achieving the desired dive rate. Between the times indicated by "C" and "D", the aircraft is decreasing the dive rate to zero while ending the turn. To end the turn, the bank angle converges toward zero. To return the dive rate to zero, the angle-of-attack α is increased. Related comments are applicable to the second half of the plotted simulation results. △

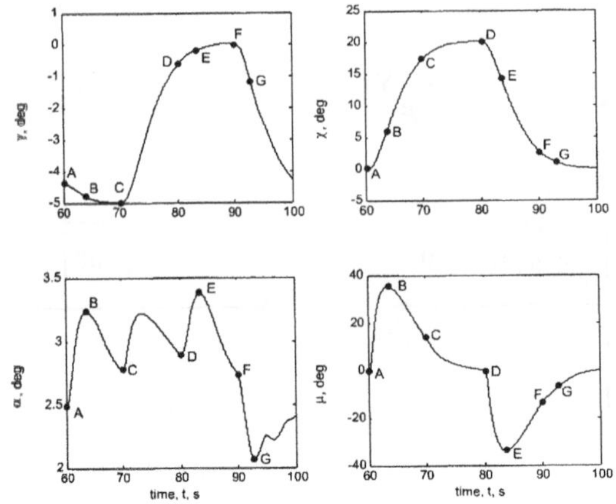

Figure 8.7: Time series plots of γ and χ in the top row and α and μ in the bottom row. The data is explained in Example 8.1.

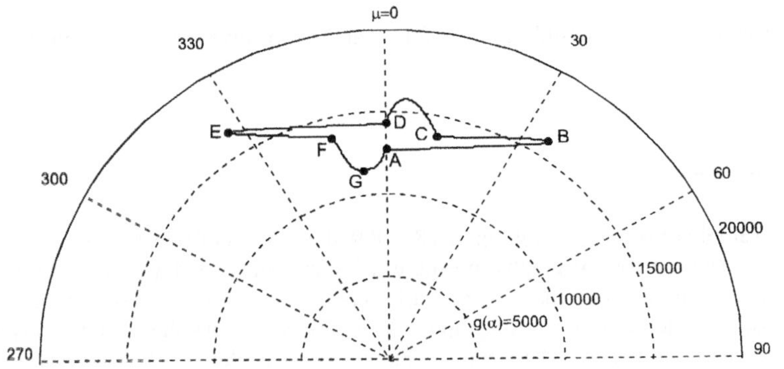

Figure 8.8: Polar plot with $\hat{g}(\alpha) = \|(X, Y)\|_2$ corresponding to the (signed) radius and $\mu = atan2(X, Y)$ defining the angle, as discussed in in Example 8.1.

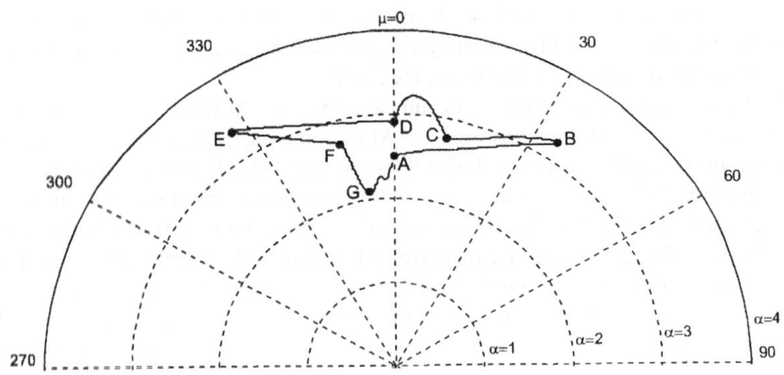

Figure 8.9: Polar plot with α corresponding to the (signed) radius and μ defining the angle, as discussed in in Example 8.1.

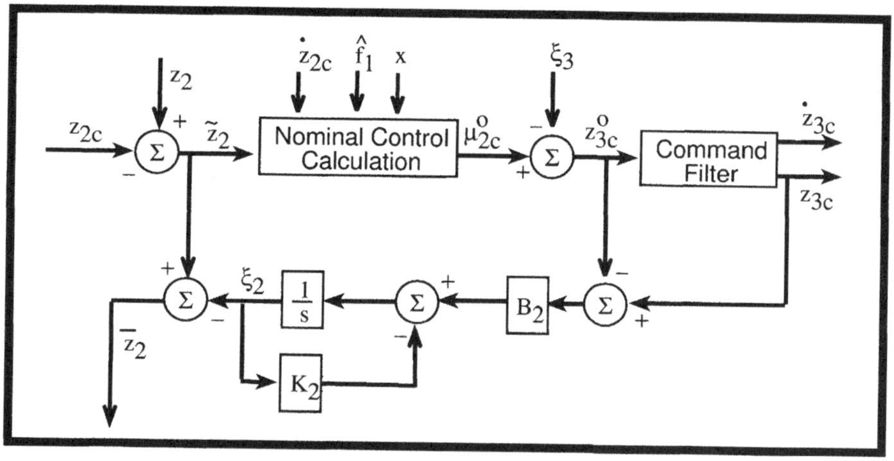

Figure 8.10: Block diagram of the wind-axes controller described in Section 8.3.2. The signal $z_2(t)$ is a subvector of $x(t)$. The function $\hat{f}_1$ is an output of the adaptive function approximation process (not shown). The nominal control calculation refers to the solution of eqn. (8.44). The signals z_{2_c} and $\dot{z}_{2_c}$ are inputs from the flight path angle controller of Section 8.3.1. The signals z_{3_c} and $\dot{z}_{3_c}$ are outputs to the angular rate controller described in Section 8.3.3. The signal ξ_3 is an input from the angular rate controller. The signal $\bar{z}_2$ is a training signal output to the function approximation process.

8.3.2 Wind-Axes Angle Control

Let z_1 be as defined in Section 8.3.1. Define $z_2 = [\mu, \alpha, \beta]^\top$. Then, the combined (z_1, z_2) dynamics are

$$\begin{aligned}\dot{z}_1 &= \bar{A}_1(x)f_1 + F_1(x) + G_1(z_2, x, T)\\ \dot{z}_2 &= A_2(x)f_1 + F_2(x) + B_2\mu_2\end{aligned}$$

where $B_2 = \begin{bmatrix} \frac{\cos\alpha}{\cos\beta} & 0 & \frac{\sin\alpha}{\cos\beta} \\ -\cos\alpha\tan\beta & 1 & -\sin\alpha\tan\beta \\ \sin\alpha & 0 & -\cos\alpha \end{bmatrix}$

$$A_2 = \frac{1}{mV}\begin{bmatrix} \sin\beta\cos\mu\tan\gamma & \cos\beta\cos\mu\tan\gamma & (\tan\beta + \tan\gamma\sin\mu) \\ 0 & 0 & -1/\cos\beta \\ \sin\beta & \cos\beta & 0 \end{bmatrix}$$

$$F_2 = \frac{1}{mV}\begin{bmatrix} (\sin\alpha\tan\gamma\sin\mu + \sin\alpha\tan\beta - \cos\alpha\tan\gamma\cos\mu\sin\beta)\,T - mg\cos\gamma\cos\mu\tan\beta \\ [-T\sin\alpha + mg\cos\gamma\cos\mu]\,\frac{1}{\cos\beta} \\ -T\sin\beta\cos\alpha + mg\cos\gamma\sin\mu \end{bmatrix}$$

are known functions and $\mu_2 = [P, Q, R]^\top$. Note that the (z_1, z_2) dynamics are not triangular, since $\bar{A}_1$, f_1, F_1 all depend on z_2. Nevertheless, the command filtered backstepping approach is applicable. The block diagram of the controller derived in this subsection is shown in Figure 8.10.

Select $\mu^o_{2_c}$ such that

$$B_2\mu^o_{2_c} = -K_2\tilde{z}_2 + \dot{z}_{2c} - A_2\hat{f}_1 - F_2 + \eta_\alpha \tag{8.44}$$

with K_2 positive definite and diagonal. The function η_α will be defined in Subsection 8.3.2.3 to ensure that α remains in $\mathcal{D}_\alpha^+$. When $\alpha \in \mathcal{D}_\alpha$, η_α will be zero. Eqn. (8.44) is always solvable for $\mu_{2_c}^o$ since B_2 is well defined and nonsingular (for $\beta \neq \pm 90°$).

To specify the angular rate control command signal z_{3_c}, we define

$$z_{3c}^o = \mu_{2c}^o - \xi_3, \tag{8.45}$$

where ξ_3 will be defined in Section 8.3.3. The signal $z_{3_c}^o$ is input to a command filter with outputs z_{3_c} and $\dot{z}_{3_c}$. The variable ξ_2 is the output of the filter

$$\dot{\xi}_2 = -K_2\xi_2 + B_2\left(z_{3_c} - z_{3_c}^o\right) \tag{8.46}$$

and the compensated tracking error is defined as $\bar{z}_2 = \tilde{z}_2 - \xi_2$. The command filter is designed to ensure that $\left|B_{22}\left(z_{3_c} - z_{3_c}^o\right)\right| \leq \frac{K_{22}}{2}\varepsilon_\alpha$ where K_{22} denotes the second diagonal element of K_2 and B_{22} is the second row of B_2. This is always possible, since the matrix B_2 is bounded (since β is near zero). Therefore,

$$|\xi_{22}| \leq \frac{\varepsilon_\alpha}{2} \tag{8.47}$$

where ξ_{22} is the second element of ξ_2. This bound is used later in the analysis.

8.3.2.1 *Tracking Error Dynamics for $\alpha \in \mathcal{D}_\alpha^+$.* Given the definitions of the previous section, for $\alpha \in \mathcal{D}_\alpha^+$, the dynamics of the z_1 and z_2 tracking errors can be derived:

$$\begin{aligned}
\dot{\tilde{z}}_1 &= \bar{A}_1 f_1 + F_1 + \hat{G}_1(\mu_1, x) - \dot{z}_{1_c} \\
&\quad + (G_1(z_2, x) - \hat{G}_1(z_2, x)) + (\hat{G}_1(z_2, x) - \hat{G}_1(\mu_1, x)) \\
&= -K_1\tilde{z}_1 - \bar{A}_1\tilde{f}_1 + (G_1(z_2, x) - \hat{G}_1(z_2, x)) + (\hat{G}_1(z_2, x) - \hat{G}_1(\mu_1, x)) \\
&= -K_1\tilde{z}_1 - A_1\tilde{f}_1 + (\hat{G}_1(z_2, x) - \hat{G}_1(\mu_1, x)),
\end{aligned} \tag{8.48}$$

where $\tilde{f}_1(x) = \hat{f}_1(x) - f_1(x)$ and algebraic manipulations result in $A_1\tilde{f}_1 = \bar{A}_1\tilde{f}_1 - (G_1(z_2, x) - \hat{G}_1(z_2, x))$ with

$$A_1 = \frac{1}{mV}\begin{bmatrix} \sin\beta\cos\mu/\cos\gamma & \cos\beta\cos\mu/\cos\gamma & \frac{\sin\mu}{\cos\gamma} \\ -\sin\beta\sin\mu & \cos\beta\sin\mu & \cos\mu \\ -V\cos(\beta) & V\sin(\beta) & 0 \end{bmatrix}. \tag{8.49}$$

Similarly, the tracking error dynamics for z_2 are

$$\begin{aligned}
\dot{\tilde{z}}_2 &= A_2 f_1 + F_2(x) + B_2 z_{3_c}^o - \dot{z}_{2_c} + B_2\left(z_3 - z_{3_c}\right) + B_2\left(z_{3_c} - z_{3_c}^o\right) \\
&= A_2 f_1 + F_2(x) + B_2\mu_{2_c}^o - B_2\xi_3 - \dot{z}_{2_c} + B_2\left(z_3 - z_{3_c}\right) + B_2\left(z_{3_c} - z_{3_c}^o\right) \\
&= -K_2\tilde{z}_2 + B_2\tilde{z}_3 - B_2\xi_3 - A_2\tilde{f}_1 + B_2\left(z_{3c} - z_{3c}^o\right) + \eta_\alpha \\
&= -K_2\tilde{z}_2 + B_2\bar{z}_3 - A_2\tilde{f}_1 + B_2\left(z_{3c} - z_{3c}^o\right) + \eta_\alpha.
\end{aligned} \tag{8.50}$$

Combining eqns. (8.41) and (8.46), respectively, with eqns. (8.48) and (8.50), the dynamics of the compensated tracking errors are

$$\begin{aligned}
\dot{\bar{z}}_1 &= \left(-K_1\tilde{z}_1 - A_1\tilde{f}_1 + (\hat{G}_1(z_2, x) - \hat{G}_1(\mu_1, x))\right) \\
&\quad - \left(-K_1\xi_1 + \left(\hat{G}_1(z_2, x) - \hat{G}_1(z_{2_c}^o, x)\right)\right) \\
&= -K_1\bar{z}_1 - A_1\tilde{f}_1
\end{aligned} \tag{8.51}$$

$$\begin{aligned}
\dot{\bar{z}}_2 &= \left(-K_2\tilde{z}_2 + B_2\bar{z}_3 - A_2\tilde{f}_1 + B_2\left(z_{3c} - z_{3c}^o\right) + \eta_\alpha\right) - \left(-K_2\xi_2 + B_2\left(z_{3_c} - z_{3_c}^o\right)\right) \\
&= -K_2\bar{z}_2 - A_2\tilde{f}_1 + B_2\bar{z}_3 + \eta_\alpha.
\end{aligned} \tag{8.52}$$

Using the notation of Section 3.1.3, modified for the application of this section, $f_1 = \left(\Theta^*_{f_1}\right)^\top \Phi_{f_1} + e_{f_1}$ where $\Theta^*_{f_1} \in \Re^{N\times 3}$ and $\Phi_{f_1} : \mathcal{D} \mapsto \Re^N$; therefore, $\tilde{f}_1 = \tilde{\Theta}^\top_{f_1}\Phi_{f_1} - e_{f_1}$, where $\tilde{\Theta}_{f_1} = \Theta_{f_1} - \Theta^*_{f_1}$ and e_{f_1} is the Minimum Functional Approximation Error (MFAE) function. Using this notation, (8.51)–(8.52) reduce to

$$\begin{aligned} \dot{\bar{z}}_1 &= -K_1\bar{z}_1 - A_1\tilde{\Theta}^\top_{f_1}\Phi_{f_1} + A_1 e_{f_1} && (8.53)\\ \dot{\bar{z}}_2 &= -K_2\bar{z}_2 - A_2\tilde{\Theta}^\top_{f_1}\Phi_{f_1} + B_2\bar{z}_3 + A_2 e_{f_1} + \eta_\alpha, && (8.54)\end{aligned}$$

which are in the form that is required to prove the desired stability properties.

8.3.2.2 Adaptive Approximation and Stability Analysis for $\alpha \in \mathcal{D}^+_\alpha$.

Let the parameter update be defined by

$$\dot{\Theta}_{f_1} = \begin{cases} 0 & \text{if } \tau(\bar{z},\varepsilon) \le 0 \\ \mathcal{P}\left(\Gamma_{f_1}\Phi_{f_1}\left(\bar{z}_1^\top A_1 + \bar{z}_2^\top A_2\right)\right) & \text{if } \tau(\bar{z},\varepsilon) > 0 \end{cases} \tag{8.55}$$

with $\varepsilon = [\varepsilon_1, \varepsilon_2, \varepsilon_3]^\top$ being a vector of designer specified constants, Γ_{f_1} a positive definite adaptation gain matrix, and the function τ defined later in (8.66).

Define the Lyapunov function as

$$\mathcal{V}_1 = \frac{1}{2}\left(\bar{z}_1^\top\bar{z}_1 + \bar{z}_2^\top\bar{z}_2 + trace\left(\tilde{\Theta}^\top_{f_1}\Gamma^{-1}_{f_1}\tilde{\Theta}_{f_1}\right)\right). \tag{8.56}$$

For $\alpha \in \mathcal{D}^+_\alpha$ and $\tau(\bar{z},\varepsilon) > 0$, when projection is not in effect, the time derivative of $\mathcal{V}_1$ along solutions of eqns. (8.53)–(8.55) is given by

$$\begin{aligned} \frac{d\mathcal{V}_1}{dt} &= \bar{z}_1^\top\left(-K_1\bar{z}_1 - A_1\tilde{\Theta}^\top_{f_1}\Phi_{f_1} + A_1 e_{f_1}\right) + trace\left(\tilde{\Theta}^\top_{f_1}\Gamma^{-1}_{f_1}\dot{\tilde{\Theta}}_{f_1}\right)\\ &\quad + \bar{z}_2^\top\left(-K_2\bar{z}_2 - A_2\tilde{\Theta}^\top_{f_1}\Phi_{f_1} + B_2\bar{z}_3 + A_2 e_{f_1} + \eta_\alpha\right)\\ &= -\bar{z}_1^\top K_1\bar{z}_1 - \bar{z}_2^\top K_2\bar{z}_2 + \bar{z}_2^\top B_2\bar{z}_3 + \left(\bar{z}_1^\top A_1 + \bar{z}_2^\top A_2\right)e_{f_1} + \bar{z}_2^\top\eta_\alpha\\ &\quad - \left(\bar{z}_1^\top A_1 + \bar{z}_2^\top A_2\right)\tilde{\Theta}^\top_{f_1}\Phi_{f_1} + trace\left(\tilde{\Theta}^\top_{f_1}\Phi_{f_1}\left(\bar{z}_1^\top A_1 + \bar{z}_2^\top A_2\right)\right)\\ &= -\bar{z}_1^\top K_1\bar{z}_1 - \bar{z}_2^\top K_2\bar{z}_2 + \bar{z}_2^\top B_2\bar{z}_3 + \bar{z}_2^\top\eta_\alpha + \left(\bar{z}_1^\top A_1 + \bar{z}_2^\top A_2\right)e_{f_1}. \end{aligned} \tag{8.57}$$

The first two terms in this expression are negative. The third term is not sign definite. The control law of Section 8.3.3 will be designed to accommodate this sign indefinite term. The right-most term, due to the inherent approximation error e_{f_1}, is also sign indefinite. It will be addressed in the overall stability analysis of Section 8.3.4. Finally, the term $\bar{z}_2^\top\eta_\alpha$ will be designed in Subsection 8.3.2.3 to ensure that $\alpha(t) \in \mathcal{D}^+_\alpha$ for all $t \ge 0$. In addition, we will show that $\bar{z}_2^\top\eta_\alpha$ is nonpositive. Therefore, this term can be dropped in subsequent analysis. The stability analysis is completed in Subsection 8.3.4.

8.3.2.3 Ensuring $\alpha \in \mathcal{D}^+_\alpha$.

Ensuring that $\alpha(t) \in \mathcal{D}^+_\alpha$ for all $t \ge 0$ is critical both for physical and for implementation reasons. Physically, if α is allowed to become too large, then the aircraft might reach a stall condition. From an implementation point of view, the approximator basis functions have α and M as inputs. The approximator will be defined to achieve accurate approximation for $\alpha \in \mathcal{D}^+_\alpha$ (defined on p. 350). For $\alpha \in \Re^1 - \mathcal{D}^+_\alpha$ the approximators are set to zero.

The portion of the control law denoted by η_α is responsible for ensuring that α remains in the region of approximation $\mathcal{D}_\alpha^+$ for all $t \geq 0$. We choose $\eta_\alpha(\alpha) = [0, \; -s_\alpha(\alpha), \; 0]^\top$ where

$$s_\alpha(\alpha) = \begin{cases} \frac{2M}{\varepsilon_\alpha}\left(\alpha - \left(\underline{\alpha} - \frac{\varepsilon_\alpha}{2}\right)\right) & \text{if } \alpha \leq \underline{\alpha} - \frac{\varepsilon_\alpha}{2} \\ \frac{2M}{\varepsilon_\alpha}\left(\alpha - \left(\bar{\alpha} + \frac{\varepsilon_\alpha}{2}\right)\right) & \text{if } \alpha \geq \bar{\alpha} + \frac{\varepsilon_\alpha}{2} \\ 0 & \text{otherwise} \end{cases}$$

as illustrated in Figure 8.11. The magnitude constraint on $M > 0$ is discussed below in (8.58). With this definition of $\eta_\alpha(\alpha)$ and (8.50) the α dynamics are

$$\dot{\alpha} = -K_{22}\tilde{\alpha} + \frac{\tilde{L}}{mV\cos\beta} + B_{22}\bar{z}_3 + B_{22}\left(z_{3_c} - z_{3_c}^o\right) + \dot{\alpha}_c - s_\alpha$$

where K_{22} denotes the second diagonal element of K_2 and B_{22} is the second row of B_2. Also, note that $\alpha\tilde{\alpha} > 0$ for $\alpha \in \mathcal{D}_\alpha^+ - \mathcal{D}_\alpha$.

Consider the time derivative of the function $V_\alpha = \frac{1}{2}\alpha^2$:

$$\dot{V}_\alpha = -\alpha K_{22}\tilde{\alpha} - \alpha s_\alpha + \alpha\left(\frac{\tilde{L}}{mV\cos\beta} + B_{22}\bar{z}_3 + B_{22}\left(z_{3_c} - z_{3_c}^o\right) - \dot{\alpha}_c\right).$$

On the set $\alpha \in \mathcal{D}_\alpha^+ - \mathcal{D}_\alpha$, the term $-\alpha K_{22}\tilde{\alpha} \leq 0$ which yields

$$\dot{V}_\alpha \leq -\alpha s_\alpha + |\alpha|\left|\frac{\tilde{L}}{mV\cos\beta} + B_{22}\bar{z}_3 + B_{22}\left(z_{3_c} - z_{3_c}^o\right) - \dot{\alpha}_c\right|.$$

We will select M to satisfy

$$M > 2\left|\frac{\tilde{L}}{mV\cos\beta} + B_{22}\bar{z}_3 + B_{22}\left(z_{3_c} - z_{3_c}^o\right) - \dot{\alpha}_c\right|. \tag{8.58}$$

The last three terms in this constraint can be directly computed. The first term $\frac{\tilde{L}}{mV\cos\beta}$ must be upper bounded.

Note that the definition of s_α ensures that the quantity $-\alpha s(\alpha)$ is negative for $\alpha \in (\bar{\alpha} + \frac{\varepsilon_\alpha}{2}, \bar{\alpha} + \varepsilon_\alpha]$ and $\alpha \in [\underline{\alpha} - \varepsilon_\alpha, \underline{\alpha} - \frac{\varepsilon_\alpha}{2})$. Constraint (8.58) ensures that

$$|\alpha s(\alpha)| > |\alpha|\left|\frac{\tilde{L}}{mV\cos\beta} + B_{22}\bar{z}_3 + B_{22}\left(z_{3_c} - z_{3_c}^o\right) - \dot{\alpha}_c\right|$$

for $\alpha \in (\bar{\alpha} + \frac{3}{4}\varepsilon_\alpha, \bar{\alpha} + \varepsilon_\alpha]$ and $\alpha \in [\underline{\alpha} - \varepsilon_\alpha, \underline{\alpha} - \frac{3}{4}\varepsilon_\alpha)$. Note that α exiting $\mathcal{D}_\alpha^+$ would require $\dot{V}_\alpha$ to be positive for either $\alpha \in (\bar{\alpha} + \frac{3}{4}\varepsilon_\alpha, \bar{\alpha} + \varepsilon_\alpha]$ or $\alpha \in [\underline{\alpha} - \varepsilon_\alpha, \underline{\alpha} - \frac{3}{4}\varepsilon_\alpha)$. Since we have just shown $\dot{V}_\alpha$ to be negative on each of these regions, α cannot exit $\mathcal{D}_\alpha^+$ (i.e., $\mathcal{D}_\alpha^+$ is positively invariant).

Finally, as discussed following (8.57), if we can show that the quantity $\bar{z}_2^\top \eta_\alpha = -(\tilde{\alpha} - \xi_\alpha)s_\alpha(\alpha)$ is always non-positive, then it can be dropped in the subsequent analysis of (8.57). We need to consider three cases:

- For $\alpha \in [\underline{\alpha} - \varepsilon_\alpha, \underline{\alpha} - \frac{\varepsilon_\alpha}{2}]$, the factor $s_\alpha(\alpha) \leq 0$ while $(\tilde{\alpha} - \xi_\alpha) \leq 0$ because $\tilde{\alpha} \leq -\frac{\varepsilon_\alpha}{2}$ while $|\xi_\alpha| \leq \frac{\varepsilon_\alpha}{2}$; therefore, $\bar{z}_2^\top \eta_\alpha \leq 0$.
- For $\alpha \in [\underline{\alpha} - \frac{\varepsilon_\alpha}{2}, \bar{\alpha} + \frac{\varepsilon_\alpha}{2}]$, the term $\bar{z}_2^\top \eta_\alpha = 0$.

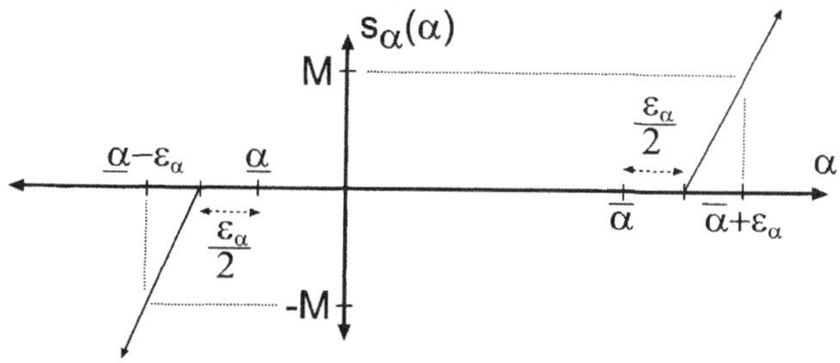

Figure 8.11: Nonlinearity used in the computation of η_α as described in Subsection 8.3.2.3. Note that this figure greatly exaggerates the size of ε_α.

- For $\alpha \in [\bar{\alpha} + \frac{\varepsilon_\alpha}{2}, \bar{\alpha} + \varepsilon_\alpha]$, the factor $s_\alpha(\alpha) \geq 0$ while $(\tilde{\alpha} - \xi_\alpha) \geq 0$ because $\tilde{\alpha} \geq \frac{\varepsilon_\alpha}{2}$ while $|\xi_\alpha| \leq \frac{\varepsilon_\alpha}{2}$; therefore, $\tilde{z}_2^\top \eta_\alpha \leq 0$.

The inequalities on $\tilde{\alpha}$ are derived using the assumed range of α and the fact that $\alpha_c(t) \in \mathcal{D}_\alpha$ for all $t \geq 0$. The inequality on $|\xi_\alpha|$ is given by (8.47).

8.3.3 Body Axis Angular Rate Control

Given the results of the previous sections, the objective of this subsection is to design a tracking controller to force z_3 to track z_{3_c} while ensuring the stability of the overall system. This controller and its derivation are very similar to that of Section 8.2. A block diagram representation of the controller derived in this section is shown in Figure 8.12.

The aircraft dynamics of eqns. (8.1)–(8.9) can be written as

$$\begin{aligned} \dot{z}_1 &= \bar{A}_1(x) f_1 + F_1(x) + G_1(z_2, x, T) \\ \dot{z}_2 &= A_2(x) f_1 + F_2(x) + B_2(x) z_3 \\ \dot{z}_3 &= A_3 f_3 + F_3(x) + B_3 G_3 \delta \end{aligned}$$

where $\delta = [\delta_1, \ldots, \delta_6]^\top$ is the control signal, $B_3 = A_3 = \begin{bmatrix} c_3 & 0 & c_4 \\ 0 & c_7 & 0 \\ c_4 & 0 & c_9 \end{bmatrix}$ are known matrices, $F_3 = \begin{bmatrix} (c_1 R + c_2 P) Q \\ c_5 PR - c_6 (P^2 - R^2) \\ (c_8 P - c_2 R) Q \end{bmatrix}$ is a known function, and $f_3 = [\bar{L}', \bar{M}', \bar{N}']^\top$ and $G_3 = \begin{bmatrix} \bar{L}_{\delta_1} & \cdots & \bar{L}_{\delta_6} \\ \bar{M}_{\delta_1} & \cdots & \bar{M}_{\delta_6} \\ \bar{N}_{\delta_1} & \cdots & \bar{N}_{\delta_6} \end{bmatrix}$ are unknown functions. The notation for the moment functions was defined in Section 8.1.2.

Select continuous δ_c^o such that

$$B_3 \hat{G}_3 \delta_c^o = -A_3 \hat{f}_3 - F_3 - K_3 \tilde{z}_3 + \dot{z}_{3_c} - B_2^\top \tilde{z}_2, \tag{8.59}$$

with K_3 positive definite. When the aircraft is over-actuated, the matrix $B_3 \hat{G}_3$ will have more columns than rows and will have full row rank. Therefore, many solutions to eqn. (8.59) exist and some form of actuator distribution [26, 68, 70] is required to select δ_c^o (see Section 8.2.2).

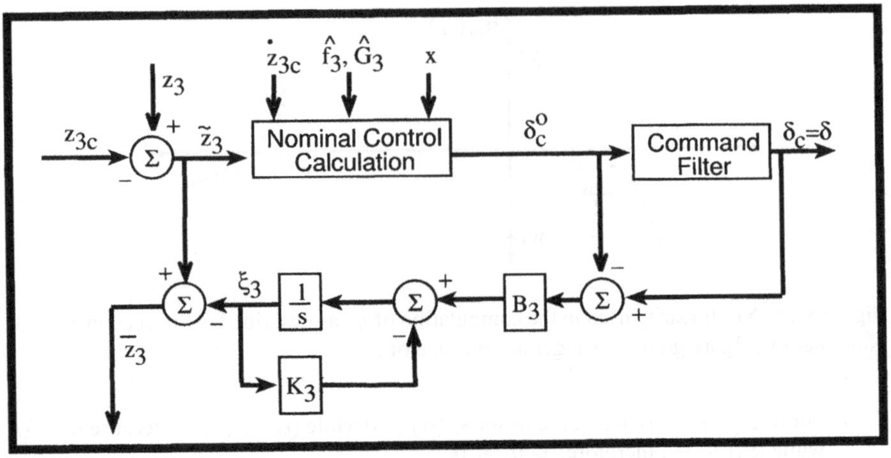

Figure 8.12: Block diagram of the angular rate controller described in Section 8.3.3. The signal $z_3(t)$ is a subvector of $x(t)$. The functions $\hat{f}_3$ and $\hat{G}_3$ are outputs of the adaptive function approximation process (not shown). The nominal control calculation refers to the solution of eqn. (8.59). The signals z_{3_c} and $\dot{z}_{3_c}$ are inputs from the wind-axes angle controller of Section 8.3.2. The signal ξ_3 is an output to the wind-axes angle controller. The signals δ_c is the surface deflection command. The signal $\bar{z}_3$ is an output training signal to be used by the adaptive function approximation process.

We pass δ_c^o through a filter, to produce δ which is within the bandwidth limitations of the actuation system.[2] The signal ξ_3 is the output of the filter

$$\dot{\xi}_3 = -K_3\xi_3 + B_3\hat{G}_3\left(\delta - \delta_c^o\right) \tag{8.60}$$

and the compensated tracing error is defined by $\bar{z}_3 = \tilde{z}_3 - \xi_3$.

Finally, select the moment function parameter adaptation laws as

$$\dot{\hat{\Theta}}_{f_3} = \dot{\Theta}_{f_3} = \begin{cases} 0 & \text{if } \tau(\bar{z},\varepsilon) \le 0 \\ \mathcal{P}\left(\Gamma_{f_3}\Phi_{f_3}\left(\bar{z}_3^\top A_3\right)\right) & \text{if } \tau(\bar{z},\varepsilon) > 0 \end{cases} \tag{8.61}$$

$$\dot{\hat{\Theta}}_{G_{3j}} = \dot{\Theta}_{G_{3j}} = \begin{cases} 0 & \text{if } \tau(\bar{z},\varepsilon) \le 0 \\ \mathcal{P}\left(\Gamma_{G_{3j}}\Phi_{G_{3j}}\delta_j\left(\bar{z}_3^\top B_3\right)\right) & \text{if } \tau(\bar{z},\varepsilon) > 0 \end{cases} \tag{8.62}$$

with $\varepsilon = [\varepsilon_1, \varepsilon_2, \varepsilon_3]^\top$ being a vector of designer-specified constants, δ_j is the j-th element of δ, Γ_{f_3} and $\Gamma_{G_{3j}}$ being positive definite matrices of appropriate dimensions, and the function τ defined in (8.66). The parameterization of $\hat{f}_3 = \Theta_{f_3}^\top\Phi_{f_3}$ and $\hat{G}_{3j} = \Theta_{G_{3j}}^\top\Phi_{G_{3j}}$ is derived in Section 8.3.5.

8.3.3.1 ***Tracking Error Dynamics and Stability Analysis*** The tracking error and compensated tracking error dynamics for z_1 are given by eqns. (8.48) and (8.53). The

[2] Alternatively, if the surface deflection is measured, then the signal δ_c^o could be used as the commanded surface positions and the measured surface deflection vector δ can be used directly to calculate ξ_3. No change is required in the notation of eqn. (8.60).

tracking error and compensated tracking error dynamics for z_2 are given by eqns. (8.50) and (8.54).

The tracking error dynamics for z_3 are

$$\begin{aligned}\dot{\tilde{z}}_3 &= A_3 f_3 + F_3(x) + B_3\hat{G}_3\delta_c^o - \dot{z}_{3_c} + B_3\hat{G}_3\left(\delta - \delta_c^o\right) + B_3\left(G_3 - \hat{G}_3\right)\delta \\ &= -K_3\tilde{z}_3 - A_3\tilde{f}_3 + B_3\tilde{G}_3\delta + B_3\hat{G}_3\left(\delta - \delta_c^o\right) - B_2^\top\tilde{z}_2 \\ &= -K_3\tilde{z}_3 - A_3\tilde{\Theta}_{f_3}^\top\Phi_{f_3} + B_3\tilde{G}_3\delta + B_3\hat{G}_3\left(\delta - \delta_c^o\right) - B_2^\top\tilde{z}_2 + A_3 e_{f_3},\end{aligned}$$

where $\tilde{f}_3 = \hat{f}_3 - f_3 = \tilde{\Theta}_{f_3}^\top\Phi_{f_3} - e_{f_3}$ and

$$\tilde{G}_3\delta = \left(\hat{G}_3 - G_3\right)\delta = \sum_{j=1}^{m}\left(\tilde{\Theta}_{G_{3_j}}^\top\Phi_{G_{3_j}} - e_{G_{3_j}}\right)\delta_j.$$

The compensated tracking error dynamics for z_3 are

$$\dot{\bar{z}}_3 = -K_3\bar{z}_3 - A_3\tilde{\Theta}_{f_3}^\top\Phi_{f_3} - B_3\sum_{j=1}^{m}\tilde{\Theta}_{G_{3_j}}^\top\Phi_{G_{3_j}}\delta_j - B_2^\top\bar{z}_2 + \rho_3$$

where $\rho_3 = A_3 e_{f_3} + B_3\sum_{j=1}^{m} e_{G_{3_j}}\delta_j$.

Define the Lyapunov function

$$\begin{aligned}\mathcal{V} &= \frac{1}{2}\left(\sum_{i=1}^{3}\bar{z}_i^\top\bar{z}_i + trace\left(\tilde{\Theta}_{f_1}^\top\Gamma_{f_1}^{-1}\tilde{\Theta}_{f_1}\right) + trace\left(\tilde{\Theta}_{f_3}^\top\Gamma_{f_3}^{-1}\tilde{\Theta}_{f_3}\right)\right. \\ &\qquad\left. + \sum_{j=1}^{m} trace\left(\tilde{\Theta}_{G_{3_j}}^\top\Gamma_{G_{3_j}}^{-1}\tilde{\Theta}_{G_{3_j}}\right)\right).\end{aligned} \tag{8.63}$$

The time derivative of $\mathcal{V}$ is

$$\begin{aligned}\dot{\mathcal{V}} &= \bar{z}_1^\top\dot{\bar{z}}_1 + \bar{z}_2^\top\dot{\bar{z}}_2 + \bar{z}_3^\top\dot{\bar{z}}_3 + trace\left(\tilde{\Theta}_{f_1}^\top\Gamma_{f_1}^{-1}\dot{\hat{\Theta}}_{f_1}\right) + trace\left(\tilde{\Theta}_{f_3}^\top\Gamma_{f_3}^{-1}\dot{\hat{\Theta}}_{f_3}\right) \\ &\quad + \sum_{j=1}^{m} trace\left(\tilde{\Theta}_{G_{3_j}}^\top\Gamma_{G_{3_j}}^{-1}\dot{\hat{\Theta}}_{G_{3_j}}\right) \\ &= -\bar{z}_1^\top K_1\bar{z}_1 - \bar{z}_2^\top K_2\bar{z}_2 - \bar{z}_3^\top K_3\bar{z}_3 \\ &\quad + \left(\bar{z}_1^\top A_1 + \bar{z}_2^\top A_2\right)e_{f_1} + \bar{z}_2^\top\eta_\alpha + \bar{z}_3^\top\rho_3.\end{aligned} \tag{8.64}$$

The term $B_2^\top\bar{z}_2$ in eqn. (8.59) results in the cancellation of one of the sign indefinite terms of eqn. (8.57). Also, the discussion of p. 359 shows that $\bar{z}_2^\top\eta_\alpha$ is negative, hence this term is dropped in the subsequent discussion. Eqn. (8.64) will be used in Section 8.3.4 to prove the stability properties of the UAV adaptive approximation based controller. Further manipulation of (8.64) is needed to determine the appropriate structure for the parameter estimation dead-zones. To continue the analysis, we express (8.64) in matrix form:

$$\dot{\mathcal{V}} \leq -\bar{z}^\top K\bar{z} + \bar{z}^\top\rho$$

where $K = diag(K_1, K_2, K_3)$ is block diagonal, $\bar{z} = [\bar{z}_1, \bar{z}_2, \bar{z}_3]^\top$ and $\rho = [\rho_1, \rho_2, \rho_3]^\top$ with $\rho_1 = A_1 e_{f_1}$, $\rho_2 = A_2 e_{f_1}$,and $\rho_3 = A_3 e_{f_3} + B_3\sum_{j=1}^{m} e_{G_{3_j}}\delta_j$. Each of the ρ_i

are bounded. The bounds are unknown, but can be made arbitrarily small by appropriate selection of the function approximator structure. However, once the designer specifies the structure, the bounds $\bar{\rho}_i$ on each ρ_i are fixed. Since K is positive definite, there exist positive definite D such that $K = D^\top D$. Therefore,

$$\begin{aligned}
\dot{\mathcal{V}} &\leq -\tilde{z}^\top D^\top D\tilde{z} + \tilde{z}^\top D^\top \left(D^\top\right)^{-1} \rho \\
&= -y^\top y + y^\top v \\
&\leq -\|y\|_2 \left(\|y\|_2 - \|v\|_2\right) \\
&\leq \frac{1}{2}\left(-\|y\|_2^2 + \|v\|_2^2\right) \\
&\leq \frac{1}{2}\left(-\underline{\lambda}_K\|\tilde{z}\|_2^2 + \bar{\lambda}_{K^{-1}}\|\rho\|_2^2\right) \\
&\leq \frac{\underline{\lambda}_K}{2}\left(-\|\tilde{z}\|_2^2 + \frac{1}{\underline{\lambda}_K^2}\|\rho\|_2^2\right)
\end{aligned} \tag{8.65}$$

where $y = D\tilde{z}$, $v = \left(D^\top\right)^{-1}\rho$, the symbol $\underline{\lambda}_K$ indicates the minimum eigenvalue of K, and $\bar{\lambda}_{K^{-1}} = \frac{1}{\underline{\lambda}_K}$ is the maximum eigenvalue of K^{-1}. Therefore, $\dot{\mathcal{V}}$ is negative definite if

$$\tau(\tilde{z}, \rho) = \left(\|\tilde{z}\|_2 - \frac{1}{\underline{\lambda}_K}\|\rho\|_2\right) > 0. \tag{8.66}$$

The theorems in the next subsection summarize the stability properties for the closed-loop system that can be proven based on the relationship of ρ to the dead-zone size parameter ε.

8.3.4 Control Law and Stability Properties

The previous subsections intermix the design of the control law equations with the analysis. This section presents in an organized summary fashion the control law implementation equations and states the stability properties that apply.

For the input signals z_{1_c} and $\dot{z}_{1_c}$ the control law is given by the following:

1. Select the control signals μ_1 so that

$$\hat{G}_1(\mu_1, x) = -K_1\tilde{z}_1 + \dot{z}_{1_c} - \bar{A}_1\hat{f}_1 - F_1 \tag{8.67}$$

   where $\tilde{z}_1 = z_1 - z_{1_c}$. Define $z_{2_c}^o = \mu_1$. Command filter $z_{2_c}^o$ to produce the signals z_{2_c} and $\dot{z}_{2_c}$.

2. Select $\mu_{2_c}^o$ such that

$$B_2\mu_{2_c}^o = -K_2\tilde{z}_2 + \dot{z}_{2c} - A_2\hat{f}_1 - F_2 + \eta_\alpha \tag{8.68}$$

   where $\tilde{z}_2 = z_2 - z_{2_c}$. Since B_2 is square and invertible, this solution is unique and straightforward. Define $z_{3_c}^o = \mu_{2_c}^o - \xi_3$. Command filter $z_{3_c}^o$ to produce the signals z_{3_c} and $\dot{z}_{3_c}$.

3. Select δ_c^o such that

$$B_3\hat{G}_3\delta_c^o = -K_3\tilde{z}_3 + \dot{z}_{3_c} - A_3\hat{f}_3 - F_3 - B_2^\top\tilde{z}_2 \tag{8.69}$$

where $\tilde{z}_3 = z_3 - z_{3_c}$. If $p > 3$, then the system is over-actuated and some form of actuator distribution process will be implemented. This actuator distribution can be used to limit the extent and rate of the commanded actuator deflections.

4. Implement the following bank of filters to compute ξ_i for $i = 1, 2, 3$:

$$\dot{\xi}_1 = -K_1\xi_1 + \left(\hat{G}_1(z_2, x) - \hat{G}_1(z^o_{2_c}, x)\right), \tag{8.70}$$

$$\dot{\xi}_2 = -K_2\xi_2 + B_2\left(z_{3_c} - z^o_{3_c}\right), \text{ and} \tag{8.71}$$

$$\dot{\xi}_3 = -K_3\xi_3 + B_3\hat{G}_3\left(\delta - \delta^o_c\right). \tag{8.72}$$

The controller includes adaptive approximation of the unknown force and moment functions using the following parameter estimation equations:

$$\dot{\Theta}_{f_1} = \mathcal{P}\left(\Gamma_{f_1}\Phi_{f_1}\left(\bar{z}_1^\top A_1 + \bar{z}_2^\top A_2\right)\right) \tag{8.73}$$

$$\dot{\Theta}_{f_3} = \mathcal{P}\left(\Gamma_{f_3}\Phi_{f_3}\left(\bar{z}_3^\top A_3\right)\right) \tag{8.74}$$

$$\dot{\Theta}_{G_{3j}} = \mathcal{P}\left(\Gamma_{G_{3j}}\Phi_{G_{3j}}\left(\bar{z}_3^\top \delta_j B_3\right)\right), \text{ for } j = 1\ldots, m \tag{8.75}$$

when $\tau(\bar{z}, \varepsilon) > 0$. Otherwise the derivatives of the approximator parameters are zero.

Such adaptive approximators are especially useful on UAVs, where the aerodynamics may change during flight, for example, due to battle damage. For the controller summarized above, the following three theorems summarize the stability properties under different application conditions. Theorem 8.3.1 is concerned with the most ideal case.

Theorem 8.3.1 *Assuming that the functions Φ_{f_1}, Φ_{f_3}, and $\Phi_{G_{3j}}$ are bounded and that perfect approximation can be achieved (i.e., $\varepsilon = \rho = 0$), the adaptive approximation based controller summarized in eqns. (8.67)–(8.75) has the following properties:*

1. *The estimated parameters Θ_{f_1}, Θ_{f_3}, $\Theta_{G_{3j}}$ and parameter errors $\tilde{\Theta}_{f_1}$, $\tilde{\Theta}_{f_3}$, $\tilde{\Theta}_{G_{3j}}$ are bounded.*

2. *The compensated tracking errors $\bar{z}_1$, $\bar{z}_2$, and $\bar{z}_3$ are bounded.*

3. *$\|\bar{z}_i(t)\| \to 0$ as $t \to \infty$ for $i = 1, 2, 3$.*

4. *$\bar{z}_i(t) \in \mathcal{L}_2$ for $i = 1, 2, 3$.*

Proof: Boundedness of the parameter error vector is due to the fact that $\mathcal{V}$ of eqn. (8.63) is positive definite in the parameter error vectors and $\frac{d\mathcal{V}}{dt}$ of eqn. (8.64) is negative semidefinite when $\varepsilon = \rho = 0$. Therefore, $\mathcal{V}(t) \leq \mathcal{V}(0)$ for $t > 0$. This implies that for any $t > 0$,

$$trace\left(\tilde{\Theta}_{f_1}^\top\Gamma_{f_1}^{-1}\tilde{\Theta}_{f_1}\right) + trace\left(\tilde{\Theta}_{f_3}^\top\Gamma_{f_3}^{-1}\tilde{\Theta}_{f_3}\right) + \sum_{j=1}^{m} trace\left(\tilde{\Theta}_{G_{3j}}^\top\Gamma_{G_{3j}}^{-1}\tilde{\Theta}_{G_{3j}}\right) \leq \mathcal{V}(0) < \infty.$$

This completes the proof of item 1. The boundedness of the compensated tracking errors is shown similarly, to complete the proof of item 2. Since the parameter errors are bounded and the optimal parameters are bounded, we also know that the estimated parameters are bounded (i.e., Θ_{f_1}, Θ_{f_3}, $\Theta_{G_{3j}} \in \mathcal{L}_\infty$). By the definition of the approximators, this also implies that $\hat{f}_1$, $\hat{f}_3$, and $\hat{G}_{3_j}$ are bounded functions, as are $\tilde{f}_1$, $\tilde{f}_3$, and $\tilde{G}_{3_j}$ on $\mathcal{D}$.

The second time derivative of the Lyapunov function is

$$\begin{aligned}\frac{d^2\mathcal{V}}{dt^2} &= -\bar{z}_1^\top\left(K_1+K_1^\top\right)\left(-K_1\bar{z}_1 - A_1\tilde{f}_1\right)\\ &\quad -\bar{z}_2^\top\left(K_2+K_2^\top\right)\left(-K_2\bar{z}_2 - A_2\tilde{f}_1 + B_2\bar{z}_3\right)\\ &\quad -\bar{z}_3^\top\left(K_3+K_3^\top\right)\left(-K_3\bar{z}_3 - A_3\tilde{f}_3 - B_3\tilde{G}\delta - B_2^\top\bar{z}_2\right)\end{aligned}$$

which is bounded. Therefore, the function $\frac{d\mathcal{V}}{dt}$ is uniformly continuous. Barbălat's Lemma (See p. 388 in Appendix A) implies that $\frac{d\mathcal{V}}{dt}\to 0$ as $t\to\infty$. This requires that $\bar{z}_i^\top K_i\bar{z}_i\to 0$ for $i=1,2,3$ as $t\to\infty$ and because $\bar{z}_i^\top K_i\bar{z}_i \geq \underline{\lambda}(K_i)\|\bar{z}_i\|^2$, where $\underline{\lambda}(K_i)$ is the minimum eigenvalue of the positive definite matrix K_i, we see that $\|\bar{z}_i\|^2\to 0$ as $t\to\infty$ for $i=1,2,3$. This completes the proof of item 3.

Integrating both sides of eqn. (8.64) with $\rho=0$ yields

$$\mathcal{V}(t)-\mathcal{V}(0) \leq \int_0^t \left(-\bar{z}_i^\top(\tau)K_i\bar{z}_i(\tau)\right)d\tau, \quad \forall\, t\geq 0 \tag{8.76}$$

$$\mathcal{V}_\infty-\mathcal{V}(0) \leq -\int_0^\infty \left(\bar{z}_i^\top(\tau)K_i\bar{z}_i(\tau)\right)d\tau \tag{8.77}$$

$$\mathcal{V}(0) \geq \int_0^\infty \left(\bar{z}_i^\top(\tau)K_i\bar{z}_i(\tau)\right)d\tau \tag{8.78}$$

where $0\leq\mathcal{V}(t)\leq\mathcal{V}(0)$ for all $t\geq 0$ and $\frac{d\mathcal{V}}{dt}\leq 0$ implies that $\lim_{t\to\infty}\mathcal{V}(t)=\mathcal{V}_\infty$ is well defined. This completes the proof of item 4. ■

Theorem 8.3.1 considered a very idealized case where $\varepsilon=\rho=0$. Theorem 8.3.2 will consider a more reasonable situation that corresponds to the dead-zone design assumption $\rho<\varepsilon$ being satisfied. Theorem 8.3.2 corresponds to the typical situation.

The proof is not included, but follows the same procedures as presented in the robustness analysis of Section 7.3.3.

Theorem 8.3.2 *Assuming that the functions Φ_{f_1}, Φ_{f_3}, and $\Phi_{G_{3j}}$ are bounded and that $\|\varepsilon\|_2>\|\rho\|_2$, the adaptive approximation based controller summarized in eqns. (8.67)–(8.75) has the following properties:*

1. *The estimated parameters Θ_{f_1}, Θ_{f_3}, $\Theta_{G_{3j}}$ and parameter errors $\tilde{\Theta}_{f_1}$, $\tilde{\Theta}_{f_3}$, $\tilde{\Theta}_{G_{3j}}$ are bounded.*

2. *The compensated tracking error vector $\bar{z}$, as $t\to\infty$, is ultimately bounded by $\|\bar{z}\|\leq\frac{1}{\underline{\lambda}_K}\|\varepsilon\|_2$. In fact, the total time spent outside the dead-zone is finite.*

3. *$\bar{z}(t)$ is small-in-the-mean-squared sense satisfying:*

$$\int_t^{t+T}\|\bar{z}\|_2^2 d\tau \leq \frac{2}{\underline{\lambda}_K}V(t)+\frac{\|\varepsilon\|_2^2}{\underline{\lambda}_K^2}T.$$

The following theorem presents stability results applicable in the worst-case scenario where the dead-zone is not large enough and the model error ρ sometimes exceeds the dead-zone size ε.

Theorem 8.3.3 *Assuming that the functions Φ_{f_1}, Φ_{f_3}, and $\Phi_{G_{3j}}$ are bounded and there exist regions of the state space where $\|\varepsilon\|_2 \leq \|\rho\|_2$, the adaptive approximation based controller summarized in eqns. (8.67)–(8.75) has the following properties:*

1. *The estimated parameters Θ_{f_1}, Θ_{f_3}, $\Theta_{G_{3j}}$ and parameter errors $\tilde{\Theta}_{f_1}$, $\tilde{\Theta}_{f_3}$, $\tilde{\Theta}_{G_{3j}}$ are bounded.*

2. *The compensated tracking error vector $\bar{z} \in \mathcal{L}_\infty$.*

3. *$\bar{z}(t)$ is small-in-the-mean-squared sense satisfying:*

$$\int_t^{t+T} \|\bar{z}\|_2^2 d\tau \leq \frac{2}{\underline{\lambda}_K} V(t) + \frac{1}{\underline{\lambda}_K^2} \int_t^{t+T} \|\rho(\tau)\|_2^2 d\tau.$$

If a nominal design model were known and used to define the functions $\hat{f}_1$, $\hat{f}_2$, and $\hat{f}_3$, then the above controller could be used without adaptive approximation. This would be similar to the baseline control approach presented in Section 8.2.2. The stability and tracking performance would be affected by the errors between the design model and the actual system as indicated in the tracking error equations (8.53), (8.54), and (8.63). In fact, if the command filters were replaced by analytic computation of the command derivatives, then the ξ_i filters could be removed (i.e., $\xi_i(t) = 0$). The remaining controller would be a backstepping controller for the aircraft designed using the nominal model. We mention this only to point out that the approximation based approach can be considered as a retrofit to a baseline nominal controller designed by the backstepping method. The retrofit would add in command filtering, adaptive approximation, and the ξ filters. Due to the adaptive approximation, the retrofit would attain both stability and performance robustness to model error.

Note that the bound on $\bar{z}$ provable in item 2 of Theorem 8.3.3 is not very reassuring. The bound would be related to the maximum value of the Lyapunov function evaluated on the boundary of the parameter set defined in the projection. Although this bound is potentially huge, it should be considered in the light of the discussion following Theorem 8.2.3 on page 344. The bound on the tracking error in Theorem 8.3.2 is much smaller and defined completely by the design parameters. It pays for the designer to be conservative in specifying the dead-zone size and the function approximator.

8.3.5 Approximator Definition

The aircraft dynamics involve three moments $(\bar{L}, \bar{M}, \bar{N})$ and three forces (D, Y, L) that define the functions f_1, f_3, and G_3. The nondimensional coefficient function approach to defining the structure of these functions has been discussed in Sections 8.1.2 and 8.2.3.1. Due to a change in subscript notation, a small portion of the material from Section 8.2.3.1 is repeated here. The objective of this section is to demonstrate that the approximators can be manipulated into the form required for the preceding theoretical analysis:

$$\hat{f}_1 = \Theta_{f_1}^\top \Phi_{f_1}, \quad \hat{f}_3 = \Theta_{f_3}^\top \Phi_{f_3}, \quad \hat{G}_{3j} = \Theta_{G_{3j}}^\top \Phi_{G_{3j}}$$

for $j = 1, \ldots, 6$. The form of the equations shown above which is convenient for analysis is not the most efficient for implementation. For implementation, it is much more efficient to manipulate the parameter adaptation equations into separate equations suitable for each nondimensional coefficient function.

Each of the coefficient functions C_* is an unknown function that is implemented as $C_*(\alpha, M) = \theta_*^\top \phi(\alpha, M)$ (e.g., $C_{D_o}(\alpha, M) = \theta_{D_o}^\top \phi(\alpha, M)$), where $\phi(\alpha, M)$ is a regressor vector that is selected by the designer and θ_* is estimated online. Note that different regressors can be used for the different functions. This section uses a single regressor vector $\phi(\alpha, M)$ for all the approximations for notational simplicity.

The drag force approximator uses the coefficient functions $\hat{C}_{D_o}$ and $\hat{C}_{D_{\delta_1}}, \ldots, \hat{C}_{D_{\delta_6}}$. By defining the matrix $\theta_D = [\theta_{D_o}, \theta_{D_{\delta_1}}, \ldots, \theta_{D_{\delta_6}}] \in \Re^{N \times 7}$, which contains in each column the parameter vector used to approximate one of the coefficient functions, we have that

$$\begin{bmatrix} \hat{C}_{D_o}(\alpha, M) \\ \hat{C}_{D_{\delta_1}}(\alpha, M) \\ \vdots \\ \hat{C}_{D_{\delta_6}}(\alpha, M) \end{bmatrix} = \theta_D^\top \phi(\alpha, M).$$

The drag force of (8.11) is then represented as

$$\hat{D} = \Psi_D \theta_D^\top \phi \tag{8.79}$$

where $\Psi_D = \bar{q}S[1, \delta_1, \ldots, \delta_m]$. Similarly, for the other forces and moments:

$$\left.\begin{array}{lllll} \hat{\bar{L}} & = & \Psi_{\bar{L}} \theta_{\bar{L}}^\top \phi & \hat{Y} = & \Psi_Y \theta_Y^\top \phi \\ \hat{\bar{M}} & = & \Psi_{\bar{M}} \theta_{\bar{M}}^\top \phi & \hat{L} = & \Psi_L \theta_L^\top \phi \\ \hat{\bar{N}} & = & \Psi_{\bar{N}} \theta_{\bar{N}}^\top \phi & & \end{array}\right\} \tag{8.80}$$

with

$$\begin{array}{lll} \Psi_{\bar{L}} & = & \bar{q}Sb\left[1, \frac{bP}{2V}, \frac{bR}{2V}, \beta, \delta_1, \ldots, \delta_m\right], \\ \Psi_{\bar{M}} & = & \bar{q}S\bar{c}\left[1, \frac{\bar{c}Q}{2V}, \delta_1, \ldots, \delta_m\right], \\ \Psi_{\bar{N}} & = & \bar{q}Sb\left[1, \frac{bP}{2V}, \frac{bR}{2V}, \beta, \delta_1, \ldots, \delta_m\right], \end{array} \qquad \begin{array}{lll} \Psi_Y & = & \bar{q}S\left[1, \beta, \frac{bP}{2V}, \delta_1, \ldots, \delta_m\right], \\ \Psi_L & = & \bar{q}S\left[1, \alpha, \delta_1, \ldots, \delta_m\right], \end{array}$$

and

$$\begin{array}{lll} \theta_Y & = & \left[\theta_{Y_o}, \theta_{Y_\beta}, \theta_{Y_P}, \theta_{Y_{\delta_1}}, \ldots, \theta_{Y_{\delta_m}}\right] \in \Re^{N \times 9} \\ \theta_L & = & \left[\theta_{L_o}, \theta_{L_\alpha}, \theta_{L_{\delta_1}}, \ldots, \theta_{L_{\delta_m}}\right] \in \Re^{N \times 8} \\ \theta_{\bar{L}} & = & \left[\theta_{\bar{L}_o}, \theta_{\bar{L}_P}, \theta_{\bar{L}_R}, \theta_{\bar{L}_\beta}, \theta_{\bar{L}_{\delta_1}}, \ldots, \theta_{\bar{L}_{\delta_m}}\right] \in \Re^{N \times 10} \\ \theta_{\bar{M}} & = & \left[\theta_{\bar{M}_o}, \theta_{\bar{M}_Q}, \theta_{\bar{M}_{\delta_1}}, \ldots, \theta_{\bar{M}_{\delta_m}}\right] \in \Re^{N \times 8} \\ \theta_{\bar{N}} & = & \left[\theta_{\bar{N}_o}, \theta_{\bar{N}_P}, \theta_{\bar{N}_R}, \theta_{\bar{N}_\beta}, \theta_{\bar{N}_{\delta_1}}, \ldots, \theta_{\bar{N}_{\delta_m}}\right] \in \Re^{N \times 10}. \end{array}$$

Each of θ_D, θ_Y, θ_L, $\theta_{\bar{L}}$, $\theta_{\bar{M}}$, and $\theta_{\bar{N}}$ is a matrix of unknown parameters.

Each of the equations (8.79)–(8.80) is linear with respect to the matrix of unknown parameters; therefore, each approximator can be rewritten into the standard vector form

$$\hat{D} = \Theta_D^\top \Phi_D \quad \hat{Y} = \Theta_Y^\top \Phi_Y \quad \hat{L} = \Theta_L^\top \Phi_L. \tag{8.81}$$

For example,

$$\begin{array}{lll} \Theta_D^\top & = & \left[\theta_{D_o}^\top, \theta_{D_{\delta_1}}^\top, \ldots, \theta_{D_{\delta_6}}^\top\right] \in \Re^{7N} \\ \Phi_D^\top & = & \bar{q}S\left[\phi^\top, \delta_1\phi^\top, \ldots, \delta_6\phi^\top\right] \in \Re^{7N}. \end{array}$$

Finally, using the above definitions,

$$\hat{f}_1 = \begin{bmatrix} \hat{D} \\ \hat{Y} \\ \hat{L} \end{bmatrix} = \begin{bmatrix} \Theta_D^\top \Phi_D \\ \Theta_Y^\top \Phi_Y \\ \Theta_L^\top \Phi_L \end{bmatrix} = \Theta_{f_1}^\top \Phi_{f_1}. \tag{8.82}$$

The moments of (8.80) require slightly more effort, because the control derivations utilize f_3 and G_3 separately. The portion of the moment equations that is independent of the surface deflections can be represented as

$$\hat{\bar{L}}' = \Theta_{\bar{L}'}^\top \Phi_{\bar{L}'} \quad \hat{\bar{M}}' = \Theta_{\bar{M}'}^\top \Phi_{\bar{M}'} \quad \hat{\bar{N}}' = \Theta_{\bar{N}'}^\top \Phi_{\bar{N}'}$$

where, for example,

$$\begin{aligned} \Theta_{\bar{L}'}^\top &= \left[\theta_{\bar{L}_o}^\top, \theta_{\bar{L}_P}^\top, \theta_{\bar{L}_R}^\top, \theta_{\bar{L}_\beta}^\top\right] \in \Re^{4N} \\ \Phi_{\bar{L}'}^\top &= \bar{q}Sb\left[\phi^\top, \frac{bP}{2V}\phi^\top, \frac{bR}{2V}\phi^\top, \beta\phi^\top\right] \in \Re^{4N}. \end{aligned}$$

Therefore,

$$\hat{f}_3 = \begin{bmatrix} \hat{\bar{L}}' \\ \hat{\bar{M}}' \\ \hat{\bar{N}}' \end{bmatrix} = \begin{bmatrix} \Theta_{\bar{L}'}^\top \Phi_{\bar{L}'} \\ \Theta_{\bar{M}'}^\top \Phi_{\bar{M}'} \\ \Theta_{\bar{N}'}^\top \Phi_{\bar{N}'} \end{bmatrix} = \Theta_{f_3}^\top \Phi_{f_3}. \tag{8.83}$$

Finally, using the above definitions,

$$\hat{G}_{3j} = \begin{bmatrix} \hat{\bar{L}}_{\delta_j} \\ \hat{\bar{M}}_{\delta_j} \\ \hat{\bar{N}}_{\delta_j} \end{bmatrix} = \begin{bmatrix} \theta_{\bar{L}_{\delta_j}}^\top \phi\delta_j \\ \theta_{\bar{M}_{\delta_j}}^\top \phi\delta_j \\ \theta_{\bar{N}_{\delta_j}}^\top \phi\delta_j \end{bmatrix} = \Theta_{G_{3j}}^\top \Phi_{G_{3j}} \tag{8.84}$$

for $j = 1, \ldots, 6$. Eqns. (8.82)–(8.84) are compatible with the approximator form used throughout the previous sections of this chapter. Therefore, the approximator parameters can be adapted according to eqns. (8.73)–(8.75).

8.3.6 Simulation Analysis

This section presents simulation results from the application of the control algorithms summarized in Section 8.3.4 to a nonlinear 6-DOF model of a flying-wing UAV. The model has previously been described in Section 8.2.4.

The scenario for this section is that the UAV is in flight, when at the time indicated by $t = 0$ some event occurs that causes substantial model error. The adaptive approximation algorithms are running throughout the simulation and must maintain stable flight and trajectory following. The bounded commands $(\chi_c^o, \gamma_c^o, V_c^o)$ as functions of time are generated outside the controller. Those signals are filtered by the controller using techniques similar to those described in Section 8.2 to generate the bounded and continuous signals (χ_c, γ_c, V_c) that the controller will track and their derivatives.

The state and state commands for the γ, χ, α, μ, Q, and P variables are shown versus time in Figure 8.13. The aircraft is commanded to simultaneously change altitude (i.e.,

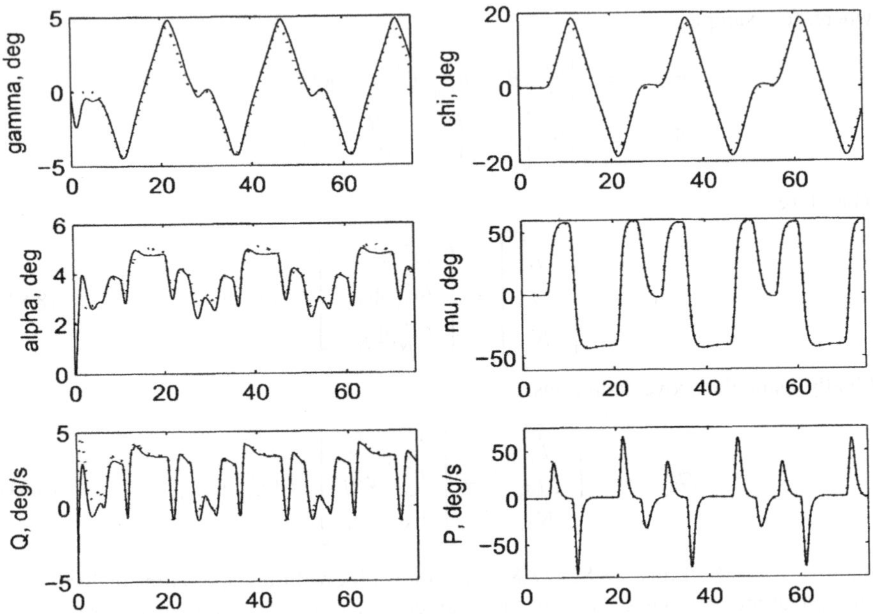

Figure 8.13: Aircraft state data for Section 8.3.6. The commanded state trajectory z_c is shown as a dotted line. The actual state trajectory is shown as a solid line. The horizontal axis shows the time, t, in seconds.

nonzero γ) and turn (i.e., time-varying χ) while holding airspeed constant and regulating sideslip to zero, i.e., coordinated turns. This type of command is relatively challenging for the autopilot because it induces significant amounts of coupling between all three channels and requires flight at high roll angles. In Figure 8.13, the commanded state is plotted as a dotted curve while the actual state is plotted as a solid curve. The tracking error is clearly evident near $t = 0$ for the variables γ, α, and Q.

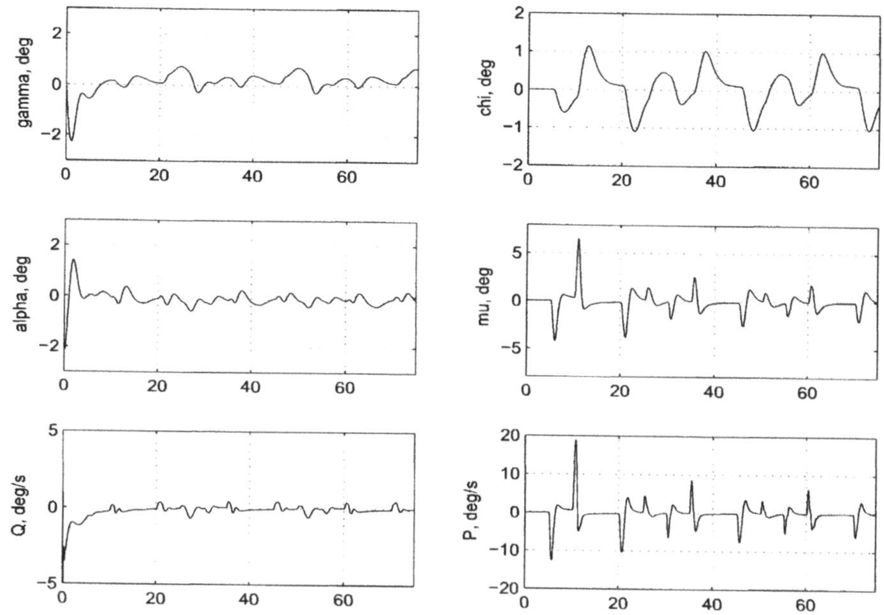

Figure 8.14: Aircraft compensated tracking error $\bar{z}$ for Section 8.3.6. The horizontal axis shows the time, t, in seconds.

Figure 8.14 plots the compensated tracking error $\bar{z}$ for the γ, χ, α, μ, Q, and P variables. Within about 10 s, the controller has learned the lift and Q-moment functions sufficiently well so that it can command the correct α and achieve that α via Q so that the γ command is tracked accurately. The μ and P tracking errors are initially large, but decreased dramatically over the first 75 s of the simulation. For this time period, $\alpha(t) \in [2.0, 5.0]$ degrees and $M \in [0.445, 0.465]$. Therefore, learning has only occurred over a small part of the operating envelope defined by $\mathcal{D}$.

Figure 8.15 shows the surface positions measured in degrees. The main purpose of these graphs is to illustrate the reasonableness in terms of magnitude and bandwidth of the control signals. Accurate tracking has been achieved, in spite of large modeling error, via adaptive approximation methods without resorting to high-gain or switching control methods.

The control gains were $K_1 = diag(0.3, 0.3, 0.2)$, $K_2 = diag(2, 2, 2)$, and $K_3 = diag(10, 30, 10)$. Therefore, $\underline{\lambda}_K = 0.2$. In the parameter adaptation dead-zone, $\|\epsilon\|_2 = 0.02$; therefore, parameter adaptation stops when $\|\bar{z}\|_2 < 0.1$. Projection was used to enforce sign constraints on the elements of $\hat{G}$, but not upper bound constraints. The region $\mathcal{D}_\alpha = [-6, 14]$ deg, $\epsilon_\alpha = 1.0°$, and $\mathcal{D}_\alpha^+ = [-7, 15]$ deg. The quantity $\frac{2M}{\epsilon_\alpha} = 0.2\frac{rad}{s}$.

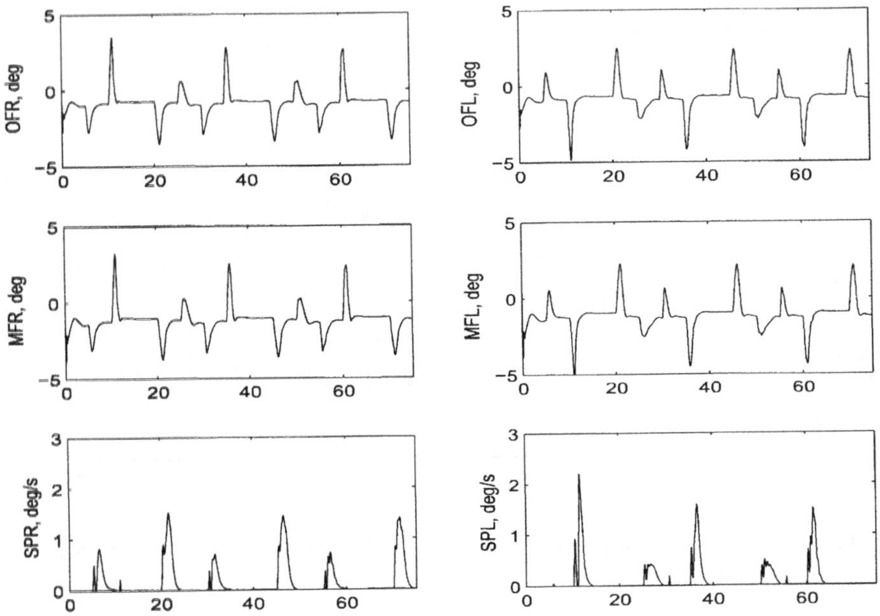

Figure 8.15: Aircraft surface deflection δ for Section 8.3.6. OF, MF, and SP denote outer-flap, mid-flap, and spoiler, respectively. The symbols R and L denote right and left, respectively. The horizontal axis shows the time, t, in seconds.

Table 8.1: Command Filter Parameters

Command variable	χ	γ	V	μ	α	β	P	Q	R
Filter bandwidth, ω_n	1.3	1.3	0.2	6	6	6	100	100	100

The purpose of the command filters for this simulation is only to compute a command and its derivative; however, the bandwidth of the command filter will influence the state trajectory. If, for example, γ_c^o were a step command, then as the bandwidth of the γ-command filter is increased, the magnitude of $\dot{\gamma}_c$ will increase. This will result in larger changes in α_c and hence Q_c and δ. Similar comments apply to χ_c^o, μ_c, P, and R. The command filter parameters used for the simulation in this section are given in Table 8.1. The damping factor in each filter was 1.0.

8.3.7 Conclusions

This section has been concerned with the problem of designing an aircraft control system capable of tracking ground track, climb rate, and speed commands from a mission planner while being robust to initial model error as well as changes to the nonlinear model that might occur during flight due to failures and battle damage. This section derives the aircraft controller using the command filtered backstepping approach with adaptive approximation to achieve robustness to unmodeled nonlinear effects, even if those effects change during flight. The stability properties are proved using Lyapunov methods. The control law and its stability properties are summarized in Section 8.3.4.

8.4 AIRCRAFT NOTATION

This section is provided as a resource to the reader. Table 8.2 defines the meaning of the constants that appear in the dynamic equations of the aircraft. Table 8.3 defines the interpretation of the symbols used to represent the state and control variables. Table 8.4 defines the unknown and approximate force and moment functions that appear in the model and control equations.

Figures 8.16 and 8.17 illustrate the definitions of the state related variables.

Symbol	Meaning
m	Mass
g	Vertical gravity component
c_i	Rotational inertia parameters defined on p. 80 in [258]
b	Reference wing span
$\bar{c}$	Mean geometric chord
S	Wing reference area

Table 8.2: Definitions of Constants

Variable	Definition
α	Angle-of-attack
β	Side slip
γ	Climb angle
θ	Pitch angle
χ	Ground track angle
μ	Roll angle
M	Mach number
P	Body axis roll rate
Q	Body axis pitch rate
R	Body axis yaw rate
V	Speed
δ_i	Deflection of the i-th control surface.
P_s	Stability axis roll rate
R_s	Stability axis yaw rate
T	Thrust

Table 8.3: Definitions of Variables

Symbol	Definition
D	Stability axis drag force. This function is unknown.
Y	Stability axis side force. This function is unknown.
L	Stability axis lift force. This function is unknown.
$\bar{L}$	Body axis roll moment. This function is unknown.
$\bar{M}$	Body axis pitch moment. This function is unknown.
$\bar{N}$	Body axis yaw moment. This function is unknown.
$\hat{D}$	Approximated stability axis drag force
$\hat{Y}$	Approximated stability axis side force
$\hat{L}$	Approximated stability axis lift force
$\hat{\bar{L}}$	Approximated body axis roll moment
$\hat{\bar{M}}$	Approximated body axis pitch moment
$\hat{\bar{N}}$	Approximated body axis yaw moment

Table 8.4: Definitions of Functions

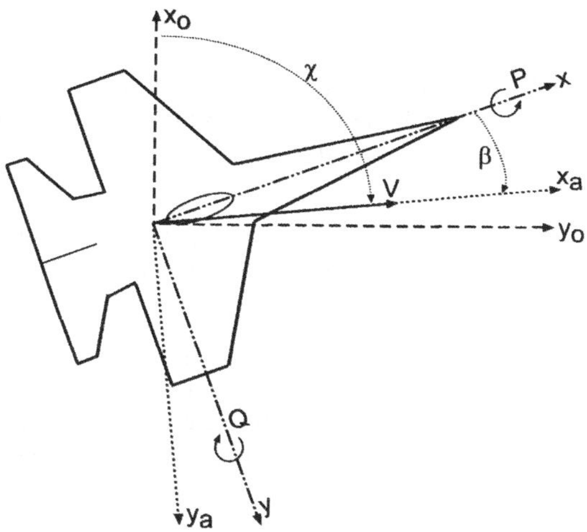

Figure 8.16: Illustration of selected aircraft variables defined in Table 8.3. For this figure, the viewer is directly above the aircraft, looking along the gravity vector. The illustration is valid for $\theta = \mu = 0$. The angular rates P and Q are defined in a right-hand sense with respect to the x and y axes, respectively.

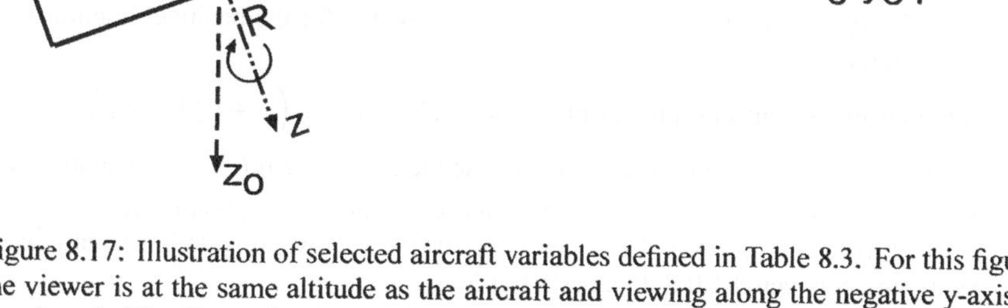

Figure 8.17: Illustration of selected aircraft variables defined in Table 8.3. For this figure, the viewer is at the same altitude as the aircraft and viewing along the negative y-axis of the aircraft. The illustration is valid for $\mu = \beta = 0$. The angular rates P and R are defined in a right-hand sense with respect to the x and z axes, respectively.

Problems

Exercise 8.1 Consider the very simple system $\dot{x} = f^* + u$ with $x \in \mathcal{D} = [-2, 2] \subset \Re^1$ and $f^* = |x| - 1$ being unknown at the design stage. Even though it is obviously not a good choice, assume that the designer has selected the approximator be $\hat{f} = \theta_0 + \theta_1 x = \theta^\top \phi(x)$ where $\phi(x) = [1, x]^\top$.

1. Use (2.35) to show that the least squares optimal parameter estimate over $\mathcal{D}$ is $\theta^* = [0, 0]^\top$. The $\mathcal{L}_\infty$ optimal estimate over $\mathcal{D}$ is also $\theta^* = [0, 0]^\top$. Therefore, $\max_{x \in \mathcal{D}} (|e_f(x)|) = 1.0 = \bar{e}_f$ where $e_f(x) = f^*(x) - \hat{f}(x)$.

2. Show that for any operating point $x_+ \in \mathcal{D}$ such that $x_+ > 0$, there is a closed neighborhood of x_+ on which f^* is perfectly approximated with $\theta^*_+ = [-1, 1]$. Similarly, for $x_- \in \mathcal{D}$ such that $x_- < 0$, there is a closed neighborhood of x_- on which f^* is perfectly approximated with $\theta^*_- = [-1, -1]$. For later use, define $\tilde{\theta}_+ = \theta - \theta^*_+$ and $\tilde{\theta}_- = \theta - \theta^*_-$.

3. Following the basic procedure of Section 8.2.3, define the control law and parameter update equations as

$$\begin{aligned} u &= -\hat{f} - K\tilde{x} + \dot{x}_c + \nu_{\mathcal{D}} \\ \dot{\theta}_f &= \Gamma \bar{x} \begin{bmatrix} 1 \\ x \end{bmatrix} \\ \bar{x} &= \begin{cases} \tilde{x} & \text{if } K|\tilde{x}| > \varepsilon \\ 0 & \text{otherwise} \end{cases} \end{aligned}$$

where x_c and $\dot{x}_c$ are the commanded state trajectory and its time derivative.

   (a) Show that the tracking error dynamics are

$$\dot{\tilde{x}} = -\tilde{\theta}^\top \phi - K\tilde{x} + e_f(x) + \nu_{\mathcal{D}}.$$

   (b) Show that, for $K|\tilde{x}| > \varepsilon$ and $x \in \mathcal{D}$, the time derivative of the Lyapunov function $V = \frac{1}{2}\left(\tilde{x} + \tilde{\theta}^\top \Gamma^{-1} \tilde{\theta}\right)$ satisfies

$$\dot{V} \leq -|\tilde{x}| \left(K|\tilde{x}| - |e_f|\right).$$

   Therefore, when $\varepsilon < K|\tilde{x}| < |e_f|$ it is possible for the positive function V to increase.

4. Simulate the system and generate plots of $V(t)$, $V_+(t) = \frac{1}{2}\left(\tilde{x} + \tilde{\theta}_+^\top \Gamma^{-1} \tilde{\theta}_+\right)$, $V_-(t) = \frac{1}{2}\left(\tilde{x} + \tilde{\theta}_-^\top \Gamma^{-1} \tilde{\theta}_-\right)$, and $(K|\tilde{x}| - |e_f|)$. Use the control gain $K = 4$, adaptation rate $\Gamma = 5I$, dead-zone radius $\varepsilon = 0.01$, and command filter parameters $\omega_n = 15$, $\zeta = 1.0$ (see eqn. (8.14)). Let $x_c^o = 0.8 sin(t) + r(t)$ where $r(t)$ is a square wave switching between ± 1.0 with a period of 50 s.

   From these plots, you should notice the following: (a) $V(t)$ is decreasing when $(K|\tilde{x}| - |e_f|)$ is positive and increasing otherwise; (b) $V_+(t)$ is decreasing when x is positive and $V_-(t)$ is decreasing when x is negative.

   The main conclusion of this exercise is that when the approximation structure is not sufficient to guarantee learning (i.e., $\bar{e}_f > \varepsilon$) then the approximator parameters will be adapted to temporarily meet this condition in the vicinity of the present operating point. Evidence that the approximator structure is not sufficient includes (i) a graph of θ versus t exhibiting clear convergence toward different parameter vectors for different regions of $\mathcal{D}$ and (ii) the tracking error not retaining improved performance in subregions of $\mathcal{D}$ for which training experience has already been obtained. If the approximation structure is sufficient to allow learning over $\mathcal{D}$, then the training error should eventually enter and remain within the dead-zone. Observation of the tracking error in this problem makes clear that the tracking error will not ultimately stay within the dead-zone.

5. Define an alternative approximator sufficient to allow learning. In doing this, the approximator structure will often be over-specified, since f^* is not known.

APPENDIX A

SYSTEMS AND STABILITY CONCEPTS

This appendix presents certain necessary concepts that are used in the main body of the book. This material is presented in the form of an appendix, as it may be familiar to many readers and will therefore not interrupt the main flow of the text. Proofs are not included. Proofs can be found in [119, 134, 169, 249], which are the main references for this appendix.

A.1 SYSTEMS CONCEPTS

Many dynamic systems (all those of interest herein) can be conveniently represented by a finite number of coupled first-order ordinary differential equations:

$$\dot{x}(t) = f_o(x(t), u(t), t) \tag{A.1}$$

$$y(t) = h_o(x(t), u(t), t) \tag{A.2}$$

where $x \in \Re^n, u \in \Re^m, y \in \Re^p, f_o : \Re^n \times \Re^m \times \Re^1 \mapsto \Re^n$, and $h_o : \Re^n \times \Re^m \times \Re^1 \mapsto \Re^p$. The parameter n is referred to as the *system order*. The vector x is referred to as the system *state*. The vector space $\Re^n$ over which the state vector is defined is the *state space*.

In the special case where $u(t)$ is a constant and f_o is not an explicit function of t, then eqn. (A.1) simplifies to

$$\dot{x}(t) = f_1(x(t)). \tag{A.3}$$

This equation, which is independent of time, is said to be *autonomous*.

Adaptive Approximation Based Control: Unifying Neural, Fuzzy and Traditional Adaptive Approximation Approaches. By Jay A. Farrell and Marios M. Polycarpou

Solutions. The analyst is often interested in qualitative properties of the solutions of the system of equations defined by eqn. (A.1). For a given signal $u(t)$, a solution to eqn. (A.1) over an interval $t \in [t_0, t_1]$ is a continuous function $x(t) : [t_0, t_1] \mapsto \Re^n$ such that $\dot{x}(t)$ is defined and $\dot{x}(t) = f_o(x(t), u(t), t)$ for all $t \in [t_0, t_1]$. The solution $x(t)$ traces a curve in $\Re^n$ as t varies from t_0 to t_1. This curve is the *state trajectory*.

Existence and Uniqueness of Solutions. The two questions of whether a differential equation has a solution and, if so, whether it is unique are fundamental to the study of differential equations. Discussion of the uniqueness of a solution requires introduction of the concept of the Lipschitz condition.

Definition A.1.1 *A function f satisfies a Lipschitz condition on $\mathcal{D}$ with Lipschitz constant γ if*

$$\|f(t, x) - f(t, y)\| \leq \gamma \|x - y\|$$

for all points (t, x) and (t, y) in $\mathcal{D}$.

Lipschitz continuity is a stronger condition that continuity. For example, $f(x) = x^p$, for $0 < p < 1$, is a continuous function on $\mathcal{D} = [0, \infty]$, but it is not Lipschitz continuous on $\mathcal{D}$ since its slope approaches infinity as x approaches zero.

The following theorem summarizes the conditions required for local existence and uniqueness of solutions.

Theorem A.1.1 *[134] If $f(t, x)$ is piecewise continuous in t and satisfies a Lipschitz condition on a compact set containing $x(t_0)$, then there exists some $\delta > 0$ such that the initial value problem*

$$\dot{x} = f(t, x), \text{ with } x(t_0) = x_0$$

has a unique solution on $[t_0, t_0 + \delta]$.

Consider, as an example, the initial value problem

$$\dot{x} = x^p, \text{ with } x(0) = 0 \text{ and } 0 < p < 1.$$

The previous discussion has already shown that $f(x) = x^p$ is not Lipschitz. Therefore, the previous theorem does not guarantee uniqueness of the solution to the initial value problem. In fact,

$$x(t) = 0 \text{ and } x(t) = ((1 - p)\, t)^{\frac{1}{1-p}}$$

are both valid solutions to the initial value problem.

Throughout the main body of this text, it will often be the case that solutions of the system equations can be shown to lie entirely in a compact set $\mathcal{D}$. In such cases, the following theorem is applicable.

Theorem A.1.2 *[134] Let $f(t, x)$ be piecewise continuous in t and locally Lipschitz in x for all $t > t_o$ and all $x \in \mathcal{A} \subset \Re^n$, where $\mathcal{A}$ is a domain containing the compact set $\mathcal{D}$. If for $x_o \in \mathcal{D}$ it is known that every solution of*

$$\dot{x} = f(t, x), \text{ with } x(t_0) = x_0$$

lies entirely within $\mathcal{D}$, then there is a unique solution defined for all $t \geq t_0$.

Equilibrium Point. Let $u(t) = 0$ for all $t \in \Re^+$. Any point $x_e \in \Re^n$ such that $f(x_e, 0, t) = 0$ for all $t \in \Re^+$ is an equilibrium point of eqn. (A.1). Conceptually, an equilibrium point is a point such that $x(t) = x_e$ solves the differential equation for $t \geq 0$. Other names for an equilibrium point include stationary point, singular point, critical point, and rest position. A differential equation can have zero, many, or an infinite number of equilibria. If x_e is an equilibrium point and there is an $r > 0$ such that $B(x_e, r)$ contains no other equilibria, then x_e is said to be an isolated equilibrium point. For example, the pendulum system described by $\dot{x} = -\sin(x)$ has an infinite number of isolated equilibria defined by $x = \pm k\pi$ where k is any integer.

Translation to the Origin. Many of the results to follow will state properties of the equilibrium solution $x(t) = 0$. The purpose of this paragraph is to show that there is no loss of generality in these statements. Let $x_o(t)$ denote the solution to eqn. (A.1) that is of interest. Let $x(t)$ be any other solution. Define $v(t) = x(t) - x_o(t)$. Then,

$$\dot{v} = f_o(x(t), u(t), t) - f_o(x_o(t), u(t), t) \tag{A.4}$$

$$\dot{v} = f(v(t), u(t), t) \tag{A.5}$$

where $f(v(t), u(t), t) = f_o(v(t) + x_o(t), u(t), t) - f_o(x_o(t), u(t), t)$. First, note that $f(v, u, t)$ has an equilibrium point at $v = 0$. Therefore, the following definitions will refer without loss of generality to properties of the solution $v(t) = 0$. Second, note that even if the original system was autonomous, the translated system of eqn. (A.6) may not be autonomous. Therefore, for generality, the subsequent definitions discuss properties of nonautonomous systems.

Operating Point. An operating point is a generalization of the idea of an equilibrium point. An operating point is any state space location at which the system can forced into equilibrium by choice of the control signal. By eqn. (A.1), the pair (x_o, u_o) is an operating point if $f(x_o, u_o, t) = 0$ for all $t \in \Re^+$. Typically, operating points are not isolated. Instead, there will exist a surface of operating points that can be selected by the value of the control signal. Note that operating points, like equilibrium points, may be either stable or unstable. For example, the system

$$\begin{aligned} \dot{x}_1 &= x_2 \\ \dot{x}_2 &= x_1^3 + u \end{aligned}$$

has an operating point at $x_o = [1, 0]^\top$ with $u_o = -1$. In fact, the surface of operating points is $x_o = [a, 0]^\top$ with $u_o = -a^3$. Every operating point on this surface is unstable. The system could be forced to operate at any point on this surface only if a stabilizing controller (e.g., $u = -x_1^3 - (x_1 - a) - x_2$) was defined.

A.2 STABILITY CONCEPTS

Based on the discussion of Section A.1, this section will discuss stability properties and analysis methods for the system

$$\dot{x} = f(x(t), t) \tag{A.6}$$

which (without loss of generality) is non-autonomous and assumed to have an equilibrium at the origin.

A.2.1 Stability Definitions

We are interested in analyzing the stability properties of the equilibrium point $x_e = 0$. By the previous phrase we mean that we want to know what happens to a solution $x(t)$ for

$t > t_o$ corresponding to the initial condition $x(t_o) = \zeta \neq 0$. This initial value problem may arise since in a physical application the system may not initially be at the origin or at some later time may become perturbed from the origin. The following definitions of stability (referred to as *stability in the sense of Lyapunov* or *internal stability*) have been shown to be useful for the rigorous classification of the stability properties of an equilibrium point.

Definition A.2.1 *The equilibrium $x_e = 0$ of eqn. (A.6) is*

stable *if for any $\epsilon > 0$ and any $t_o > 0$, there exists $\delta(\epsilon, t_o) > 0$ such that $\|x(t_o)\| < \delta(\epsilon, t_o) \Rightarrow \|x(t)\| < \epsilon$ for all $t \geq t_o$;*

uniformly stable *if for any $\epsilon > 0$ and any $t_o > 0$, there exists $\delta(\epsilon) > 0$ such that $\|x(t_o)\| < \delta(\epsilon) \Rightarrow \|x(t)\| < \epsilon$ for all $t \geq t_o$;*

unstable *if it is not stable;*

asymptotically stable *if it is stable and for any $t_o \geq 0$ there exist $\eta(t_o) > 0$ such that $\|x(t_o)\| < \eta(t_o) \Rightarrow \|x(t)\| \rightarrow 0$ as $t \rightarrow \infty$;*

uniformly asymptotically stable *if (1) it is uniformly stable, and (2) there exist $\delta > 0$ independent of t such that $\forall \epsilon > 0$ there exists $T(\epsilon) > 0$ such that $\|x(t_o)\| < \delta \Rightarrow \|x(t)\| < \epsilon$ for all $t > t_o + T(\epsilon)$;*

exponentially stable *if for any $\epsilon > 0$ there exist $\delta(\epsilon) > 0$ such that $\|x(t_o)\| < \delta \Rightarrow \|x(t)\| < \epsilon e^{-\alpha(t-t_o)}$, $\forall t > t_o \geq 0$ for some $\alpha > 0$.*

The definition of stability in the sense of Lyapunov includes the intuitive idea that solutions are bounded, but also requires that the bound on the solution can be made as small as desired by restriction of the size of the initial condition. The property of instability implies that there is some $\epsilon > 0$ such that, no matter how small the bound on the initial condition is required to be, there will exist some initial condition for which the corresponding solution always grows larger than ϵ. Note that instability does *not* require the solution to blow up (i.e., $\|x(t)\| \rightarrow \infty$). The main distinction between stability and uniform stability is that in the later case δ is independent of t_o. In either case, stability is a local property of the origin. Asymptotic stability requires solutions to converge to the origin. Exponential stability requires at least an exponential rate of convergence to the origin.

The set of initial conditions $\mathcal{D} = \{x_o \in \Re^n | x(t_o) = x_o \text{ and } \|x(t)\| \rightarrow 0 \text{ as } t \rightarrow \infty\}$ is the domain of attraction of the origin. If $\mathcal{D}$ is equal to $\Re^n$, then the origin is said to be *globally* asymptotically stable.

In some cases of importance in the main text, it will not be possible to prove stability of the origin due to certain perturbations. In such cases concepts related to boundedness are important.

Definition A.2.2 *The equilibrium $x_e = 0$ is*

uniformly ultimately bounded *if there exist positive constants R, $T(R)$, b such that $\|x(t_o)\| \leq R$ implies that $\|x(t)\| < b$ for all $t > t_o + T$;*

globally uniformly ultimately bounded *if $R = \infty$.*

The constant b is referred to as the *ultimate bound.* There are several important distinctions between the classification as *stable* or *uniformly ultimately bounded* (UUB). For stability, δ will be less that ϵ and $\|x(t)\| < \epsilon$ for all t. For UUB, R is normally larger that b and

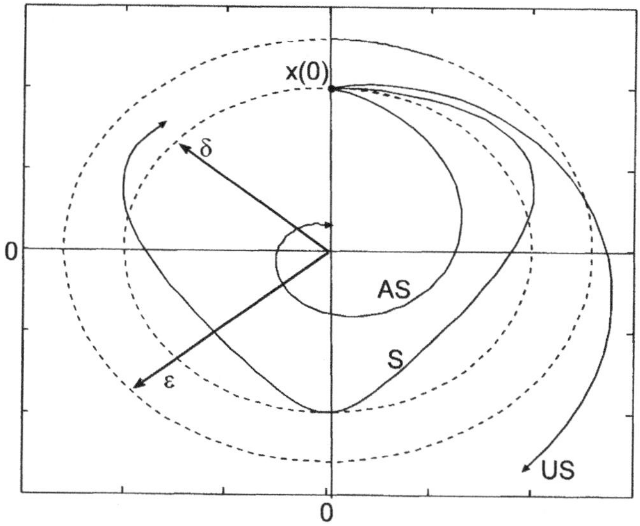

Figure A.1: Example trajectories for systems with different stability properties. Trajectory US is unstable. Trajectory S is stable. Trajectory AS is asymptotically stable. Also shown are the ϵ and δ contours of the stability definitions.

$\|x(t)\| < b$ only after an intervening time denoted by T. Also, for stability, the quantity ϵ can be made arbitrarily small. For UUB the quantity b is determined by physical aspects of the system. For example, b may be a function of the control parameters and a bound on the disturbances. The form of the functionality can be important as it provides the designer guidance on how control performance can be affected by choice of the design parameters. The UUB classification is uniform in the sense that the constants R, T, b do not depend on t_o.

A.2.2 Stability Analysis Tools

The previous section presented the technical definitions of various forms of stability. As written, the definitions are not easily applicable to the classification of systems. This section presents various results that have been found useful for classifying systems according to the definitions of the previous section.

A.2.2.1 Lyapunov Functions Figure A.1 shows trajectories for stable, unstable, and asymptotically stable systems. For the stable system, given any $\epsilon > 0$, it is possible to find a δ such that starting within the $\delta > 0$ contour ensures that the solution is always inside the ϵ contour. This is also true for the asymptotically stable system, with the added property that the trajectory of the AS system eventually converges to zero. For the unstable system, for the given ϵ, there is not $\delta > 0$ that yields trajectories within the ϵ contour for all $t > 0$.

Figure A.1 illustrates the stability definitions in a two-dimensional plane. The two-dimensional case is special since it is the highest order system that can be conveniently and completely illustrated graphically. Since most physical systems have state dimension greater than two, an analysis tool is required to allow the application of the stability definitions, the key ideas of which are illustrated in Figure A.1, in higher dimensions where graphical analysis of trajectories is not possible. Lyapunov's *direct* method provides these tools, without the need to explicitly solve for the solution of the differential equation.

The key idea of Lyapunov's direct approach is that the analyst define closed contours in $\Re^n$ that correspond to the level curves of a sign definite function. The analysis then focuses on the behavior of the system trajectories relative to these contours. The ideas of Lyapunov's direct method are rigorously summarized by the Theorem A.2.1. Before presenting that theorem, we introduce a few essential concepts. In the following definition, $B(r)$ denotes an open set containing the origin.

Definition A.2.3

1. *A continuous function $V(x)$ is positive definite on $B(r)$ if*
   (a) $V(0) = 0$, *and*
   (b) $V(x) > 0$, $\forall x \in B(r)$ *such that* $x \neq 0$.
2. *A continuous function $V(x)$ is positive semidefinite on $B(r)$ if*
   (a) $V(0) = 0$, *and*
   (b) $V(x) \geq 0$, $\forall x \in B(r)$ *such that* $x \neq 0$.
3. *A continuous function $V(x)$ is negative (semi-)definite on $B(r)$ if $-V(x)$ is positive (semi-)definite.*
4. *A continuous function $V(x)$ is radially unbounded if*
   (a) $V(0) = 0$,
   (b) $V > 0$ *on* $\Re^n - \{0\}$, *and*
   (c) $V(x) \to \infty$ *as* $\|x\| \to \infty$.
5. *A continuous function $V(t,x)$ is positive definite on $\Re \times B(r)$ if there exists a positive definite function $w(x)$ on $B(r)$ such that*
   (a) $V(t,0) = 0$, $\forall t \geq 0$, *and*
   (b) $V(t,x) \geq w(x)$, $\forall t \geq 0$ *and* $\forall x \in B(r)$.
6. *A continuous function $V(t,x)$ is radially unbounded if there exists a radially unbounded function $w(x)$*
   (a) $V(t,0) = 0$, $\forall t \geq 0$, *and*
   (b) $V(t,x) \geq w(x)$, $\forall t \geq 0$ *and* $\forall x \in \Re^n$.
7. *A continuous function $V(t,x)$ is decrescent on $\Re \times B(r)$ if there exists a positive definite function $w(x)$ on $B(r)$ such that*

$$V(t,x) \leq w(x), \ \forall t \geq 0 \ and \ \forall x \in B(r).$$

The concept of positive definiteness is important since positive definite functions characterize closed contours around the origin in $\Re^n$. If the time derivative of a positive definite function V along the system's trajectories can be shown to always be negative, then the trajectories are only crossing contours in the direction towards the origin. This fact can be used through the Lyapunov theorems to rigorously prove asymptotic stability. Similar theorems will show the conditions sufficient to prove the other forms of stability.

Before presenting the Lyapunov theorems, it is necessary to state that the rate of change of $V(t,x)$ along solutions of eqn. (A.6) is defined by

$$\left.\frac{dV}{dt}\right|_{(A.6)} = \frac{\partial V}{\partial t} + \nabla V(t,x)^\top \frac{dx}{dt} \tag{A.7}$$

$$= \frac{\partial V}{\partial t} + \nabla V(t,x)^\top f(x,t) \tag{A.8}$$

where $\nabla V(t,x)$ denotes the gradient of V with respect to x. The gradient of V is a vector pointing in the direction of maximum increase of V. The vector $f(x,t)$ is tangent to the solution $x(t)$. Therefore, if $\frac{\partial V}{\partial t} = 0$, the condition $\nabla V(t,x)^\top f(x,t) < 0$ implies that the solutions $x(t)$ always cross the contours of V with an angle greater than 90 degrees relative to the outward normal. Therefore, the direct method of Lyapunov replaces the n-dimensional analysis problem that is difficult to visualize with a lower dimensional problem that is easy to interpret. The difficulty of the Lyapunov approach is the specification of a suitable Lyapunov function V. In the following theorem, D is an open region containing the origin.

Theorem A.2.1 *Let $V(t,x) : \Re^+ \times D \mapsto \Re^1$ be a continuously differentiable and positive definite function.*

1. *If $\left.\frac{dV}{dt}\right|_{(A.6)} \leq 0$ for $x \in D$, then the equilibrium $x = 0$ is stable.*

2. *If $V(t,x)$ is decrescent and $\left.\frac{dV}{dt}\right|_{(A.6)} \leq 0$ for $x \in D$, then the equilibrium $x = 0$ is uniformly stable.*

3. *If $\left.\frac{dV}{dt}\right|_{(A.6)}$ is negative definite for $x \in D$, then the equilibrium $x = 0$ is asymptotically stable.*

4. *If $V(t,x)$ is decrescent and $\left.\frac{dV}{dt}\right|_{(A.6)}$ is negative definite for $x \in D$, then the equilibrium $x = 0$ is uniformly asymptotically stable.*

5. *If there exist three positive constants c_1, c_2, and c_3 such that $c_1\|x\|^2 \leq V(t,x) \leq c_2\|x\|^2$ and $\left.\frac{dV}{dt}\right|_{(A.6)} \leq -c_3\|x\|^2$ for all $t \geq 0$ and for all $x \in D$, then the equilibrium $x = 0$ is exponentially stable.*

A key advantage of this theorem is that it can be applied without finding the solutions of the differential equation. A key disadvantage is that there is no systematic method for generating the Lyapunov function V. In addition, if a particular choice of Lyapunov function does not yield the desired definiteness properties for its time derivative, then *no conclusion* can be made about the stability properties of the system by use of that Lyapunov function; instead, another Lyapunov function candidate must be evaluated.

■ EXAMPLE A.1

Consider the linear system

$$\dot{x} = A\,x. \tag{A.9}$$

Let $V(x) = x^T P\, x$, where P is a symmetric and positive definite matrix.[1] Then,

$$\begin{aligned} \left.\frac{dV}{dt}\right|_{(A.9)} &= x^\top A^\top P\, x + x^\top P\, A\, x \\ &= x^\top \left(A^\top P + P\, A\right) x \\ &= -x^\top Q\, x \end{aligned}$$

where

$$Q = -\left(A^\top P + P\, A\right) \tag{A.10}$$

is a symmetric matrix. If Q is positive definite, then the linear system is globally[2] exponentially stable.

If Q is not positive definite, then nothing can be said about the stability properties of the system. The fact that Q is not positive definite may be the result of a poor choice of P for the problem of interest. Therefore, the method of selecting P and calculating $Q = -\left(A^\top P + P\, A\right)$ is not the preferred approach.

The equation $Q = -\left(A^\top P + P\, A\right)$ is referred to as the *continuous-time Lyapunov equation.* Note that if Q is specified and A is known, then the Lyapunov equation is linear in P. Therefore the preferred approach is (1) to specify a positive definite Q, and (2) to solve the Lyapunov equation for P. If the resulting P is positive definite, then the linear system is exponentially stable. If the resulting P has any negative eigenvalue, then the linear system is unstable. △

EXAMPLE A.2

Consider the example of the pendulum described by

$$\left.\begin{aligned} \dot{x}_1 &= x_2 \\ \dot{x}_2 &= -\sin(x_1) - x_2. \end{aligned}\right\} \tag{A.11}$$

The total energy for this system is

$$E(x_1, x_2) = \int_0^{x_1} \sin(v)dv + \frac{1}{2}x_2^2.$$

For solutions of the eqn. (A.11),

$$\frac{dE}{dt} = (\nabla E)^\top \cdot \begin{bmatrix} \dot{x}_1 \\ \dot{x}_2 \end{bmatrix} \tag{A.12}$$

$$= [\sin(x_1), x_2] \begin{bmatrix} x_2 \\ -\sin(x_1) - x_2 \end{bmatrix} \tag{A.13}$$

$$= -x_2^2 \leq 0, \ \forall x_1, \ x_2. \tag{A.14}$$

Therefore, the function E is positive definite for $x_1 \in (-\pi, \pi), \forall x_2 \in \Re^1$, with a negative semidefinite derivative. The conclusion by Theorem A.2.1 is that the system is uniformly stable.

[1] A symmetric matrix has all real eigenvalues. If all these eigenvalues are positive, then the matrix is positive definite.

[2] The region $B(r)$ in Theorem A.2.1 has $r = \infty$.

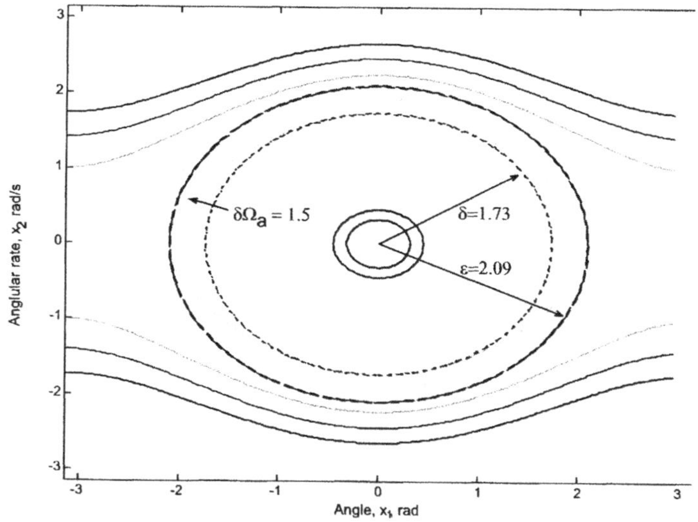

Figure A.2: Energy contours discussed in Example A.2 with the $\|x\| = \varepsilon$ and δ contours of the Lyapunov stability definitions shown.

Although Theorem A.2.1 will not be proved herein, the flavor of the proof is illustrated in this paragraph and Figure A.2. To relate the function E back to the definition of stability, consider a specific value of ε. Figure A.2 shows the contour $\|x\| = \varepsilon$ for $\varepsilon = 2.09$. First, find $a = inf_{\|x\|=\varepsilon} E(x)$. Let $\Omega_a = \{x \in \Re^2 | E(x) < a\}$ and let $\delta\Omega_a = \{x \in \Re^2 | E(x) = a\}$. Figure A.2 shows the boundary of Ω_a for $a = 1.5$. Note that by the properties of E and the definition of Ω_a, $\Omega_a \subset B(0, \varepsilon)$. Second, find $\delta = inf_{x \in \delta\Omega_a} \|x\|$. The contour $\|x\| = \delta$ for $\delta = 1.73$ is shown in Figure A.2. Note that $B(0, \delta) \subset \Omega_a$. By the definitions of δ and Ω_a of this paragraph, if $\|x_0\| < \delta$, then $E(x_0) < a$. Since $\frac{dE}{dt} \leq 0$ along solutions of the system, $E(x(t)) < a$, $\forall t > 0$ (i.e., $x(t) \in \Omega_a \forall t > 0$). Since $\Omega_a \subset B(0, \varepsilon)$, $\|x(t)\| < \varepsilon, \forall t > 0$. Therefore, the $\varepsilon - \delta$ definition of uniform stability is satisfied.
$\triangle$

■ EXAMPLE A.3

Consider the system

$$\dot{x}(t) = -a\, x(t) + b\, u(t) \tag{A.15}$$

where $a > 0$ is known and the unknown constant parameter b is to be estimated. Define the parameter estimation system to be

$$\dot{v}(t) = -a\, v(t) + c(t)\, u(t) \tag{A.16}$$

$$\dot{c}(t) = g(u, v, x) \tag{A.17}$$

where $c(t)$ is the estimate of b. The remaining step in the estimator design requires specification of the function $g(u, v, x)$ so that $c(t) \rightarrow b$.

Define the error variables $e(t) = x(t) - v(t)$ and $\theta(t) = c(t) - b$. Then,

$$\dot{\theta} = \dot{c} = g(u, v, x)$$

and

$$\begin{aligned} \dot{e} &= \dot{x}(t) - \dot{v}(t) \\ &= (-a\, x(t) + b\, u(t)) - (-a\, v(t) + c(t)\, u(t)) \\ \dot{e} &= -a\, e(t) - \theta(t)\, u(t). \end{aligned}$$

To analyze this system, let $V(e, \theta) = \frac{1}{2}(e^2 + \theta^2)$. The time derivative of V along solutions of eqn. (A.22) is

$$\left.\frac{dV}{dt}\right|_{(A.22)} = e\dot{e} + \theta\dot{\theta} \tag{A.18}$$

$$= -a\, e^2 + \theta(-e\, u + \dot{\theta}) \tag{A.19}$$

$$= -a\, e^2 + \theta\, (-e\, u + g\, (u, v, x))\, . \tag{A.20}$$

If the designer selects $g(u, v, x) = e\, u$, then

$$\left.\frac{dV}{dt}\right|_{(A.22)} = -a\, e^2 \tag{A.21}$$

which is negative semidefinite. Therefore, we know that the origin of the (e, θ) system is uniformly stable.

Due to the choice of $g(u, v, x) = e\, u$, the dynamics of the error variables are defined by a linear time varying (LTV) system:

$$\begin{bmatrix} \dot{e} \\ \dot{\theta} \end{bmatrix} = \begin{bmatrix} -a & u(t) \\ -u(t) & 0 \end{bmatrix} \begin{bmatrix} e \\ \theta \end{bmatrix}. \tag{A.22}$$

Several interesting observations related to this simple example have direct relevance to the main topic of the text.

1. Even though the original system of eqn. (A.15) is linear time invariant (LTI), the corresponding parameter estimation system of eqn. (A.22) is LTV.

2. If the parameter a were also unknown, then that parameter estimation problem would involve a nonlinear system of equations.

3. The time rate of change of c depends on the signal $u(t)$. In particular, if $u(t) = 0, \forall t > t_o$, then b cannot be estimated.

4. The above analysis shows that the solutions of eqn. (A.22) never increase V. However, either one of e or θ can increase, as long as the other decreases at least as fast.

△

Note that although Examples A.2 and A.3 have only demonstrated uniform stability, stronger forms of stability may be provable either by an alternate choice of Lyapunov function or by more advanced forms of analysis.

A.2.2.2 Invariance Theory Analysis of dynamic systems often results in situations where the derivative of the Lyapunov function is only negative semidefinite. For autonomous systems, it is sometimes possible to conclude asymptotic stability, even when the time derivative of the Lyapunov function is only negative semidefinite. This extension of Lyapunov Theory is referred to as LaSalle's Theory and relies on the concept of invariant sets.

Definition A.2.4 *A set Γ is a positively invariant set of a dynamic system if every trajectory starting in Γ at $t = 0$ remains in Γ for all $t > 0$.*

Regarding the invariant sets of a dynamic system, consider the following observations:

- Any equilibrium of a system is an invariant set.
- The set of all equilibria of a system is an invariant set.
- Any solution of an initial value problem related to the dynamic system is an invariant set.
- The domain of attraction of an equilibrium is an invariant set.
- A system can have many invariant sets.
- An invariant set need not be connected.
- The union of invariant sets yields an invariant set.

Using the concept of invariant sets, the local and global invariant set theorems can be stated.

Theorem A.2.2 (Local Invariant Set Theorem) *For an autonomous system $\dot{x} = f(x)$, with f continuous on domain $\mathcal{D}$, let $V(x) : \mathcal{D} \mapsto \Re^1$ be a function with continuous first partial derivatives on $\mathcal{D}$. If*

1. *the compact set $\Omega \subset \mathcal{D}$ is a positively invariant set of the system, and*
2. $\dot{V} \leq 0 \;\; \forall x \in \Omega$,

then every solution $x(t)$ originating in Ω converges to M as $t \to \infty$, where $R = \{x \in \Omega \mid \dot{V}(x) = 0\}$ and M is the union of all invariant sets in R.

Theorem A.2.3 (Global Invariant Set Theorem) *For an autonomous system, with f continuous, let $V(x)$ be a function with continuous first partial derivatives. If*

1. $V(x) \to \infty$ *as* $\|x\| \to \infty$, *and*
2. $\dot{V} \leq 0 \;\; \forall x \in \Re^n$,

then all solutions $x(t)$ converge to M as $t \to \infty$, where $R = \{x \in \Re^n \mid \dot{V}(x) = 0\}$ and M is the union of all invariant sets in R.

Note that neither theorem requires V to be positive definite. Also, in the local theorem, when the set M contains a single equilibrium point, the set Ω provides an estimate of the domain of attraction of the equilibrium point.

EXAMPLE A.4

Consider the system described by

$$\left.\begin{aligned}\dot{x}_1 &= x_2 \\ \dot{x}_2 &= -f(x_1)x_2 - g(x_1)\end{aligned}\right\} \tag{A.23}$$

where f and g are differentiable on $\Re^1$, $g(0) = 0$, $x_1 g(x_1) > 0\ \forall x_1 \neq 0$, and $f(x_1) > 0\ \forall x_1 \in \Re^1$. This system is a state space representation of the Lienard equation. The only equilibrium point of this system is the origin.

Consider the function

$$V(x) = \int_0^{x_1} g(v)dv + \frac{1}{2}x_2^2.$$

The time derivative of V along solutions of eqn. (A.23) is

$$\begin{aligned}\dot{V} &= g(x_1)\dot{x}_1 + x_2\dot{x}_2 \\ &= g(x_1)x_2 - f(x_1)x_2^2 - g(x_1)x_2 \\ \dot{V} &= -f(x_1)x_2^2 \leq 0 \quad \forall x \in \Re^2.\end{aligned}$$

Therefore, $R = \{(x_1, x_2) \in \Re^2 \mid x_2 = 0\}$. The only invariant set in R is $\{(0,0)\}$; therefore, M is the set containing the origin.

Since this V happens to be positive definite, there does exist $l > 0$ such that $\Omega_l = \{x \in \Re^2 \mid V(x) \leq l\}$ is bounded. Therefore, the local invariant set theorem shows that the origin of the system is locally asymptotically stable.

If g has the property that $\int_0^{x_1} g(v)dv \rightarrow \infty$ as $x_1 \rightarrow \infty$, then $V(x) \rightarrow \infty$ as $\|x\| \rightarrow \infty$. In this case, the global invariant set theorem shows that the origin is globally asymptotically stable. △

A.2.2.3 Barbălat's Lemma LaSalle's Theorem is applicable to the analysis of autonomous systems. For nonautonomous systems, it may be unclear how to define the sets R and M. Following are various useful forms of *Barbălat's Lemma* that are useful for nonautonomous systems.

Lemma A.2.4 *Let* $\phi(t) : \Re^+ \mapsto \Re^1$ *be in* $\mathcal{L}_\infty$, $\frac{d\phi}{dt} \in \mathcal{L}_\infty$ *and* $\frac{d\phi}{dt} \in \mathcal{L}_2$, *then* $\lim_{t\rightarrow\infty} \phi(t) = 0$.

Lemma A.2.5 *Let* $\phi(t) : \Re^+ \mapsto \Re^1$ *be uniformly continuous on* $[0, \infty]$. *If*

$$\lim_{t\rightarrow\infty} \int_0^t \phi(s)ds = M < \infty,$$

then $\lim_{t\rightarrow\infty} \phi(t) = 0$.

Note that the uniform continuity of ϕ needed for these Lemmas can be proven by showing either that $\dot{\phi} \in \mathcal{L}_\infty([0,\infty))$ or that $\phi(t)$ is Lipschitz on $[0,\infty)$. The importance of Barbălat's Lemma is highlighted by the following two examples [119, 249]. The application of Barbălat's Lemma is demonstrated in the third example.

EXAMPLE A.5

Consider the function $f(t) = \sin(\log(t))$, which does not have a limit as $t \to \infty$. The derivative of f is

$$\frac{df}{dt} = \frac{\cos(\log(t))}{t},$$

which approaches zero as $t \to \infty$. This function f demonstrates the main conclusion of this example which is

$$\dot{\mathbf{f}}(\mathbf{t}) \to \mathbf{0} \;\; \textbf{does not imply that } \mathbf{f}(\mathbf{t}) \;\; \textbf{converges to a constant.}$$

The fact that $\lim_{t\to\infty} \dot{f} = 0$ only implies that as t increases the rate of change of f becomes increasingly slow. Similar examples exist for which $f(t)$ is unbounded. △

EXAMPLE A.6

Consider the function

$$f(t) = \frac{\sin((1+t)^n)}{1+t}$$

for $n \geq 2$, which converges to zero as $t \to \infty$. The derivative of f is

$$\frac{df}{dt} = -\frac{\sin((1+t)^n)}{(1+t)^2} + n(1+t)^{n-2}\cos(1+t)^n,$$

which has no limit as $t \to \infty$. In fact, for $n > 2$ the function $\dot{f}$ is unbounded. This function f demonstrates the main conclusion of this example which is

$$\mathbf{f}(\mathbf{t}) \to \mathbf{c} \;\; \textbf{does not imply that } \dot{\mathbf{f}}(\mathbf{t}) \;\; \textbf{converges to zero.}$$

△

Before proceeding to the last example of this section, the following lemma is introduced. The lemma is used in the example and in the main body of the text.

Lemma A.2.6 *If $f(t) : \Re^1 \mapsto \Re^1$ is bounded from below and $\dot{f} \leq 0$, then $\lim_{t\to\infty} f(t) = f_\infty$ exists.*

EXAMPLE A.7

Example A.3 (beginning on page 385) analyzed the system described by

$$\begin{bmatrix} \dot{e} \\ \dot{\theta} \end{bmatrix} = \begin{bmatrix} -a & u(t) \\ -u(t) & 0 \end{bmatrix} \begin{bmatrix} e \\ \theta \end{bmatrix}$$

using the Lyapunov function $V(x) = \frac{1}{2}(e^2 + \theta^2)$. The analysis of that example showed that

$$\left.\frac{dV}{dt}\right|_{(A.22)} = -a\,e^2. \qquad \text{(A.24)}$$

Based on the basic Lyapunov theorems, the origin of the (e, θ) system was shown to be uniformly stable.

Consider the function $\phi(t) = \dot{V}(t)$. The derivative of $\phi(t)$ is

$$\dot{\phi} = 2a\left(a\, e^2 - e\, u(t)\, \theta\right).$$

Since $V(x) = \frac{1}{2}(e^2 + \theta^2)$ and $\dot{V} = -a\, e^2 \leq 0$, we have that

$$\frac{1}{2}e^2(t) \leq V(t) \leq V(0), \text{ and } \frac{1}{2}\theta^2(t) \leq V(t) \leq V(0),$$

which shows that e and θ are in $\mathcal{L}_\infty([0,\infty))$. Therefore, if $u(t) \in \mathcal{L}_\infty([0,\infty))$, then $\dot{\phi}(t) \in \mathcal{L}_\infty([0,\infty))$. This shows that $\phi(t)$ is uniformly continuous.

By Lemma A.2.6, $lim_{t\to\infty} V(t)$ exists. Therefore,

$$\lim_{t\to\infty} \int_0^t \phi(s)ds = \lim_{t\to\infty} V(t) - V(0)$$

exists and is finite. Then, by Barbălat's Lemma A.2.5,

$$\phi(t) = \dot{V}(t) \to 0 \text{ as } t \to \infty.$$

Therefore, for $u(t) \in \mathcal{L}_\infty([0,\infty))$ we have that $e(t) \to 0$ as $t \to \infty$.

Note that this example has still only proven that $\theta \in \mathcal{L}_\infty([0,\infty))$, not convergence of θ to zero. △

A.2.2.4 *Stable in the Mean Squared Sense*

In many adaptive applications, asymptotic stability of certain error variables can only be proven in idealized settings. In realistic situations involving disturbance signals, robust parameter estimation approaches are required and stability can only be proven in an input-output sense. The concept of mean square stability (MSS) will be frequently referred to in the main body of the text.

Definition A.2.5 *The signal $x : [0,\infty) \mapsto \Re^n$ is μ-small in the mean squared sense if and only if $x \in \mathcal{S}(\mu)$ where*

$$\mathcal{S}(\mu) = \left\{ x \in \mathcal{L}_{2e} \,\middle|\, \int_t^{t+T} x(\tau)^\top x(\tau)d\tau \leq c_0\, \mu\, T + c_1, \forall t, T \geq 0 \right\}$$

where c_0 and c_1 are finite, positive constants with c_0 independent of μ.

For example, let the dynamics of e be

$$\dot{e} = -ke - \tilde{\theta}^\top \phi(t) + \epsilon(t)$$

for $k > 0$, $\epsilon(t) < \bar{\epsilon}$ and $\phi(\cdot) : [0,\infty) \mapsto \Re^N$ with

$$\dot{\tilde{\theta}} = \begin{cases} 0 & \text{for} \quad |e| \leq \bar{\epsilon} \\ \phi e & \text{for} \quad |e| > \bar{\epsilon}. \end{cases}$$

Choosing the Lyapunov function

$$V(e, \tilde{\theta}) = \frac{1}{2}\left(e^2 + \tilde{\theta}^\top \tilde{\theta}\right).$$

The time derivative of V along solutions of the above system while $|e| > \bar{\epsilon}$ is

$$\dot{V} = -ke^2 + e\epsilon.$$

To show MSS, we choose $\gamma \in (0, k)$ and complete the square on the right hand side:

$$\begin{aligned}
\dot{V} &= -(k-\gamma)e^2 - \left(\sqrt{\gamma}e - \frac{\epsilon}{2\sqrt{\gamma}}\right)^2 + \frac{\epsilon^2}{4\gamma} \\
&\leq -(k-\gamma)e^2 + \frac{\epsilon^2}{4\gamma} \\
(k-\gamma)e^2 &\leq -\dot{V} + \frac{\epsilon^2}{4\gamma} \\
\int_t^{t+T} e(\tau)d\tau &\leq \frac{1}{(k-\gamma)}V(t) + \frac{\epsilon^2}{4\gamma(k-\gamma)}T \\
\int_t^{t+T} e(\tau)d\tau &\leq \frac{2}{k}V(0) + \frac{\bar{\epsilon}^2}{k^2}T
\end{aligned}$$

where we have assumed that $|e(0)| > \bar{\epsilon}$ (without loss of generality). From this we can conclude that $e \in \mathcal{S}\left(\frac{\bar{\epsilon}^2}{k^2}\right)$.

A.2.3 Strictly Positive Real Transfer Functions

The concepts of Positive Real (PR) and Strictly Positive Real (SPR) , which are useful in some forms of stability analysis, are derived from network theory, where a rational transfer function is the driving point impedance of a network that does not generate energy if and only if it is PR. A network that does not generate energy is known as a passive or dissipative network, and it consists mainly of resistors, inductors and capacitors.

Specifically, a rational transfer function $W(s)$ of the complex variable $s = \sigma + j\omega$ is PR if $W(s)$ is real for real s, and $\text{Re}[W(s)] \geq 0$ for all $\text{Re}[s] \geq 0$. A transfer function $W(s)$ is SPR if for some $\epsilon > 0$, $W(s-\epsilon)$ is PR.

The following result of Ioannou and Tao [120] provide frequency domain conditions for SPR transfer functions.

Lemma A.2.7 *A strictly proper transfer function $W(s)$ is SPR if and only if*

1. *$W(s)$ is stable;*
2. *$Re[W(j\omega)] > 0$, for all $\omega \in (-\infty, \infty)$; and*
3. *$\lim_{|\omega|\to\infty} \omega^2 Re[W(j\omega)] > 0$.*

It is clear that the class of SPR transfer functions is a special class of stable transfer functions, which also satisfy a minimum phase condition. The following example illustrates the class of SPR transfer functions for a second order system.

EXAMPLE A.8

Consider the transfer function

$$W(s) = \frac{s + k_1}{(s+k_2)(s+k_3)}.$$

Using Lemma A.2.7, $W(s)$ is SPR if and only if following conditions hold:

- $k_1 > 0, \quad k_2 > 0, \quad k_3 > 0$
- $k_1 < k_2 + k_3$.

The details of the proof are left as an exercise (see Exercise A.3). △

An important result concerning SPR transfer functions is the Kalman-Yakubovich-Popov (KYP) Lemma. This lemma provides a useful property, which is employed extensively in parameter estimation texts [119, 179, 235, 268].

Lemma A.2.8 (Kalman-Yakubovich-Popov Lemma) [6] *Given a strictly proper, stable, rational transfer function $W(s)$. Assume that*

$$W(s) = C(sI - A)^{-1}B$$

where $(A,\ B,\ C)$ is a minimal state-space realization of $W(s)$ with (A, B) controllable and (A, C) observable. Then, $W(s)$ is SPR if and only if there exist symmetric positive definite matrices P, Q such that

$$\begin{aligned} A^{\top}P + PA &= -Q \\ B^{\top}P &= C. \end{aligned}$$

The KYP Lemma is particularly useful in adaptive systems where the dynamics of an error vector z are defined as

$$\dot{z} = Az + B\tilde{\theta}\phi(t)$$

where $\tilde{\theta}$ is an unknown vector to be estimated and $\phi(t)$ is known. See, for example, Section 7.2.2.1. Estimation of $\tilde{\theta}$ involves a training error $e = Cz$. The KYP Lemma provides a direct method to define a vector C such that the transfer function from $\tilde{\theta}\phi(t)$ to e is SPR.

The above definitions and lemmas for PR and SPR transfer functions are applicable to scalar transfer functions. The extension to matrix transfer functions is omitted. The interested reader is referred to [119] for SPR conditions for matrix transfer functions.

A.3 GENERAL RESULTS

This section presents and proves a set of theorems referenced from the main body of the text. The theorems of this section are generalizations of the basic results presented previously in this appendix.

Lemma A.3.1 *Given the system*

$$\begin{aligned} \dot{x}_1 &= f_1(x_1, x_2) \\ \dot{x}_2 &= f_2(x_1, x_2) \end{aligned}$$

with an equilibrium at $x_1 = 0 \in \Re^{n_1}$ and $x_2 = 0 \in \Re^{n_2}$, where f_1 and f_2 are Lipschitz functions of (x_1, x_2). If there exists a continuously differentiable function $\mathcal{V}(x_1, x_2)$ such that

$$\alpha_1\|x_1\|_2^2 + \alpha_2\|x_1\|_2^2 \ \leq V(x_1, x_2) \leq \ \beta_1\|x_1\|_2^2 + \beta_1\|x_1\|_2^2$$

where α_1, α_2, β_1, β_2 *are positive constants and if*

$$\frac{\partial \mathcal{V}}{\partial x_1} f_1 + \frac{\partial \mathcal{V}}{\partial x_2} f_2 \le -\gamma_1 \|x_1\|_2^2 \tag{A.25}$$

with $\gamma_1 > 0$, *then*

1. *the system is uniformly stable (i.e.,* $x_1, x_2 \in \mathcal{L}_\infty$*);*
2. $x_1 \in \mathcal{L}_2$*; and,*
3. *If* $\dot{x}_1 \in \mathcal{L}_\infty$ *(i.e.,* $f_1(x_1, x_2)$ *is bounded), then* $x_1 \to 0$ *as* $t \to \infty$.

Proof: The fact that the system is uniformly stable is immediate from Theorem A.2.1. By Lemma A.2.6, $V_\infty = lim_{t\to\infty} V(t)$ exists and is finite. From eqn. (A.25), we have that

$$\begin{aligned}
\frac{d\mathcal{V}}{dt} &\le -\gamma_1 \|x_1\|_2^2 \\
-\int_0^\infty \frac{d\mathcal{V}(\tau)}{d\tau} d\tau &\ge \gamma_1 \int_0^\infty \|x_1(\tau)\|_2^2 d\tau \\
\int_0^\infty \|x_1(\tau)\|_2^2 d\tau &\le \frac{1}{\gamma_1} \left(\mathcal{V}(0) - V_\infty\right),
\end{aligned}$$

which shows that $x_1 \in \mathcal{L}_2$. Finally, using Barbălat's Lemma A.2.5 with $\phi = \frac{d\mathcal{V}}{dt}$ and using the fact that f_1 is bounded, we have that $x_1 \to 0$. ∎

Lemma A.3.1 is a special case of results by LaSalle and Yoshizawa. This lemma is useful in the proofs related to stability of adaptive approximation systems. In such proofs, x_1 will denote the tracking error of the closed-loop system and x_2 will denote the estimated parameters of the approximator.

Lemma A.3.2 *Suppose* $v(t) \ge 0$ *satisfies the inequality*

$$\dot{v}(t) \le -cv(t) + \lambda$$

where $c > 0$ *and* $\lambda > 0$ *are constants. Then* $v(t)$ *satisfies*

$$v(t) \le \left(v(0) - \frac{\lambda}{c}\right) e^{-ct} + \frac{\lambda}{c}.$$

Proof: Since $\dot{v}(t) \le -cv(t) + \lambda$, there exists a function $k(t) \ge 0$ such that $\dot{v}(t) = -cv(t) + \lambda - k(t)$. Therefore $v(t)$ satisfies

$$\begin{aligned}
v(t) &= e^{-ct} v(0) + \int_0^t e^{-c(t-\tau)} \left(\lambda - k(\tau)\right) d\tau \\
&= e^{-ct} v(0) + \lambda \int_0^t e^{-c(t-\tau)} d\tau - \underbrace{\int_0^t e^{-c(t-\tau)} k(\tau) d\tau}_{\le 0} \\
&\le \left(v(0) - \frac{\lambda}{c}\right) e^{-ct} + \frac{\lambda}{c}.
\end{aligned}$$

This concludes the proof. ∎

According to the above lemma, if $\dot{v}(t) \le -cv(t) + \lambda$ then given any $\mu > \frac{\lambda}{c}$ there exists a time T_μ such that for all $t \ge T_\mu$ we have $v(t) \le \mu$. Figure A.3 illustrates a possible plot for $v(t)$ versus t.

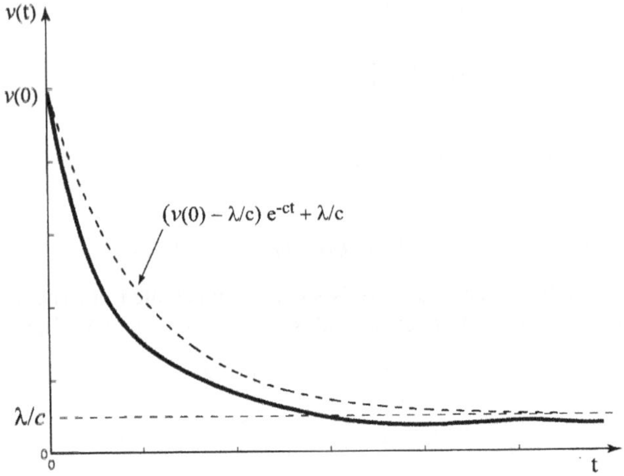

Figure A.3: Plot of a possible $v(t)$ versus time t.

A.4 TRAJECTORY GENERATION FILTERS

Advanced control approaches often assume the availability of a continuous and bounded desired trajectory $y_d(t)$ and its first r derivatives $y_d^{(r)}(t)$. The first time that this assumption is encountered it may seem unreasonable, since a user will often only specify a command signal $y_c(t)$. However, this assumption can always be satisfied by passing the commanded signal $y_c(t)$ through a single-input, multi-output prefilter of the form

$$\dot{z}(t) = \begin{bmatrix} 0 & 1 & 0 & \dots & 0 \\ 0 & 0 & 1 & \dots & 0 \\ \vdots & \vdots & \vdots & \ddots & \vdots \\ 0 & 0 & 0 & \dots & 1 \\ -a_0 & -a_1 & -a_2 & \dots & -a_{n-1} \end{bmatrix} z(t) + \begin{bmatrix} 0 \\ 0 \\ \vdots \\ 0 \\ a_0 \end{bmatrix} y_c(t) \quad \text{(A.26)}$$

$$\begin{bmatrix} y_d \\ y_d^{(1)} \\ \vdots \\ y_d^{(r)} \end{bmatrix} = \begin{bmatrix} I_{(r+1)\times(r+1)} & 0_{(r+1)\times(n-r-1)} \end{bmatrix} z(t) \quad \text{(A.27)}$$

where $z \in \Re^n$, $r < n$, and

$$s^n + \sum_{i=0}^{n-1} a_i s^i = 0$$

is a stable (Hurwitz) polynomial. If $y_c(t)$ is bounded, then this prefilter will provide as its output vector the bounded and continuous signals $y_d^{(i)}(t)$, $i = 0, \dots, r$. Each $y_d^{(i)}(t)$, $i =$

$0, \ldots, r$ is continuous and bounded as it is a state of a stable linear filter with a bounded input. Note that $y_d(t)$ and its first r derivatives are produced without differentiation[3].

The transfer function from y_c to y_d is

$$\frac{Y_d(s)}{Y_c(s)} = H(s) = \frac{a_0}{s^n + a_{n-1}s^{n-1} + \ldots + a_1 s + a_0}$$

which has unity gain at low frequencies. Therefore, the error $|y_d(t) - y_c(t)|$ is small if the bandwidth of $Y_c(s)$ is less than the bandwidth of $H(s)$. If the bandwidth of y_c is specified and the only goal of the filter is to generate y_d and its necessary derivatives with $|y_d - y_c|$ small, then the designer simply chooses the $H(s)$ as a stable filter with sufficiently high bandwidth. However, there are typically additional constraints.

■ EXAMPLE A.9

In the case that $n = 2$ and $r = 1$, a block diagram of a suitable command prefilter is shown in Figure A.4 where $a_0 = \omega_n^2$ and $a_1 = 2\zeta\omega_n$ with ζ being the damping ratio and ω_n being the undamped natural frequency. If the only design objectives are

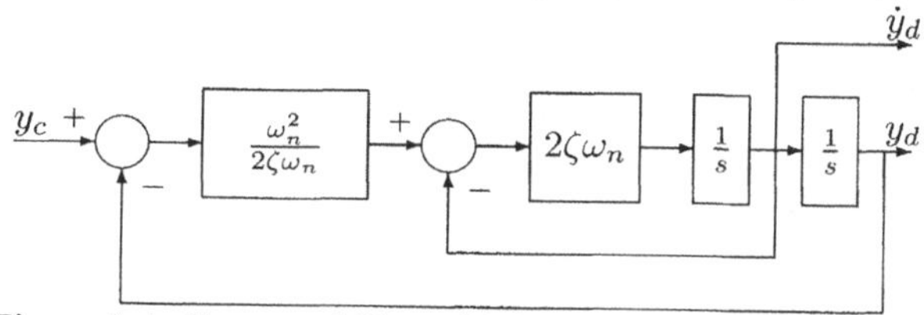

Figure A.4: Command filter with input y_c and outputs y_d and $\dot{y}_d$.

derivative generation and $|y_d - y_c|$ small when the maximum bandwidth of $Y_c(s)$ is specified to be 5 Hz, then any positive value of ζ and $\omega_n > 30\frac{rad}{s}$ should suffice. △

For many advanced control methods the objective is to design the feedback control law so that the plant state $x(t) \in \Re^r$ will track the reference trajectory $z_r(t) = [y_d(t), \ldots, y_d^{(r)}]$ perfectly. Perfect tracking has two conditions. First, if $x(0) = z_r(0)$, then $x(t) = z_r(t)$ for any $t \geq 0$. Second, if $x(0) \neq z_r(0)$, then $e(t) = x(t) - z_r(t)$ should converge to zero exponentially (see p. 217 in [249]). If x tracks z_r perfectly, then the transfer function $H(s)$ of the prefilter defined in eqn. (A.26) largely determines the bandwidth required for the plant actuators and does determine the transient response to changes in y_c. In some applications, this transient response is critically important. For example, in aircraft control it is referred to as *handling qualities* and has its own literature. Therefore the choice of the parameters $[a_0, a_1, a_2, \ldots, a_{n-1}]$, and the pole locations of $H(s)$ that they determine, should be carefully considered.

[3]Note that the approach described herein is essentially the same as that described in eqns. (7.31) and (7.42). For example, in eqn. (7.31):

$$\dot{x}_d = Ax_d + Br,$$

with A and B is defined on page 295. If r is selected as

$$r = a_0 y_c - \sum_{i=1}^{n} a_{i-1} x_{d_i}$$

then both approaches yield identical results.

EXAMPLE A.10

In the case that $n = 2$ and $r = 1$ that was considered in Example A.9, if the control specification is for y_c defined as a step function to be tracked with less than 5% overshoot with rise time $T_r \in [0.1, 0.2]$ s and settling time to within 1% of its final value in $T_s < 0.5$ s, then appropriate pole locations are $p = -10 \pm j5$, which are achieved for $a_0 = 125$ and $a_1 = 20$. The selection of pole locations to achieve time domain specifications is discussed in for example Section 3.4 in [86].

The trajectory output by the prefilter will achieve the time domain tracking specification. The prefilter is outside of the feedback control loop. If the feedback controller achieves perfect tracking, then the state of the plant will also achieve the tracking specification. △

Finally, in adaptive approximation based control, the desired trajectory is assumed to remain in the operating region denoted by $\mathcal{D}$. This assumption can also be enforced by a suitable trajectory prefilter design, as shown in the following example.

EXAMPLE A.11

In the case that $n = 2$ and $r = 1$ that was considered in Example A.9, assume that $\mathcal{D} = [\underline{y}, \overline{y}] \times [\underline{v}, \overline{v}]$ and an additional constraint on the prefilter is to ensure that $(y_d(t), \dot{y}_d(t)) \in \mathcal{D}\ \forall t \geq 0$, assuming that $(y_d(0), \dot{y}_d(0)) \in \mathcal{D}$. A filter designed to help enforce this constraint is

$$\begin{aligned}
\begin{bmatrix} \dot{z}_1 \\ \dot{z}_2 \end{bmatrix} &= \begin{bmatrix} z_2 \\ a_1 \left(v_{c_l}(t) - z_2 \right) \end{bmatrix} \\
\begin{bmatrix} y_d \\ \dot{y}_d \end{bmatrix} &= \begin{bmatrix} z_1 \\ z_2 \end{bmatrix}
\end{aligned}$$

where

$$\begin{aligned}
v_{c_l}(t) &= g\left(\frac{a_0}{a_1} \left(y_{c_l}(t) - z_1(t) \right), \underline{v}, \overline{v} \right) \\
y_{c_l}(t) &= g\left(y_c(t), \underline{y}, \overline{y} \right).
\end{aligned}$$

The saturation function indicated by g is defined as

$$g(x, \underline{x}, \overline{x}) = \begin{cases} \overline{x} & \text{if } x \geq \overline{x} \\ x & \text{if } \underline{x} \leq x \leq \overline{x} \\ \underline{x} & \text{if } x \leq \underline{x}. \end{cases}$$

This filter is depicted as a block diagram format in Figure A.5.

The signal $y_{c_l}(t)$ is a magnitude limited version of $y_c(t)$. This ensures that the user does not inadvertently command y_d to leave $[\underline{y}, \overline{y}]$. The signal y_{c_l} is interpreted as the commanded value for $z_1 = y_d$. The error $(y_{c_l} - y_d)$ is multiplied by the gain $\frac{a_0}{a_1}$ and limited to the range $[\underline{v}, \overline{v}]$ to produce the signal v_{c_l} that is treated as the commanded value for $z_2 = \dot{y}_d$.

Note that even such a filter does not guarantee that $(y_d(t), \dot{y}_d(t)) \in \mathcal{D}\ \forall t \geq 0$, because z_2 will lag v_{c_l}. Therefore, the region enforced by the command filter should be selected as a proper subset of the physical operating envelope. △

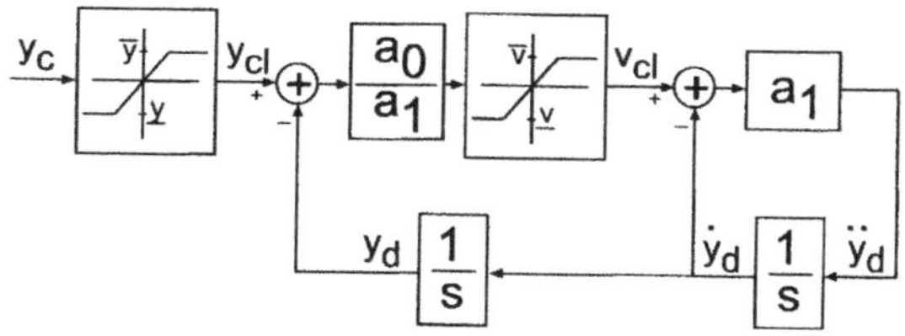

Figure A.5: Trajectory generation prefilter for Example A.11 that ensures $y_d(t) \in [\underline{y}, \overline{y}]$ and $\dot{y}_d(t) \in [\underline{v}, \overline{v}]$.

A.5 A USEFUL INEQUALITY

Most of the bounding techniques that are developed and used throughout the book require the use of a switching, or signum function sgn(ξ), which is discontinuous at $\xi = 0$. In order to avoid hard switching in the control or adaptive laws, it is desirable to use a smooth approximation of the signum function. The following result presents a useful inequality for using the function $\tanh(\xi/\varepsilon)$ as a smooth approximation to sgn(ξ).

Lemma A.5.1 *The following inequality holds for any* $\varepsilon > 0$ *and for any* $\xi \in \Re^1$

$$0 \leq |\xi| - \xi \tanh\left(\frac{\xi}{\varepsilon}\right) \leq \kappa\varepsilon \tag{A.28}$$

where κ *is a constant that satisfies* $\kappa = e^{-(\kappa+1)}$*; i.e.,* $\kappa = 0.2785$.

Proof: By dividing throughout by ε, proving (A.28) is in fact equivalent to

$$0 \leq |z| - z\tanh(z) \leq \kappa, \qquad z \in \Re^1 \tag{A.29}$$

where $z = \xi/\varepsilon$. Let

$$M(z) = |z| - z\tanh(z).$$

Since $M(-z) = M(z)$ (i.e., M is an even function), we only need to consider the case of $z \geq 0$. Moreover, we note that $M(0) = 0$, so for $z = 0$ (A.29) holds trivially. Hence, it is left to show that for positive z we have $0 \leq M(z) \leq \kappa$. For $z > 0$,

$$M(z) = z(1 - \tanh(z)),$$

and therefore $M(z) \geq 0$. To prove that $M(z) \leq \kappa$ we note that $M(z)$ has a well-defined maximum (see Figure A.6). To determine the maximum, we take the derivative and set it to zero, which yields

$$\begin{aligned} \frac{dM}{dz} &= \frac{d}{dz}\left\{z\left(1 - \frac{e^z - e^{-z}}{e^z + e^{-z}}\right)\right\} \\ &= \frac{2}{(e^z + e^{-z})^2}\left(1 - 2z + e^{-2z}\right) \equiv 0. \end{aligned}$$

Hence, the value $z = z^*$ that achieves the maximum satisfies

$$e^{-2z^*} = 2z^* - 1.$$

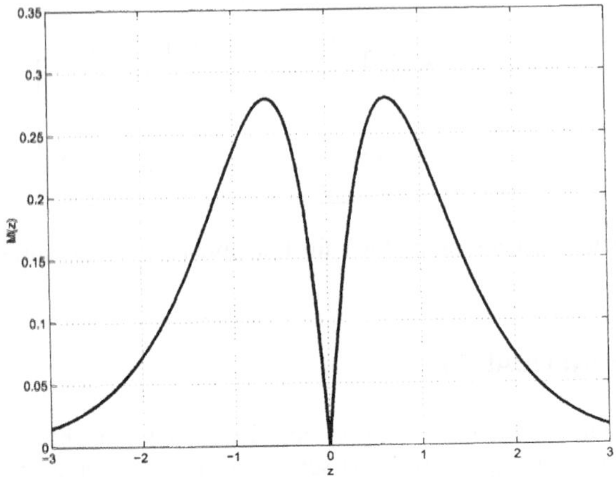

Figure A.6: Plot of $M(z) = |z| - z\tanh(z)$.

After some algebraic manipulation, it can be shown that

$$\begin{aligned} M(z^*) &= \frac{2z^* e^{-z^*}}{e^{z^*} + e^{-z^*}} \\ &= \frac{(e^{-2z^*} + 1)e^{-z^*}}{e^{z^*} + e^{-z^*}} \\ &= e^{-2z^*} \\ &= 2z^* - 1 \end{aligned}$$

Therefore, the maximum value of $M(z)$ is $2z^* - 1$ and it occurs at $z = z^*$ satisfying $2z^* - 1 = e^{-2z^*}$. If we let $\kappa = 2z^* - 1$ then $M(z) \leq \kappa$, where κ satisfies $\kappa = e^{-(\kappa+1)}$. By numerical methods, it can be readily shown deduced that $\kappa = e^{-(\kappa+1)}$ is satisfied for $\kappa = 0.278464...$; therefore, we take $\kappa = 0.2785$ as an upper bound. ■

A.6 EXERCISES AND DESIGN PROBLEMS

Exercise A.1 For the linear system $\dot{x} = A\,x$:

1. Show that if A is nonsingular, then the system has a single equilibrium point.
2. Show that if A is singular, then the system has an (uncountably) infinite number of equilibria. Are these equilibria isolated?

Exercise A.2 For the system

$$\ddot{\theta} + 2\dot{\theta} + \sin(\theta) = 0,$$

find all equilibria. Are any of the equilibria isolated?

Exercise A.3 Consider the Example A.8 on page 391. Show that the second-order system is Strictly Positive Real (SPR) if and only if the listed conditions hold.

APPENDIX B

RECOMMENDED IMPLEMENTATION AND DEBUGGING APPROACH

The approach to implementation and debugging presented in this appendix has been defined based on interactions with numerous students and colleagues. The objective is to correctly implement a working adaptive approximation based controller.

1. Derive a state space model for the plant that is of interest. Relative to the model, clearly record which portions are known and which are not. Denote the unknown functions by f_i where i counts over the number of unknown functions.

2. Choose a control design approach. For this approach, assume for a moment the all portions of the model are known. Derive a control law applicable to this known system that is provably stable. Note the stability properties that are expected.

3. Implement a simulation of the state space system. Also, implement the controller equations. In the controller, let the symbol $\hat{f}_i$ represent the approximation to f_i. Make sure that the controller implements $\hat{f}_i$ as a clearly distinguishable entity as it will be replaced later. For this step in the debugging process, assume some reasonable function for each f_i and let $\hat{f}_i = f_i$. With this perfect modeling, the stability properties provable in the previous step should hold exactly.

4. Run the simulation from various initial conditions and with various commanded trajectories. Make sure that all proven stability properties hold. For example, if

Adaptive Approximation Based Control: Unifying Neural, Fuzzy and Traditional Adaptive Approximation Approaches. By Jay A. Farrell and Marios M. Polycarpou

you have proven that the derivative of a function V is negative definite, then make sure that it is in the simulation. If any proven stability properties do not hold, even intermittently, then debugging is required. If any bugs are not removed at this step, then they may lead to misinterpretations or instability later.

5. Parameterize each unknown function: $f_i = (\theta^*)^\top \phi(x, \sigma^*) + e_i(x)$.

6. Derive parameter adaptation laws for θ and σ such that the adaptive closed-loop system has the desired set of stability properties required for the application conditions.

7. Modify the simulation from Step 3 so that $\hat{f}_i = \theta^\top \phi(x, \sigma)$ where θ and σ are estimated by the methods determined in Step 6. It is particularly important that relative to the working simulation from Step 3, the only changes should be those required to change the $\hat{f}_i$ functions to the form required for adaptive approximation.

8. Run the simulation from various initial conditions and with various commanded trajectories. Make sure that all proven stability properties hold. Assuming that the simulation was properly debugged in Step 3, this step should only involve tuning and debugging of the approximator and parameter estimation routines.

9. Translate the adaptive approximation based controller resulting from the above process to the platform required for actual implementation.

It is important to not skip Steps 3 and 4. Skipping those steps can result in bugs in the basic control law implementation being misinterpreted as problems or bugs in the adaptive approximation process. The above stepwise derivation and debugging approach decomposes the problem into pieces that can be separably solved, analyzed, and debugged.

REFERENCES

1. J. Albus. Data storage in the cerebellar model articulation controller (CMAC). *Trans. ASME J. Dynamic Syst. Meas. and Contr.*, 97:228–233, 1975.
2. J. Albus. A new approach to manipulator control: The cerebellar model articulation controller (CMAC). *Trans. ASME J. Dynamic Syst. Meas. and Contr.*, 97:220–227, 1975.
3. A. Alessandri, M. Baglietto, T. Parisini, and R. Zoppoli. A neural state estimator with bounded errors for nonlinear systems. *IEEE Transactions on Automatic Control*, 44(11):2028–2042, 1999.
4. P. An, W. Miller, and P. Parks. Design improvements in associative memories for cerebellar model articulation controllers (CMAC). In *International Conference on Artificial Neural Networks*, pages 1207–1210, 1991.
5. B. D. O. Anderson. Adaptive systems, lack of persistency of excitation and bursting phenomena. *Automatica*, 21:247–258, 1985.
6. B. D. O. Anderson and S. Vongpanitlerd. *Network Analysis and Synthesis*. Prentice-Hall, Englewood Cliffs, NJ, 1973.
7. Anonymous. Recommended practice for atmospheric and space flight vehicle coordinate systems. Technical Report R-004-1992, AIAA/ANSI, 1992.
8. M. Anthony and P.L. Bartlett. *Neural Network Learning: Theoretical Foundations*. Cambridge University Press, Cambridge, UK, 1999.
9. P. J. Antsaklis, W. Kohn, A. Nerode, and S. Sastry. *Hybrid Systems II*, volume 999 of *Lecture Notes in Computer Science*. Springer-Verlag, New York, 1995.
10. P. J. Antsaklis and A. N. Michel. *Linear Systems*. McGraw-Hill, Reading, MA, 1997.
11. K. Astrom and B. Wittenmark. *Adaptive Control*. Addison-Wesley, Reading, MA, 2nd edition, 1995.

Adaptive Approximation Based Control: Unifying Neural, Fuzzy and Traditional Adaptive Approximation Approaches. By Jay A. Farrell and Marios M. Polycarpou

12. C. G. Atkeson. Using modular neural networks with local representations to control dynamical systems. Technical Report AFOSR-TR-91-0452, MIT AI Lab, Cambridge, MA, 1991.

13. C. G. Atkeson, A. W. Moore, and S. Schaal. Locally weighted learning. *Artificial Intelligence Review*, 11:11–73, 1997.

14. M. Azam and S. N. Singh. Invertibility and trajectory control for nonlinear maneuvers of aircraft. *AIAA Journal of Guidance, Control, and Dynamics*, 17(1):192–200, 1994.

15. R. Babueska. *Fuzzy Modeling for Control.* Kluwer Academic Publishers, Boston, 1998.

16. W. Baker and J. Farrell. Connectionist learning systems for control. In *SPIE OE/Boston '90*, 1990.

17. A. Barron. Universal approximation bounds for superpositions of a sigmoidal function. *IEEE Transactions on Information Theory*, 39(3):930–945, 1993.

18. R. L. Barron, R. L. Cellucci, P. R. Jordan, N. E. Beam, P. Hess, and A. R. Barron. Applications of polynomial neural networks to FDIE and reconfigurable flight control. In *Proc. National Aerospace and Electronics Conference*, page 507–519, 1990.

19. J. S. Bay. *Fundamentals of Linear State Space Systems*. McGraw-Hill, Boston, MA, 1998.

20. R. Bellman. *Adaptive Control Processes*. Princeton University Press, Princeton, NJ, 1961.

21. H. Berenji. Fuzzy logic controllers. In R. Yager and L. Zadeh, editors, *An Introduction to Fuzzy Logic Applications and Intelligent Systems*. Kluwer Academic Publisher, Boston, MA, 1992.

22. C. P. Bernard and J.-J. E. Slotine. Adaptive control with multiresolution bases. In *Proceedings of the 36th IEEE Conference on Decision and Control*, pages 3884–3889, 1997.

23. D. Bertsekas and J. Tsitsiklis. *Neuro-dynamic Programming.* Athena Scientific, Belmont, MA, 1996.

24. S. Billings and W. Voon. Correlation based model validity tests for nonlinear models. *International Journal of Control*, 44:235–244, 1986.

25. S. A. Billings and H.-L. Wei. A new class of wavelet networks for nonlinear system identification. *IEEE Transactions on Neural Networks*, 16(4):862–874, 2005.

26. M. Bodson. Evaluation of optimization methods for control allocation. *AIAA Journal of Guidance, Control, and Dynamics*, 25(4):703–711, 2002.

27. S. A. Bortoff. Approximate feedback linearization using spline functions. *Automatica*, 33(8):1449–1458, 1997.

28. G. Box, G. M. Jenkins, and G. Reinsel. *Time Series Analysis: Forecasting and Control*. Prentice-Hall, Englewood Cliffs, NJ, 3rd edition, 1994.

29. W. Brogan. *Modern Control Theory*. Prentice-Hall, Englewood Cliffs, NJ, 1991.

30. D. Broomhead and D. Lowe. Multivariable functional interpolation and adaptive networks. *Complex Systems*, 1988.

31. D. Broomhead and D. Lowe. Radial basis functions, multivariable functional interpolation and adaptive networks. Technical Report 4148, Royal Signals and Radar Establishment, March 1988.

32. M. Brown and C. Harris. *Neurofuzzy Adaptive Modelling and Control*. Prentice-Hall, Englewood Cliffs, NJ, 1994.

33. A. E. Bryson and Y. C. Ho. *Applied Optimal Control*. Blaisdell, Waltham, MA, 1969.

34. D. J. Bugajski, D. F. Enns, and M. R. Elgersma. A dynamic inversion based control law with application to high angle of attack research vehicle. In *AIAA Guidance, Navigation and Control Confernece*, number AIAA-90-3407-CP, pages 826–839, 1990.

35. M. D. Buhmann. *Radial Basis Functions: Theory and Implementation*. Cambridge University Press, Cambridge, UK, 2003.

36. A. J. Calise and R. T. Rysdyk. Nonlinear adaptive flight control using neural networks. *IEEE Control Systems Magazine*, 18(6):14–25, 1998.

37. M. Cannon and J.-J. E. Slotine. Space-frequency localized basis function networks for nonlinear system estimation and control. *Neurocomputing*, 9:293–342, 1995.

38. M. Carlin, T. Kavli, and B. Lillekjendlie. A comparison of four methods for nonlinear data modeling. *Chemometrics and Intelligent Laboratory Systems*, 23:163–178, 1994.

39. C.-T. Chen. *Linear System Theory and Design*. Oxford University Press, Oxford, UK, 3rd edition, 1998.

40. F.-C. Chen and H. K. Khalil. Adaptive control of nonlinear systems using neural networks. *International Journal of Control*, 55(6):1299–1317, 1992.

41. F.-C. Chen and H. K. Khalil. Adaptive control of a class of nonlinear discrete-time systems using neural networks. *IEEE Transactions on Automatic Control*, 40:791–801, 1995.

42. F.-C. Chen and C. C. Liu. Adaptively controlling nonlinear continuous-time systems using multilayer neural networks. *IEEE Transactions on Automatic Control*, 39(6):1306–1310, 1994.

43. S. Chen and S. Billings. Neural networks for nonlinear dynamic system modelling and identification. In *Advances in Intelligent Control*. Taylor and Francis, London, 1994.

44. S. Chen, S. Billings, C. Cowan, and P. Grant. Practical identification of NARMAX models using radial basis functions. *International Journal of Control*, 52(6):1327–1350, 1990.

45. S. Chen, S. Billings, and P. Grant. Non-linear system identification using neural networks. *International Journal of Control*, 51:1191–1214, 1990.

46. S. Chen, S. Billings, and P. Grant. Recursive hybrid algorithm for non-linear system identification using radial basis function networks. *International Journal of Control*, 55(5):1051–1070, 1992.

47. S. Chen, C. F. N. Cowan, and P. M. Grant. Orthogonal least squares learning algorithm for radial basis function networks. *IEEE Transactions on Neural Networks*, 2(2):302–309, 1991.

48. E. W. Cheney. *Introduction to Approximation Theory*. McGraw-Hill, New York, 1966.

49. J.Y. Choi and J.A. Farrell. Nonlinear adaptive control using networks of piecewise linear approximators. *IEEE Transactions on Neural Networks*, 11(2):390–401, 2000.

50. M.-Y. Chow. *Methodologies of Using Neural Network and Fuzzy Logic Technologies for Motor Incipient Fault Detection*. World Scientific, London, 1998.

51. C. Chui. *An Introduction to Wavelets*. Academic Press, San Diego, CA, 1992.

52. C. W. Clenshaw. A comparison of "best" polynomial approximations with truncated chebyshev series expansions. *Journal of the Society for Industrial and Applied Mathematics: Series B, Numerical Analysis*, 1:26–37, 1964.

53. C. W. Clenshaw. Curve and surface fitting. *J. Inst. Math. Appl.*, 1:164–183, 1965.

54. T. F. Coleman and Y. Li. A globally and quadratically convergent affine scaling method for l_1 problems. *Mathematical Programming*, 56, Series A:189–222, 1992.

55. M. Cox. Practical spline approximation. In *Topics in Numerical Analysis*, pages 79–112. Springer-Verlag, Berlin, 1981.

56. M. Cox. Algorithms for spline curves and surfaces. Technical report, NPL Report DITC 166/90, 1990.

57. M. G. J. Cox. Curve fitting with piecewise polynomials. *Inst. Math. Appl.*, 8:36–52, 1971.

58. G. Cybenko. Approximation by superposition of a sigmoidal function. *Mathematics of Control, Signals, and Systems*, 2(4):303–314, 1989.

59. M. Daehlen and T. Lyche. Box splines and applications. In H. Hagen and D. Roller, editors, *Geometric Modeling: Methods and Applications*. Springer-Verlag, Berlin, 1991.

60. I. Daubechies. *Ten Lectures on Wavelets*. SIAM, Philadelphia, PA, 1992.

61. J. D'Azzo and C. Houpis. *Linear Control System Analysis and Design: Conventional and Modern*. McGraw-Hill, New York, 1995.

62. C. de Boor. *A Practical Guide to Splines*, volume 27 of *Applied Mathematical Sciences*. Springer-Verlag, New York, 1978.

63. C. De Silva. *Intelligent Control: Fuzzy Logic*. CRC Press, Boca Raton, FL, 1995.

64. J. D. DePree and C. W. Swartz. *Introduction to Real Analysis*. John Wiley and Sons, New York, 1988.

65. Y. Diao and K. M. Passino. Stable fault-tolerant adaptive fuzzy/neural control for a turbine engine. *IEEE Transactions on Control Systems Technology*, 9:494–509, 2001.

66. R. Dorf and R. Bishop. *Modern Control Systems*. Addison-Wesley, Reading, MA, 9th edition, 1998.

67. D. Driankov, H. Hellendoorn, and M. Reinfrank. *An Introduction to Fuzzy Control*. Springer-Verlag, Berlin, 1993.

68. W. C. Durham. Computationally efficient control allocation. *AIAA Journal of Guidance, Control, and Dynamics*, 24(3):519–524, 2001.

69. B. Egardt. *Stability of Adaptive Controllers*. Spinger-Verlag, Berlin, 1979.

70. D. Enns. Control allocation approaches. In *AIAA Guidance, Navigation and Control Conference*, number AIAA-98-4109, pages 98–108, 1998.

71. R. Eubank. *Spline Smoothing and Nonparametric Regression*. Marcel Dekker, New York, 1988.

72. S. Fabri and V. Kadirkamanathan. Dynamic structure neural networks for stable adaptive control of nonlinear systems. *IEEE Transactions on Neural Networks*, 7(5):1151–1167, 1996.

73. J. A. Farrell. Neural control systems. In W. Levine, editor, *The Controls Handbook*, pages 1017–1030. CRC Press, Boca Raton, FL, 1996.

74. J. A. Farrell. Persistancy of excitation conditions in passive learning control. *Automatica*, 33(4):699–703, 1997.

75. J. A. Farrell. Stability and approximator convergence in nonparametric nonlinear adaptive control. *IEEE Transactions on Neural Networks*, 9(5):1008–1020, 1998.

76. J. A. Farrell and M. M. Polycarpou. Neural, fuzzy, and approximation-based control. In T. Samad, editor, *Perspectives in Control Engineering Technologies, Applications, and New Directions*, pages 134–164. IEEE Press, Piscataway, NJ, 2001.

77. J. A. Farrell, M. M. Polycarpou, and M. Sharma. Longitudinal flight path control using on-line function approximation. *AIAA Journal of Guidance, Control, and Dynamics*, 26(6):885–897, 2003.

78. J. A. Farrell, M. Sharma, and M. M. Polycarpou. Backstepping-based flight control with adaptive function approximation. *AIAA Journal of Guidance, Control, and Dynamics*, 28(6):1089–1102, 2005.

79. G. E. Fasshauer. Meshfree methods. In M. Rieth and W. Schommers, editors, *Handbook of Theoretical and Computational Nanotechnology*. American Scientific Publ., Stevenson Ranch, CA, 2005.

80. S. P. Fears, H. M. Ross, and T. M. Moul. Low-speed wind-tunnel investigation of the stability and control characteristics of a series of flying wings with sweep angles of 50°. Technical Memorandum 4640, NASA, 1995.

81. A. F. Filippov. Differential equations with discontinuous right hand sides. *American Mathematical Society Translations*, 42:199–231, 1964.

82. T. B. Fomby, R. C. Hill, and S. R. Johnson. *Advanced Econometric Models*. Springer-Verlag, New York, 1984.

83. R. Franke. Locally determined smooth interpolation at irregularly spaced points in several variables. *J. Inst. of Math. Appl.*, 19:471–482, 1977.

84. R. Franke. Scattered data interpolation: Tests of some methods. *Mathematics of Computation*, 38(157), 1982.

85. R. Franke and G. Nielson. Scattered data interpolation and applications: A tutorial and survey. In H. Hagen and D. Roller, editors, *Geometric Modeling*. Springer-Verlag, Berlin, 1991.

86. G. F. Franklin, J. D. Powell, and A. Emani-Naeini. *Feedback Control of Dynamic Systems*. Addison-Wesley, Reading, MA, 3rd edition, 1994.

87. M. French, C. Szepesvári, and E. Rogers. *Performance of Nonlinear Approximate Adaptive Controllers*. John Wiley, Hoboken, NJ, 2003.

88. K. Funahashi. On the approximate realization of continuous mappings by neural networks. *Neural Networks*, 2:183–192, 1989.

89. V. Gazi, K. M. Passino, and J. A. Farrell. Adaptive control of discrete time nonlinear systems using dynamic structure approximators. In *Proceedings of the American Control Conference*, pages 3091–3096, 2001.

90. S. Ge, C. Hang, T.H. Lee, and T. Zhang. Adaptive neural network control of nonlinear systems by state and output feedback. *IEEE Transactions on Systems, Man, and Cybernetics, Part B: Cybernetics*, 29(6):818–828, 1999.

91. S. Ge, C. Hang, T.H. Lee, and T. Zhang. *Stable Adaptive Neural Network Control*. Kluwer, Boston, MA, 2001.

92. S. Ge, T.H. Lee, and C. Harris. *Adaptive Neural Network Control of Robotic Manipulators*. World Scientific, London, 1998.

93. S. Ge, G. Li, and T.H. Lee. Adaptive neural network control for a class of strict feedback discrete-time nonlinear systems. *Automatica*, 39:807–819, 2003.

94. S. Ge and C. Wang. Adaptive neural control of uncertain MIMO nonlinear systems. *IEEE Transactions on Neural Networks*, 15(3):674–692, 2004.

95. S. Ge, J. Zhang, and T.H. Lee. State feedback NN control of a class of discrete MIMO nonlinear systems with disturbances. *IEEE Transactions on Systems, Man, and Cybernetics, Part B: Cybernetics*, 34(4):1634–1645, 2004.

96. W. L. Gerrard, D. F. Enns, and A. Snell. Nonlinear longitudinal control of a supermaneuverable aircraft. In *Proceedings of the American Control Conference*, pages 142–147, 1989.

97. F. Girosi and T. Poggio. Networks and the best approximation property. *Biological Cybernetics*, 63:169–176, 1990.

98. S. T. Glad and O. Härkegård. Backstepping control of a rigid body. In *Proceedings of the 41st IEEE Conference on Decision and Control*, pages 3944–3945, 2002.

99. G. Golub and C. Van Loan. *Matrix Computations*. John Hopkins University Press, Baltimore MD, 1996.

100. D. Gorinevsky. On the persistency of excitation in radial basis function network identification of nonlinear-systems. *IEEE Transactions on Neural Networks*, 6(5):1237–1244, 1995.

101. M. M. Gupta and N. K. Sinha, editors. *Intelligent Control Systems: Theory and Applications*. IEEE Press, New York, 1996.

102. F. M. Ham and I. Kostanic. *Principles of Neurocomputing for Science and Engineering*. McGraw-Hill, New York, 2000.

103. J. D. Hamilton. *Time Series Analysis*. Princeton University Press, Princeton, NJ, 1994.

104. R. L. Hardy. Multiquadratic equations of topography and other irregular surfaces. *J. Geographical Res.*, 76:1905–1915, 1971.

105. R. L. Hardy. Research results in the application of multiquadratic equations to surveying and mapping problems. *Surveying and Mapping*, 35:321–332, 1975.

106. O. Härkegård. *Backstepping and Control Allocation with Applications to Flight Control*. Ph. D. dissertation 820, Linkoping Studies in Science and Technology, 2003.

107. O. Härkegård and S. T. Glad. A backstepping design for flight path angle control. In *Proceedings of the 39th IEEE Conference on Decision and Control*, pages 3570–3575, 2000.

108. C. Harris, C. Moore, and M. Brown. *Intelligent Control: Some Aspects of Fuzzy Logic and Neural Networks*. World Scientific Press, Hackensack, NJ, 1993.

109. S. Haykin. *Neural Networks: A Comprehensive Foundation*. Prentice-Hall, Englewood Cliffs, NJ, 2nd edition, 1999.

110. K. Hornik, M. Stinchcombe, and H. White. Multilayer feedforward networks are universal approximators. *Neural Networks*, 2:359–366, 1989.

111. N. Hovakimyan, F. Nardi, A. Calise, and N. Kim. Adaptive output feedback control of uncertain nonlinear systems using single-hidden-layer neural networks. *IEEE Transactions on Neural Networks*, 13(6):1420–1431, 2002.

112. N. Hovakimyan, R. Rysdyk, and A. Calise. Dynamic neural networks for output feedback control. *International Journal of Robust and Nonlinear Control*, 11(1):23–29, 2001.

113. D. Hrovat and M. Tran. Application of gain scheduling to design of active suspension. In *Proc. of the IEEE Conf. on Decision and Control*, pages 1030–1035, December 1993.

114. L. Hsu and R. Costa. Bursting phenomena in continuous-time adaptive systems with a σ-modification. *IEEE Transactions on Automatic Control*, 32(1):84–86, 1987.

115. K. Hunt, G. Irwin, and K. Warwick, editors. *Neural Network Engineering in Dynamic Control Systems*. Springer, Berlin, 1995.

116. K. Hunt and D. Sbarbaro-Hofer. Neural networks for nonlinear internal model control. *IEE Proc. D*, 138(5):431–438, 1991.

117. D. Hush and B. Horne. Progress in supervised neural networks: What's new since Lippman? *IEEE Signal Processing Magazine*, 10:8–39, 1993.

118. R. A. Hyde and K. Glover. The application of H_∞ controllers to a VSTOL aircraft. *IEEE Transactions on Automatic Control*, 38:1021–1039, 1993.

119. P. A. Ioannou and J. Sun. *Robust Adaptive Control*. Prentice Hall, Upper Saddle River, NJ, 1996.

120. P. A. Ioannou and G. Tao. Frequency domin conditions for strictly positive functions. *IEEE Transactions on Automatic Control*, 32(1):53–54, 1987.

121. A. Isidori. *Nonlinear Control Systems*. Springer-Verlag, Berlin, 1989.

122. R. A. Jacobs and M. I. Jordan. A modular connectionist architecture for learning piecewise control strategies. In *Proceedings of the American Control Conference*, 1991.

123. S. Jagannathan and F. L. Lewis. Multilayer discrete-time neural net controller with guaranteed performance. *IEEE Transactions on Neural Networks*, 7(1):107–130, 1996.

124. D. James. Stability of a model reference control system. *AIAA Journal*, 9(5), 1971.

125. M. Jamshidi, N. Vadiee, and T. Ross, editors. *Fuzzy Logic and Control: Software and Hardware Applications*. Prentice Hall, Englewood Cliffs, NJ, 1993.

126. J. Jiang. Optimal gain scheduling controllers for a diesel engine. *IEEE Control Systems Magazine*, 14(4):42–48, 1994.

127. R. Johansson. *System Modeling and Identification*. Prentice Hall, Englewood Cliffs, NJ, 1993.

128. J. Judd. *Neural Network Design and the Complexity of Learning*. MIT Press, Cambridge, MA, 1990.

129. J. Kacprzyk. *Multistage fuzzy control: a model-based approach to fuzzy control and decision making*. Wiley, Chichester, 1997.

130. T. Kailath. *Linear Systems*. Prentice-Hall, Englewood-Cliffs, NJ, 1980.

131. A. Kandel and G. Langholz, editors. *Fuzzy control systems*. CRC Press, Boca Raton, FL, 1994.

132. T. Kavli. ASMOD-an algorithm for adaptive spline modelling of observation data. *International Journal of Control*, 58(4):947–967, 1993.

133. S. M. Kay. *Fundamentals of Statistical Signal Processing*. Prentice Hall Signal Processing Series, Englewood Cliffs, NJ, 1993.

134. H. Khalil. *Nonlinear Systems*. Prentice Hall, Englewood Cliffs, NJ, 1996.

135. M. A. Khan and P. Lu. New technique for nonlinear control of aircraft. *AIAA Journal of Guidance, Control, and Dynamics*, 17(5):1055–1060, 1994.

136. J. Kindermann and A. Linden. Inversion of neural networks by gradient descent. *Parallel Computing*, 14:277–286, 1990.

137. E. Kosmatopoulos, M. Polycarpou, M. Christodoulou, and P. Ioannou. High-order neural network structures for identification of dynamical systems. *IEEE Transactions on Neural Networks*, 6(2):422–431, 1995.

138. G. Kreisselmeier. Adaptive observers with exponential rate of convergence. *IEEE Transactions on Automatic Control*, 22(1):2–8, 1977.

139. M. Krstic, I. Kanellakopoulos, and P. Kokotovic. *Nonlinear and Adaptive Control Design*. Wiley, New York, 1995.

140. B. C. Kuo. *Automatic Control Systems*. Prentice-Hall, Englewood Cliffs, NJ, 6th edition, 1991.

141. A. J. Kurdila, F. J. Narcowich, and J. D. Ward. Persistency of excitation in identification using radial basis function approximants. *SIAM Journal of Control and Optimization*, 33(2):625–642, 1995.

142. S. Lane, D. Handelman, and J. Gelfand. Theory and development of higher-order CMAC neural networks. *IEEE Control Systems Magazine*, pages 23–30, 1992.

143. S. H. Lane and R. F. Stengel. Flight control design using nonlinear inverse dynamics. *Automatica*, 31(4):781–806, 1988.

144. E. Lavretsky, N. Hovakimyan, and A. Calise. Upper bounds for approximation of continuous-time dynamics using delayed outputs and feedforward neural networks. *IEEE Transactions on Automatic Control*, 48(9):1606–1610, 2003.

145. Y. LeCun. Une procedure d'apprentissage pour reseau a seuil assymetrique. *Cognitiva*, 85:599–604, 1985.

146. M. Leshno, V. Y. Lin, A. Pinkus, and S. Schocken. Multilayer feedforward networks with a nonpolynomial activation function can approximate any function. *Neural Computation*, 6:861–867, 1993.

147. F. L. Lewis, J. Campos, and R. R. Selmic. *Neuro-Fuzzy Control of Industrial Systems With Actuator Nonlinearities*. SIAM Press, Philadelphia, PA, 2002.

148. F. L. Lewis, S. Jagannathan, and A. Yesildirek. *Neural Network Control of Robot Manipulators and Nonlinear Systems*. Taylor & Francis, London, 1999.

149. F. L. Lewis, A. Yesildirek, and K. Liu. Multilayer neural-net robot controller with guaranteed tracking performance. *IEEE Transactions on Neural Networks*, 7:1–12, 1996.

150. H. Lewis. *The Foundations of Fuzzy Control*. Plenum Press, New York, 1997.

151. C. Lin. *Neural Fuzzy Control Systems with Structure and Parameter Learning*. World Scientific, Singapore, 1994.

152. Lipmann. A critical overview of neural network pattern classifiers. In *Proceedings of the IEEE Workshop on Neural Networks for Signal Processing*, pages 266–275, 1991.

153. L. Ljung. *System Identification: Theory for the User*. Prentice-Hall, Englewood Cliffs, NJ, 2nd edition, 1999.

154. L. Ljung and T. Söderström. *Theory and Practice of Recursive Identification*. MIT Press, Cambridge, MA, 1983.

155. G. Lorentz. *Approximation of Functions*. Holt, Rinehart, and Winston, New York, 1966.

156. D. Lowe. On the iterative inversion of RBF networks: A statistical interpretation. In *IEE 2nd International Conference on Artificial Neural Networks*, pages 29–39, 1991.

157. D. G. Luenberger. *Linear and Nonlinear Programming*. Addison-Wesley, Reading, MA, 2nd edition, 1984.

158. I. Mareels and R. Bitmead. Nonlinear dynamics in adaptive control: Chaotic and periodic stabilization. *Automatica*, 22:641–655, 1986.

159. R. Marino and P. Tomei. *Nonlinear Control Design: Geometric, Adaptive and Robust*. Prentice-Hall, Englewood Cliffs, NJ, 1995.

160. W. D. Maurer and T. G. Lewis. Hash table methods. *Computing Surveys*, 7(1):5–19, 1975.

161. D. McRuer, I. Ashkenas, and D. Graham. *Aircraft Dynamics and Automatic Control*. Princeton University Press, Princeton, NJ, 1973.

162. M. Mears and M. Polycarpou. Stable neural control of uncertain multivariable systems. *International Journal of Adaptive Control and Signal Processing*, 17:447–466, 2003.

163. J. M. Mendel. *Discrete Techniques of Parameter Estimation: The Equation Error Formulation*. Marcel Dekker, New York, 1973.

164. J. M. Mendel. *Lessons in Estimation Theory for Signal Processing, Communications, and Control*. Prentice Hall, Englewood Cliffs, NJ, 1995.

165. P. K. A. Menon, M. E. Badget, R. A. Walker, and E. L. Duke. Nonlinear flight test trajectory controllers for aircraft. *AIAA Journal of Guidance, Control, and Dynamics*, 10(1):67–72, 1987.

166. G. Meyer, R. Su, and L. R. Hunt. Application of nonlinear transformations to automatic flight control. *Automatica*, 20(1):103–107, 1984.

167. C. A. Micchelli. Interpolation of scattered data: Distance matrices and conditionally positive definite functions. *Constructive Approximation*, pages 11–22, 1986.

168. A. N. Michel and D. Liu. *Qualitative Analysis and Synthesis of Recurrent Neural Networks*. Marcel Dekker, New York, 2002.

169. R. K. Miller and A. N. Michel. *Ordinary Differential Equations*. Academic Press, New York, 1982.

170. W. T. Miller, F. Glanz, and G. Kraft. CMAC: An associative neural network alternative to backpropagation. *Proc. IEEE*, 78(10):1561–1567, 1990.

171. W. T. Miller, F. Glanz, and G. Kraft. Real-time dynamic control of an industrial manipulator using a neural-network based learning controller. *IEEE Transactions on Robotics and Automation*, 6(1):1–9, 1990.

172. W. T. Miller, R. S. Sutton, and P. J. Werbos. *Neural Networks for Control*. MIT Press, Cambridge, MA, 1990.

173. P. Millington. Associative reinforcement learning for optimal control. Master's thesis, Department of Aeronautics and Astronautics, MIT, Cambridge, MA, 1991.

174. R. S. Minhas and S. A. Bortoff. Robustness considerations in spline-based adaptive feedback linearization. In *Proceedings of the 1996 IFAC World Congress*, volume E, pages 191–196, 1996.

175. J. Moody and C. Darken. Fast learning in networks of locally-tuned processing units. *Neural Comput.*, 1:281–294, 1989.

176. F. K. Moore and E. M. Greitzer. A theory of post-stall transients in axial compression systems — part I: development of equations. *Journal of Turbomachinery*, 108:68–76, 1986.

177. A. S. Morse. Global stability of parameter adaptive control systems. *IEEE Transactions on Automatic Control*, 25:433–439, 1980.

178. J. Nakanishi, J. A. Farrell, and S. Schaal. Composite adaptive control with locally weighted statistical learning. *Neural Networks*, 18(1):71–90, 2005.

179. K. S. Narendra and A. M. Annaswamy. *Stable Adaptive Systems*. Prentice Hall, Englewood Cliffs, NJ, 1989.

180. K. S. Narendra, Y. H. Lin, and L. S. Valavani. Stable adaptive controller design, part II: Proof of stability. *IEEE Transactions on Automatic Control*, 25:440–448, 1980.

181. K. S. Narendra and K. Parthasarathy. Identification and control of dynamical systems using neural networks. *IEEE Transactions on Neural Networks*, 1(1):4–27, 1990.

182. H. T. Nguyen, editor. *Theoretical Aspects of Fuzzy Control*. Wiley, New York, 1995.

183. R.A. Nichols, R.T. Reichert, and W.J. Rugh. Gain scheduling for H_∞ controllers: A flight control example. *IEEE Transactions on Control Systems Technology*, 1:69–75, 1993.

184. J. Nie and D. Linkens. *Fuzzy-Neural Control: Principles, Algorithms, and Applications*. Prentice Hall, New York, 1995.

185. H. Nijmeijer and A. van der Schaft. *Nonlinear Dynamical Control Systems*. Spinger-Verlag, New York, 1990.

186. O. Omidvar and D. L. Elliott, editors. *Neural Systems for Control*. Academic Press, San Diego, 1997.

187. J. Ozawa, I. Hayashi, and N. Wakami. Formulation of CMAC-fuzzy system. In *Proc. IEEE Intern. Conf. Fuzzy Systems*, pages 1179–1186, 1992.

188. G. Page, J. Gomm, , and D. Williams, editors. *Application of Neural Networks to Modeling and Control*. Chapman & Hall, London, 1993.

189. R. Palm, D. Driankov, and H. Hellendoorn. *Model Based Fuzzy Control: Fuzzy Gain Schedulers and Sliding Mode Fuzzy Controllers*. Springer, Berlin, 1997.

190. Y. Pao. *Adaptive Pattern Recognition and Neural Networks*. Addison-Wesley, Reading, MA, 1989.

191. T. Parisini and R. Zoppoli. A receding-horizon regulator for nonlinear systems and a neural approximation. *Automatica*, 31(10):1443–1451, 1995.

192. T. Parisini and R. Zoppoli. Neural approximations for infinite-horizon optimal control of nonlinear stochastic systems. *IEEE Transactions on Neural Networks*, 9(6):1388–1408, 1998.

193. J. Park and I.W. Sandberg. Universal approximation using radial basis function networks. *Neural Computation*, 3(2):246–257, 1991.

194. D.B. Parker. Learning-logic: Casting the cortex of the human brain in silicon. Technical Report TR-47, Center for Computational Research in Economics and Management Science, MIT, Cambridge, MA, 1985.

195. P. Parks and J. Militzer. A comparison of five algorithms for the training of CMAC memories for learning control systems. *Automatica*, 28(5):1027–1035, 1992.

196. P. C. Parks. Lyapunov redesign of model reference adaptive control systems. *IEEE Transactions on Automatic Control*, 11:362–367, 1966.

197. K. Passino. *Biomimicry for Optimization, Control, and Automation*. Springer-Verlag, London, 2005.

198. K. Passino and S. Yurkovich. *Fuzzy Control*. Addison-Wesley, Menlo Park, CA, 1998.

199. Y. C. Pati and P. S. Krishnaprasad. Analysis and synthesis of feedforward neural networks using discrete affine wavelet transform. *IEEE Transactions on Neural Networks*, 4(1):73–85, 1993.

200. A. Patrikar and J. Provence. Nonlinear system identification and adaptive control using polynomial networks. *Mathematical & Computer Modelling*, 23:159–173, 1996.

201. W. Pedrycz. *Fuzzy Control and Fuzzy Systems*. Wiley, New York, 2nd edition, 1993.

202. R. Penrose. A generalized inverse for matrices. In *Proceedings of the Cambridge Philosophical Society*, volume 51, Part 3, pages 406–413, 1955.

203. D. Pham and L. Xing. *Neural Networks for Identification, Prediction, and Control*. Springer-Verlag, London, 1995.

204. T. Poggio and F. Girosi. A theory of networks for approximation and learning. Technical Report AIM 1140, AI Laboratory, MIT, Cambridge, MA, 1989.

205. T. Poggio and F. Girosi. Networks for approximation and learning. *Proceedings of the IEEE*, 78(9):1481–1497, 1990.

206. R. Policar. The engineer's ultimate guide to wavelet analysis. http://users.rowan.edu/ polikar/WAVELETS/WTtutorial.html.

207. M. Polycarpou and A. Helmicki. Automated fault detection and accommodation: A learning system approach. *IEEE Transactions on Systems, Man, and Cybernetics*, 25(11):1447–1458, 1995.

208. M. Polycarpou and P. Ioannou. Modeling, identification and stable adaptive control of continuous-time nonlinear dynamical systems using neural networks. In *Proc. 1992 American Control Conference*, pages 36–40, 1992.

209. M. M Polycarpou. Stable adaptive neural control scheme for nonlinear systems. *IEEE Transactions on Automatic Control*, 41(3):447–451, 1996.

210. M. M. Polycarpou. On-line approximators for nonlinear system identification: A unified approach. In C. Leondes, editor, *Control and Dynamic Systems: Neural Network Systems Techniques and Applications*, pages 191–230. Academic Press, New York, NY, 1998.

211. M. M. Polycarpou and P. A. Ioannou. Identification and control of nonlinear systems using neural network models: Design and stability analysis. Technical Report 91-09-01, University of Southern California, Dept. Electrical Engineering – Systems, September 1991.

212. M. M. Polycarpou and P. A. Ioannou. Neural networks as on-line approximators of nonlinear systems. In *Proceedings of the 31st IEEE Conference on Decision and Control*, pages 7–12, 1992.

213. M. M. Polycarpou and P. A. Ioannou. On the existence and uniqueness of solutions in adaptive control systems. *IEEE Transactions on Automatic Control*, 38(3):474–479, 1993.

214. M. M. Polycarpou and P. A. Ioannou. Stable nonlinear system identification using neural network models. In G. Bekey and K Goldberg, editors, *Neural Networks for Robotics*, pages 147–164. Kluwer Acedemic Publishers, 1993.

215. M. M. Polycarpou and P. A. Ioannou. A robust adaptive nonlinear control design. *Automatica*, 32(3):423–427, 1996.

216. M. M. Polycarpou and M. Mears. Stable adaptive tracking of uncertain systems using nonlinearly parametrized on-line approximators. *International Journal of Control*, 70(3):363–384, 1998.

217. M. M. Polycarpou, M. J. Mears, and S. E. Weaver. Adaptive wavelet control of nonlinear systems. In *Proceedings of the 36th IEEE Conference on Decision and Control*, pages 3890–3895, 1997.

218. M. Powell. *Approximation Theory and Methods*. Cambridge University Press, Cambridge, UK, 1981.

219. M. Powell. Radial basis functions for multivariable interpolation: A review. In J. Mason and M. Cox, editors, *Algorithms for Approximation of Functions and Data*, pages 143–167. Oxford University, Oxford, UK, 1987.

220. D. V. Prokhorov and D. C. Wunsch. Adaptive critic designs. *IEEE Transactions on Neural Networks*, 8(5):997–1007, 1997.

221. Shorten R. and Murray-Smith R. Side effects of normalising radial basis function networks. *International Journal of Neural Systems*, 7(2):167–179, 1996.

222. H. Ritter, T. Martinez, and K. Schulten. Topology conserving maps for learning visuo-motorcoordination. *Neural Networks*, 2(2):159–168, 1989.

223. F. Rosenblatt. *Principles of Neurodynamics:Perceptrons and the Theory of Brain Mechanisms*. Spartan Books, Washington, DC, 1961.

224. G. A. Rovithakis and M. A. Christodoulou. Adaptive control of unknown plants using dynamical neural networks. *IEEE Trans. Systems, Man, and Cybernetics*, 24(3):400–412, 1994.

225. W. J. Rugh. *Linear System Theory*. Prentice-Hall, Englewood Cliffs, NJ, 2nd edition, 1995.

226. D. Rumelhart and J. McClelland (Eds.). *Parallel Distributed Processing: Explorations in the Microstructure of Cognition*. MIT Press, Cambridge, MA, 1986.

227. D. E. Rumelhart, G. E. Hinton, and R. J. Williams. Learning representations of backpropagation errors. *Nature*, 323:533–536, 1986.

228. E. W. Saad, D. V. Prokhorov, and D. C. Wunsch. Comparative study of stock trend prediction using time delay, recurrent and probabilistic neural networks. *IEEE Transactions on Neural Networks*, 9(6):1456–1470, 1998.

229. N. Sadegh. A perceptron network for functional identification and control of nonlinear systems. *IEEE Transactions on Neural Networks*, 4(6):982–988, 1993.

230. A. Saffiotti, E. H. Ruspini, and K. Konolige. Using fuzzy logic for mobile robot control. In H. Prade, D. Dubois, and H. J. Zimmermann, editors, *International Handbook of Fuzzy Sets and Possibility Theory*, volume 5. Kluwer Academic Publishers Group, Norwell, MA, and Dordrecht, The Netherlands, 1997.

231. A. Samuel. Some studies in machine learning using the game of checkers. *IBM Journal of Research and Development*, 3:210–229, 1959.

232. R. Sanner and J. Slotine. Gaussian networks for direct adaptive ccontrol. *IEEE Transactions on Neural Networks*, 3(6):837–863, 1992.

233. R. M. Sanner and J.-J. E. Slotine. Stable recursive identification using radial basis function networks. In *Proceedings of the American Controls Conference*, volume 3, pages 1829–1833, 1992.

234. S. Sastry. *Nonlinear Systems: Analysis, Stability, and Control*. Springer-Verlag, New York, 1999.

235. S. Sastry and M. Bodson. *Adaptive Control: Stability, Convergence and Robustness*. Prentice Hall, Englewood Cliffs, NJ, 1989.

236. S. Schaal and C. G. Atkeson. Receptive field weighted regression. Technical Report TR-H-209, ATR Human Information Processing Laboratories, Kyoto, Japan, 1997.

237. S. Schaal and C. G. Atkeson. Constructive incremental learning from only local information. *Neural Computation*, 10(8):2047–2084, 1998.

238. I. J. Schoenberg. Spline functions and the problem of graduation. *Proceedings of the National Academy of Sciences*, 52:947–950, 1964.

239. B. Schölkopf and A.J. Smola. *Learning with Kernels*. The MIT Press, Cambridge, MA, 2002.

240. L. Schumaker. *Spline Functions Basic Theory*. John Wiley, New York, 1981.

241. R. R. Selmic and F. L. Lewis. Neural network approximation of piecewise continuous functions: application to friction compensation. *IEEE Transactions on Neural Networks*, 13(3):745–751, 2002.

242. J. S. Shamma and M. Athans. Analysis of gain scheduled control for nonlinear plants. *IEEE Transactions on Automatic Control*, 35(8):898–907, 1990.

243. J. S. Shamma and M. Athans. Gain scheduling: Potential hazards and possible remedies. *IEEE Control Systems Magazine*, 12:101–107, 1992.

244. M. Sharma and A. J. Calise. Neural-network augmentation of existing linear controllers. *AIAA Journal of Guidance, Control, and Dynamics*, 28(1):12–19, 2005.

245. M. Sharma, J. A. Farrell, M. M. Polycarpou, N. D. Richards, and D. G. Ward. Backstepping flight control using on-line function approximation. In *Proc. of the AIAA Guidance, Navigiation, and Control Conference*, 2003.

246. S. Shekhar and M. Amin. Generalization by neural networks. *IEEE Transactions on Knowledge and Data Engineering*, 4(2):177–185, 1992.

247. J. Si, A. Barto, Powell W, and D. Wunsch, editors. *Handbook of Learning and Approximate Dynamic Programming*. Wiley-Interscience, Hoboken, NJ, 2004.

248. G. R. Slemon and A. Straughen. *Electric Machines*. Addison-Wesley, Reading, MA, 1980.

249. J. J. Slotine and W. Li. *Applied Nonlinear Control*. Prentice Hall, Englewood Cliffs, NJ, 1991.

250. S. A. Snell, D. F. Enns, and W. L. Garrard. Nonlinear inversion flight control for a supermaneuverable aircraft. *AIAA Journal of Guidance, Control, and Dynamics*, 14(4):976–984, 1992.

251. T. Soderstrom and P. Stoica. *System Identification*. Prentice Hall, New York, 1989.

252. D. Specht. A general regression network. *IEEE Transactions on Neural Networks*, 2(6):568–576, 1991.

253. M. Spivak. *Calculus on Manifold*. W.A.Benjamin, New York, 1965.

254. J. Spooner, M. Maggiore, R. Ordonez, and K. Passino. *Stable Adaptive Control and Estimation for Nonlinear Systems: Neural and Fuzzy Approximator Techniques*. Wiley-Interscience, New York, 2002.

255. J. Spooner and K. Passino. Stable adaptive control using fuzzy systems and neural networks. *IEEE Transactions on Fuzzy Systems*, 4(3):339–359, 1996.

256. G. Stein. Adaptive flight control – a pragmatic view. In K. S. Narendra and R. V. Monopoli, editors, *Applications of Adaptive Control*. Academic Press, New York, 1980.

257. G. Stein, G. Hartmann, and R. Hendrick. Adaptive control laws for F-8 flight test. *IEEE Transactions on Automatic Control*, 22:758–767, 1977.

258. B. L. Stevens and F. L. Lewis. *Aircraft Control and Simulation*. Wiley Interscience, New York, 1992.

259. M. Stinchcombe and H. White. Universal approximation using feedforward networks with non-sigmod hidden layer activation functions. In *Proceedings of the International Joint Conference on Neural Networks*, volume 1, pages 613–617, 1989.

260. G. Strang. Wavelet transforms versus Fourier transforms. *Bulletin of the American Mathematical Society*, 28(2):288–305, 1993.

261. M. Sugeno and M. Nishida. Fuzzy control of model car. *Fuzzy Sets and Systems*, 16:103–113, 1985.

262. N. Sureshbabu and J.A. Farrell. Wavelet based system identification for nonlinear control applications. *IEEE Transactions on Automatic Control*, 44(2):412–417, 1999.

263. H. J. Sussmann and P. V. Kokotovic. The peaking phenomenon and the global stabilization of nonlinear systems. *IEEE Transactions on Automatic Control*, 36(4):424–440, 1991.

264. J. Suykens, J. Vandewalle, and B. De Moor. *Artificial neural networks for modelling and control of non-linear systems*. Kluwer Academic Publishers, Boston, MA, 1996.

265. D. Sworder and J. Boyd. *Estimation Problems in Hybrid Systems*. Cambridge University Press, Cambridge, UK, 1999.

266. T. Takagi and M. Sugeno. Fuzzy identification of systems and its application to modeling and control. *IEEE Trans. Systems, Man and Cybernetics*, 15(1):116–132, 1985.

267. T. Takagi and M. Sugeno. Stability analysis and design of fuzzy control systems. *Fuzzy Sets Syst.*, 45:135–156, 1992.

268. G. Tao. *Adaptive Control Design and Analysis*. Wiley-Interscience, Hoboken, NJ, 2003.

269. H. Tolle and E. Ersu. *Neurocontrol: Learning Control Systems Inspired by Neuronal Architectures and Human Problem Solving*, volume 172 of *Lecture Notes in Control and Information Sciences*. Springer-Verlag, New York, 1992.

270. H. Tolle, P. Parks, E. Erus, M. Hormel, and J. Militzer. Learning control with interpolating memories. In C. Harris, editor, *Advances in Intelligent Control*. Taylor and Francis, London, 1994.

271. A. Trunov and M. Polycarpou. Automated fault diagnosis in nonlinear multivariable systems using a learning methodology. *IEEE Transactions on Neural Networks*, 11(1):91–101, 2000.

272. E. Tzirkel-Hancock and F. Fallside. A direct control method for a class of nonlinear systems using neural networks. In *Proc. 2nd IEE Int. Conf. on Artificial Neuural Networks*, pages 134–138, 1991.

273. E. Tzirkel-Hancock and F. Fallside. Stable control of nonlinear systems using neural networks. *International Journal Robust Nonlinear Control*, 2(2):67–81, 1992.

274. V. Vapnik. *Statistical Learning Theory*. Wiley, New York, NY, 2001.

275. A. Vemuri and M. Polycarpou. Neural network based robust fault diagnosis in robotic systems. *IEEE Transactions on Neural Networks*, 8(6):1410–1420, 1997.

276. A. Vemuri and M. Polycarpou. Robust nonlinear fault diagnosis in input-output systems. *International Journal of Control*, 68(2):343–360, 1997.

277. A. Vemuri, M. Polycarpou, and S. Diakourtis. Neural network based fault detection and accommodation in robotic manipulators. *IEEE Transactions on Robotics and Automation*, 14(2):342–348, 1998.

278. G. K. Venayagamoorthy, R. G. Harley, and D. C. Wunsch. Comparison of heuristic dynamic programming and dual heuristic programming adaptive critics for neurocontrol of a turbogenerator. *IEEE Transactions on Neural Networks*, 13(3):764–773, 2002.

279. M. Vidyasagar. *Nonlinear Systems Analysis*. Prentice-Hall, Englewood Cliffs, NJ, 2nd edition, 1993.

280. M. Vidyasagar. *A Theory of Learning and Generalization: with Applications to Neural Networks and Control Systems*. Springer-Verlag, London, 1997.

281. G. Walter. *Wavelets and Other Orthogonal Systems with Applications*. CRC Press, Boca Raton, FL, 1994.

282. L.-X Wang. Stable adaptive fuzzy control of nonlinear systems. *IEEE Transactions on Fuzzy Systems*, 1(2):146–155, 1993.

283. L.-X Wang. *Adaptive Fuzzy Systems and Control: Design and Stability Analysis*. Prentice Hall, Englewood Cliffs, NJ, 1994.

284. L.-X. Wang. *A Course in Fuzzy Systems and Control*. Prentice Hall, Upper Saddle River, NJ, 1997.

285. L.-X Wang and J. Mendel. Fuzzy basis functions, universal approximation, and orthogonal least-squares learning. *IEEE Transactions on Neural Networks*, 3(5):807–814, 1992.

286. L.-X. Wang and J. M. Mendel. Generating fuzzy rules by learning from examples. *IEEE Transactions on Systems, Man, and Cybernetics*, 22:1414–1427, 1992.

287. K. Warwick, G. Irwin, and K. Hunt, editors. *Neural Networks for Control and Systems*. P. Peregrinus/IEE, London, 1992.

288. S. Weaver, L. Baird, and M. Polycarpou. An analytical framework for local feedforward networks. *IEEE Transactions on Neural Networks*, 9(3):473–482, 1998.

289. S. Weaver, L. Baird, and M. Polycarpou. Using localized learning to improve supervised learning algorithms. *IEEE Transactions on Neural Networks*, 12(5):1037–1046, 2001.

290. H. Wendland. Piecewise polynomial, positive definite and compactly supported radial functions of minimal degree. *Adv. in Comput. Math*, 4:389–396, 1995.

291. P. Werbos. Beyond regression: New tools for prediction and analysis in the behavioral sciences. Master's thesis, Harvard University, Cambridge, MA, 1974.

292. P. Werbos. Backpropagation through time: What it does and how to do it. *Proceedings of the IEEE*, 78:1550–1560, 1990.

293. H. Werntges. Partitions of unity improve neural function approximation. In *Proc. IEEE Int. Conf. Neural Networks*, pages 914–918, San Francisco, CA, 1993.

294. E. Weyer and T. Kavli. Theoretical properties of the ASMOD algorithm for empirical modelling. *International Journal of Control*, 67(5):767–790, 1997.

295. H. P. Whitaker, J. Yamron, and A. Kezer. Design of model reference adaptive control systems for aircraft. Technical Report R-164, Instrumentation Lab, Massachusetts Institute of Technology, 1958.

296. D. White and D. Sofge, editors. *Handbook of Intelligent Control: Neural, Fuzzy, and Adaptive Approaches*. Van Nostrand Reinhold, New York, 1992.

297. B. Widrow and M. Hoff. Adaptive switching circuits. In *IRE WESCON Convention Record*, pages 96–104, 1960.

298. B. Widrow and M. Lehr. 30 years of adaptive neural networks: Perceptron, Madaline and Backpropagation. *Proc. IEEE*, 78(9):1415–1441, 1990.

299. B. Widrow and S. Stearns. *Adaptive Signal Processing*. Prentice Hall, Englewood Cliffs, NJ, 1985.

300. D. Wolpert. A mathematical theory of generalization: Part I. *Complex Systems*, 4:151–200, 1990.

301. D. Wolpert. A mathematical theory of generalization: Part II. *Complex Systems*, 4:201–249, 1990.

302. R. Yager and D. Filev. *Essentials of Fuzzy Modeling and Control*. Wiley, New York, 1994.

303. L. Zadeh. Fuzzy sets. *Information and Control*, 8:338–353, 1965.

304. L. Zadeh. Outline of a new approach to the analysis of complex systems and decision processes. *IEEE Transactions on Systems, Man, and Cybernetics*, 3(1):28–44, 1973.

305. J. Zhang, J. Raczkowsky, and A. Herp. Emulation of spline curves and its applications in robot motion control. In *Procs. of the IEEE Int. Conf. on Fuzzy Systems*, pages 831–836, 1994.

306. Q. Zhang and A. Benveniste. Wavelet networks. *IEEE Transactions on Neural Networks*, 3(6):889–898, 1992.

307. X. Zhang, T. Parisini, and M. Polycarpou. A unified methodology for fault diagnosis and accommodation for a class of nonlinear uncertain systems. *IEEE Transactions on Automatic Control*, 49(8):1259–1274, 2004.

308. X. Zhang, M. Polycarpou, and T. Parisini. A robust detection and isolation scheme for abrupt and incipient faults in nonlinear systems. *IEEE Transactions on Automatic Control*, 47(4):576–593, 2002.

309. Y. Zhang. A primal-dual interior point approach for computing the l_1 and l_∞ solutions of overdetermined linear systems. *J. Optimization Theory and Applications*, 77:592–601, 1993.

INDEX

Actuator, 1
Adaptation, 19
Adaptive approximation, 116
Adaptive approximation based control, robust, 286
Adaptive approximation problem, 124
Adaptive bounding, 220, 241, 252
Adaptive function approximation, 33
Adaptive linear control, 6
Adaptive nonlinear control, 222
Affine function, 46
Algebra, 48
Approximable by linear combinations, 44
Approximation based backstepping, 309
Approximation based backstepping, command filtered, 323
Approximation based feedback linearization, 288, 289
Approximation based input-output feedback linearization, 306
Approximation based input-state feedback linearization, 294
Approximation error, inherent, 75
Approximation error, residual, 74, 75
Approximation theory, 23
Approximation, degree of, 44
Approximation, nonparametric, 74
Approximation, scattered data, 84
Approximation, structure free, 74
Asymptotically stable, 380
Atomic fuzzy proposition, 101

Backpropagation, 95, 148, 152
Backpropagation through time, 154
Backpropagation, dynamic, 154
Backstepping control design, 203
Backstepping, approximation based, 309
Banach space, 43
Barbălat's Lemma, 260, 388
Basis-Influence functions, 57
Batch function approximation, 31
Best approximation, 44
Best approximator, 52
Boundedness, uniform ultimate, 380
Bounding control, 211, 239
Break point, 78
Bursting phenomenon, 292

Cardinal B-splines, 80
Cauchy sequence, 43
Cerebellar Model Articulation Controller, 87
Certainty equivalence principle, 222
Chattering, 211, 212, 215
Chebyshev space, 29
Class $\mathcal{K}$ function, 216, 219
CMAC, 87
Collocation matrix, 29, 77
Command filter, 208, 336, 352, 356
Command filtered approximation based backstepping, 323
Command filtering formulation, 207
Companion form, 190, 295

Adaptive Approximation Based Control: Unifying Neural, Fuzzy and Traditional Adaptive Approximation Approaches. By Jay A. Farrell and Marios M. Polycarpou

Condition number, matrix, 31
Continuous-time parameter estimation, 126, 141
Control system, 1
Control system design objectives, 3
Control terminology, 1
Controllable, 194, 253
Coordinate transformation, 193, 203
Corrective control law, 219
Covariance matrix, 151
Covariance resetting, 151
Covariance wind-up, 151
Cruise control example, 2
Curse of dimensionality, 136

Daubechies wavelets, 111
Dead-zone, 221
Dead-zone modification, 170
Definiteness, 382
Defuzzification, 104
Degree of approximation, 44
Dense, 45
Density, 108
Diffeomorphism, 193
Dilation, 83
Dilation parameter, 83
Direct adaptive control, 222
Discontinuous control law, 239
Discrete-time parameter estimation, 126
Discrete-time parametric modeling, 134
Distributed information processing, 96
Distribution of training data, 26
Disturbance, 2

Embedding function, 89
Epsilon-modification, 169
Equilibrium, 378
Error backpropagation algorithm, 148, 152
Error filtering online learning, 116
Estimator, 138
Excitation, sufficient, 42
Exponentially stable, 380

Feedback linearization, 180, 188, 237, 253
Feedback linearization, approximation based, 288, 289
Feedback linearization, input-output, 196
Feedback linearization, input-state, 190
Filtering techniques, 129
Finite escape time, 192
Function approximation, 30
Functional approximation error, 116
Fuzzification, 101
Fuzzy approximation, 96
Fuzzy implication, 101
Fuzzy inference, 103
Fuzzy logic, 96
Fuzzy rule base, 101
Fuzzy singleton, 97

Gain scheduling, 6, 186
Generalization, 29, 36, 54, 74
Generalization parameter, 64
Global approximation structure, 56
Global stability, 235
Global support, 56
Globally asymptotically stable, 204, 380
Gradient algorithm, normalized, 150
Gradient descent, 148
Guaranteed learning algorithm, 43

Haar space, 29, 66, 77
Haar wavelet, 110
Handling qualities, 395
Hidden layer, 94
High-gain feedback, 211, 214, 220
Hurwitz matrix, 295
Hurwitz polynomial, 191, 394
Hybrid systems, 127

Ill-conditioned, 32
Indirect adaptive control, 222
Inherent approximation error, 344
Input-output feedback linearization, 196
Input-output feedback linearization, approximation based , 306
Input-state linearization, 190, 194
Instability mechanisms, 264
Integrator backstepping, 203
Integrators, appended, 190
Internal dynamics, 198, 306
Interpolation, 28, 30
Interpolation matrix, 29
Interpolation, Lagrange, 29
Interpolation, scattered data, 84
Invariant set, 387
Involutivity, 195

Kalman-Yakubovich-Popov Lemma, 392
Knot, 78
Knots, nonuniformly spaced, 81
KYP Lemma, 392

Lagrange interpolation, 29
LaSalle's Theorem, 387
Lattice, 63, 86, 88
Learning, 19
Learning algorithms, robust, 163, 164
Learning interference, 89
Learning scheme, 124
Learning, supervised, 24
Least squares with forgetting, 152, 175
Least squares, batch recursive, 33
Least squares, batch weighted, 31
Least squares, continuous-time, 38, 150
Least squares, continuous-time recursive, 151
Least squares, discrete-time recursive, 33
Least squares, discrete-time weighted, 33
Legendre polynomials, 76

Lie derivative, 196
Linear control design, 4
Linearization, feedback, 180
Linearization, small-signal, 180, 253
Linearly parameterized approximators, 41, 126, 131
LIP approximators, 41
Lipschitz condition, 378
Local approximation structure, 56
Local function, 48
Local stability, 182, 235
Local support, 56
Locally weighted learning, 161, 177
Lyapunov equation, 296, 384
Lyapunov function, 381
Lyapunov redesign method, 215
Lyapunov's direct method, 382

Maar wavelet, 106
Mass-spring-damper model, 73
Matching condition, 216, 307
Matrix Inversion Lemma, 34
Measurement noise, 2, 26
Membership function, 96
Memoryless system, 116
Metric space, 43
Mexican hat wavelet, 106
MFAE, 75, 116, 128, 243, 267, 278, 286
Minimum functional approximation error, 75, 116, 121, 128
Minimum phase, 198
Model structure, 72
Model, physically based, 72
Modeling errors, 232
Modeling simplifications, 232
Modified control input, 204
Moore-Penrose pseudo-inverse, 32
Mother wavelet, 106
Multi-layer perceptron, 93
Multiresolution analysis, 108

Nearest neighbor matching, 25
Network, feedforward, 94
Network, recurrent, 94
Neural network training, 17
Nodal address, 65
Nodal processor, 40, 48
Noise, 2, 26
Nominal model, 128
Nonlinear control design, 9
Nonlinear damping, 219
Nonlinear state transformation, 193
Nonlinear systems, 3
Nonlinearly parameterized approximators, 126, 278
Nonuniformly spaced knots, 81
Normal form, 198

Offline function approximation, 31
Offline parameter estimation, 126
Online learning schemes, 116
Operating envelope, 2, 226, 286, 350
Operating point, 5, 180, 186, 344, 379
Order, system, 377
Orthogonal wavelet, 111
Orthonormality, 108
Output layer, 94
Over-constrained solution, 31

Parameter adaptive law, 125
Parameter convergence, 127, 145, 161
Parameter drift, 164, 242, 247, 261, 262, 265, 291
Parameter estimation, 115
Parameter estimation, Lyapunov based, 143
Parameter estimation, optimization based, 148
Parameter uncertainty, 116
Parametric model, 124
Parametric modeling, 127
Partition of unity, 57, 176
Peaking phenomenon, 238
Pendulum model, 72
Perceptron, 93
Perfect tracking, 21, 395
Persistency of excitation, 8, 35, 124, 127, 145, 159, 161
Persistently exciting signal, 120, 161, 162
Physically based models, 72
Plant, 1
Polynomial precision, 84
Polynomials, 75
Positive real, 391
Positively invariant set, 247
Predictor-corrector, 35
Prefilter, 2
Projection modification, 165, 221, 261
Projection, boundedness, 288
Projection, stabilizability, 288
Pseudo-inverse, 32

Radial basis function network, 123
Radial basis functions, 84
Rank, matrix, 31
RBF networks, 84
Receptive field weighted regression, 161, 176
Recursive parameter estimation, 126
Reference input, 1
Regional stability, 235
Regressor filtering online learning, 116
Regulation, 2
Relative degree, 197
Residual approximation error, 74
RFWR, 176
Richness condition, 162
Robotic manipulator model, 195
Robust learning algorithms, 116, 163
Robust nonlinear control, 211

Satellite model, 185
Scaling, 108
Scattered data approximation, 17, 54, 84

Scattered data interpolation, 29, 84
Self-organizing, 19
Semi-global stability, 235
Sensor, 1
Separation, 108
Sigma-modification, 168, 221
Sigmoidal neural network, 40
Sign function, 213
Singular values, matrix, 31
Sliding manifold, 213
Sliding mode control, 212
Sliding surface, 213
Small signal linearization, 253
Small-in-the-mean-square sense, 292, 364, 390
Small-signal linearization, 180, 238
Smoothing the control law, 239
Solution existence, 378
Solution uniqueness, 378
Splines, 78
Splines, B-splines, 80
Splines, natural, 78
SPR, 391
SPR filtering, 131
Squashing function, 40, 48, 94
Stability, 379
Stabilizability, 181, 189, 253, 261
Stabilization, 236
Stable, 380
Stable, asymptotically, 380
Stable, exponentially , 380
Stable, uniformly , 380
Stable, uniformly asymptotically, 380
State, 377
State space, 377
State transformation, 193
State-space parametric modeling, 133
Static system, 116
Statistical learning theory, 54
Steepest descent, 148
Stone-Weierstrass theorem, 51
Strictly positive real, 391
Structure free approximation, 74
Sufficiently exciting, 35
Sufficiently rich, 162
Supervised learning, 24, 95, 152
Support, 176, 264
Support, global, 56
Support, local, 56
Switching control, 211
Systems terminology, 1

Takagi-Sugeno fuzzy system, 104
Taylor series approximation, 75
Tchebycheff set, 53
Time constant, 4
Tracking, 2, 253
Translation, 83

Under-constrained solution, 32
Uniform ultimate boundedness, 380
Uniformly completely controllable, 184
Universal approximator, 50, 51
Universe of discourse, 96

Vandermonde matrix, 32
Vanishing perturbation, 338
Virtual control input, 203, 205, 310, 315, 317

Wavelet transform, 106
Wavelet, mother, 106
Wavelets, 106
Weierstrass theorem, 44, 45, 77

Zero dynamics, 197, 198

Adaptive and Learning Systems for Signal Processing, Communications, and Control

Editor: Simon Haykin

Beckerman / ADAPTIVE COOPERATIVE SYSTEMS

Candy / MODEL-BASED SIGNAL PROCESSING

Chen and Gu / CONTROL-ORIENTED SYSTEM IDENTIFICATION: An $\mathcal{H}_\infty$ Approach

Cherkassky and Mulier / LEARNING FROM DATA: Concepts, Theory, and Methods

Diamantaras and Kung / PRINCIPAL COMPONENT NEURAL NETWORKS: Theory and Applications

Farrell and Polycarpou / ADAPTIVE APPROXIMATION BASED CONTROL: Unifying Neural, Fuzzy and Traditional Adaptive Approximation Approaches

Hänsler and Schmidt / ACOUSTIC ECHO AND NOISE CONTROL: A Practical Approach

Haykin / UNSUPERVISED ADAPTIVE FILTERING: Blind Source Separation

Haykin / UNSUPERVISED ADAPTIVE FILTERING: Blind Deconvolution

Haykin and Puthussarypady / CHAOTIC DYNAMICS OF SEA CLUTTER

Haykin and Widrow / LEAST-MEAN-SQUARE ADAPTIVE FILTERS

Hrycej / NEUROCONTROL: Towards an Industrial Control Methodology

Hyvärinen, Karhunen, and Oja / INDEPENDENT COMPONENT ANALYSIS

Kristić, Kanellakopoulos, and Kokotović / NONLINEAR AND ADAPTIVE CONTROL DESIGN

Mann / INTELLIGENT IMAGE PROCESSING

Nikias and Shao / SIGNAL PROCESSING WITH ALPHA-STABLE DISTRIBUTIONS AND APPLICATIONS

Passino and Burgess / STABILITY ANALYSIS OF DISCRETE EVENT SYSTEMS

Sánchez-Peña and Sznaier / ROBUST SYSTEMS THEORY AND APPLICATIONS

Sandberg, Lo, Fancourt, Principe, Katagiri, and Haykin / NONLINEAR DYNAMICAL SYSTEMS: Feedforward Neural Network Perspectives

Spooner, Maggiore, Ordóñez, and Passino / STABLE ADAPTIVE CONTROL AND ESTIMATION FOR NONLINEAR SYSTEMS: Neural and Fuzzy Approximator Techniques

Tao / ADAPTIVE CONTROL DESIGN AND ANALYSIS

Tao and Kokotović / ADAPTIVE CONTROL OF SYSTEMS WITH ACTUATOR AND SENSOR NONLINEARITIES

Tsoukalas and Uhrig / FUZZY AND NEURAL APPROACHES IN ENGINEERING

Van Hulle / FAITHFUL REPRESENTATIONS AND TOPOGRAPHIC MAPS: From Distortion- to Information-Based Self-Organization

Vapnik / STATISTICAL LEARNING THEORY

Werbos / THE ROOTS OF BACKPROPAGATION: From Ordered Derivatives to Neural Networks and Political Forecasting

Yee and Haykin / REGULARIZED RADIAL BIAS FUNCTION NETWORKS: Theory and Applications

Printed and bound by CPI Group (UK) Ltd, Croydon, CR0 4YY
16/09/2026
14963448-0002